T0336391

17th Edition

Mary Hansen

Australia • Brazil • Japan • Korea • Mexico • Phillippines • Singapore • Spain • United Kingdom • United States

Business Math, Seventeenth Edition
Mary Hansen

Vice President of Editorial, Business:
Jack W. Calhoun

Vice President/Editor-in-Chief: Karen Schmohe

Vice President/Marketing: Bill Hendee

Executive Editor: Eve Lewis

Senior Developmental Editor: Enid Nagel

Editorial Assistant: Virginia Wilson

Assistant Marketing Manager: Linda Kuper

Content Project Manager: Darrell E. Frye

Media Editor: Mike Jackson

Website Project Manager: Ed Stubenrauch

Senior Manufacturing Buyer: Kevin Kluck

Production Service: Pre-PressPMG

Senior Art Director: Tippy McIntosh

Internal and Cover Designer: Lou Ann Thesing

Cover Illustration: Lou Ann Thesing

Photography Manager: John Hill

ISBN-13: 978-0-538-44873-4

ISBN-10: 0-538-44873-3

South-Western Cengage Learning
5191 Natorp Boulevard
Mason, OH 45040
USA

Cengage Learning products are represented in Canada by Nelson Education, Ltd.

For your course and learning solutions, visit **www.cengage.com/school**
Visit our company website at **www.cengage.com**

Printed in the United States of America
Print Number: 17 Print Year: 2025

Reviewers

Ronald E. Anderson
Business Education Instructor
Winchester Community High School
Winchester, Indiana

Norma Brown
Marketing Instructor, NBCT
Dutch Fork High School
Irmo, South Carolina

Rebecca L. Dufour
Business Educator
Tioga High School
Tioga, Louisiana

Ryan Harrison
Marketing and Business Education Teacher
Rosemount High School
Rosemont, Minnesota

Robin Levasseur
Middle/High School Math
Wisdom Middle/High School
St. Agatha, Maine

Robert A. Lever
Business Education Chair
Catonsville High School
Catonsville, Maryland

Greg Malkin
Director, Entrepreneur Institute
University School
Cleveland, Ohio

Fred Pimentel
Teacher, Business Department
New Bedford High School
New Bedford, Massachusetts

Jennifer B. Reinhardt
Teacher, Business Department
Prior Lake High School
Savage, Minnesota

John S. Salerno
Teacher, Career/Technical Education Department
Newington High School
Newington, Connecticut

Pamela Powell Sales
Teacher, Business Education
Westside High School
Macon, Georgia

Sandy Schaefer
Business Education Teacher
Sleepy Eye High School
Sleepy Eye, Minnesota

About the Author

Mary Hansen taught mathematics, money management, and special education at the high school and college level in Kansas, North Carolina, and Texas. An educational consultant, she is the coauthor of three high school mathematics textbooks.

Contents

Chapter 5

Loans 168

Chapter 6

Own a Home or a Car 206

Chapter 7

Insurance and Investments 270

Chapter 8

Budgets 334

Chapter 9

Business Costs 380

Chapter 10

Sales and Marketing 416

Chapter 13

International Business 546

Inside the Student Edition

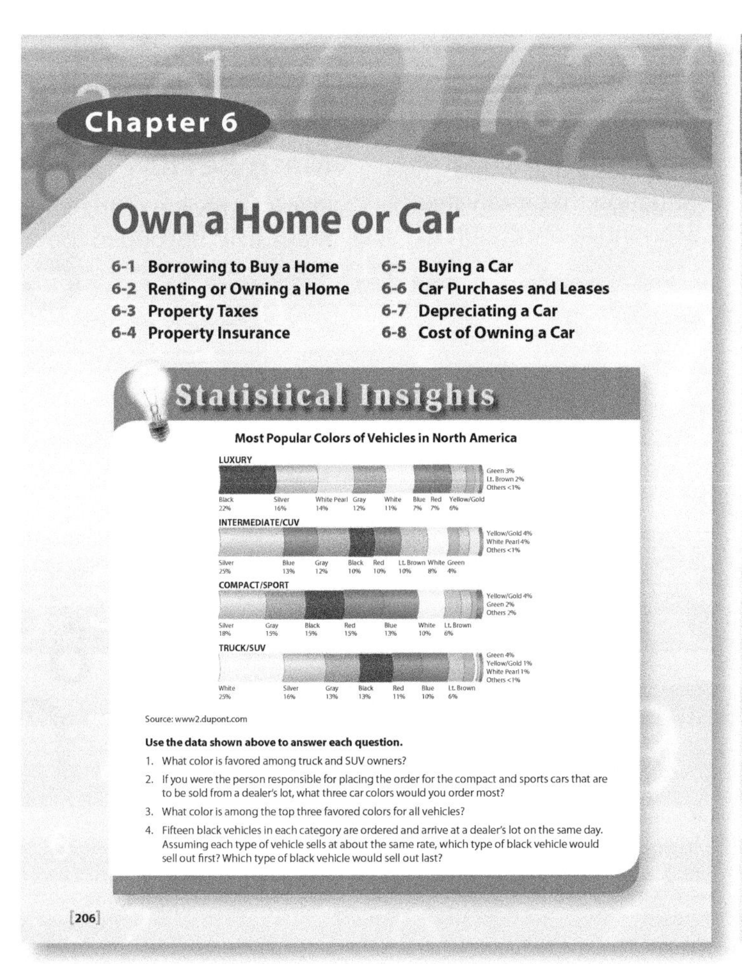

Chapter 6

Own a Home or Car

Statistical Insights

Most Popular Colors of Vehicles in North America

Source: www2.dupont.com

Use the data shown above to answer each question.

1. What color is favored among truck and SUV owners?
2. If you were the person responsible for placing the order for the compact and sports cars that are to be sold from a dealer's lot, what three car colors would you order most?
3. What color is among the top three favored colors for all vehicles?
4. Fifteen black vehicles in each category are ordered and arrive at a dealer's lot on the same day. Assuming each type of vehicle sells at about the same rate, which type of black vehicle would sell out first? Which type of black vehicle would sell out last?

[206]

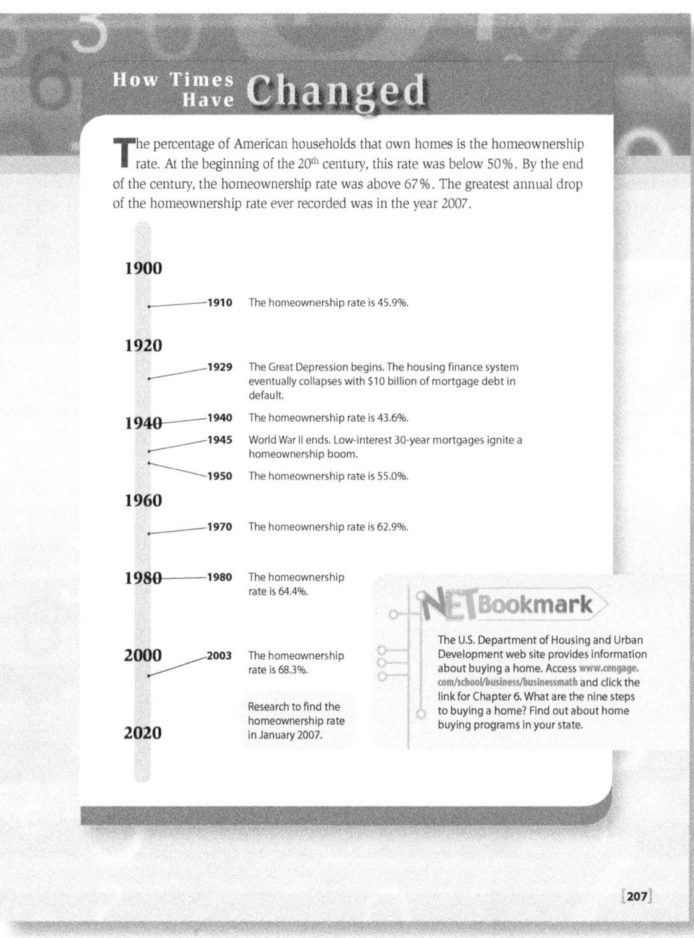

How Times Have Changed

The percentage of American households that own homes is the homeownership rate. At the beginning of the 20th century, this rate was below 50%. By the end of the century, the homeownership rate was above 67%. The greatest annual drop of the homeownership rate ever recorded was in the year 2007.

1900
1910 The homeownership rate is 45.9%.
1920
1929 The Great Depression begins. The housing finance system eventually collapses with $10 billion of mortgage debt in default.
1940
1940 The homeownership rate is 43.6%.
1945 World War II ends. Low-interest 30-year mortgages ignite a homeownership boom.
1950 The homeownership rate is 55.0%.
1960
1970 The homeownership rate is 62.9%.
1980
1980 The homeownership rate is 64.4%.
2000
2003 The homeownership rate is 68.3%.
Research to find the homeownership rate in January 2007.
2020

NET Bookmark

The U.S. Department of Housing and Urban Development web site provides information about buying a home. Access www.cengage.com/school/business/businessmath and click the link for Chapter 6. What are the nine steps to buying a home? Find out about home buying programs in your state.

[207]

Statistical Insights

Data Analysis

Presents real-world data that is relevant to the business and math topics of the chapter.

How Times Have Changed

Timeline Feature

Provides historical details relevant to the chapter. Looking at events, inventions, and technology from the past can give you an understanding and appreciation for the way business is done today.

Net Bookmark

Internet Feature

Incorporates Internet activities into every chapter.

Excellent Organization

Goals

Objectives

Clear and concise description of what you will learn in the lesson

Terms

New Vocabulary

Lists the new terms defined in the lesson.

Start Up

Engages Students

Provides a real-world situation that motivates learning in the chapter.

Math Skill Builder

Integrated Basic Math Review

Provides review of the basic math skills needed in the lesson

6-6

Car Purchases and Leases

GOALS

- Calculate the total amount paid and the finance charge for installment loan car purchases
- Calculate the cost of leasing cars
- Compare the costs of leasing and buying cars

KEY TERMS

- lease

Start Up

Bobbi tells you that she doesn't like her family's cars because they are kept too long, for 8–10 years. She claims that she will always lease cars so she can drive a new car all the time. She asks you whether you agree with her. What would you say?

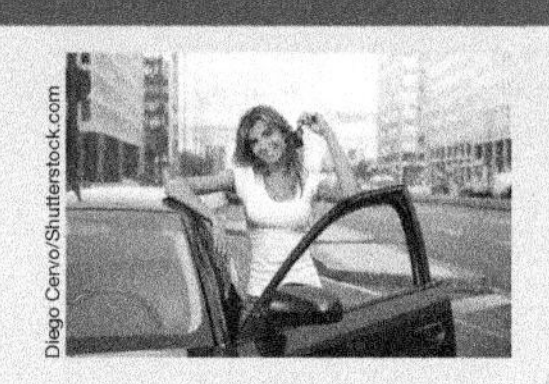
Diego Cervo/Shutterstock.com

Math Skill Builder

Review these math skills and solve the exercises that follow.

1. **Add** money amounts.
   Find the sum. $24,875 + $3,100 + $450 = $28,425

   1a. $180 + $570 + $43

   1b. $24,765 + $1,250

   1c. $562 + $19 + $1,235

2. **Subtract** money amounts.
   Find the difference. $34,279 − $33,892 = $387

   2a. $18,367 − $17,907

   2b. $38,431 − $8,400

   2c. $64,395 − $16,338

3. **Multiply** money amounts by whole numbers and percents.
   Find the product. 24 × $582 = $13,968
   Find the product. 13% × $16,300 = 0.13 × $16,300 = $2,119

   3a. 48 × $648 **3b.** 60 × $610.45

   3c. 21% × $19,250 **3d.** 17% × $9,670

   3e. 64 × $19.95 **3f.** 35% × $13,842

244 **Chapter 6** Own a Home or Car

Examples

Show Work Step-By-Step

Each lesson has worked out examples. The example explains step-by-step how to solve the problem. The text in blue helps you identify what you calculate.

Check Your Understanding

Practice for Example

Allows you to practice before continuing. The questions are like the example and will help you know if you understand the skills being taught

Wrap Up

Concludes Start Up

You will already have formed an opinion and answer about the Start Up question. This gives you an opportunity to check your answer and your reasoning.

Rate of Depreciation

When the straight-line method of finding depreciation is used, the average annual depreciation may be shown as a percent of the original cost. The percent is called the *rate of depreciation.*

Rate of Depreciation = Average Annual Depreciation ÷ Original Cost

EXAMPLE 2

A $12,000 car is sold 3 years later for $6,960. What is the rate of depreciation?

SOLUTION

Find the total depreciation.

$12,000 − $6,960 = $5,040

Find the average annual depreciation.

$5,040 ÷ 3 = $1,680

Divide the annual depreciation by the original cost to find the rate of depreciation.

$1,680 ÷ $12,000 = 0.14, or 14%   rate of depreciation

✔CHECK YOUR UNDERSTANDING

C. A new car that cost $23,000 is worth $16,100 a year later. What was the rate of depreciation for the one year?

D. Billy Macon sold his car for $368. He paid $9,200 for the car when he bought it 12 years ago. What was the annual rate of depreciation?

Wrap Up ▸▸▸

The value of a used car depends not only on its age and the miles it has been driven, but also on its overall condition and how it has been maintained. After the two cars are seen and their service records examined, their true value is determined by what price they will bring in the market. The Kelley Blue Book is one source that provides used car pricing information. You may also want to look at the want ad prices for similar cars or check the prices posted on used car web sites or the Blue Book site.

TEAM Meeting

Meet with three to four members of your class and select one specific make and model that would fit into each of these vehicle categories: luxury car, luxury SUV, economy car, economy SUV. Find the approximate selling price of each vehicle a year ago and what each vehicle would sell for today as a one-year-old vehicle. Calculate the percent of depreciation for each vehicle. Study the depreciation percents and write a paragraph about any differences you find.

[252] **Chapter 6** Own a Home or Car

Features Enhance Learning

Consumer Alert

Payday Loans = HIGH Interest

Payday loans are usually short term loans designed to provide money until the next payday. While payday loans might seem like a convenient way to get out of a jam, the interest rate for these loans can range from 400% – 800%.

A common payday loan is for $500 with $25 per $100 due for interest. If the loan is to be repaid in two weeks, what is the interest rate for this loan?

Companies offering payday loans may ask borrowers to leave a check for the principal and the interest that will be cashed on the day the loan is due. Online companies may ask for electronic access to the borrower's bank account to electronically withdraw funds on the due date. The Consumer Federation of America recommends never transmitting bank account numbers, Social Security numbers, or other personal financial information over the Internet or by fax to unknown companies.

Consumer Alert

Provides you information about scams and schemes targeting consumers.

Financial Responsibility

Shows you how to be fiscally prudent with your money.

Financial Responsibility

When making travel plans, whether it is for business travel or personal travel, you should consider buying *trip insurance*. Without insurance, if you become ill, have a family emergency, or some other reason that you cannot make your trip as planned, any nonrefundable payments made are forfeited. You cannot get your money back.

With insurance, the money that you have prepaid can be refunded. There are many different types of trip insurance ranging in price. You should buy a policy that meets the level of coverage you desire.

The cost of insurance is minimal compared to the cost of the trip.

Communication

Write a brief paragraph you might include in an e-mail to a friend who works explaining why it is important for the friend to file a tax return.

There are three important guidelines that should be followed when sending e-mail.

1. Your e-mail should cover only one topic.
2. Your message should be brief.
3. Be courteous and professional in your message.

Remember, once the e-mail is sent, you cannot get it back.

Communication

Provides you opportunities to use various methods of communication.

Team Meeting

Provides group activities where you can put into practice the topics you are learning

TEAM Meeting

With two other students, investigate the advantages and disadvantages of health maintenance organization health plans, or HMOs, and indemnity health plans. You should use Internet search tools and talk to at least one health insurance agent to obtain your information.

You need to define a health maintenance organization plan. Name the requirements and the general premise under which they operate. Also explain an indemnity health plan, its requirements, and how they function.

List the advantages and disadvantages of each. It is a good idea to question adults who participate in each type of program and get their opinions.

Tips Provide Extra Help

Algebra *Tip*

To find the interest on a bond, you can use the simple interest formula:.

$$I = P \times R \times T$$

where P is the principal or the par value of the bond, r is the rate of interest, and t is the time in years.

Math *Tip*

A mill is one-tenth of a cent and one thousandth of a dollar.

Problem Solving *Tip*

To find the employee share of the premium, deduct the employer's percentage share from 100%. The difference is the employee's percentage share. Then multiply the total premium by the employee's percentage share.

Calculator *Tip*

Find the total using the memory keys. For each line, after you find the product, press [M+] to store the product. Repeat for each line of multiplication. When all of the multiplication is done, press [MR]. The total will be displayed, and you can divide by the number of items. To clear the memory before a new problem, press [MC].

Spreadsheet *Tip*

Spreadsheets can be used to find the days between dates. Enter the first date in B1 and the other date in B2. Then in B3 enter: = B2 − B1. Next format B3 to display a *number* with *no decimal places*.

Integrated Technology

Technology Workshop

Task 1 Calculating Mortgage Payments

Enter data into a template that calculates the monthly mortgage loan payments and the total interest paid on the loan. You may use the template to compare the effects of changes in the interest rate and loan term on the total interest paid.

Open the spreadsheet for Chapter 6 (tech6-1.xls) and enter the data shown in blue (cells B3-5) into the spreadsheet. The spreadsheet will calculate the monthly loan payment, total amount paid on the loan, and the total interest paid on the loan.

Your computer screen should look like the one shown below when you are done.

	A	B
1	MORTGAGE LOAN CALCULATOR	
2	Mortgage Loan Data	
3	Amount	$110,000.00
4	Interest Rate (%)	8.160
5	Term (in years)	30
6	Mortgage Payment Data	
7	Mortgage Factor	0.9128142
8	Number of Payments	360
9	Monthly Payment	$819.44
10	Total Amount Paid	$294,998.40
11	Less Original Mortgage	$110,000.00
12	Total Interest Paid	$184,998.40

Task 2 Analyze the Spreadsheet Output

Answer these questions about the mortgage l

1. For how many years was the loan made?
2. What amount was borrowed?
3. What was the monthly payment?
4. What total amount was paid on the mort
5. What total amount of interest was paid o

Now move the cursor to cell B4, which holds rate 8.66%, which is $\frac{1}{2}$% higher than the rate 8.66% without the percent symbol.

Answer these questions.

6. What total interest would be paid on the
7. Approximately how much more would b $\frac{1}{2}$% higher interest rate?

8. Assume you changed the loan term to 25 years and kept the rate at 8.66%. Over which term, 25 years or 30 years, do you think you would pay the greatest total amount of interest? Now, change the term to 25 years and check your thinking.

Task 3 Design an Insurance Loss Payment Spreadsheet

You are to design a spreadsheet that will calculate the amount of loss paid by an insurance company under a coinsurance policy. Also calculate the required amount of coinsurance.

The spreadsheet should have two sections, one for input data, and another for calculated data. Design your spreadsheet so the amount of loss paid is never greater than the insurance carried on the property.

SITUATION: Dewayne Clayton owns a home worth $50,000. He insures it for $35,000 under an 80% coinsurance policy. His roof was damaged by high winds, and its repair will cost $2,000. Find the amount of this loss that will be paid by the insurance company.

Task 4 Analyze the Spreadsheet Output

Answer these questions about your completed spreadsheet:

9. What amount of loss did the insurance company pay?
10. What coinsurance amount should have been carried on the home?
11. How much of the loss will Dewayne have to pay?
12. If the policy had a deductible, what change would you have to make in your spreadsheet?

Technology Workshop

Spreadsheet Activities

A template task is provided for you to enter information and learn how spreadsheets work. You answer questions about the results and the template. Then you design your own spreadsheet to fit a situation or data given.

Career Planning

Planning a Career In...

Career Cluster Feature

Includes career awareness information that introduce career options you may want to consider in the future.

Planning a Career in Architecture and Construction

Careers in architecture and construction can be as varied as an engineer building a bridge to a surveyor verifying boundary lines in a home sale. If you choose a career in architecture you can specialize in designs for homes, buildings, ships, planes, bridges, highways, or landscaping. A job in construction can mean you build skyscrapers, install plumbing, run wires for electricity, manage construction projects or crews, or inspect buildings or structures. If you have the ability to envision an idea that does not yet exist, or the skills to bring that idea into existence, a career in architecture and construction may be a good avenue for you.

Job Titles

- Civil engineer
- Brick layer
- Drafter
- Architect
- Aerospace engineer
- Landscape architect
- Electrician
- Boilermaker mechanic

Needed Skills

- strong organizational and leadership skills
- an exceptional eye for detail

- mathematical and scientific skills
- excellent problem-solving skills
- technical and computer skills
- the disciple to work independently, as well as in a team

What's it like to work in Architecture?

Architects specialize in designs of structures of either interiors or exteriors. Landscape architects design outdoor areas around houses, shopping centers, roadways, schools, and buildings. Landscape architects can work with engineers, scientists, and surveyors to plan the locations for buildings, roads, and walkways. These plans include sketches, reports, cost estimates, material lists, and the use of specialized software programs. Many landscape architects have their own businesses. Forty-nine states require landscape architects to be licensed. This is achieved by examination.

What About You?

What aspect of architecture and construction appeals to you? How might you best prepare for a career in this field?

How Times Have Changed

For Questions 1–2, refer to the timeline on page 207 as needed.

1. It is common for a homebuyer to pay a down payment of 10%, 15% or 20%. If a home in Baltimore, Maryland cost 315,000 in 2007, what is the range of money a buyer would likely pay as a down payment?
2. The number of households consists of the number of homeowners and the number of renters. The number of households in 2007 was about 111 million. About how many households owned homes in 2007? About how many 2007 households were renters?

Chapter 6 Assessment [267]

Ongoing Assessment and Review

Exercises

Practice for Basic and Business Math

Exercises start with basic math skills followed by the business math in the examples.

Financial Decision Making

Ask you to apply financial decision making skills.

Stretching Your Skills

The numbers may be larger or your answer may not be as easy to arrive at, but the skills you use are the same.

Exercises

Write as a decimal.

1. 5%
2. 2.5%
3. 10.35%
4. 0.7%
5. 105%
6. $6\frac{1}{2}\%$

Find the product. Round to the nearest cent.

7. 2% × $563.98
8. $3\frac{1}{4}\%$ × $762.37

For the car insurance problems in this textbook, use the premiums table to find the cost of insurance. If the insurance coverage is not given, assume it is one of these standard coverages: bodily injury, $25/50,000; property damage, $25,000; collision, $100 deductible; comprehensive, $50 deductible. The same rate will apply to all types of motor vehicles unless otherwise indicated.

9. What is the total premium for standard insurance coverage on a car driven to work?
10. On a truck he drives to work, Norbert carries bodily injury insurance of $50/100,000 and $250 deductible on collision. Other coverage is standard. Find his annual premium.

June Driscoll uses her truck for business and insures the truck with standard coverage.

11. What annual premium does she pay?
12. If June took the highest deductibles, what amount would she save annually on her total car insurance bill?

23. **FINANCIAL DECISION MAKING** A car that you drive to work is worth about $1,200. If you did not insure your car for comprehensive and collision coverage, you would save about $500 a year in insurance costs. Should you drop these two coverages to save money?

24. **STRETCHING YOUR SKILLS** The IRS allows taxpayers to deduct the expense of operating a vehicle for business purposes. The taxpayer can keep records documenting the miles driven for business and the annual expenses for the vehicle and deduct the percentage of expenses based on the percentage of miles that were driven for business. Or, the taxpayer can use the standard deduction of 55 cents per mile. Jason Selvidge calculated his annual deductible expenses for his car to be $4,800. He drove his car a total of 16,000 miles during the year, and 10,000 of those miles were for business. How much of his actual expenses can he deduct? How much could he deduct if he takes the standard deduction?

Exercises

Best Buy

You make calculations for two different situations and then determine the better choice.

Critical Thinking

Asks you to apply critical thinking to the business math concepts.

Integrating Your Knowledge

You will use skills that were taught in more than one lesson.

Mixed Review

You will work exercises from the previous lessons that provide ongoing review and assessment.

BEST BUY Jorge Conseco can work for ABM, Inc. for $437 per week or Zeda, Inc. for $1,408 per month. Benefits average 19% of yearly wages at ABM and 25% at Zeda. Job expenses are estimated to be $1,096 per year at ABM and $636 per year at Zeda.

18. Which job would give Jorge more net job benefits for a year?

19. How much more?

20. **CRITICAL THINKING** The value of some benefits, such as paid holidays, can be figured very accurately. However, the value of other benefits, such as free recreation facilities, can only be estimated. If you were offered a benefit package that included use of a free gymnasium, how would you estimate its dollar value?

INTEGRATING YOUR KNOWLEDGE Nora Bertram works at Radnor Products, Inc. and is paid a salary of $25,000 plus 5% commission on all her sales. Last year her sales were $200,000. Nora's benefits were: paid pension, $3,150; health insurance, $2,400; paid vacations and holidays, $3,365. Her job expenses are $3,007. She is considering a job offer from B-Tree, Inc. that pays a salary of $30,000 plus 6% commission on all sales over $100,000. She estimates her benefits at B-Tree to be $8,489 and her job expenses to be $2,050.

Photodisc/Getty Images

21. If Nora's sales at B-Tree were $200,000, which job would give her more net job benefits?

22. Use your answer from Exercise 22 to determine how much more the net job benefits would be.

Mixed Review

Change to decimals.

23. $\frac{3}{8}$ **24.** 87.6% **25.** 0.5% **26.** $\frac{1}{4}$%

Change to fractions or mixed numbers and simplify.

27. 25% **28.** 250% **29.** 10%

30. Cromwell, Inc. employs 5 people at a branch office. Their weekly wages are: Fred, $423.34; Erin, $479.14; Bob, $378.98; Susan, $528.20; and James, $462.93. What is the average weekly wage at the branch office?

[62] **Chapter 2** Net Pay

Chapter Review

Vocabulary and Lesson Review

Provides a summary list of vocabulary terms and review questions for each lesson in the chapter.

Chapter *Review*

Vocabulary Review

Find the term, from the list at the right, that completes each sentence. Use each term only once.

1. The money paid to purchase an insurance policy is called the _?_.
2. A contract that allows you to use property, such as a car, for a certain period is known as a(n) _?_.
3. A type of car insurance that covers your liability for injury to other persons is called _?_.
4. The money paid in addition to a down payment to complete the purchase of a home is _?_.
5. The gradual reduction in the value of a home due to aging and use is referred to as _?_.
6. A home's estimated worth used for tax purposes is called _?_.
7. The insurance usually obtained by tenants to cover the things they own and to provide personal liability protection is known as a(n) _?_.
8. The protection you get from car insurance that covers possible damage to your car is known as _?_.
9. An amount of money kept by a landlord to cover any damage you may cause when renting property is called the _?_.
10. The money you pay to a government unit based on the value of the property you own is called _?_.

assessed value
bodily injury
closing costs
collision
comprehensive damage
depreciation
down payment
homeowners insurance
lease
manufacturer's suggested retail price (MSRP)
negotiation
mortgage loan
premium
principal
property damage
property taxes
renters policy
resale value
security deposit
trade-in value

6-1 Borrowing to Buy a Home

11. Chester Thornton plans to buy a home for $105,700 with a 15% down payment. He estimates his closing costs as inspections, $360; property survey, $250; legal fees, $1,300; title insurance, $220; loan administration fee, $115; and recording fee, $180. What amount will Chester have to borrow to buy the home? What amount of cash will he need?
12. LuAnne Wiggins is buying a home for $167,000. She will make a 10% down payment and borrow the balance for 30 years at 8.23%. Her monthly mortgage payments will be $1,127.04. What total amount of interest will she pay over 30 years?
13. Robbie Whitaker's monthly mortgage payment is $823. His new monthly payment will be $694 if he refinances the mortgage loan. The refinancing costs are closing costs of $1,074 and a prepayment penalty of $421. How much will Robbie save in the first year by refinancing his mortgage?

Chapter 6 Review 261

Chapter *Assessment*

Chapter Test

Answer each question.

1. $1,057 + $186.20 + $595.86
2. $248,112 − $162,9
3. 360 × $918.47
4. 2.45% × $137,00
5. $1\frac{3}{5} \times \$24,000$
6. $6,840 ÷ 0.5%
7. 18 is what percent of 360?
8. $542 increased by
9. $6\frac{1}{2} + 1\frac{3}{5} + 4\frac{3}{4}$

Solve.

10. The home that Hilda Vaughan wants to buy sells for $213,000. She plans to make a 5% down payment and borrow the balance at 7.67% for 25 years. Her monthly mortgage payments will be $1,517.80. What total interest will Hilda pay over 25 years?
11. Eldon Hudspeth estimates that the cost of taxes, insurance, and maintenance on a home he bought is $4,980 a year. His other yearly costs would be $12,760 in mortgage loan interest and $2,050 in estimated depreciation. For a year, he would lose $1,030 interest on his down payment and save $2,400 on his income taxes. Eldon could have rented a similar home for $1,400 a month and had annual expenses of $302 for insurance and $2,760 for utilities. The security deposit on the rented home would have been $500. For the first year, which would have been less expensive, buying or renting a home? How much less?
12. Find the property tax due on a building assessed at $360,000 if the tax rate is $37.40 per $1,000.
13. Wanda Frey insured her home for $156,400. Her annual insurance rate is $0.83 per $100. What annual premium did she pay?
14. Kerry Molloy's homeowners policy has a face value of $78,200 and a $250 deductible. How much of a $4,783 loss will the insurance company pay?
15. Janese Mosby bought a new car for $27,437. She made a down payment of $4,200. Janese must pay 5% sales tax on the purchase and $121 in registration costs. She received a $500 rebate from the manufacturer when she bought the car. Find the delivered price and the balance due on this car purchase.
16. A car bought for $18,216 was sold for $12,870 after 2 years of use. What was the car's rate of depreciation, to the nearest tenth percent?
17. Edward Laffin uses his car for pleasure driving. The insurance coverages he wants and their costs are: collision, $368.12; comprehensive, $146.89; bodily injury, $164.14; property damage, $186.34. Edward receives a 5% discount on the total premium because he drives the car less than 7,500 miles a year. What annual premium will Edward pay?
18. A car costing $34,000 can be leased for $473 monthly over 5 years with a $2,420 down payment. The car's residual value is estimated to be $16,400. If the car is purchased with a $5,000 down payment, the monthly loan payments will be $623 for 60 months. Is buying or leasing less expensive, and how much less?

266 **Chapter 6** Own a Home or Car

Chapter Assessment

Chapter Test

Provides questions about basic math skills and business math topics covered in the chapter.

Chapters 5–6 *Cumulative Review*

MULTIPLE CHOICE

Select the best choice for each question.

1. What is the finance charge per $100 on a loan of $5,300 with a finance charge of $954?

   A. $0.18 B. $1.80 C. $5.50
   D. $18 E. $555

2. What is the interest due at maturity for a $900 note borrowed for 1.5 years at a rate of 16.5%?

   A. $222.75 B. $148.50 C. $99
   D. $22.75 E. $14.85

3. You paid $55 interest on a 6-month promissory note of $1,000. What rate of interest, to the nearest percent, did you pay?

   A. 5.5% B. 5.8% C. 9.1%
   D. 11% E. 12.2%

4. How many days are between March 2 and May 9?

   A. 64 B. 65 C. 66
   D. 67 E. 68

5. You are thinking of buying a home
   payment. You estimate closing cost
   How much cash will you need to b

   A. $16,166.50 B. $
   D. $21,381.50 E. $

6. A home worth $105,000 is insured
   How much of a $28,000 loss would

   A. $20,000 B. $
   D. $25,200 E. $

7. Vicky Zielinski bought a truck on t
   payment, $2,500; rebate, $1,200; sa
   What was the balance due?

   A. $20,579 B. $
   D. $17,805 E. $

8. Charles Codwell bought a used car
   and paid the balance in 24 paymen
   charge on the loan?

   A. $448.28 B. $
   D. $351.72 E. $

9. A car bought new for $27,118 is so
   average annual depreciation?

   A. $1,869.14 B. $
   D. $2,854.52 E. $

OPEN ENDED

10. Larry Pons has a 3-year, $10,000 installment loan at 8%. His monthly payment is $313.36. After making 22 payments, his balance is $4,175.32. If Larry decides to pay off the loan with his next payment, how much should he pay?
11. Yoko Soga borrowed $450 on a loan with a finance charge of $94.50. Find the finance charge per $100 of the amount financed.

Nathan Bogart borrows $1,540 on a simple interest installment loan at $11\frac{1}{2}$% agreeing to repay it in 12 equal monthly payments of $136.47.

12. What was the total finance charge on the loan?
13. What was the interest paid for the first month?
14. What was the amount applied to principal at the end of the first month?
15. The basic annual cost of an auto insurance policy on Kara Malgren's car is $726. She gets a 2% discount for having a theft alarm and side impact air bags. She also gets a 4% discount for her safe driving record. What annual premium will Kara pay?
16. A home may be bought for $72,000 with a 15% down payment. The monthly payments on a 20-year loan at 7.61% will be $497.15. What total interest will be paid on the loan?
17. A town needs $580,000 to maintain its parks. Park use fees raise $124,000 of that amount. The total assessed value of property in the town is $10,900,000. What tax rate is needed to provide enough money to maintain the parks? Round to four decimal places.
18. A home's assessed value is $185,410. If the property tax rate is $29.18 per $1,000 of assessed value, what tax is due on the home?
19. A tax rate of 24 mills per $1 of assessed value is equivalent to what rate in dollars?
20. Bess Ambrose owns a car with an average monthly cost of gas, oil, and repairs of $184. The annual costs include insurance, $842; depreciation, $1,080; license plates, $68; and lost interest, $48. What is the total annual cost of operating the car?
21. Yong's Flower Shop purchased a delivery van for $32,599. The salesperson guaranteed that the dealership would buy back the van after four years for $12,550. What is the expected rate of depreciation?

CONSTRUCTED RESPONSE

22. A car can be leased for $289 a month for 36 months with no money down. The same car could be bought for $15,680 with $1,700 down and 36 monthly payments of $453. Write a note to a friend explaining why even with the $164 monthly payment difference, leasing the car may be more expensive than buying.

Cumulative Review

Multiple Choice

Presents question in the five-choice format found on most standardized tests.

Quantitative Comparison

Presents questions in a testing format where you must calculate two different problems and then compare those answers.

Constructed Response

Presents a question where your response will include both a calculation and an explanation.

Skills Workshop 1

Place Value and Order

EXAMPLE 1 Write 3,758,162,345,678.9123 in words.

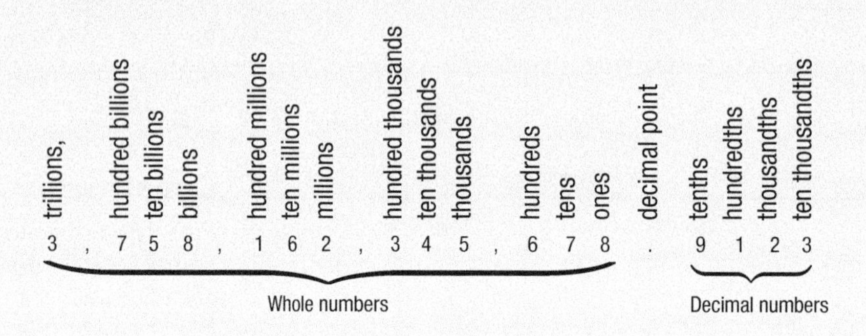

SOLUTION The number shown is three trillion, seven fifty-eight billion, one hundred sixty-two million, three hundred forty-five thousand, six hundred seventy-eight and nine thousand one hundred twenty-three ten-thousandths.

EXAMPLE 2 Use $<$ or $>$ to make this sentence true: 6 ■ 2

SOLUTION Numbers can be graphed on a number line. The number farther to the right is the greater number. Remember, $<$ means "less than" and $>$ means "greater than."
Six is greater than two. $6 > 2$

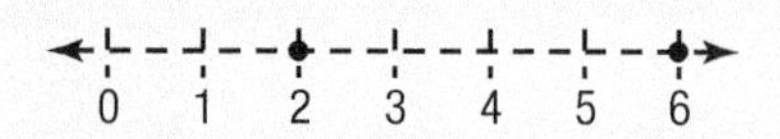

Write each number in words.

1. 3,647 **2.** 6,004,300.002 **3.** 0.9001

4. 17.049 **5.** 40,372 **6.** 6,071.435

Write each of the following as a number.

7. two million, one hundred fifty thousand, four hundred seventeen

8. five thousand, one hundred twenty and five hundred two thousandths

9. nine million, ninety thousand, nine hundred and ninety-nine ten-thousandths

10. four hundred sixty-eight thousand, forty-six and fourteen thousandths

Use $<$ or $>$ to make each sentence true.

11. 9 ■ 8 **12.** 164 ■ 246 **13.** 63,475 ■ 6,435

14. 52 ■ 50 **15.** 5.39 ■ 9.02 **16.** 43.94 ■ 53.69

17. 1.75 ■ 1.25 **18.** 1,476 ■ 1,467 **19.** 847.05 ■ 846.75

Skills Workshop 2

Rounding Whole Numbers and Decimals

When rounding whole numbers, first locate the digit in the place value to which you are rounding. If the digit to the right of it is 5 or greater, increase the digit in the specified place by 1. If the digit to the right of it is less than 5, the digit in the specified place remains the same. All digits to the right of the place value to which you are rounding become zero.

EXAMPLE 1 Round 4,782 to the nearest hundred.

SOLUTION Find the hundreds place. 4,<u>7</u>92 7 is in the hundreds place.

The digit to the right of 7 is 9.

Since 9 is greater than 5, round 7 up to 8.

So, 4,792 rounds to 4,800. All digits to the right of the hundreds place become zero.

When rounding decimals, follow the same procedure for rounding whole numbers except no zeros are needed if they are to the right of the place to which you are rounding and to the right of the decimal point. This is because they are not significant digits.

EXAMPLE 2 Round 8.243 to the nearest tenth.

SOLUTION Find the tenths place. 8.<u>2</u>43 2 is in the tenths place.

The digit to the right of 2 is 4.

Since 4 is less than 5, 2 remains the same.

So, 8.243 rounds to 8.200. All digits to the right of the tenths place are dropped.

8.200 can be written as 8.2

Round each number to the indicated place value.

1. 429 to the nearest ten
2. 9,058 to the nearest thousand
3. 36,815 to the nearest hundred
4. 85,726 to the nearest thousand
5. 48,280 to the nearest ten thousand
6. 392,682 to the nearest ten thousand
7. 6,329,451 to the nearest hundred thousand
8. 9,583,507 to the nearest thousand
9. 93,487,991 to the nearest hundred thousand
10. 4,540,597 to the nearest million
11. 0.38 to the nearest tenth
12. 6.849 to the nearest hundredth
13. 62.9042 to the nearest hundredth
14. 601.584 to the nearest tenth
15. 3.9015 to the nearest thousandth
16. 42.3952 to the nearest thousandth
17. 9.00391 to the nearest ten-thousandth
18. 0.92403 to the nearest ten-thousandth
19. 18.670308 to the nearest hundred-thousandth
20. 3.0040916 to the nearest hundred-thousandth

Skills Workshop 3

Add and Subtract Whole Numbers and Decimals

To add or subtract whole numbers and decimals, write the digits so the place values line up. Add from right to left, renaming when necessary. When adding or subtracting decimals, be sure to place the decimal point in the answer directly below the aligned decimals in the problem.

EXAMPLE 1

Find the sum of .058; 25.39; 6,346; and 1.57.

The answer is called the **total** or **sum.**

SOLUTION

```
 1 1 2
    0.058
   25.39
6,346.
 +  1.57
---------
6,373.018
```

The red zero is used to show there are no ones. The decimal point is at the end of whole numbers. Add from right to left.

EXAMPLE 2

Find the difference between 10,049 and 5,364.

The answer is called the **difference**.

SOLUTION

```
    9
  9 1014
 10,049
- 5,364
-------
  4,685
```

Rename as needed to subtract.

EXAMPLE 3

Subtract 6.37 from 27.

The lesser number is subtracted from the greater.

SOLUTION

```
   9
 6 1010
 27.00
 -6.37
------
 20.63
```

Add a decimal point and zeros to help you complete the subtraction.

Add or subtract.

1. 23.146 + 17.215

2. 46.48 − 6.57

3. 52 − 1.95

4. 0.86 + 0.75

5. 83 − 82.743

6. 9.45 + 13.2

7. 14.5 − 9.684

8. 913.03 − 79

9. 0.8523 − 0.794

10. 1,765.36 + 1,587.50 + 1,400

11. 6.4 + 54.2 + 938.05 + 3.7 + 47.3

12. 51,876.36 + 48,156.95 + 1,417.86

13. 17,347.85 − 12,516.90

14. 76.2 + 80 + 56 + 9.321

15. 107,285 − 61,500.25

16. 567.1 + 6 + 13.452 + 100

17. 6.013 + 39 + 14.09

18. 4.621 + 372.14

19. 1,468,329 − 370,418.5

20. 472.13 + 1,695.006 + 27.127 + 50,040 + 0.578

Skills Workshop 4

Multiply Whole Numbers and Decimals

To multiply whole numbers, find each partial product and then add.

The red zeros are added to help align the answers.

Add. Then rewrite the answer with commas.

EXAMPLE 1 Multiply 5,754 by 236.

SOLUTION

5,754	factor
× 236	factor
34524	5,754 × 6
172620	5,754 × 30
+ 1150800	5,754 × 200
1357944	Add.
1,357,944	product

When multiplying decimals, locate the decimal point in the product so that there are as many decimal places in the product as the total number of decimal places in the factors.

EXAMPLE 2 Multiply 2.6394 by 3,000.

SOLUTION

2.6394	4 decimal places
× 3,000	+ 0 decimal places
7,918.2000	4 decimal places
or 7,918.2	

Zeros at the end (far right) *after* the decimal point can be dropped because they are not *significant digits*.

EXAMPLE 3 Multiply 3.92 by 0.023.

SOLUTION

3.92	2 decimal places
× 0.023	+3 decimal places
1176	
+ 7840	
0.09016	5 decimal places

The red zero is added *before* the nine, so the product will have five decimal places.

Multiply.

1. 36 × 45 **2.** 500 × 30 **3.** 17,000 × 230 **4.** 6.2 × 8

5. 950 × 1.6 **6.** 3.652 × 20 **7.** 179 × 83 **8.** 257 × 320

9. 8,560 × 275 **10.** 467 × 0.3 **11.** 2.63 × 183 **12.** 0.758 × 321.8

13. 49.3 × 1.6 **14.** 6.859 × 7.9 **15.** 794.4 × 321.8 **16.** 0.08 × 4

17. 0.062 × 0.5 **18.** 0.0135 × 0.003 **19.** 21.6 × 3.1 **20.** 8.76 × 0.005

Skills Workshop 5

Divide Whole Numbers and Decimals

Dividing whole numbers and decimals involves a repetitive process of estimating a quotient, multiplying, and subtracting.

Math *Tip*

In Example 1, 239 is the dividend, 7 is the divisor, 34 is the quotient, and 1 is the remainder.

EXAMPLE 1 Find 239 ÷ 7.

SOLUTION

```
     34
  7)239
   −21↓      3 × 7   Subtract. Bring down the 9.
     29
    −28      4 × 7
      1
```

EXAMPLE 2 Find 283.86 ÷ 5.7.

SOLUTION When dividing decimals, move the decimal point in the divisor to the right until it is a whole number. Move the decimal point in the dividend the same number of places that you moved the decimal point in the divisor. Then place the decimal point in the answer directly above the new location of the decimal point in the dividend.

```
                          49.8
5.7)283.86   →    57)2838.6
                    −228
                      558
                     −513
                       45 6
                      −45 6
                          0
```

If answers do not have a remainder of 0, you can add 0's after the last digit and continue dividing.

Divide.

1. 72 ÷ 6	**2.** 6,000 ÷ 2	**3.** 26,568 ÷ 8
4. 5.6 ÷ 7	**5.** 120 ÷ 0.4	**6.** 936 ÷ 12
7. 3.28 ÷ 4	**8.** 0.1960 ÷ 5	**9.** 1,968 ÷ 0.08
10. 16 ÷ 0.04	**11.** 1,525 ÷ 0.05	**12.** 109.94 ÷ 0.23
13. 0.6 ÷ 24	**14.** 7.924 ÷ 0.28	**15.** 32.6417 ÷ 9.1
16. 24 ÷ 0.6	**17.** 1,784.75 ÷ 29.5	**18.** 0.01998 ÷ 0.37
19. 7.8 ÷ 0.3	**20.** 12,000 ÷ 0.04	**21.** 820.94 ÷ 0.02
22. 121.55 ÷ 18.7	**23.** 29,000 ÷ 1,450	**24.** 5,929.52 ÷ 9.4
25. 618.03 ÷ 12.6	**26.** 22.1616 ÷ 34.2	**27.** 235,083.36 ÷ 67.09

Skills Workshop 6

Average of a Group of Numbers

An **average**, or **mean**, is a measure of central tendency. To find the average of a group of numbers, find the sum of the numbers, and then divide the sum by the number of items in the group.

EXAMPLE 1 Find the average of $36, $49, $22, $48, $39, $40, and $18.

SOLUTION Find the sum of the numbers.

$36 + $49 + $22 + $48 + $39 + $40 + $18 = $252

Count how many items are in the group.

Divide the sum by 7 since there are 7 items.

$$\frac{\$252}{7} = \$36$$

The average of the set of numbers is $36.

EXAMPLE 2 Find the average of the following set of numbers.

3.2, 4.2, 6.05, 9.25, 5, 7.1, 9.8, 12.4, 3.3

SOLUTION Find the sum of the numbers.

$3.2 + 4.2 + 6.05 + 9.25 + 5 + 7.1 + 9.8 + 12.4 + 3.3 = 60.3$

Count how many items are in the group.

Divide the sum by 9 since there are 9 items.

$$\frac{60.3}{9} = 6.7$$

The average of the set of numbers is 6.7.

Find the average of each set of numbers.

1. 85, 60, 72, 68, 95, 83, 97, 84

2. $125, $149, $135, $146

3. 260, 362, 302, 381, 295, 332, 280

4. 37, 28, 33, 30, 25, 15

5. 32,052; 33,559; 30,129; 34,058

6. 1.1, 1.5, 1.8, 2.7, 1.6

7. 6, 8, 9, 6, 3, 4, 6, 8, 7, 8, 6, 5, 9, 9, 8

8. 24.5, 20.5, 22.4, 28.2, 25, 23.7

9. 2.5, 3.5, 2.6, 2.8, 2.4, 3.2, 2.3, 3.1

10. $545, $425, $600, $562, $399, $457

11. $17, $24, $28, $16, $28, $17, $18, $24, $22, $25, $29, $25

12. $625, $501, $399, $572, $459, $680, $377, $540

13. $10.25, $12.32, $11.24, $13.08, $14.48, $10.76, $15.30

14. 135.05, 241.62, 452.13, 105.95, 261.48

15. 9.1, 9.4, 9.9, 8.9, 9.3, 9.1, 9.1, 9.4, 8.8, 9.4

16. $135.06, $132.29, $145.92, $162.37, $127.55, $144.26, $150.05, $138.42

Skills Workshop 7

Multiply and Divide Fractions

To multiply fractions, multiply the numerators and then multiply the denominators. Write the answer in simplest form.

EXAMPLE 1 Multiply $\frac{2}{5}$ and $\frac{7}{8}$.

SOLUTION

$$\frac{2}{5} \times \frac{7}{8} = \frac{2 \times 7}{5 \times 8} = \frac{14}{40} = \frac{7}{20}$$

To divide by a fraction, multiply by the reciprocal of that fraction. To find the reciprocal of a fraction, invert (turn upside down) the fraction. The product of a fraction and its reciprocal is 1. Since $\frac{2}{3} \times \frac{3}{2} = \frac{6}{6}$ or 1, $\frac{2}{3}$ and $\frac{3}{2}$ are reciprocals of each other.

EXAMPLE 2 Divide $1\frac{1}{5}$ by $\frac{2}{3}$.

SOLUTION

$1\frac{1}{5} \div \frac{2}{3} = \frac{6}{5} \div \frac{2}{3}$ Write the mixed number as a fraction.

$= \frac{6}{5} \times \frac{3}{2}$ Rewrite division as multiplication by the inverse of the divisor.

$= \frac{6 \times 3}{5 \times 2}$ Multiply and simplify.

$= \frac{18}{10}$

$= 1\frac{4}{5}$

Multiply or divide. Write each answer in simplest form.

1. $\frac{2}{3} \div \frac{5}{6}$

2. $\frac{3}{5} \times \frac{10}{12}$

3. $\frac{5}{8} \div \frac{1}{4}$

4. $\frac{1}{2} \times \frac{2}{3}$

5. $\frac{2}{3} \times \frac{1}{2}$

6. $\frac{2}{3} \times \frac{1}{2}$

7. $\frac{1}{2} \div \frac{2}{3}$

8. $\frac{2}{3} \div \frac{1}{2}$

9. $\frac{3}{4} \div \frac{5}{8}$

10. $2\frac{2}{3} \div 1\frac{3}{5}$

11. $1\frac{1}{5} \times 2\frac{1}{4}$

12. $3\frac{1}{3} \times 1\frac{1}{10}$

13. $5\frac{2}{5} \div 2\frac{4}{7}$

14. $2\frac{4}{7} \div 5\frac{2}{5}$

15. $2\frac{4}{7} \times 5\frac{2}{5}$

16. $1\frac{7}{8} \div 1\frac{7}{8}$

17. $\frac{3}{4} \times \frac{2}{3} \times 1\frac{5}{8} \times 2\frac{2}{3}$

18. $\frac{1}{8} \times \frac{3}{4} \times 1\frac{2}{3} \times \frac{7}{10}$

19. $5\frac{1}{8} \times 3\frac{4}{5}$

20. $1\frac{4}{7} \div 7\frac{5}{6}$

21. $6\frac{3}{4} \times 9\frac{7}{8}$

Skills Workshop 8

Add Fractions

To add fractions with a common denominator, add the numerators and write the sum over the denominator they have in common. Then write the answer in simplest form.

EXAMPLE 1 Add $\frac{7}{8}$ and $\frac{3}{8}$.

SOLUTION

$$\begin{array}{r} \frac{7}{8} \\ +\frac{3}{8} \\ \hline \frac{10}{8} \end{array}$$

Add the numerators.
Use the common denominator.

Rewrite as a mixed number. $\frac{10}{8} = 1\frac{2}{8} = 1\frac{1}{4}$

To add fractions without a common denominator, first find a common denominator by finding the least common multiple of the denominators. Next, rename each fraction with an equivalent fraction using the common denominator. Then add the numerators and write the sum over their common denominator. Write the answer in simplest form.

EXAMPLE 2 Add $\frac{3}{4}$ and $\frac{5}{6}$.

SOLUTION

$$\begin{array}{rcl} \frac{3}{4} = \frac{3}{4} \times \frac{3}{3} = & & \frac{9}{12} \\ +\frac{5}{6} = \frac{5}{6} \times \frac{2}{2} = & + & \frac{10}{12} \\ \hline & & \frac{19}{12} \end{array}$$

Add the numerators.
Use the common denominator.

Rewrite as mixed number. $\frac{19}{12} = 1\frac{7}{12}$

Add. Write each answer in simplest form.

1. $\frac{1}{5} + \frac{2}{5}$

2. $\frac{2}{3} + \frac{1}{3}$

3. $\frac{8}{9} + \frac{4}{9}$

4. $\frac{11}{15} + \frac{4}{15}$

5. $\frac{1}{5} + \frac{1}{10}$

6. $\frac{5}{8} + \frac{3}{4}$

7. $\frac{16}{21} + \frac{2}{21}$

8. $\frac{6}{7} + \frac{1}{3}$

9. $\frac{11}{14} + \frac{3}{4}$

10. $2\frac{1}{2} + 3\frac{1}{2}$

11. $6\frac{5}{8} + 3\frac{7}{8}$

12. $3\frac{2}{3} + 4\frac{1}{2}$

13. $6\frac{1}{2} + 5\frac{7}{9}$

14. $7\frac{2}{3} + 6\frac{1}{5}$

15. $11\frac{4}{5} + 9\frac{1}{4}$

16. $4\frac{3}{8} + 2\frac{1}{6}$

17. $11\frac{2}{5} + 9\frac{4}{9}$

18. $5\frac{3}{10} + 13\frac{7}{8}$

19. $3\frac{1}{2} + 9\frac{4}{5} + 2\frac{2}{5}$

20. $1\frac{1}{5} + 2\frac{1}{3} + 5\frac{1}{4}$

21. $10\frac{7}{8} + 3\frac{3}{4} + 6\frac{1}{2} + 2\frac{5}{8}$

Skills Workshop 9

Subtract Fractions

To subtract fractions with a common denominator, subtract the numerators and write the difference over the common denominator. Then write the answer in simplest form.

EXAMPLE 1 Find the difference of $\frac{7}{8} - \frac{3}{8}$.

SOLUTION

$$\begin{array}{r} \frac{7}{8} \\ -\frac{3}{8} \\ \hline \frac{4}{8} \end{array}$$

To subtract fractions without a common denominator, first find a common denominator by finding the least common multiple of the denominators. Next, rename each fraction with an equivalent fraction using the common denominator. Then subtract the numerators and write the difference over the common denominator. Write the answer in simplest form.

EXAMPLE 2 Subtract $1\frac{3}{5}$ from $5\frac{1}{2}$.

SOLUTION

$$\begin{array}{rcrcr} 5\frac{1}{2} & = & 5\frac{5}{10} & = & 4\frac{15}{10} \\ -\ 1\frac{3}{5} & = & -\ 1\frac{6}{10} & = & -\ 1\frac{6}{10} \\ \hline & & \uparrow & & 3\frac{9}{10} \end{array}$$

You can not subtract $\frac{6}{10}$ from $\frac{5}{10}$, so rename again.

Subtract. Write each answer in simplest form.

1. $\frac{6}{7}-\frac{2}{7}$

2. $\frac{4}{9}-\frac{2}{9}$

3. $\frac{7}{16}-\frac{3}{16}$

4. $\frac{9}{11}-\frac{6}{11}$

5. $\frac{3}{4}-\frac{1}{3}$

6. $\frac{5}{8}-\frac{1}{4}$

7. $\frac{7}{12}-\frac{1}{6}$

8. $\frac{9}{10}-\frac{3}{4}$

9. $\frac{15}{16}-\frac{5}{8}$

10. $\frac{7}{8}-\frac{1}{5}$

11. $\frac{5}{8}-\frac{1}{12}$

12. $\frac{3}{5}-\frac{7}{12}$

13. $2\frac{3}{4}-1\frac{1}{4}$

14. $5\frac{1}{8}-3\frac{7}{8}$

15. $1\frac{1}{3}-\frac{2}{3}$

16. $8\frac{1}{10}-5\frac{2}{3}$

17. $6\frac{1}{2}-5\frac{3}{5}$

18. $10\frac{5}{8}-9\frac{3}{4}$

19. $24\frac{4}{7}-8\frac{5}{12}$

20. $15\frac{5}{6}-13\frac{2}{3}$

21. $5\frac{1}{2}-3\frac{4}{11}$

22. $24\frac{4}{7}-8\frac{5}{12}$

23. $8\frac{4}{9}-8\frac{3}{10}$

24. $37\frac{7}{8}-33\frac{8}{9}$

Skills Workshop 10

Fractions, Decimals, and Percents

Percent means *per hundred*. Thus, 35% means 35 out of 100. Percents can be written as equivalent decimals and fractions.

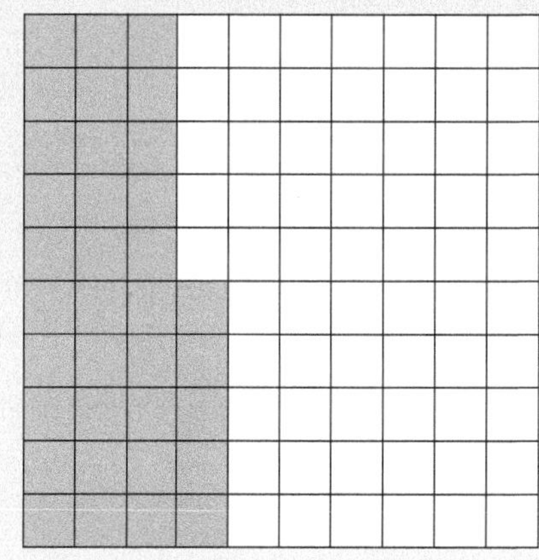

$35\% = 0.35$ Move the decimal point two places to the left.

$35\% = \frac{35}{100}$ Write the fraction with a denominator of 100.

$= \frac{7}{20}$ Then simplify.

EXAMPLE Write $\frac{3}{8}$ as a decimal and as a percent.

SOLUTION $\frac{3}{8} = 0.375$ To change a fraction to a percent, first divide and write the answer as a decimal.

$0.375 = 37.5\%$ Change the decimal to a percent by moving the decimal point two places to the right and adding a percent symbol.

Percents greater than 100% represent whole numbers or mixed numbers.

$200\% = 2$ or 2.00 $350\% = 3.5$ or $3\frac{1}{2}$

Complete each table. Write all fractions in simplest form.

	Fraction	Decimal	Percent
1.	$\frac{1}{2}$		
2.		0.63	
3.			10%
4.	$\frac{1}{4}$		
5.		0.15	
6.			12%
7.			100%

	Fraction	Decimal	Percent
8.	$\frac{3}{4}$		
9.		0.4	
10.			150%
11.		2.35	
12.	$3\frac{7}{8}$		
13.			160%
14.		10.125	

Skills Workshop 11

Percent

To find a percent of a number, first write the percent as a decimal by moving the decimal point two places to the left. Then multiply the given number by this decimal.

EXAMPLE 1 Find 20% of 275.

SOLUTION Write 20% as a decimal.

$20\% = 0.2$

Multiply.

$$\begin{array}{r} 275 \\ \times\ 0.2 \\ \hline 55.0 \end{array}$$

So 20% of 275 is 55.

To find what percent a number is of another number, write a fraction where the numerator is the *part* and the denominator is the *whole*. Reduce the fraction if possible. Then divide to change the fraction to a decimal. Finally, change the decimal to a percent.

EXAMPLE 2 What percent of 50 is 16?

SOLUTION Write a fraction and reduce. $\frac{16}{50} = \frac{8}{25}$

Divide.

$$\begin{array}{r} 0.32 \\ 25\overline{)8.00} \\ -7\,5 \\ \hline 50 \\ -\ 50 \\ \hline 0 \end{array}$$

$0.32 = 32\%$

So, 16 is 32% of 50.

Math Tip

The *whole* is the number that comes after the word "of."

Find each percent.

1. 2% of 18

2. 21% of 54

3. 85% of 400

4. 25% of 384

5. 27% of 12.2

6. 82% of 55

7. 34% of 2,200

8. 72.5% of 340

9. 6% of 12,500

Find the following.

10. What percent of 60 is 42?

11. What percent of 5 is 2?

12. 9 is what percent of 36?

13. What percent of 45 is 36?

14. What percent of 125 is 32?

15. 81 is what percent of 3,600?

16. 28 is what percent of 32?

17. What percent of 63 is 63?

18. What percent of 112 is 21?

19. 114 is what percent of 475?

Skills Workshop 12

Estimation Skills

You can round numbers to find the estimated sum, difference, product, or quotient. Round the numbers to the highest place value they have in common and then perform the operation.

EXAMPLE 1 Find the estimated difference of 6,452 − 2,806.

SOLUTION Round each number to the nearest thousandth.
6,000 − 3,000 = 3,000 The difference is about 3,000.

You can use **front-end estimation** to find an estimated sum, difference, product, or quotient. Use only the front digit of each number and the rest of the digits become zero.

EXAMPLE 2 Find the estimated product of 842 × 67 using front-end estimation.

SOLUTION

$$\begin{array}{r} 842 \\ \times\ 67 \\ \hline \end{array} \longrightarrow \begin{array}{r} 800 \\ \times\ 60 \\ \hline 48{,}000 \end{array}$$

The product is about 48,000.

You can use **adjusted front-end estimation** by looking at more than the front digits to get an answer that is more accurate.

EXAMPLE 3 Find the estimated sum of 4,367 + 5,735 using adjusted front-end estimation.

SOLUTION

$$\begin{array}{r} 4{,}\boxed{367} \\ +\ 5{,}\boxed{735} \\ \hline 9{,}000 \end{array} \longrightarrow \text{about } 1{,}000$$

9,000 + 1,000 = 10,000 The sum is about 10,000.

Estimate by rounding the numbers.

1. 395 + 842
2. 9,940 − 2,504
3. 459 × 34
4. 837 ÷ 38
5. 34,651 + 84,648
6. 668,345 − 229,048
7. $338 × 22
8. $39,648 ÷ 196
9. $75,045 − $8,654

Use front-end estimation to perform each operation.

10. 248 × 49
11. 6,482 + 8,248
12. 946 ÷ 305
13. 8,459 − 6,218
14. 84,516 + 3,811
15. $735 × 63
16. $245,364 − $19,563
17. 946 + 358 + 205
18. $6480 ÷ 231

Use adjusted front-end estimation to perform each operation.

19. 5,638 + 3,281
20. 867 − 311
21. 326 + 284
22. 2,942 + 9,133
23. 649 − 48
24. 81,506 − 9,408
25. $3,945 − $3,108
26. $1,235 + 72
27. 948 + 629 + 755 + 371

Skills Workshop 13

Elapsed Time

The amount of time that passes between two given times is called elapsed time. You can find elapsed time by finding the difference in the earlier time and the later time.

EXAMPLE 1 Find the elapsed time from 9:15 A.M. to 10:55 A.M.

SOLUTION Subtract the earlier time from the later time.

$$\begin{array}{r} 10{:}55 \\ -\ 9{:}15 \\ \hline 1{:}40 \end{array}$$

The elapsed time is 1 hour and 40 minutes.

Sometimes you are not able to subtract the number of minutes in the earlier time from the number of minutes in the later time. When this happens, rewrite the later time with 1 less hour and 60 more minutes.

If one of the times is A.M. and the other is P.M., add 12 hours to the later time before finding the difference.

EXAMPLE 2 Find the elapsed time from 11:40 A.M. to 3:30 P.M.

SOLUTION

Subtract the earlier time from the later time.	Rewrite 3:30 as 2:90	Add 12 hours to 2:90 since one time is A.M. and the other is P.M.	
$\begin{array}{r} 3{:}30 \\ -\ 11{:}40 \\ \hline \end{array}$	$\begin{array}{r} 2{:}90 \\ -\ 11{:}40 \\ \hline \end{array}$	$\begin{array}{r} 2{:}90 + 12{:}00 \\ -\ 11{:}40 \\ \hline \end{array}$	$\begin{array}{r} 14{:}90 \\ -\ 11{:}40 \\ \hline 3{:}50 \end{array}$

The elapsed time is 3 hours and 50 minutes.

Find the elapsed time.

1. from 5:00 P.M. to 11:00 P.M.
2. from 6:15 A.M. to 9:15 A.M.
3. from 2:10 P.M. to 7:33 P.M.
4. from 12:30 p.m. to 8:45 P.M.
5. from 8:00 A.M. to 5:00 P.M.
6. from 1:42 P.M. to 3:17 P.M.
7. from 2:55 A.M. to 9:45 A.M.
8. from 8:15 P.M. to 2:30 A.M.
9. from 11:00 A.M. to 9:00 P.M.
10. from 11:47 A.M. to 12:45 P.M.
11. from 4:12 P.M. to 1:33 A.M.
12. from 7:10 A.M. to 9: 30 P.M.
13. from 3:30 P.M. to 3:50 A.M.
14. from 9:33 P.M. to 4:08 A.M.
15. from 7:34 A.M. to 4:18 P.M.
16. from 1:42 P.M. to 3:37 A.M.
17. from 4:18 P.M. to 2:03 A.M.
18. from 7:56 A.M. to 5:12 P.M.

Skills Workshop 14

Problem Solving: 4-Step Plan

When solving word problems it is helpful to follow a 4-step plan.

1. **Understand** Read the problem and determine what information is given and what it is you are to find.
2. **Plan** Determine the method or strategy you will use to solve the problem.
3. **Solve** Carry out your plan to find an answer to the problem.
4. **Look Back** Go back over the problem and your answer to determine if your answer makes sense and make sure your computations are correct.

EXAMPLE 1 Jessica is buying two sheets of \$0.34 stamps from a vending machine. There are 20 stamps on each sheet. If she needs to put the exact amount of money in the vending machine, how much money does she need?

SOLUTION **Understand** It is given that Jessica is buying 2 sheets of 20 stamps. Each stamp costs \$0.34. You are to find the total cost.

Plan First find the cost of one sheet. Then double this amount to find the cost of both sheets.

Solve $\$0.34 \times 20 = \$6.80$ The cost of one sheet of stamps is \$6.80

$\$6.80 \times 2 = \$13.60$

The exact amount of money Jessica needs for the stamps is \$13.60.

Look Back You can use estimation to determine if your answer makes sense. Round the amount of one stamp to \$0.30.

$\$0.30 \times 20 = \$6.00$

$\$6.00 \times 2 = \$12.00$

Because the stamps are slightly more than \$0.30, the answer \$13.60 makes sense.

Use the 4-step plan to solve each problem.

1. Latoya earns \$1.75 an hour for each child she baby-sits. How much does she earn if she baby-sits 3 children for 4 hours?
2. James works at a department store and receives a 25% discount on his purchases. He purchases some new clothes at the store and his total before the discount is \$145.20. What is James's total after the discount?
3. Kegan borrows \$78 from his sister. He will pay her back over a 4-week period. If he pays the same amount each week, how much will he have paid back after the third week?
4. Ashley stops by the grocery to pick up a few items. She buys a loaf of bread for \$1.09, a pound of turkey for \$4.59, 2 cans of soup for \$0.79 each, and 4 oranges for \$0.27 each. How much money does Ashley spend at the grocery?
5. Jackie buys a one-way ticket to Nevada and a one-way ticket back home. Each way costs \$118. Joe buys a round-trip ticket to Nevada for \$227. Whose ticket is less? By how much?

Skills Workshop 15

Metric Measures

The basic metric units are meter (length), liter (capacity), and gram (mass or weight). All measurements can be expressed in terms of these three basic units. However, prefixes are used with the basic units to avoid dealing with very large and very small numbers.

The same prefixes are used for length, capacity, and mass.

1,000 m	100 m	10 m	1 m	0.1 m	0.01 m	0.001 m
kilo-meter	hecto-meter	deca-meter	meter	deci-meter	centi-meter	milli-meter
km	hm	dcm	m	dm	cm	mm

1,000 L	100 L	10 L	1 L	0.1 L	0.01 L	0.001 L
kilo-liter	hecto-liter	deca-liter	liter	deci-liter	centi-liter	milli-liter
kL	hL	dcL	L	dL	cL	mL

1,000 g	100 g	10 g	1 g	0.1 g	0.01 g	0.001 g
kilo-gram	hecto-gram	deca-gram	gram	deci-gram	centi-gram	milli-gram
kg	hg	dcg	g	dg	cg	mg

EXAMPLE 1 Change 0.68 meters to centimeters

SOLUTION Think: 1 m = 100 cm
0.68 m = 68 cm Move the decimal point right 2 spaces.

EXAMPLE 2 Change 8,000 grams to kilograms

SOLUTION Think: 1,000 g = 1 kg
8,000 g = 8 kg Move the decimal point left 3 spaces.

EXAMPLE 3 Change 5.2 liters to milliliters

SOLUTION Think: 1 L = 1,000 mL
5.2 L = 5,200 mL Move the decimal point right 3 spaces.

Change each measurement to the named unit.

1. 76 g = __ cg
2. 88 mL = __ L
3. 200 m = __ mm
4. 34 kL = __ cL
5. 123 mg = __ g
6. 7,065 L = __ kL
7. 4.35 g = __ hg
8. 0.98 m = __ cm
9. 12.5 kg = __ g
10. 44 dcm = __ mm
11. 600 kg = __ cg
12. 0.025 L = __ kL

Skills Workshop 16

Customary Measures

The only major industrial nation that uses the Customary System of Measurement is the United States. If you are converting from a larger unit to a smaller unit, multiply. If you are converting from a smaller unit to a larger unit, divide.

Length
1 foot (ft) = 12 inches (in.)
1 yard (yd) = 3 ft
1 mile (mi) = 5,280 ft or 1,760 yd

Mass or Weight
1 pound (lb) = 16 ounces (oz)
1 ton (T) = 2,000 lb

Capacity
1 tablespoon (tbsp) = 3 teaspoons (tsp)
1 fluid ounce (fl oz) = 6 tsp
1 cup (c) = 16 tbsp or 8 fl oz
1 pint (pt) = 2 c
1 quart (qt) = 2 pt
1 gallon (gal) = 4 qt

EXAMPLE 1 Change 3 feet to inches

SOLUTION Think: 1 ft = 12 in.

You are converting from a larger unit to a smaller unit, so multiply.

3 ft = 3 × 12 = 36 in. Multiply 3 by 12.

EXAMPLE 2 Change 6 cups to pints

SOLUTION Think: 1 pt = 2 c

You are converting from a smaller unit to a larger unit, so divide.

6 c = 6 ÷ 2 = 3 pt Divide 6 by 2.

EXAMPLE 3 Change 1.5 tons to pounds

SOLUTION Think: 1 T = 2,000 lb

You are converting from a larger unit to a smaller unit, so multiply.

1.5 T = 1.5 × 2,000 = 3,000 lb Multiply 1.5 by 2,000.

Change each measurement to the named unit.

1. 14 yd = __ ft
2. 7 qt = __ pt
3. 40 tbsp = __ c
4. 16 c − __ fl oz
5. 18 gal = __ qt
6. 5 mi = __ yd
7. 11,000 lb = __ T
8. 0.25 T = __ oz
9. 12 tsp = __ tbsp
10. 18 ft = __ yd
11. 7,920 ft = __ mi
12. 8 pt = __ gal
13. 252 in. = __ yd
14. 129 tsp = __ fl oz
15. 48 oz = __ lb

Skills Workshop 17

Sales Tax

Sales tax is a percentage of the price of an item or a percentage of the total of all taxable items. Sales tax is rounded to the nearest cent.

Sales Tax = Price of Item × Sales Tax Rate

The buyer pays the price of the item plus the sales tax.

Total Cost of Item = Price of Item + Sales Tax

EXAMPLE 1 Ellie Kramer purchased a chain saw priced at $285.99. She paid 6.5% sales tax. What amount did she pay in sales tax? What was the total amount Ellie paid for the chain saw?

SOLUTION 6.5% = 0.065 Write the sales tax rate as a decimal.
$285.99 × 0.065 = $18.58935, or $18.59 Multiply price by rate.
$285.99 + $18.59 = $304.58 Add price and sales tax.

To find the sales tax when some items of a purchase are not taxable, first find the subtotal of all taxable items. Then calculate the sales tax on that portion of the bill only. To find the total bill, add the subtotals of nontaxable and taxable items, with the amount of sales tax.

Sales Tax = Subtotal of Taxable Items × Sales Tax Rate

Total = Subtotal of Taxable Items + Subtotal of Nontaxable Items + Sales Tax

EXAMPLE 2 A service station mechanic took 3 hours to repair a car. The service charge was $65 an hour. Two parts were replaced at a cost of $117.98 and $49.39. A sales tax of 4% is charged on goods, but not on labor. Find the total bill.

SOLUTION 3 × $65 = $195 cost of nontaxable labor
$117.98 + $49.39 = $167.37 Add the cost of the taxable parts.
$167.37 × 0.04 = $6.694, or $6.69 Multiply rate by subtotal.
$195 + $167.37 + $6.69 = $369.06 nontaxable labor + taxable items + sales tax

Find the sales tax on each of the items below.

1. computer scanner, $99.99, 5.8%
2. garden tractor, $1,568.89, 3.5%
3. golf club set, $635.18, 2.9%
4. antique dresser, $498.89, 4.6%
5. Mona Allen wants to buy a bed that costs $695. The city sales tax rate is 7%. In a nearby city the sales tax rate is 4%. How much less would the bed cost if Mona bought it in the nearby city?
6. Rollie Gusewelle has his auto dealer install auto seat covers on his used car. The seat covers costs $189.99 and installation cost $45.89. The sales tax rate is 4.7% but is not applied to labor. What is the total cost of the seat covers to Rollie?

Skills Workshop 18

Bar Graphs

Business firms use graphs to show data about their companies or industries. Graphs often show facts and trends more clearly than do numbers in tables.

The vertical bar graph shown at the right displays the daily sales of The Building Center for a week. The height of each bar shows the sales for each day. Each vertical block on the graph equals $100 of sales. The daily sales are rounded to the nearest $50.

The Building Center
Daily Sales
Week Ending May 7, 20--

Sales in Dollars
1500
1000
500
0
M T W T F S
Days of Week

EXAMPLE 1 Use the vertical graph for The Building Center to find the day on which sales were the greatest. What was the amount of sales for that day?

SOLUTION Determine the sales for each bar. Select the bar that shows the greatest sales amount. The greatest sales amount is $1,650 for Saturday.

The horizontal bar graph at the right, with bars running left to right, shows the sales by department of The Building Center. Each horizontal block on the graph equals $2,000. The amounts for each bar were rounded to the nearest $1,000.

EXAMPLE 2 Use the horizontal graph above to find the department in which the sales for the second quarter were greater than $29,000. Find the total quarterly sales in that department.

SOLUTION Determine the sales represented by each bar. Identify the department with sales greater than $29,000. The lumber department's sales were $38,000.

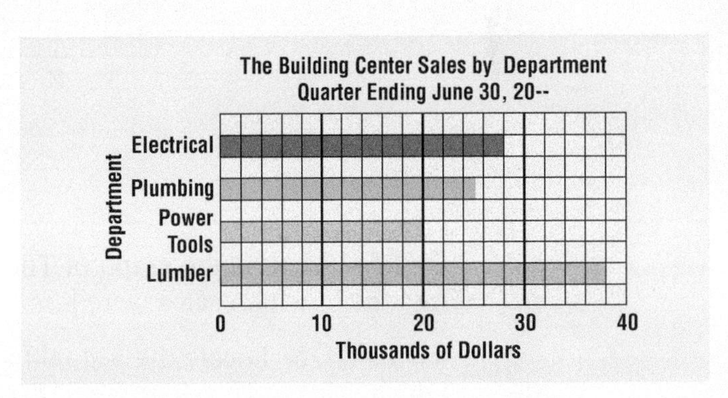

Refer to the vertical bar graph of The Building Center for Exercises 1–3.

1. On which two days was the difference in sales the greatest?
2. On which days were sales below $1,200?
3. What were the total sales for Wednesday through Friday?

Refer to the horizontal bar graph of The Building Center for Exercises 4–5.

4. For which two departments were the sales most nearly the same during the quarter?
5. How much greater were the sales of lumber than the sales of power tools?

Skills Workshop 19

Line Graphs

The line graph shown displays the sales of The Building Center by months. The time scale runs from left to right and is at the bottom of the graph. The dollar scale runs from bottom to top and is at the left.

The monthly sales were rounded to the nearest $1,000. The line graph is made by first placing dots showing each month's sales. The dots are then connected by drawing a line using a ruler.

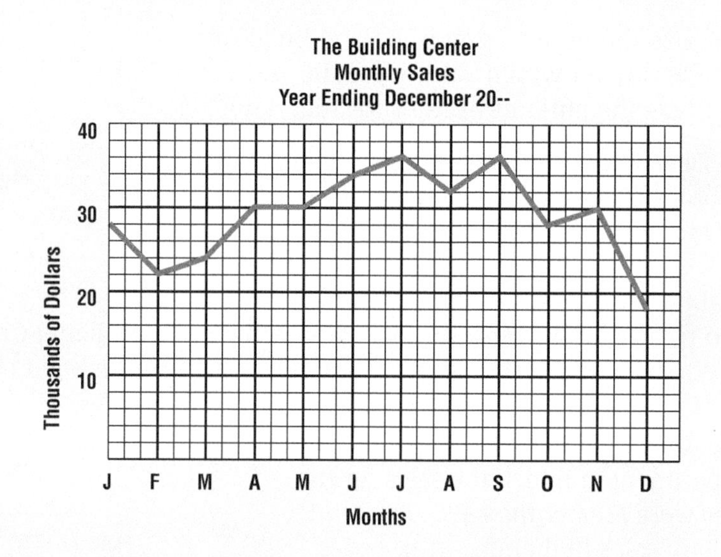

EXAMPLE Use the graph of The Building Center's monthly sales to find the months in which sales were less than $24,000.

SOLUTION Locate the $24,000 mark on the dollar scale. Locate all the months in which sales were below the $24,000 mark. February and December were months with sales less than $24,000.

Refer to the line graph of The Building Center for Exercises 1–4.

1. In the graph of The Building Center's monthly sales, what were the approximate sales for each month from October through December?
2. Between which two consecutive months did sales increase the most?
3. Between which months was there less than a $2,000 difference in sales?
4. What were the total sales for the second quarter of the year?

Skills Workshop 20

Circle Graphs

Circle graphs are used to show how parts relate to the whole and to each other. Circle graphs are based on a whole circle, or 100%. A circle has (360°). A circle graph is divided into parts, called sectors.

EXAMPLE Marvin earns $250 net pay per month. He plans to spend these amounts monthly in each category: Entertainment, $50; Clothes, $50; Meals, $30; School, $20; Miscellaneous, $25; Savings, $75. Display Marvin's budget in a circle graph.

SOLUTION Show the budget amounts as percents by dividing the amount budgeted for each category by the total budget, rounded to the nearest percent. Multiply each percent by 360°, rounded to the nearest whole degree.

Budget Category	Amount	Expressed as Percent	Degrees in Sector
Entertainment	$50	$\frac{\$50}{\$250} = 20\%$	20% of 360° = 72°
Clothes	$50	$\frac{\$50}{\$250} = 20\%$	20% of 360° = 72°
Meals	$30	$\frac{\$30}{\$250} = 12\%$	12% of 360° = 43°
School	$20	$\frac{\$20}{\$250} = 8\%$	8% of 360° = 29°
Miscellaneous	$25	$\frac{\$25}{\$250} = 10\%$	10% of 360° = 36°
Savings	$75	$\frac{\$75}{\$250} = 30\%$	30% of 360° = 108°
TOTALS	$250	$\frac{\$250}{\$250} = 100\%$	100% of 360° = 360°

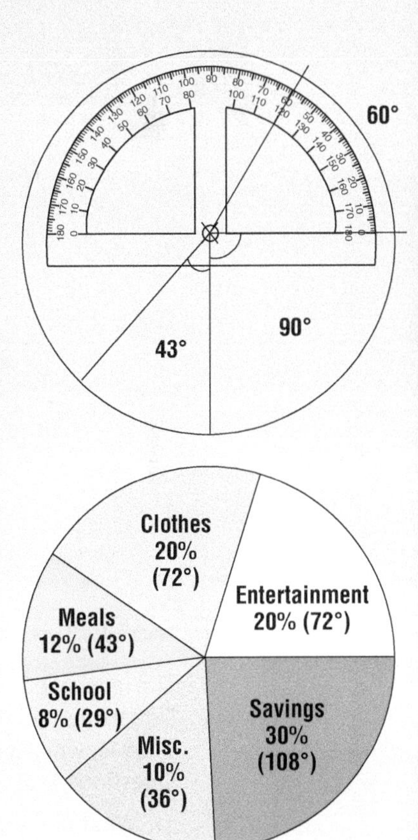

Use a compass to draw a circle. Mark the center of the circle. Use a protractor to draw the angles from the center of the circle that match each degree calculated above. Label each category.

The table shows how LaRowe Company spent $60,000 on advertising.

1. Find the percent of advertising spent on each type.
2. Make a circle graph of the LaRowe Company's advertising costs.

Newspaper	$18,000
Direct mail	$15,000
Internet	$12,000
Coupons	$9,000
Product samples	$3,000
Other	$3,000

Chapter 1

Gross Pay

Statistical Insights

Median Weekly Income Based on Level of Education

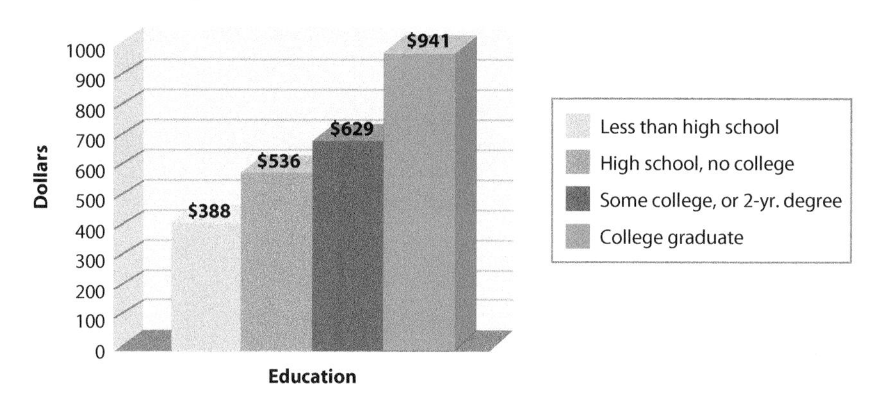

Source: U.S. Dept of Labor, Bureau of Labor Statistics

The level of education people receive consistently affects their earning power. Statistics from the Department of Labor show an American with a college degree earns a median weekly income of $941; while a person without a high school diploma earns $388. Use the bar graph to answer Questions 1–5.

1. What percent of a college graduate's earnings does a high school graduate make?
2. About how many weeks would a person without a high school diploma have to work to earn what a person with some college does in one week?
3. How much less each week does a person that did not finish high school earn than a college graduate?
4. If the average person earns a 4-yr college degree by the age of 22, about how old would that person be when he or she earns in excess of $1,000,000 in gross wages?
5. If the average age of a person that drops out of high school is 17 years old, is it likely that a person that does not finish high school will earn in excess of $1,000,000 in gross wages before retirement at the age of 65? Explain.

How Times Have Changed

The minimum wage does not automatically rise with inflation. The United States Congress determines when and how much to raise the minimum wage. The 25-cent minimum wage set by Congress in 1938 seems very low by today's standards, but when taking inflation into account, 25-cents in 1938 would have the buying power of $3.81 in 2008.

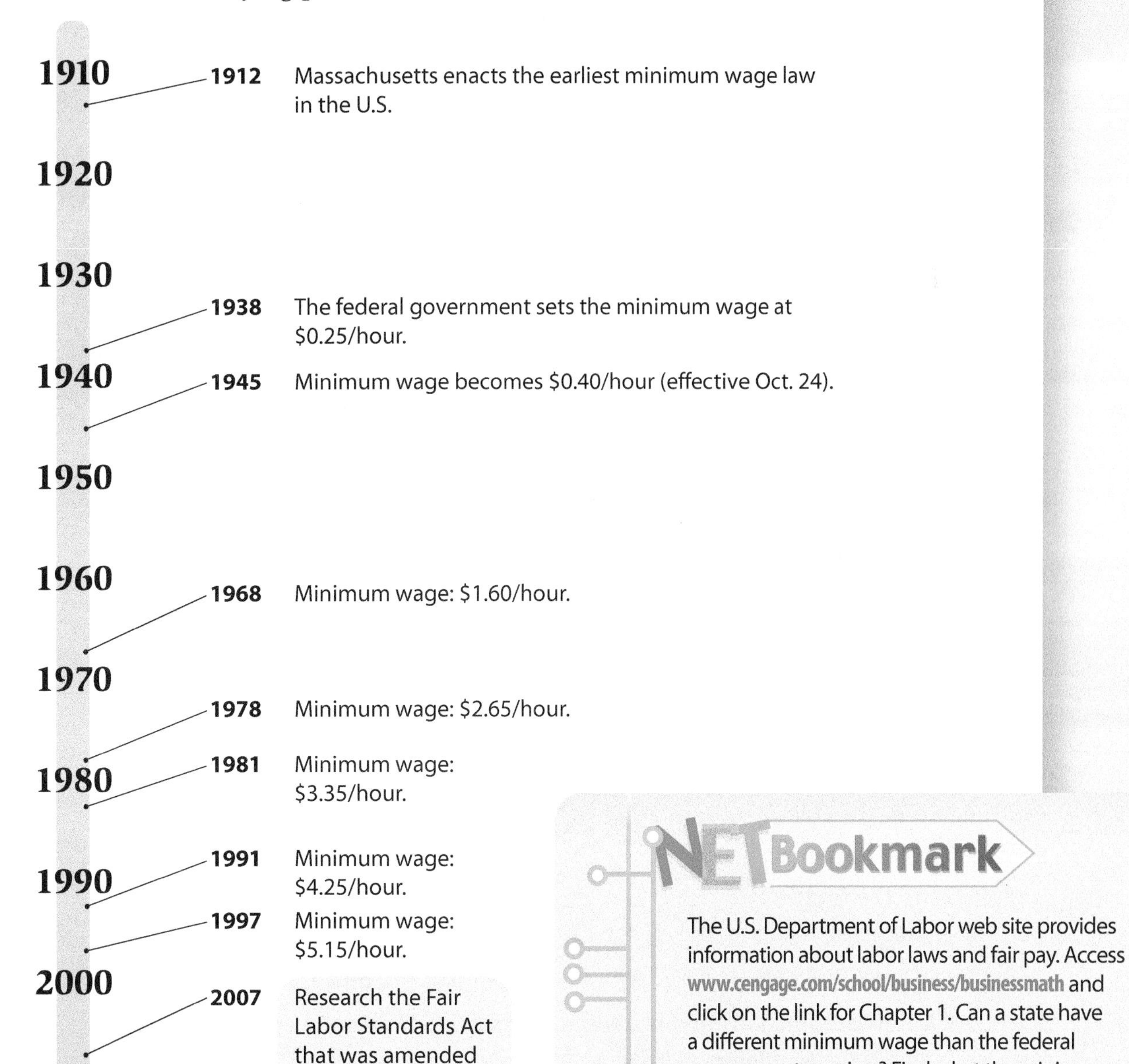
1910
1920
1930
1940
1950
1960
1970
1980
1990
2000
2010
1912 Massachusetts enacts the earliest minimum wage law in the U.S.
1938 The federal government sets the minimum wage at $0.25/hour.
1945 Minimum wage becomes $0.40/hour (effective Oct. 24).
1968 Minimum wage: $1.60/hour.
1978 Minimum wage: $2.65/hour.
1981 Minimum wage: $3.35/hour.
1991 Minimum wage: $4.25/hour.
1997 Minimum wage: $5.15/hour.
2007 Research the Fair Labor Standards Act that was amended in 2007 to find the changes made to the minimum wage laws

NET Bookmark

The U.S. Department of Labor web site provides information about labor laws and fair pay. Access **www.cengage.com/school/business/businessmath** and click on the link for Chapter 1. Can a state have a different minimum wage than the federal government requires? Find what the minimum wage is in your state.

1-1

Hourly Pay

GOALS
- Calculate gross pay for hourly-rate employees
- Compute overtime pay rates
- Calculate regular and overtime pay

KEY TERMS
- employee
- employer
- hourly rate
- gross pay
- overtime
- time-and-a-half pay
- double-time pay

Start Up

Sarah has been offered a new job as an assistant manager in a shoe department. She will work 54 hours each week and earn \$15 an hour. For this work schedule and rate of pay, is it possible for Sarah's gross pay for one month to be at least \$3,750?

Photodisc/Getty Images

Math Skill Builder

Review these math skills and solve the exercises that follow.

1 **Round** to the nearest cent.
Round. \$1,874.898 = \$1,874.90

1a. \$24.373 **1b.** \$99.995 **1c.** \$537.307

2 **Multiply** decimals. Remember to round to the nearest cent.
Multiply. $\$18.30 \times 6.5 = \$118.95$

2a. $\$12.80 \times 1.5$ **2b.** $\$11.63 \times 40.5$ **2c.** $\$9.87 \times 45.7$

3 **Add** decimals.
Add. $6.25 + 4.75 + 7.5 = 18.5$

3a. $5.5 + 7.25 + 8 + 7 + 6.75$ **3b.** $8.4 + 9.1 + 7.7$

3c. $3.2 + 6.45 + 1.35 + 9$ **3d.** $7.55 + 6.82 + 2.8 + 4.3$

4 **Rewrite** fractions and mixed numbers as their decimal equivalents.
Rewrite as a decimal. $1\frac{1}{2} = 1.5$

4a. $\frac{3}{4}$ **4b.** $\frac{1}{2}$

4c. $6\frac{3}{4}$ **4d.** $48\frac{3}{4}$

Gross Pay for Hourly-Rate Employees

Most people earn money by working for others. Those who work for others are called **employees**. The person or company an employee works for is called an **employer**.

An employee who is paid by the hour works for an **hourly rate**, which is a certain amount for each hour worked. The total amount of money that an employee is paid is called **gross pay** or *gross wages*. Gross pay may also be called total earnings or total pay.

The gross pay earned by employees who are paid by the hour is found by multiplying the pay per hour by the hours worked.

Gross Pay = Number of Hours Worked × Hourly Rate

Algebra *Tip*

Gross pay is calculated using the formula

$$G = H \times R$$

where G represents gross pay, H represents the number of hours, and R represents the rate of pay per hour.

EXAMPLE 1

Mark Willow works as a customer service representative and is paid $9.10 per hour. He worked 38 hours last week. Find his gross pay.

SOLUTION

Substitute the known values in the formula.

$G = 38 \times \$9.10 = \$345.80$ $\quad G = H \times R$

Mark's gross pay for last week was $345.80.

Calculator *Tip*

When you input decimal numbers into a calculator, the decimal point in the answer will be placed correctly automatically. This feature is called *floating decimal point.* Your calculator will show 345.8. To write the answer, add the 0 and a dollar sign.

✔ CHECK YOUR UNDERSTANDING

A. Kenny Walker works as a shopping mall security guard and is paid $7.50 an hour. Find his gross pay when he works 46 hours a week.

B. Cassie Boland earns $16.25 an hour. What gross wages did she earn last week by working 25 hours?

EXAMPLE 2

Sharon Medal is paid $7 an hour. Last week Sharon worked 8 hours a day for 5 days. Find her gross pay for last week.

SOLUTION

$5 \times 8 = 40$ $\quad$ hours worked in last week

$G = 40 \times \$7 = \$280$ $\quad$ gross pay for last week

✔ CHECK YOUR UNDERSTANDING

C. Rosa Mendez works 8 hours Monday through Friday as a legal assistant. Her hourly wage is $18.75. Find how many hours she worked in one week and her gross pay for that week.

D. Vincent O'Malley worked the following schedule one week: Monday, 8 hours; Tuesday, 6 hours; Wednesday, 7 hours; Thursday, 6 hours; Friday, 5 hours. He was paid $9 an hour. How many hours did Vincent work that week and what was his gross pay?

Recording Hours Worked

Many companies keep an exact record of the number of hours their employees work. They record the times people arrive at work, take breaks, and leave for the day.

Companies may use electronic methods such as a magnetic stripe card or a time card to track employees' time. Other companies use a time sheet where employees write the hours they worked.

The hours that employees work are typically recorded by the quarter hour (15 minute segments) or the tenth of an hour (6 minute segments). Parts of the hour recorded in quarters shown as $\frac{1}{4}$, $\frac{1}{2}$, or $\frac{3}{4}$ of an hour, or 0.25, 0.5, or 0.75 hours. Parts of any hour recorded in tenths are shown as 0.1, 0.2, and so on.

Photodisc/Getty Images

Employees who arrive late or leave early may be penalized by a quarter or tenth of an hour. For example, a company that records time by the quarter hour may penalize an employee 15 minutes for arriving 3 or more minutes late.

Overtime and Overtime Pay Rates

Companies record regular hours of work and **overtime**, which is time worked beyond the regular working day or week. Daily overtime is based on a regular working day, such as an 8-hour day. So, an employee who works 10 hours in one day will be paid for 8 hours regular time and 2 hours overtime.

If the regular working week is 40 hours, then an employee who works 45 hours in a week is paid for 40 regular hours and 5 overtime hours.

Overtime pay is often figured at one and a half times (× 1.5) the regular-time rate and is called **time-and-a-half pay**. Sometimes **double-time pay** is given for work over a certain number of hours, or for work on weekends and holidays. Double-time pay is twice (× 2) the regular-time pay rate.

Time-and-a-Half-Rate = 1.5 × Regular Pay Rate (Do NOT round.)

Double-Time Rate = 2 × Regular Pay Rate (Do NOT round.)

Algebra *Tip*

To find the pay rate for overtime hours, whatever the multiplying factor is, use the formula,

$$O = F \times R$$

where O is the overtime pay rate, F is the number of times the regular pay rate is increased, and R is the regular pay rate.

EXAMPLE 3

Paul Mears' regular pay is \$10.73 an hour. His employer pays overtime at 1.5 of the regular rate and double time at twice the regular rate of pay. What are Paul's time-and-a-half and double-time pay rates?

SOLUTION

1.5 × \$10.73 = \$16.095 time-and-a-half rate

2 × \$10.73 = \$21.46 double-time rate

✔ CHECK YOUR UNDERSTANDING

E. Find time-and-a half and double-time rates for these regular-time pay rates.

1) $7.51 **2)** $8.76 **3)** $13.67

F. Emilio's regular-time pay rate is $11.25 per hour. What time-and-a-half and double-time rates would he earn for overtime work?

Regular and Overtime Wages

To find gross wages for an employee who has worked both regular time and overtime, use these steps.

Step 1: Find the number of regular-time and overtime hours worked.

Step 2: Find the regular time pay.
Regular Pay = Hourly Rate × Regular Hours Worked

Step 3: Find the overtime pay.
Overtime Pay = Overtime Rate × Overtime Hours

Step 4: Find the gross wages.
Gross Pay = Regular Pay + Overtime Pay

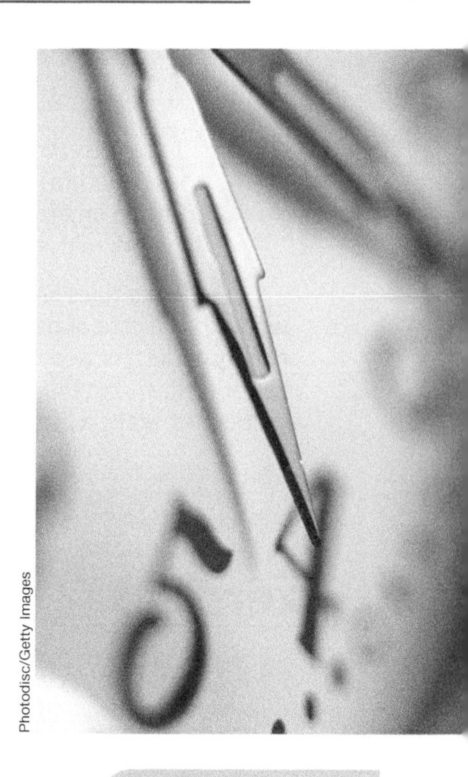
Photodisc/Getty Images

EXAMPLE 4

Stanley Bartlett worked these hours last week: Monday, $7\frac{1}{4}$; Tuesday, 10; Wednesday, $9\frac{1}{2}$; Thursday, 8; Friday, 9. Stanley is paid based on an 8-hour day with time-and-a-half for daily overtime. If Stanley's regular-time pay rate is $9.50 per hour, what gross wages did he earn last week?

SOLUTION

Step 1: $7.25 + 8 + 8 + 8 + 8 = 39.25$ regular-time hours
$2 + 1.5 + 1 = 4.5$ time-and-a-half overtime hours

Step 2: $39.25 \times \$9.50 = 372.875 \approx \$372.88$ regular-time pay

Step 3: $4.5 \times \$9.50 \times 1.5 = \$64.125 \approx \$64.13$ time-and-a-half pay

Step 4: $\$372.88 + \$64.13 = \$437.01$ gross wages

Calculator *Tip*

It is often easier to work with decimals than fractions. To change a fraction such as $\frac{3}{4}$ to a decimal, use $3 \div 4 = 0.75$.

✔ CHECK YOUR UNDERSTANDING

G. Xavier Centor works an 8-hour day. He is paid $13.69 an hour for regular-time work and time-and-a-half for any hours over 8 hours a day. Xavier worked these hours last week: Monday, $9\frac{3}{4}$; Tuesday, 8; Wednesday, 6; Thursday, $8\frac{1}{2}$; Friday, 8. Complete all steps to find Xavier's gross wages for last week.

H. Diedra McKenney works on a 40-hour week basis with time-and-a-half paid for overtime work. Her regular-time hourly rate is $17.50. Last week she worked 45.3 hours from Monday through Thursday and 8.1 hours on Friday. Complete all steps to find Diedra's gross wages for last week.

Wrap Up ▶▶▶

If Sarah is paid regular time for the 54 hours per week, her weekly gross pay is \$810. A month has anywhere from 4 weeks to about $4\frac{1}{2}$ weeks. So, Sarah's monthly pay will be between \$3,240 and \$3,645. If Sarah earns time-and-a-half overtime pay for any hours over 40 hours a week, then she would make between \$3,660 and \$4,117.50 per month.

TEAM Meeting

Imagine you are responsible for supervising a small group of workers and have been assigned to write guidelines for them. The company you work for expects employees to be on time for work and not leave early. The penalty for employees who break this work rule is a loss of pay.

What rules would you have about paying employees who are ill or need time off from work to take care of personal business? Will the rules be different for new employees or the same as those rules for people who have been with the company longer? Be fair to your employees, to yourself, and to the company when writing the guidelines. As a class, discuss everyone's guidelines. Make a list of the rules the class generally agrees on.

Exercises

Find each sum.

1. $8 + 7 + 6.5 + 7 + 8$

2. $7 + 6.5 + 7.5 + 8 + 6$

3. \$1,500 + \$723

4. \$680 + \$72

Find each product.

5. 38 × \$12

6. 40 × \$9.80

7. 25 × \$412

8. 7.5 × \$8.10

9. 7.75 × \$418

10. 10 × \$1,200

Rewrite as decimals.

11. $40\frac{1}{4}$

12. $\frac{3}{4}$

13. $43\frac{1}{2}$

Round to nearest cent.

14. \$78.438

15. \$298.987

16. \$419.097

Solve.

17. Francesco Jardin earns \$7 an hour at his part-time job. Last week he worked 16 hours. What was his gross pay for the week?

18. Alberta Doan worked 6 hours at time-and-a-half pay and $3\frac{1}{4}$ hours at double-time pay. Her regular pay rate was \$9.72 an hour. What was Alberta's total overtime pay for the week?

Jaci Welk is paid $11.95 an hour with time-and-a-half pay for all hours she works over 40 hours a week. Last week she worked $45\frac{1}{2}$ hours.

19. How many overtime hours did Jaci work?

20. What was her overtime rate?

21. What was her overtime pay?

Steve Gaimes is paid overtime for all time worked past 40 hours in a week. His regular-time pay rate is $12 an hour, and his overtime pay rate is $18 an hour. Last week Steve worked 47.3 hours.

22. How many regular-time hours did Steve work last week?

23. How many overtime hours did he work last week?

24. What was Steve's regular-time pay last week?

25. What was his overtime pay last week?

26. What was Steve's gross or total pay last week?

Photodisc/Getty Images

Ike Phillips worked 6.7 hours at time-and-a-half pay and 3.4 hours at double-time pay last week. His regular earnings for the week were $454.80 figured on a regular pay rate of $11.37 an hour.

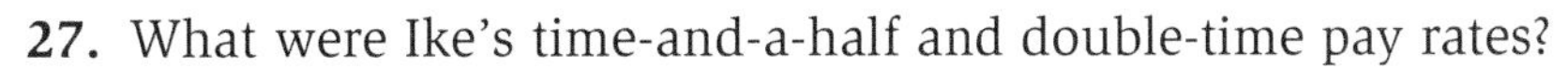

27. What were Ike's time-and-a-half and double-time pay rates?

28. What amounts did he earn for time-and-a-half and double-time work?

29. What was Ike's total gross pay for the week?

The chart below shows the hourly pay and hours worked for Trilton Company's three employees. Employees get paid time-and-a-half for any hours worked over 40 hours per week. Copy and complete the chart.

	Employee	Hourly Pay	Hours Worked	Gross Pay
30.	Rick Wilson	$9.25	40	
31.	Art Dillart	$9.45	52	
32.	Letitia Reed	$9.75	42	
33.	Total			

34. **CRITICAL THINKING** At the end of an interview you are offered a job. You would start at $7.50 an hour and work an average of 45 hours a week. At the end of six months with a positive evaluation, your hourly pay would increase to $8. The job pays overtime at a time-and-a-half rate, based on a $37\frac{1}{2}$ hour regular work week. What total earnings could you expect to make by working a full year?

Mixed Review

35. $680 + $302

36. 52 × $826

37. 40 × $7.80

38. 7.5 × $9.42

39. 115% × $30

40. $34,000 ÷ $10,000

1-2

Salary

GOALS

- Compare hourly pay and salary
- Calculate gross pay for salaried employees

KEY TERM

- salary

Start Up

Riel is offered two jobs. One job pays $11.25 per hour and requires 40 hours a week, 50 weeks a year. The other job has a salary of $22,000 per year and includes 2 weeks of vacation. What should he consider in order to decide which job to take?

Blend Images/Jupiter Images

Math Skill Builder

Review these math skills and then answer the questions that follow.

1. **Multiply** money amounts by whole numbers.
   Find the product. $385 × 12 = $4,620

   1a. $520 × 4 **1b.** $782 × 2 **1c.** $500 × 52

2. **Divide** money amounts by whole numbers.
   Find the quotient. $27,000 ÷ 12 = $2,250

   2a. $42,000 ÷ 24 **2b.** $65,000 ÷ 52 **2c.** $18,000 ÷ 12

3. **Round** to the nearest cent.
   Round this amount to the nearest cent. $842.875 = $842.88

   3a. $675.891 **3b.** $893.996 **3c.** $258.333

Comparing Hourly Pay and Salary

Some employees are paid a **salary**, which is a fixed amount of money for each pay period worked. These employees are referred to as *salaried employees* and have an agreement with their employers about how much the job pays, their responsibilities, and benefits.

Salaried employees get paid a fixed amount, regardless of the number of hours worked. They do not get paid for overtime hours. A full-time salaried employee may work more than 40 hours per week on a regular basis.

Unlike hourly employees, salaried employees still get paid the same amount even if they miss work due to illness or a vacation, as long as they are within the guidelines specified by the employer.

Typically, salaried employees earn more money than hourly employees and often more education and skills are required. For example, a bank manager would likely be paid a salary, while a bank teller would likely be paid an hourly rate.

Gross Pay for Salaried Employees

Salaries are often stated as a yearly amount, although they can be stated by day, week, or month. Salaried employees are most often paid monthly, bi-monthly (2 times/mo), bi-weekly (every other week), or weekly. For a salaried employee, if a weekly or monthly salary is known, find the annual gross pay by multiplying the salary for a pay period by the number of periods worked. If a yearly salary is known, find the gross pay per pay period by dividing the annual salary by the number of time periods in a year.

Math Tip

1 year = 12 months
1 year = 52 weeks
1 year = 365 days

EXAMPLE 1

Albert Meyer is paid a salary of $465 a week. How much gross pay does Albert receive for 4 weeks of work?

SOLUTION

$465	pay for 1 week
× 4	weeks worked
$1,860	gross pay for 4 weeks

✔CHECK YOUR UNDERSTANDING

A. Tek Research pays its office manager a weekly salary of $680. What gross pay will the office manager receive every 2 weeks?

B. Tom works as a dispatcher and is paid a weekly salary of $540. What gross pay will Tom earn for one year of work?

EXAMPLE 2

Janice Compton is paid a yearly salary of $51,000. What are her gross monthly wages?

SOLUTION

Divide the yearly salary by 12, and round to the nearest cent, if necessary.

$51,000 ÷ 12 = $4,250 monthly wages; There are 12 months in a year.

✔CHECK YOUR UNDERSTANDING

C. How much gross pay will a salaried worker receive each week if he is paid $85,000 yearly?

D. A bank manager makes $95,000 per year. If she is paid bi-monthly, what are her gross wages each paycheck?

Wrap Up ▸▸▸

Riel should compare the gross earnings for each job. The hourly rate job pays $22,500 per year, assuming that he works 40 hours per week for 50 weeks of the year. The salaried job will pay $22,000, even if he takes sick leave, but he may have some weeks that he has to work more than 40 hours. Riel could consider the hours required for the salaried position are not fixed. He could also consider which job has the work that he likes best or that will help him in his future.

Consumer Alert

Work at Home Scams

"Make BIG $$ working from the comfort of your own home. Work a few hours a day, and make $10,000 a month!"

Work from home job listings are everywhere—on television, radio, Internet, and hanging on the local bulletin board. Some claim you can make big bucks stuffing envelopes, doing data entry, typing ads, or any number of ways. Typically these jobs come with no hourly wage or salary, only a promise you will make money. Often they require you to pay fees or make purchases in order to get started.

Most of these work-at-home opportunities are scams that are designed to take your money, and not make you money.

Tips for Avoiding Work From Home Scams:

1. Avoid listings that guarantee you wealth or financial success or that will help you get rich fast. If it sounds too good to be true, it probably is.
2. Check out the company with the Federal Trade Commission, the Better Business Bureau, state Attorney General, or your local consumer protection agency.
3. Check references. Ask for names of contractors or employees to speak with about the company.
4. Ask how you will be paid, and how often.
5. Ask what equipment you need to provide.
6. Do not send money. Legitimate employers don't charge you to get started.
7. Do not purchase work-at-home directories or start-up kits. Free information is available online.
8. Do not give out your personal information to a person or company you don't know.

Exercises

Find the product or quotient.

1. $15,000 ÷ 12
2. $24,000 ÷ 24
3. $1,875 × 24
4. $584 × 52

Round to the nearest cent.

5. $1,435.783
6. $589.355

Solve.

7. Tom Page earns $368 a week. Tom is paid every two weeks. What gross pay does he receive each payday?

8. Eldon Cavanaugh is paid a weekly salary of $562. How much would Eldon earn in 4 weeks of work?

Emily Casper earns a weekly salary of $785. How much will she make after:

9. two weeks
10. four weeks?
11. one year?

Jorge Rodriguez earns an annual salary of $48,000. Find his gross wages for each given pay period.

12. monthly
13. bi-weekly
14. weekly

Photodisc/Getty Images

15. **STRETCHING YOUR SKILLS** Ryo Akita currently earns a monthly salary of $2,200. She has been offered a raise of $250 per month. How much more will she earn per year at her new salary?

16. **STRETCHING YOUR SKILLS** An employer has three employees who are each paid a salary of $1,250 per month. He wants to give each of the three employees a $125 per month raise and he wants to hire an additional salaried employee. If his salary budget for these employees is $5,325 per month, how much can he afford to offer the new employee?

17. **CRITICAL THINKING** What are the advantages of being paid a salary instead of an hourly rate? What are the disadvantages of being paid a salary instead of an hourly rate?

18. **FINANCIAL DECISION MAKING** You are offered two jobs that you like. One job pays an hourly rate of $15.00 per hour for a 40-hour work week. Employees can take a two-week paid vacation. The employer says you will work about 5 hours of overtime per week at time-and-a-half pay. The other job pays $38,000 per year. You will be expected to work at least 55 hours per week, and you will have two weeks vacation and sick leave. Which job would you choose? Why?

Mixed Review

19. $835 × 52
20. $1,435.60 × 12
21. $15,000 ÷ 12

22. Andrea Marshall is paid $10 per hour for a 40-hour work week, and time-and-a-half for hours over 40 per week. She worked the following hours last week: Monday 9 hours, Tuesday 7 hours, Wednesday $8\frac{1}{2}$ hours, Thursday 6 hours, Friday 9 hours, Saturday 3 hours. What were her gross wages?

23. Joanna Grimshaw makes an hourly rate of $12.50, and she works 40 hours per week. Her boss offers her a promotion to a salaried position that pays $675 per week. How much more will she earn per week with the salaried position?

1-3

Commission

GOALS

- Calculate straight commission earnings
- Calculate commission earnings based on quota
- Calculate graduated commission earnings
- Find the rate of commission

KEY TERMS

- commission
- straight commission
- quota
- graduated commission

Start Up

Two sales jobs are advertised in the newspaper. The first job pays a commission of 3.5% on all sales, with expected sales of $52,000 a month. The second job pays a commission of 4% on sales up to $5,000 and 12% on sales over $5,000, with expected sales of $25,000 a month. Based on the expected sales, which job pays more?

Photodisc/Getty Images

Math Skill Builder

Review these math skills and solve the exercises that follow.

1. **Rewrite** a percent as a decimal. Rewrite this percent. 50% = 0.5

1a. 12% **1b.** 18.6% **1c.** 2.5%

2. Find a **percent of a number**.
Find the percent. 1% of $1,956 = 0.01 × $1,956 = $19.56

2a. 14% of $500 **2b.** 3.5% of $1,200

2c. 12% of $600 **2d.** 2% of $38.50

3. **Rewrite** a decimal as a percent. Rewrite this decimal. 0.25 = 25%

3a. 0.08 **3b.** 0.75

3c. 0.01 **3d.** 0.1825

4. Find **what percent** one number is of another number. Find the percent.
$10 ÷ $80 = 0.125 or 12.5%

4a. $15 ÷ $75 **4b.** $4,800 ÷ $32,000

4c. $34 ÷ $68 **4d.** $8 ÷ $200

Straight Commission

Some salespeople earn a commission instead of a fixed salary or hourly pay. A **commission** may be an amount for each item sold, or it may be a percent of the dollar value of sales. A higher commission may be earned for goods that are harder to sell than for goods that are easy to sell. Both a salary and a commission may be earned.

Salespeople whose earnings come only from commission work on a **straight commission** basis. When the rate of commission is an amount for each item sold, multiply the number of items by the rate to find the commission.

Commission = Quantity Sold × Rate of Commission

> **Algebra Tip**
>
> A verbal model can be helpful when solving problems.
>
> Commission Paid = Amount × Rate of Commission
>
> $C = A \times R$
>
> *A* represents a quantity or sales. *R* is a rate and when it is a percent it needs to be converted to a decimal.

EXAMPLE 1

Maxwell Lytle sells decorative notepads and is paid a straight commission of $0.80 on each notepad he sells. During December, he sold 750 notepads. Find his commission.

SOLUTION

$C = 750 \times \$0.80 = \$600$ commission

✔ CHECK YOUR UNDERSTANDING

A. Lorraine Wilk is paid a commission of $1.30 for each hand-painted tile she sells. What commission did she earn by selling 74 tiles last week?

B. Leo Margolis receives a $0.075 commission for each newspaper he sells at his newsstand. What commission would he earn by selling 1,200 newspapers?

Photodisc/Getty Images

When the rate of commission is a percent, multiply the amount of the sales by the rate to find the commission.

Commission = Sales × Rate of Commission

EXAMPLE 2

Huey Gaines is paid a straight commission of 6% on his sales. During February, his sales were $38,000. What was his commission?

SOLUTION

$C = \$38,000 \times 0.06 = \$2,280$ commission

✔ CHECK YOUR UNDERSTANDING

C. Melvin's sales last month of a new tile cleaner were $9,500. If he receives a commission rate of 15% of all sales, what commission did he earn?

D. Jacqueline earned 15% commission on her monthly sales of $2,870, $3,150, and $3,940. What was her total commission for the three months?

Commission Based on Quota

Some salespeople may be paid a commission that is a percent of their sales above a certain amount. This fixed amount is called a **quota**. Salespersons may also be paid a salary in addition to commission.

EXAMPLE 3

Leona Bahr is paid a commission of 12% on all sales above $7,000 for the week. She is also paid a weekly salary of $380. What are her total earnings for a week in which her sales were $9,800?

SOLUTION

Sales	$9,800
Quota	−7,000
Sales over Quota	$2,800

Commission: $2,800 × 0.12 = $336

Salary	$380
Commission	+$336
Total Earnings	$716

Leona's total earnings were $716.

✔CHECK YOUR UNDERSTANDING

E. Colby Richards is paid a salary of $125 a week and a 3% commission on all sales he makes above $2,000 for the week. What total earnings did he have for a week in which his sales were $2,890?

F. Lula Krobo is paid a 7% commission on all sales over $15,000 in a month and a monthly salary of $2,300. Her last month's sales were $29,700. What were Lula's total earnings for the month?

Graduated Commission

Some salespeople are paid a **graduated commission**. This means their rate of commission increases as their sales increase. For example, the rate may be 3% on the first $12,000 of sales; 4% on the next $6,000; and 5% on sales over $18,000. Graduated commissions may also be based on the number of units sold.

EXAMPLE 4

Lamont Cotton is paid 4% commission on the first $10,000 of monthly sales and 10% on all sales over $10,000. Last month his sales were $38,000. What was his commission?

SOLUTION

Commission on first $10,000: $10,000 × 0.04 = $400

Sales over $10,000: $38,000 − $10,000 = $28,000

Commission on sales over $10,000: $28,000 × 0.10 = $2,800

Total commission: $400 + $2,800 = $3,200

✔ CHECK YOUR UNDERSTANDING

G. Morgan Lee is paid a commission of 3% on the first $100,000 of monthly sales and 5% on any sales above that amount. What commission did he earn if his sales for a month were $120,000?

H. Janice Corrudo is paid a commission of 15% of her yearly sales up to $85,000 and 18% of any sales above $85,000. Her total sales for last year were $112,000. What total commission did she earn last year?

Rate of Commission on Sales

To find the rate of commission, divide the amount of commission paid on total sales by the total sales amount.

Rate of Commission = Amount of Commission ÷ Sales

EXAMPLE 5

A salesperson sold a laptop computer and software for $3,000 and received a $120 commission. What percent commission did the salesperson receive?

SOLUTION

$120 ÷ $3,000 = 0.04 = 4%

Calculator *Tip*

If your calculator has a % key, you can find the percent directly by following these steps: key 120, press ÷, key 3000, press %. Be sure to add the percent symbol to your answer.

✔ CHECK YOUR UNDERSTANDING

I. Marc received a commission of $448 for selling $6,400 in goods in the past two weeks. What rate of commission did he earn?

J. Nedra's sales last month were $54,000 for which she received a commission of $3,240. What rate of commission was Nedra paid?

Wrap Up ▸▸▸

The first sales job would pay monthly commission of $1,820. The second job would pay commission of $200 on the first $5,000 of sales and $2,400 on the $20,000 of sales over $5,000 in a month. Total monthly commission for the second job would be $2,600. The second sales job pays more.

Communication

The BBB Company now pays a straight commission of 15% of all monthly sales. Expected sales are $24,000 a month per salesperson. The company's new plan will pay a monthly salary of $1,500 and a 22% commission on all sales over $15,000 a month. Write a statement that either accepts or rejects the new plan based on an increase or decrease in the annual wages of salespeople.

Exercises

Rewrite as a decimal.

1. 9.25%
2. 16.2%
3. 0.5%

Find the amount.

4. 12% of $800
5. 7.5% of $13,000
6. 4.2% of $569

Find the percent.

7. $1.26 ÷ $7
8. $1.20 ÷ $8
9. $90 ÷ $2,000

Rewrite as a percent.

10. 0.08
11. 0.11625
12. 0.0025

Solve.

13. Dan Pawlik is paid a straight commission of $5.75 for each item he sells. Last month he sold 103 items. Find his estimated and exact commissions.

14. A student who sells subscriptions for a magazine that costs $35 a year makes a commission of $5.25 on each subscription. What percent commission does the student make?

15. Paul Batik earns a commission of 9% on sales. Last week he had sales of $646.70, $237.58, $1,984.89, $658.66, and $953.73. Find his total commission to the nearest cent.

16. Jo Ann White is paid a salary of $410 a week and a commission of 5.6% on all sales. Her sales last week were $6,700. Find her total earnings for the week.

17. Sheldon Cole earns a salary of $150 a week and a commission of 7% on all sales. If Cole's sales for one week were $6,890, what were his total weekly earnings?

18. Roosevelt Quinn receives a weekly salary of $600 plus $\frac{1}{2}$% commission on all sales in excess of $12,500 a week. Last week his sales were $48,370. What were his total earnings for the week?

19. Alice Miller works for a paint manufacturer. She is paid 3% commission on her first $20,000 of monthly sales and 8% commission on all sales over $20,000. In March her sales were $54,500; in April, her sales totaled $47,300. What were the total commissions she earned for the 2 months?

20. Olivia Thoms sells surplus books to bookstores. She is paid a weekly commission of $1.50 each on the first 50 books she sells, $1.75 each on the next 100 books, and $2 on any books she sells over 150. Last week she sold 225 books. What was her commission for the week?

21. Ludmilla Pavel is paid a commission on all sales over $3,000 a week. Last week she earned a commission of $420 on sales of $16,000. What rate of commission was she paid to the nearest tenth percent?

22. Nola Potter earns a salary of $1,200 a month and a commission of 7.5% on all sales over $4,000. This month her sales were $21,400. Find her total earnings for the month.

Martin Ellis sells a line of cooking pots. He is paid a salary of $1,150 a month plus a commission on all sales. Last month his sales were $35,000, and he earned a total salary and commission of $3,250.

23. How much commission was Martin paid?

24. What rate of commission was he paid?

25. **CRITICAL THINKING** When a new inkjet printer model is introduced salespeople may be paid a larger commission by their store for each old printer model they sell. Why would the store's manager offer such an incentive to salespeople?

26. **CRITICAL THINKING** Write two paragraphs that explain the advantages and disadvantages of working on commission. The first paragraph should be from an employee's viewpoint, the second from the viewpoint of an employer.

27. **FINANCIAL DECISION MAKING** You see ads on an Internet job listing service from two companies looking for salespeople. Both companies market a weight-loss system. The Slo-Loss Company pays a weekly salary of $100 and a commission of 14.5% on sales. The Slim-Now Company pays a straight commission of 30% of sales. Both companies expect you to be able to have sales of $1,000 in the first month and reach sales of $5,000 at the end of six months. List the reasons in outline form why you want to get one job over the other.

Mixed Review

28. Find $\frac{1}{4}$% of $28,000

29. $97,398 × 1%

30. 60 is 15% of what number?

31. 20% more than $18 is what amount?

32. $250 decreased by what percent of itself is $235?

33. Marc Bullard has two part-time jobs. At one job he worked 12 hours last week and was paid $8.15 an hour. Marc worked 6.25 hours last week at his second job that pays $7 an hour. What was his gross pay last week from both jobs?

34. Carolyn Mills does maintenance work at a golf course. She worked these hours last season: April, 150; in each of the next four months, 200; September, 100. If she gets paid $10.50 an hour, what was her gross pay for the season?

Photodisc/Getty Images

1-4

Other Wage Plans

GOALS
- Calculate gross pay for piece-rate employees
- Calculate gross pay for per diem employees
- Calculate gross pay for tip employees

KEY TERMS
- piece-rate
- per diem
- tip

Start Up

Employees whose pay varies widely may need to estimate or project annual gross income based on current earnings. Assume that a waiter's monthly earnings from hourly wages and tips for the first quarter of the year are: January, $1,367; February, $1,845; March, $2,398. What is the projected annual gross income for the waiter?

Photodisc/Getty Images

Math Skill Builder

Review these math skills and solve the exercises that follow.

1. **Add** whole numbers and money amounts.
   Find the sum. 34 + 25 + 31 + 37 + 28 = 155

   1a. 78 + 92 + 101 + 86 **1b.** $135 + $176 + $157

2. **Multiply** money amounts by whole numbers.
   Find the product. 347 × $0.81 = $281.07

   2a. 181 × $1.24 **2b.** 5 × $98

3. **Multiply** money amounts by percents.
   Find the product. 15% × $54 = 0.15 × $54 = $8.10

   3a. 20% × $26 **3b.** 5% of $20.45 **3c.** 10% of $60

Piece-Rate Employees

Employers use a variety of ways to pay their employees. Some employees are paid for each item or *piece* they produce. Their wages are paid on a **piece-rate** basis. To figure their gross pay, you must multiply their pay per piece by the number of pieces produced. If employees are paid only for usable pieces produced, they get no pay for the pieces that are rejected.

Gross Pay = Number of Pieces Produced × Piece Rate

EXAMPLE 1

Helen Burchett is paid $1.30 for each usable picture frame she produces. What was Helen's gross pay for last week if she produced the following quantities of usable frames:

Monday	52
Tuesday	47
Wednesday	54
Thursday	50
Friday	45

SOLUTION

$52 + 47 + 54 + 50 + 45 = 248$ usable frames produced

$G = 248 \times \$1.30 = \$322.40$ gross pay

Algebra Tip

A verbal model for a piece-rate wage is:

Gross Pay = Number × Rate

$G = N \times R$

N represents the number of items produced. R is the rate paid per piece.

✔ CHECK YOUR UNDERSTANDING

A. Louise Schubert is paid $18 for each computer she installs at customer offices. She installed these numbers of computers in 5 days last week: 7, 6, 9, 8, 5. What gross pay did Louise earn for the week?

B. Trevor Sherr is paid $1.20 for each hand-painted dish he produces. He is not paid for dishes that are not acceptable. On Monday, he painted 56 dishes; on Tuesday, he painted 44 dishes. For the two days 6 dishes contained slight errors and were unacceptable. What gross pay did Trevor earn for the two days?

Photodisc/Getty Images

Per Diem Employees

Some people are paid on a per diem basis. **Per diem** means "by the day." Per diem employees are paid a fixed daily amount by their employer. Many per diem employees are temporary employees provided to a company by temporary help agencies.

Self-employed persons may charge a per diem rate for their services. These people may provide a specialized service to their clients. Self-employed persons work for themselves instead of for employers.

The gross pay of someone paid by the day is found by multiplying the per diem rate by the number of days worked.

Gross Pay = Number of Days × Per Diem Rate

EXAMPLE 2

Shawn Traylor worked 5 days last week as a temporary computer operator. His per diem pay rate was $120. What gross pay was Shawn paid for the week?

SOLUTION

$G = 5 \times \$120 = \$600$ gross pay for the week

✔ **CHECK YOUR UNDERSTANDING**

C. Sherry McCoy is a tax consultant. She charges $425 per diem for her services. If she worked 180 days last year, what was her gross income for the year?

D. Charlie's neighbors are often out of town and they hire him to house-sit. They pay Charlie $20 for each day they are gone. If they were out of town 57 days last year, what was Charlie's income from house sitting?

Tip Employees

Many workers receive income in the form of tips. A **tip** is an amount of money given to someone for services they provide. The person receiving the service pays a tip voluntarily. Many employees who earn tips are paid less than minimum wage, and some employees who receive tips must share them with other employees who assist them.

A tip, also called a *gratuity,* is calculated as a percentage when there is a dollar value attached to the service. A waiter, for example, may receive a tip of 20% of the total restaurant bill.

Tip Amount = Total Bill × Tip Percent

An airport skycap, on the other hand, may receive a specific amount for each piece of luggage handled.

Tip Amount = Number of Units × Tip Per Unit

Tipping practices vary considerably. The table below suggests guidelines for tipping certain types of workers. Most people round tips to the nearest quarter, or even dollar amount.

	Suggested Tipping Amounts
Airport skycap	$1 per bag
Hair stylist	15% of cost, minimum $1
Hotel chambermaid	$5 to $9 a night
Pizza delivery person	$1 to $5 depending on distance
Waiter/waitress	15–20% of total bill
Buffet waitstaff	5–10% of total bill
Taxi driver	15% of fare

Algebra *Tip*

Formulas for calculating hourly pay, commission, piece-rate, per diem and tips are all based on **amount × rate**. The Amount may be hours, sales, pieces, days or total bill, while the rate may be a dollar amount, percent, or unit.

Math *Tip*

A quick way to calculate a 20% tip is to double the amount of a 10% tip. A 10% tip on a bill of $25.00 is $2.50. (Move the decimal point left one place.)

Tipping guidelines for many other tip employees are available online. Use the Internet to find the tipping guidelines for three other types of employees.

EXAMPLE 3

Photodisc/Getty Images

After the Sutton family finished their meal at a local restaurant, the waiter brought them a check for $46.86. If Mrs. Sutton leaves a 20% tip, what amount of tip wages will the waiter receive for serving dinner to the Suttons? What will be the total meal cost to the Sutton's?

SOLUTION

The check amount is multiplied by the tip percentage to find the amount of the tip. The tip is added to the check amount to find the meal's total cost.

$T = 20\% \times \$46.86 = 0.2 \times \$46.86 = \$9.372$ round tip to $9.50

$\$9.50 + \$46.86 = \$56.36$ total meal cost

✔CHECK FOR UNDERSTANDING

E. Jack orders the lunch special and a beverage. His check comes to $10.20. How much will the waitress receive if a 15% tip is left? What is the total cost of the meal to Jack?

F. Lydia and Sarah share a cab ride to work. Their fare is $7.60. At 15%, how much should they tip the driver? What is their total cost to ride the cab?

Wrap Up ▶▶▶

A waiter's total earnings for the first 3 months, or one quarter year, are $5,610. Since there are 4 quarters in a year, multiply the total earnings for three months by 4 to find the total gross income for the year. So, $\$5,610 \times 4 = \$22,440$ total annual gross income.

Financial Responsibility

Reporting Tips

Stockbyte/Getty Images

If you work in a job where you receive tips, you are required to document how much you receive in tips and report the amount to your employer.

If you are an employer in the restaurant industry, the amount of tips that tip employees report must be at least 8% of your total receipts.

1. Jason is a waiter at a local restaurant. He made the following in tips last week: $65, $55, $125, $93, $75. How much tip income should he report to his employer for the week?
2. The restaurant that Jason works for had $50,000 in total receipts last week. What is the minimum amount of tip income that should be reported by all of the tip employees?

Exercises

Find the sum.

1. 40 + 38 + 39 + 45 + 41
2. \$135 + \$18.60

Find the product.

3. 87 × \$1.12
4. 3 × \$97
5. 15 × \$3 × 5

Find the product.

6. 5% × \$38
7. 20% × \$187
8. \$425 × 22

Solve.

9. An airport skycap handled 520 bags in a weekend. His average tip per bag was \$1.25. What total earnings did he have from tips for the weekend?

10. A waitress in an exclusive restaurant presented a food and beverage check in the amount of \$340 to customers at a table. The customers decided to leave a 20% tip. What tip amount did the waitress receive?

Photodisc/Getty Images

11. Sandra Mitchell worked 22 days last month as a temporary employee in the Purchasing department. Her per diem pay was \$95. What were Sandra's total earnings for the month?

12. To meet a shortage of medical staff, a doctor agreed to work 6, 24-hour shifts in the emergency room of a hospital during the next year. She is paid \$950 for each shift worked. What total pay will the doctor receive for the 6 days of emergency room work?

13. Lu Ying works at the Wilkins Bike Shop and is paid \$3.25 for every bike he assembles. The shop owner charges customers \$20 for this service. During the five working days of one week, Lu assembled these numbers of bikes: 27, 33, 29, 27, 31. What was Lu's gross pay for that week?

For each of these piece-rate employees at Dover Industries, find the total pieces produced and the gross pay for the week. Copy and complete the chart.

	Name	M	T	W	T	F	Total pieces	Rate per piece	Gross Pay
14.	Zinke, T	54	55	59	62	60		\$1.60	
15.	Bello, V.	24	28	30	31	27		\$2.80	
16.	Dixon, S.	63	69	59	62	50		\$1.55	
17.	Maier, B.	68	65	72	74	75		\$1.18	

18. **FINANCIAL DECISION MAKING** You are offered two jobs. One job pays an hourly rate of \$8.50, and requires 40 hours of work per week. The second job pays \$5.25 per hour plus tips and requires 8 hours per day, 5 days per week. Another employee who does the same job says that he usually earns about \$50 per day in tips. Based on the pay scale, which job would you accept, and why?

Photodisc/Getty Images

19. **CRITICAL THINKING** Steve took a taxi from the airport to his hotel across town. The fare came to $19.30. Steve handed the driver a $20 bill and told him to keep the change. Do you think Steve gave the driver a generous tip, an adequate tip, or not enough tip to express his appreciation for good service?

INTEGRATING YOUR KNOWLEDGE Fred must make a decision about keeping his current job, which he dislikes, or accepting an offer for a new job, which he thinks he would enjoy. His current job pays an hourly rate of $12. Fred works 40 hours a week.

At the new job, Fred would earn $0.80 for each item he produces up to 125 pieces per day. For each piece over 125 produced in a day, Fred would receive $0.85. The average production rate is 15 pieces an hour. Because of his experience and skill, Fred believes that he can produce 18 an hour. At the new job, Fred would work 8 hours a day, 5 days a week.

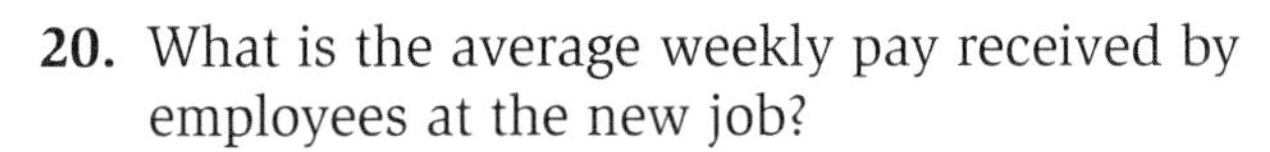

20. What is the average weekly pay received by employees at the new job?
21. How much does Fred expect to make each week at the new job?
22. What is Fred's weekly pay at his current job?
23. Create a chart, like the one shown, that will allow you to compare both jobs by estimated daily, weekly, and annual earnings based on Fred's predictions that he will produce 18 pieces an hour.

	Current Job	New Job
Daily Earnings		
Weekly Earnings		
Annual Earnings		

Mixed Review

24. $1\frac{5}{8} + 2\frac{3}{4}$

25. $15 \div \frac{5}{8}$

26. What number increased by 8% of itself equals 1,944?
27. Bob Turnquist works 42 hours a week at a pay rate of $12.50 an hour. What amount will Bob earn in 4 weeks?

Brenda Peoples earned $43,680 last year. Her usual work schedule is 50 hours a week. What were her average earnings

28. per month?
29. per week?
30. per hour?

31. Danny Mills receives a salary of $660 a month and a 7.5% commission on all sales above his monthly sales quota of $15,000. His sales for February totaled $32,000. What was Danny's total income for February?
32. A waitress at a Sunday brunch served 50 customers in a 4-hour period. The total of all the food and beverage checks she wrote for customers was $1,500. Her customers left an average tip of 8%. What is her tip income for Sunday?

1-5

Average Pay

GOALS
- Calculate simple averages
- Calculate averages from grouped data
- Find the unknown item in a set of data

KEY TERMS
- average
- mean

Start Up

A six-figure income is a total yearly earnings amount that most people will never receive. A six-figure income means that a person has annual earnings from $100,000 to $999,999. If one person earning $100,000 and another earning $999,999 annually were paid each week, what would be their gross pay each week, rounded to the nearest dollar?

Michael G. Smith/Shutterstock.com

Math Skill Builder

Review these math skills and solve the exercises that follow.

1 **Divide** money amounts.
Find the quotient. $95 ÷ 5 = $19

1a. $450 ÷ 5 **1b.** $18,000 ÷ 12

1c. $720 ÷ 15 **1d.** $4,708 ÷ 22

2 **Round** to the nearest cent.
Round this amount to the nearest cent. $9.287 = $9.29

2a. $8.765 **2b.** $9.996

2c. $7.097 **2d.** $13.602

2e. $526.889 **2f.** $0.737

3 **Multiply** money amounts by whole numbers.
Find the product. 8 @ $8.25 = $66

3a. 7 @ $7.25 **3b.** 4 @ $178

3c. 4 @ $7.25 **3d.** 11 @ $1.65

3e. 9 @ $12.20 **3f.** 100 @ $0.573

Math *Tip*

1 year = 12 months

1 year = 52 weeks

1 year = 365 days

Math *Tip*

The symbol @ means at. It means the same as multiply.

Simple Averages

An **average** is a single number used to represent a group of numbers. The most commonly used average is the **mean**.

A mean is found by adding several numbers and dividing the sum by the number of items added. Another name for a simple average is the *simple average*.

EXAMPLE 1

Tricia Willard earned these amounts for the 5 days she worked last week: Monday, \$82; Tuesday, \$91; Wednesday, \$96; Thursday, \$80; Friday, \$86. What was her average pay for the 5 days?

SOLUTION

Add daily amounts to find total pay.

$\$82 + \$91 + \$96 + \$80 + \$86 = \$435$ total pay

Divide the sum by the number of days to find the average pay per day.

$\$435 \div 5 = \$87$ average pay per day

Calculator *Tip*

Be sure you press [=] to find the total pay before you divide.

✔ CHECK YOUR UNDERSTANDING

A. Monica Wilkes earned these amounts last week at her part-time job: Friday, \$18; Saturday, \$58; Sunday, \$32. What was her average daily pay for the 3 days she worked?

B. A clothing designer earned these amounts in four consecutive months: \$2,400; \$3,200; \$1,500; \$1,700. What average monthly earnings did the designer have for these 4 months?

Photodisc/Getty Images

EXAMPLE 2

Valeria Mishkov earns \$18,000 per year as assistant manager at a local store. Find her average pay per hour (to the nearest cent) if she works 37.5 hours for 50 weeks per year and gets 2 weeks paid vacation.

SOLUTION

Find total weeks for which pay is received.

$50 + 2 = 52$ weeks

Find total hours for which she is paid in 1 year.

$37.5 \times 52 = 1{,}950$ total hours

Divide the total earnings by the number of hours. Round the final answer to the nearest cent.

$\$18{,}000 \div 1{,}950 = \$9.2308$, or \$9.23 average pay per hour

Math *Tip*

In solving problems in this book, round to the nearest cent all answers involving money amounts unless you are directed otherwise.

✔ CHECK YOUR UNDERSTANDING

C. Last month Antoine Beal earned $264 by working 38 hours at his part-time job. What average hourly pay did he earn last month, to the nearest cent?

D. LaKeisha Jones earned $35,800 in the first year of her new job. After receiving a promotion she earned $47,000 in the second year. What were her average earnings for the two years?

Averages in Grouped Data

A number or rate can occur more than once in a set of data. When that happens, you can find the sum quickly by grouping the common numbers.

EXAMPLE 3

Brandon Chin sells sarongs at the beach. During the first month of the season, he earned $600. For the next three months he earned $1,050 per month. In the last month, Brandon earned $180. What were his average earnings per month?

Simone van den Beerg/Shutterstock.com

SOLUTION

Find his total earnings for months worked.

1 month @ $600 = $600

3 months @ $1,050 = $3,150

1 month @ $180 = $180

5 total months = $3,930   total earnings

Divide the total earnings by the months worked.

$3,930 × 5 = $786   average monthly earnings

✔ CHECK YOUR UNDERSTANDING

E. The Runwell Company has 8 employees. Five of the employees earn $12 an hour, two earn $9 an hour, and 1 earns $11 an hour. What is the average amount per hour that the employees are paid?

F. Rosalind Jeszko earns a commission in addition to her salary. She earned these amounts in commission in the first 4 weeks of the year: $280, $315, $424, $265. What was her average commission for the 4 weeks, rounded to the nearest dollar?

Calculator *Tip*

Find the total using the memory keys. For each line, after you find the product, press [M+] to store the product. Repeat for each line of multiplication. When all of the multiplication is done, press [MR]. The total will be displayed, and you can divide by the number of items. To clear the memory before a new problem, press [MC].

Unknown Items in a Set of Data

If one item in a group or set of data is unknown, you can find it. Averages are used often in finding the value of the unknown item.

EXAMPLE 4

The weekly pay of four picture frame assemblers in a company averages $433 per employee. The weekly pay amounts of three of the four employees are $400, $410, and $460. What is the weekly pay of the fourth employee?

SOLUTION

Multiply the average pay by 4 to find the total pay.

$433 × 4 = $1,732   total pay

Add to find the total pay of the three employees.

$400 + $410 + $460 = $1,270   total pay of 3 employees

Subtract the totals to find the missing weekly pay.

$1,732 – $1,270 = $462   weekly pay of fourth employee

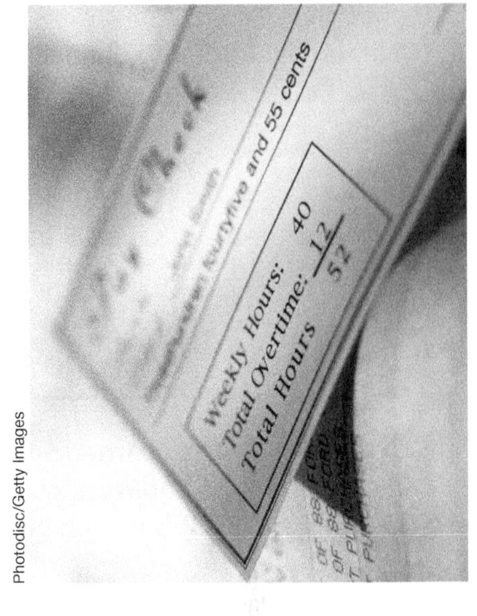

Photodisc/Getty Images

✔ CHECK YOUR UNDERSTANDING

G. To earn a $2,000 bonus, Kylene must average $25,000 in sales in a six-month period. If her sales for the first 5 months are $22,000, $30,000, $28,000, $20,000, and $18,000, how much must she sell in the last month to earn a bonus?

H. Hank Borden was paid $420 for 5 days work. For 4 of those days, his average pay was $86. Find his pay for the fifth day.

Wrap Up ▶▶▶

The six-figure income must be divided by 52 weeks to calculate the average weekly pay. A person earning $100,000 would earn $1,923 weekly. The person earning $999,999 would earn $19,231 weekly. Because the weekly pay figures are rounded, the answers are approximate.

TEAM Meeting

This class activity will simulate a company's weekly payroll for hourly-rate employees. You need two boxes. In a box marked Dollars, place a slip of paper for each amount from $5 to $11. In a box marked Cents, place paper slips marked from $0.55 to $0.95 in multiples of 0.05. Each student randomly chooses one slip from each box. The numbers drawn become the student's hourly rate. After each draw, return the slips to the boxes.

1. List all hourly rates. Record the number of students at each rate.
2. Find the average hourly pay rate for all class members.
3. Calculate the total weekly gross pay for all students for a 40-hour week.
4. Find the average weekly gross pay of the class.

Exercises

Find the quotient or product.

1. $620 ÷ 40
2. 12 × $1,578
3. 4 @ $12.75

Round to the nearest cent.

4. $406.439
5. $10.407

Two years ago Randi earned $18,400. Last year she earned $19,700. Suppose she earns $21,600 this year.

6. What total amount will Randi have earned for the three years?
7. What will be her average earnings per year?
8. How much would Randi have to earn next year so that her four year average is $21,000?

Rashad earned $662 by working five days a week at his full-time job. He earned $118 for 12 hours of work at his part-time job.

9. What average pay per day did he earn from full-time work?
10. What average amount per hour did he earn from part-time work?

Solve.

11. Ben's earnings for work he did from Monday through Saturday were: $78, $94, $115, $108, $67, $78. What was his average daily pay for the days he worked?
12. Four employees are paid a monthly salary as follows: Wilma, $1,820; Mavis, $1,615; Martha, $1,977; Tom, $1,560. What is their average salary per month?
13. Emma has been offered a job that pays $38,950 for working 52 weeks. What average amount does the job pay per month? per week?
14. Beatrix Thompson owns a craft shop and plans to sell 260 ceramic vases this year. She earns $25 for every vase sold. For the first 8 months of this year, she sold an average of 16 vases a month and had total earnings of $3,200. How many vases must she sell, on average, in each of the remaining 4 months to reach her goal?
15. At her job of grooming horses, Polly Yaskovich worked 8 hours a day on Monday and Tuesday earning $69 each day. On Wednesday, she earned $82. On Thursday, she earned $78. How much must Polly earn on Friday, to have average earnings of $75 a day?
16. The Willis Avenue Door Company gave its employees bonuses. Six employees received a bonus of $940 each; 4 employees received a bonus of $820 each; 5 employees were paid a bonus of $1,150 each. What was the average bonus paid to these employees?
17. The owner of Mid-Town Rapid Delivery plans to spend no more than $980 a day for employees' wages. The owner now has 8 employees who earn an average of $108 a day. What is the most a new employee can be paid without spending more money than planned?

Photodisc/Getty Images

Photodisc/Getty Images

Three weeks ago James worked 46 hours and earned $414. Two weeks ago he earned $333 by working 37 hours. Last week he worked 42 hours for $378.

18. What were his total earnings?
19. What were his average hourly earnings?
20. What were his average weekly earnings?
21. How much does James need to earn this week so that his average weekly earning is $380? How many hours does he have to work to earn this amount?

Yi Chin was offered a new job. For the first 3 months she works, she will be paid a monthly salary of $1,600. Her monthly pay for the next 3 months will be $1,760. For the next 6 months after that, Yi's monthly salary will be $1,936.

22. How much will Yi earn during a full year?
23. What average pay per month will she receive if she works a full year?
24. **STRETCHING YOUR SKILLS** Marsha's scores on seven tests were: 82, 78, 77, 93, 85, 91, and 86. What does she need to score on her next text to have an 86 average?
25. **STRETCHING YOUR SKILLS** The owner of a small business bought 5 cases of copier paper for $14 each, 8 cases for $13.50, and 11 cases for $12.75. What average price did the owner pay for each case, to the nearest cent?
26. **CRITICAL THINKING** The owner of a business has two skilled employees who each earn $160 a day and 12 unskilled employees who each earn $56 a day. Find the average daily pay for the 14 employees. Does the average show what the "average" worker is paid?
27. **FINANCIAL DECISION MAKING** The daily cost of public transportation to get you to and from your current job is $5. You regularly work 8 hours a day Monday through Friday and 4 hours on Saturday. You have been offered a new job close to home where you could walk to and from work each day. The new job pays $0.75 less an hour than your current job. You will work an average of 44 hours a week. Will you make or lose money by taking the new job?

Mixed Review

28. Find the quotient: $31,920 ÷ 12
29. Find the product: $235 × 4
30. What is $12.655 rounded to the nearest cent?
31. Carmella Petrocelli earns a weekly salary of $1,352 as a web site designer. If she works at this pay rate for a year, what will be her total annual earnings?
32. Tracy Voisene earns $18.15 an hour. What are her gross weekly wages if she works 40 hours at regular pay and 12 hours at time-and-a-half pay?

Chapter *Review*

Vocabulary Review

Find the term, from the list at the right, that completes each sentence. Use each term only once.

average
commission
double-time pay
employee
employer
graduated commission
gross pay
hourly rate
overtime
piece-rate
per diem
quota
salary
straight commission
time-and-a-half pay
tip

1. The total amount of an employee's earnings is called __?__.
2. Pay that is 1.5 times the regular hourly pay rate of an employee is called __?__.
3. An amount of money, often calculated as a percent, given to someone for service they provide is called a(n) __?__.
4. A fixed amount of pay for a week or a month is called __?__.
5. Salespeople who receive a specific percent of the sales they make are paid on (a, an) __?__ basis.
6. One number that represents a group of numbers is called a(n) __?__.
7. Time worked beyond the end of a usual working day is called __?__.
8. A wage rate based on the amounts produced by an employee is called __?__.
9. Employees that are paid a fixed amount daily are called __?__ employees.

1-1 Hourly Pay

10. Raphael Winston is paid $15.60 for each hour he works. What is his gross pay for a week in which he works 43 hours?
11. Monique Valla is paid an hourly rate of $17.63 for regular-time work. What will be her time-and-a-half and double-time hourly pay rates for overtime work?
12. Raul Pina worked these hours in five days: $8\frac{1}{4}$, $7\frac{1}{2}$, 10, $8\frac{3}{4}$, and 8 hours. Overtime is based on an 8-hour workday. How many regular hours and overtime hours did he work in the five days?
13. Penelope Schoenberg's overtime is figured on a 40-hour week. Last week she worked 45.6 hours. This week she worked 9.7 hours on Monday, 8.3 hours on Tuesday, 8 hours on Wednesday, 9.1 hours on Thursday, and 8.6 hours on Friday. How many overtime hours did she work in the two weeks?
14. Eddie Fantin is paid every two weeks. For the first week of his pay period he worked 42 hours, 4 of which were overtime hours. In the second week he worked 40 regular hours and 7 overtime hours. His regular pay rate is $14.40 an hour with time-and-a-half for overtime. What are his regular, overtime, and total wages for the two weeks?

1-2 Salary

15. Karolyn Yoder is paid a salary of $4,600 a month. What are her total annual earnings?

16. Marcel Ouimet earns a salary of $30,000 per year. What is his monthly salary?

17. Jason Kimee earns $450 per week. His boss gives him a raise of $30 per week. How much more will Jason earn per year at his new salary?

1-3 Commission

18. Gaston Kohl is paid a straight commission of 6.5% on all sales. In March his sales were $105,000. What were his commission earnings in March?

19. A company pays sales staff a monthly commission of 4% on the first $15,000 of sales, 6% on the next $20,000 of sales, and 7.5% on all sales above $35,000. What amount would Tony Renshaw earn if his sales for a month were $41,000?

20. Irene Ogan earned a commission of $5,130 on sales of $90,000. What rate of commission was she paid?

21. Doris Bommarito is paid a commission of 1.4% on all monthly sales above $80,000. Her sales for November were $382,000. What commission amount did she earn for the month?

1-4 Other Wage Plans

22. Steven Kahn is paid $4.25 for each wooden duck he paints that passes inspection. What is his pay on a day when he paints 38 ducks, 2 of which were rejected?

23. Tammy Scott-Hogan charges $350 a day to develop a personal training program for her clients. What amount did she earn last month if she worked 17 days?

24. What tip will Brady get if a customer adds a 15% tip to his $18.52 meal cost?

25. Amanda delivers newspapers to subscription customers. She receives an average annual tip of $12 from her 156 customers. What is her tip income for the year?

1-5 Average Pay

26. The gross earnings of 6 employees in a picture framing shop for a week were: $620, $524, $715, $670, $588, and $675. What was the average amount earned for the week by these employees?

27. For the 9 warmest months of the year Kendrick Beachom installed chain link fences and earned $3,500 a month. For the remaining 3 months of the year he earned these monthly amounts by working several part-time jobs: $2,450, $1,785. $3,025. What were his average monthly earnings for the year?

28. The average annual pay of 4 construction workers is $46,800. Three of the workers earned these annual amounts of pay: $44,200; $47,450; and $45,900. What was the annual pay of the fourth worker?

29. Sol Levin sold 22 pairs of shoes each day, Monday through Wednesday, 28 pairs on Thursday, and 31 on Friday. How many shoes must he sell on Saturday to average 29 pairs sold for the 6 days he worked during the week?

Technology Workshop

Task 1 Enter Data into a Payroll Detail Template

You are to complete a template that calculates the weekly gross wages for each employee of the Bainbridge Company. All employees receive regular hourly pay for time worked and overtime pay for hours worked beyond 40 hours in a week.

Open the spreadsheet for Chapter 1 (tech1-1.xls). Next, enter into the spreadsheet the hours worked by each employee for the three days shown in blue cells (cells G5-I14). The spreadsheet will calculate regular, overtime, and total gross wages for each employee. When finished, your spreadsheet should look like the one shown below.

	A	B	C	D	E	F	G	H	I	J	K	M	N	O	P
1							Bainbridge Company								
2							Payroll Detail Sheet for January 8, 20—								
3	Employee		Daily Hours Worked							Total Hours		Hourly	Gross Wages		
4	No.	Name	M	T	W	T	F	S	S	Reg	O.T.	Rate	Reg	O.T.	Total
5	1	Ajanaku	8.00	8.00	7.80	8.00	8.00	0.00	0.00	39.80	0.00	10.46	416.31	0.00	416.31
6	2	Bell	8.00	8.00	4.50	8.00	8.00	0.00	0.00	36.50	0.00	11.15	406.98	0.00	406.98
7	3	Cole	8.00	10.00	10.00	10.00	8.10	0.00	0.00	40.00	6.10	12.23	489.20	111.90	601.10
8	4	Dern	7.70	8.00	8.00	8.00	8.00	4.10	2.00	40.00	5.80	10.15	406.00	88.31	494.31
9	5	Evers	8.00	9.80	8.10	8.00	8.00	0.00	0.00	40.00	1.90	12.85	514.00	36.62	550.62
10	6	Ford	7.10	8.00	8.00	8.00	8.00	4.10	0.00	40.00	3.20	10.46	418.40	50.21	468.61
11	7	Gomez	8.00	8.00	8.00	8.00	8.00	3.90	2.00	40.00	5.90	12.35	494.00	109.30	603.30
12	8	Huang	8.00	9.30	8.00	8.00	8.00	0.00	0.00	40.00	1.30	11.70	468.00	22.82	490.82
13	9	Isom	8.00	8.00	5.40	8.00	8.00	0.00	0.00	37.40	0.00	11.32	423.37	0.00	423.37
14	10	Jackson	8.00	8.00	10.00	8.10	8.40	0.00	0.00	40.00	2.50	13.20	528.00	49.50	577.50
15		Totals											4,564.26	468.66	5,032.92
16															
17										Pay Raise Factor		1.000			

Task 2 Analyze the Spreadsheet Output

Answer these questions about your completed payroll sheet.

1. What hourly pay rate did Dern have?
2. Which employee had the largest gross pay for the week?
3. Which employee worked the least regular-time hours?
4. Which employee worked the most overtime hours?
5. What was the total amount paid to all employees for overtime work?
6. What total gross pay was paid to employees for the one-week pay period?

Now move the cursor to cell M17, labeled Pay Raise Factor. The current entry in the cell should be 1.000. Now enter 1.021. This change shows what would happen if the Bainbridge Company gives its employees a 2.1% pay increase. The increase would raise wages to 102.1%, or 1.021, of their current level. Notice how the hourly rate and the regular, overtime, and total gross wages figures changed for all workers.

Answer these questions about your updated payroll sheet.

7. What is the formula used in Cell N5? What does the formula calculate?
8. What is the formula used in Cell O7? What arithmetic is done in the cell?
9. What is the formula used in Cell P12? What does it do?
10. What hourly rate does Dern now earn? How much more per hour is Dern paid after the raise was calculated?
11. Find the difference between the original and the new total gross wages, then calculate the percent increase to the nearest tenth percent. What does your answer show?

Task 3 Design a Sales Commission Spreadsheet

You are to design a spreadsheet that will compute the monthly earnings of salespersons that are paid monthly on a graduated commission basis.

The spreadsheet for Task 1 includes formulas that use subtraction, multiplication, the IF function, and the SUM function. Create a spreadsheet that will use similar math operators and functions. The spreadsheet should allow you to calculate the gross pay for a month for all employees listed. Your spreadsheet should contain a row for each employee. The column formulas should calculate the amount of commission earned by each employee for each commission level and the total gross wages for all employees.

SITUATION: You are the payroll clerk in the office of the Betadyne Company. The company pays its salespeople a commission on their sales. It does not pay them any salary. The employee names and their April sales are shown below on the left. Shown below on the right are the sales levels and rates by which your company figures commission payments.

Salesperson	April Sales
Boyce, Thad	$53,000
Elkins, James	$48,000
Kubik, Lucy	$92,000
Mays, Nora	$64,000

Betadyne Company Commission Structure
1.5% on all sales
3.4% on the first $50,000 of sales
4.6% on all sales above $50,000

Task 4 Analyze the Spreadsheet Output

Answer these questions about your completed spreadsheet.

12. How did you figure the commission on all sales?
13. How did you test the spreadsheet to make sure the calculations were correct?
14. How would you change the spreadsheet to calculate the average gross commissions earned by employees?

Chapter *Assessment*

Chapter Test

Answer each question.

1. Add: $8\frac{1}{4} + 9 + 7\frac{1}{2} + 8 + 10\frac{1}{4}$
2. Subtract: $47.3 - 37.5$
3. Multiply: $38\frac{3}{4} \times \$13.35$.
4. Divide: $\$85{,}956 \div 52$
5. Rewrite 1.0567 as a percent.
6. $30 \div \frac{5}{8} = ?$
7. What is 1.25% of $34,500?
8. 120 is $\frac{1}{4}$ greater than what number?
9. Write 5.68% as a decimal rounded to the nearest hundredth.
10. Round $0.8794 to the nearest cent.
11. What number increased by 20% of itself equals $103.20?

Solve.

12. Fiona Wolfe was paid $9 an hour for 46 hours of work last week at her full-time job. She also worked 7 hours last week at a part-time job that pays $11 an hour. What total gross pay did she earn last week from both jobs?
13. Zygmund Oleksik is paid a salary of $800 a week to manage a party store. What amount will he make in one year working at this job?
14. Rosie Belin worked these total weekly hours in four weeks of work: 45, 38, 42, 43. Her job pays $14 for each hour she works. What average gross pay did she earn per week for these four weeks of work?
15. The six employees in the security department of a company average $120 gross pay a day. Three of the employees earn $125 per day. Two others earn $116.50 a day. How much does the sixth employee earn per day?
16. Frank Camp's regular hourly pay rate is $11.87 an hour. His overtime pay rate is time-and-a-half. How much is Frank paid per hour for overtime work?
17. Justine Gilbert worked these hours last week: Monday, 8; Tuesday, $10\frac{1}{4}$; Wednesday, $8\frac{1}{2}$; Thursday, 9; Friday, $7\frac{3}{4}$. She works on an 8-hour day with time-and-a-half being paid for overtime work. Her regular gross pay rate is $17.10 an hour. What was Justine's overtime pay amount for last week?
18. Barney Mullins' regular pay rate is $14.85 an hour. He is paid time-and-a-half for hours worked over 40 hours in a week, including weekend work. He worked these hours from Monday through Saturday last week: 8.2, 8.9, 10.1, 9.6, 8.8, 6.7. What was his gross pay for the week?
19. Tiffany Penfield is paid a salary of $750 a month at her sales job. She also earns a commission on her sales in this way: 2% on all sales up to $34,000 in a month and 8% on all higher sales. What were Tiffany's total earnings for a month where her total sales were $80,000?
20. All employees of the Crafton Company are paid $0.375 for each wrench set they pack. How much would Kenny Pace earn if he packed 1,740 wrench sets in one week of work?

Planning a Career in Information Technology

Photodisc/Getty Images

The opportunities that exist in information technology range from creating web pages and computer programs to the installation and repair of computer equipment. Many of these jobs are in businesses or government, where computer networks are installed and monitored to increase effectiveness and efficiency. Some positions in information technology involve creativity while others require mechanical and electrical expertise. If you are good at problem solving, are comfortable with cutting-edge technology, and work well with little supervision, a career in information technology may be for you.

Job Titles

- Computer programmer
- Network administrator
- Computer support specialist
- Technical writer
- Web designer
- Computer scientist
- Telecommunications equipment installer
- Computer software engineer
- Technical writer

Needed Skills

- Excellent computer and technology skills
- Ability to work independently, as well as with others

- Outstanding problem solving skills
- Strong mathematical and science skills
- Vivid imagination, creativity, and exceptional writing talent
- Assessment and decision making skills

What's it Like to Work in Computer Programming?

Computer programmers write programs for computers to follow to perform specific functions. Programmers usually translate a design for a computer program by coding the design into a programming language that the computer can understand. There is more than one programming language and computer programmers generally know more than one. Computer programs are written for the payroll industry, the financial industry, the entertainment industry, and many other industries. After the programmer writes the specific instructions for the program, the programmer is responsibly for testing, maintaining, and improving the program.

What About You?

What aspect of computer programming appeals to you? How might you best prepare for a career in this field?

How Times Have Changed

For Questions 1–2, refer to the timeline on page 3 as needed.

1. Opponents to increases in minimum wage believe that by increasing the wage, jobs will be lost because small business cannot afford to keep workers at a higher rate. Suppose a full-time worker who earns minimum wage in 1997 receives the increases enacted in 2007, 2008, and 2009. How much more per week will the company have to pay her at the new rate in 2007 than 1997? How much more will they pay in 2008 than 2007? How much more will they pay in 2009 than 2008?
2. How much does a full-time minimum wage worker that works 52 weeks a year earn annually at the 2007 minimum wage rate? the 2009 rate?

Chapter 2

Net Pay

2-1 **Deductions from Gross Pay**
2-2 **Federal Income Taxes**
2-3 **State and City Income Taxes**
2-4 **Benefits and Job Expenses**
2-5 **Analyze Take-Home Pay**

Statistical Insights

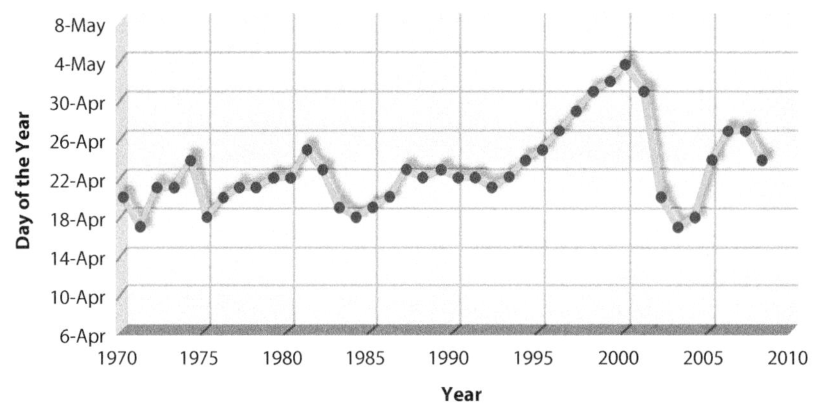

Source: www.taxfoundation.org

If American workers applied every dollar earned beginning January 1 to pay the government their annual tax obligations, all their taxes would be paid by a day in April or May. This day is referred to as Tax Freedom Day and varies from year to year. The line graph above has 39 points; one for each year from 1970 to 2008. Use the graph to answer Questions 1–5.

1. What year did Americans work the greatest number of days to pay their tax burden?
2. What two years did Americans work the least number of days to pay their tax burden?
3. Recent tax stimulus packages have resulted in Tax Freedom Day arriving earlier in the year. Which four recent years appear to be years with these tax cuts?
4. What day was Tax Freedom Day the year you were born?
5. **Explain** why the Tax Freedom Day continues to fall within the same four weeks of the year as it did in the 1970s.

How Times Have Changed

Social Security was created to keep older Americans out of poverty. It is not an insurance policy or savings account; the money deposited today is used to pay benefits today. In 1950, there were about 16 workers to cover one Social Security beneficiary. Today, there are about 3.3 workers for each beneficiary. It is predicted that by about 2042, Social Security will run out of funds.

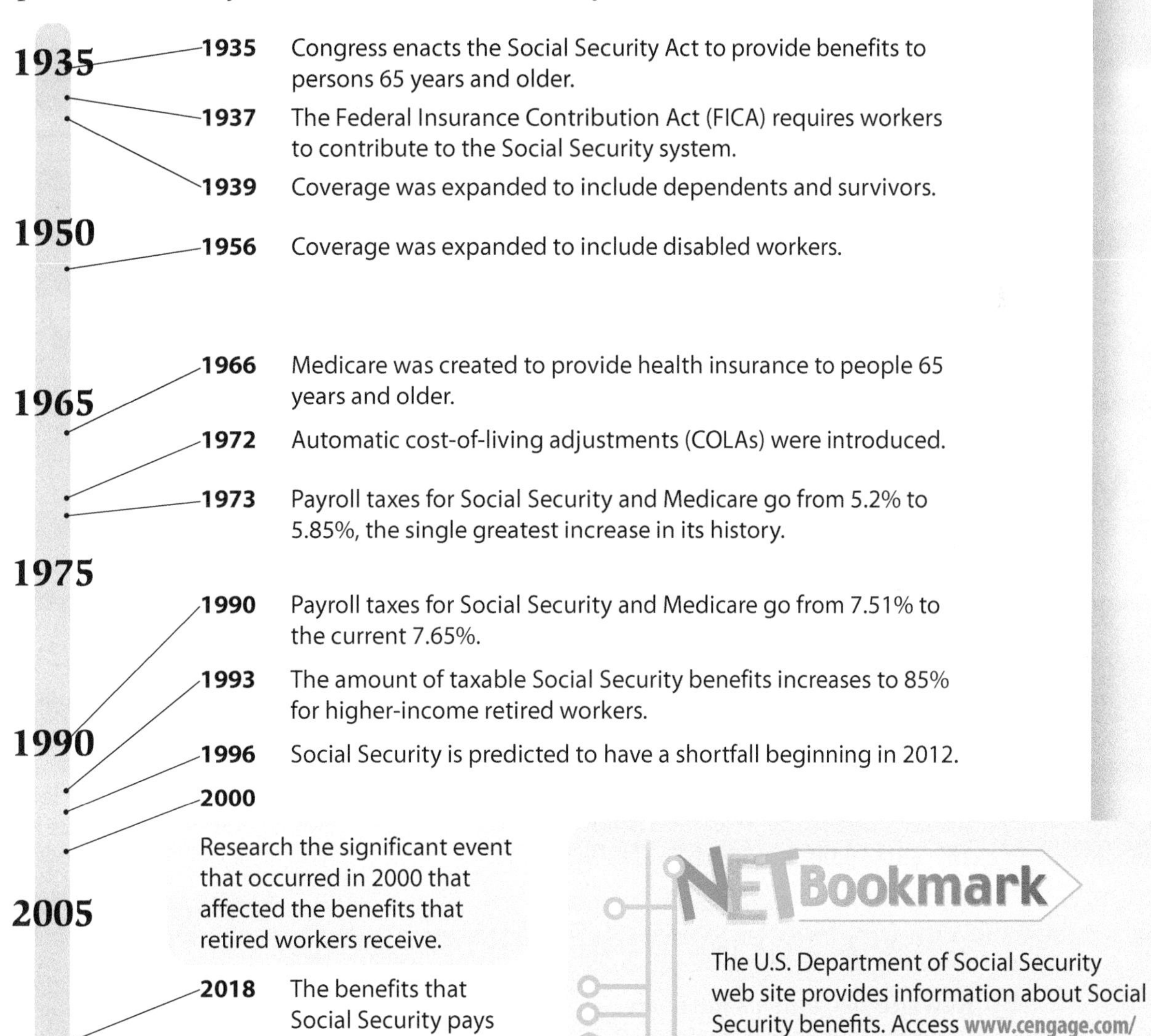

1935
1950
1965
1975
1990
2005
2020
1935 Congress enacts the Social Security Act to provide benefits to persons 65 years and older.
1937 The Federal Insurance Contribution Act (FICA) requires workers to contribute to the Social Security system.
1939 Coverage was expanded to include dependents and survivors.
1956 Coverage was expanded to include disabled workers.
1966 Medicare was created to provide health insurance to people 65 years and older.
1972 Automatic cost-of-living adjustments (COLAs) were introduced.
1973 Payroll taxes for Social Security and Medicare go from 5.2% to 5.85%, the single greatest increase in its history.
1990 Payroll taxes for Social Security and Medicare go from 7.51% to the current 7.65%.
1993 The amount of taxable Social Security benefits increases to 85% for higher-income retired workers.
1996 Social Security is predicted to have a shortfall beginning in 2012.
2000
Research the significant event that occurred in 2000 that affected the benefits that retired workers receive.
2018 The benefits that Social Security pays out are predicted to exceed the amount collected.
2042 The fund is predicted to be out of money, if changes are not made to the plan.

NET Bookmark

The U.S. Department of Social Security web site provides information about Social Security benefits. Access www.cengage.com/school/business/businessmath and click on the link for Chapter 2. You can calculate your benefits under the retirement heading. What three types of calculators are offered? Besides retirement, what other reasons might a person have to use one of these calculators?

2-1

Deductions from Gross Pay

GOALS
- Find federal withholding tax deductions
- Calculate Social Security and Medicare tax deductions
- Calculate total deductions and net pay

KEY TERMS
- deduction
- withholding tax
- withholding allowance
- net pay

Start Up ▸▸▸

Janice is single and has just graduated from community college. She needs at least $480 each week to pay for her rent and other living expenses to afford to live alone. If she earns $13 an hour and works 40 hours each week, will she earn enough to pay her expenses?

Photodisc/Getty Images

Math Skill Builder

Review these math skills and solve the exercises that follow.

1 **Add** money amounts. Find the sum.
$35.62 + $12.65 + $87.61 + $27.59 = $163.47

1a. $77.12 + $18.92 + $40.56 + $9.21

1b. $53.07 + $3.76 + $21.98 + $82.16

2 **Subtract** money amounts from money amounts.
Find the difference. $540.09 − 62.72 = $477.37

2a. $3,145.00 − $809.12

2b. $723.82 − $129.04

2c. $235.88 − $13.48

3 **Rewrite** percents as decimals.
Rewrite 7.52% as a decimal. 7.52% = 0.0752

3a. 4% **3b.** 5.2% **3c.** 10.5% **3d.** 4.34%

4 **Multiply** money amounts by percents and round the product to the nearest whole cent.
Find the product. $387.25 × 7% = $387.25 × 0.07 = $27.1075, or $27.11

4a. $1,249.00 × 4%

4b. $478.53 × 7.2%

4c. $809.42 × 1.45%

Math *Tip*

To rewrite a percent as a decimal, move the decimal point two places to the left and drop the percent sign.

Federal Withholding Tax Deduction

Deductions are subtractions from gross pay. The federal government, as well as many states and cities, require employers to deduct money from employee wages for income taxes, or **withholding taxes**, plus Social Security and Medicare taxes.

The amount of withholding tax depends on a worker's wages, marital status, and number of withholding allowances claimed. A **withholding allowance** is used to reduce the amount of tax withheld. Workers may claim one withholding allowance for themselves, one for a spouse, and one for each child or dependent.

To find the amount withheld from a worker's wages, you can use an income tax withholding table prepared by the government. Use the tables shown to find federal withholding taxes on weekly wages.

First determine whether the person is single or married. Then, using the table for the employee's marital status, read down the *if the wages are*—column at the left until you reach the correct wage line. Next, read across to the column headed by the number of withholding allowances claimed by the employee.

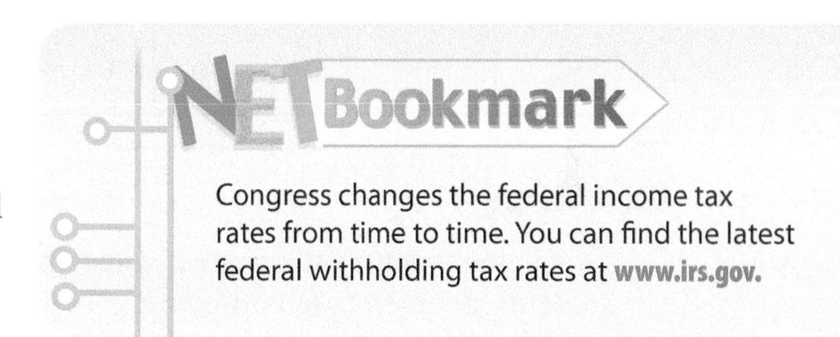

NET Bookmark
Congress changes the federal income tax rates from time to time. You can find the latest federal withholding tax rates at www.irs.gov.

Single Persons—Weekly Payroll Period

(For Wages Paid in 2008)

If the wages are—		And the number of withholding allowances claimed is—										
At least	But less than	0	1	2	3	4	5	6	7	8	9	10
		The amount of income tax to be withheld is—										
350	360	38	28	18	10	3	0	0	0	0	0	0
360	370	40	30	20	11	4	0	0	0	0	0	0
370	380	41	31	21	12	5	0	0	0	0	0	0
380	390	43	33	23	13	6	0	0	0	0	0	0
390	400	44	34	24	14	7	1	0	0	0	0	0
400	410	46	36	26	15	8	2	0	0	0	0	0
410	420	47	37	27	17	9	3	0	0	0	0	0
420	430	49	39	29	18	10	4	0	0	0	0	0
430	440	50	40	30	20	11	5	0	0	0	0	0
440	450	52	42	32	21	12	6	0	0	0	0	0
450	460	53	43	33	23	13	7	0	0	0	0	0
460	470	55	45	35	24	14	8	1	0	0	0	0
470	480	56	46	36	26	16	9	2	0	0	0	0
480	490	58	48	38	27	17	10	3	0	0	0	0
490	500	59	49	39	29	19	11	4	0	0	0	0
500	510	61	51	41	30	20	12	5	0	0	0	0
510	520	62	52	42	32	22	13	6	0	0	0	0
520	530	64	54	44	33	23	14	7	0	0	0	0
530	540	65	55	45	35	25	15	8	1	0	0	0
540	550	67	57	47	36	26	16	9	2	0	0	0

Married Persons—Weekly Payroll Period

(For Wages Paid in 2008)

If the wages are—		And the number of withholding allowances claimed is—										
At least	But less than	0	1	2	3	4	5	6	7	8	9	10
		The amount of income tax to be withheld is—										
440	450	29	22	16	9	2	0	0	0	0	0	0
450	460	30	23	17	10	3	0	0	0	0	0	0
460	470	32	24	18	11	4	0	0	0	0	0	0
470	480	33	25	19	12	5	0	0	0	0	0	0
480	490	35	26	20	13	6	0	0	0	0	0	0
490	500	36	27	21	14	7	0	0	0	0	0	0
500	510	38	28	22	15	8	1	0	0	0	0	0
510	520	39	29	23	16	9	2	0	0	0	0	0
520	530	41	31	24	17	10	3	0	0	0	0	0
530	540	42	32	25	18	11	4	0	0	0	0	0
540	550	44	34	26	19	12	5	0	0	0	0	0
550	560	45	35	27	20	13	6	0	0	0	0	0
560	570	47	37	28	21	14	7	1	0	0	0	0
570	580	48	38	29	22	15	8	2	0	0	0	0
580	590	50	40	30	23	16	9	3	0	0	0	0
590	600	51	41	31	24	17	10	4	0	0	0	0
600	610	53	43	33	25	18	11	5	0	0	0	0
610	620	54	44	34	26	19	12	6	0	0	0	0
620	630	56	46	36	27	20	13	7	0	0	0	0
630	640	57	47	37	28	21	14	8	1	0	0	0

EXAMPLE 1

A single receptionist's weekly wages are $380 with 1 withholding allowance. What federal income tax is withheld?

SOLUTION

Use the table for single persons. The wages, $380, are on the fourth line of this part of the table. Read across to find the column under 1 withholding allowance. The amount of tax is $33.

✔ CHECK YOUR UNDERSTANDING

A. Jared Brown is a single hospital technician with weekly wages of $458. He claims 1 withholding allowance. What amount should be deducted from his wages for federal withholding taxes?

B. Imy Berstein is a married worker earning $514 each week. She claims 2 withholding allowances. What amount should be deducted from her weekly earnings for federal withholding taxes?

Social Security and Medicare Tax Deductions

The tax for Social Security is part of the Federal Insurance Contributions Act and is also called the FICA tax. The FICA benefits include:

Photodisc/Getty Images

- *disability benefits* for workers who are disabled and unable to work
- *Medicare,* which provides hospital insurance for some disabled people and for people over 65
- *retirement benefits* for people who are at least 62
- *survivors' benefits,* which are paid to spouses and dependent children when a Social Security recipient dies

FICA tax rates and the maximum wages on which the taxes are charged are set by Congress and may change from time to time.

The overall tax rate of 7.65% is used in this text. This rate is made up of the Social Security tax rate of 6.2% applied to a maximum wage of $102,000 and the Medicare tax rate of 1.45%, applied to all wages.

If a person earns more than $102,000 a year from one job, the employer does not deduct Social Security tax after the wages exceed $102,000. If a person earns more than $102,000 a year from several jobs, each employer withholds 6.2% Social Security tax to the maximum $102,000 earning limit. The taxpayer must apply for a return of the overpayments when a federal income tax return is filed.

Business Tip

People who work for themselves must also pay FICA taxes on yearly net earnings. The tax rates for the self-employed are twice the rates paid by employees because the self-employed person must pay the employee and employer shares.

EXAMPLE 2

Find the total FICA tax on incomes of $35,000, $80,000, and $104,000.

SOLUTION

Income	Social Security	Medicare	Total FICA
$35,000	$35,000 × 6.2% = $2,170	$35,000 × 1.45% = $507.50	$2,677.50
$80,000	$80,000 × 6.2% = $4,960	$80,000 × 1.45% = $1,160	$6,120.00
$104,000	$102,000 × 6.2% = $6,324	$104,000 × 1.45% = $1,508	$7,832.00

✔ CHECK YOUR UNDERSTANDING

Find the FICA tax on each income.

C. $24,000 **D.** $110,000

Math Tip

If an employee has not exceeded the $102,000 limit, another way to find the total FICA tax is to multiply the income by the combined Social Security and Medicare rate of 7.65%

$35,000 × 7.65% = $2,677.50

$80,000 × 7.65% = $6,120.00

$104,000 has exceeded the $102,000 limit, so the separate rates must be used.

FICA taxes owed by workers are collected by their employers. Employers deduct the tax from each employee's earnings. Employers must also pay a FICA tax equal to the FICA taxes they deduct from their employees' earnings.

EXAMPLE 3

Sarah Fellows earned $562 during the last week of January. Find the total FICA taxes her company deducted from her wages.

SOLUTION

Since Sarah's wages are paid in January, you are sure that her wages have not exceeded the Social Security earning limit. She is taxed on both Social Security and Medicare. Use 7.65% as the tax rate.

7.65% = 0.0765 Rewrite as a decimal.

$562 × 0.0765 = $42.993 Multiply the weekly wages by the decimal rate.

The total FICA tax amount deducted from Sarah's wages was $42.99.

✔CHECK YOUR UNDERSTANDING

Find the total FICA tax amount on each weekly wage.

E. $460

F. $712.44

G. $1,087.30

H. $375.88

Total Deductions and Net Pay

In addition to withholding, Social Security, and Medicare taxes, other deductions may also be subtracted from gross pay, such as union dues, health and life insurance, and government bonds. After all deductions are subtracted from total wages, or gross pay, an amount remains that is called **net pay**, or *take-home pay.*

Net Pay = Gross Pay − Deductions

Calculator *Tip*

On many calculators, you can multiply by a percent directly without rewriting it as a decimal. For example, to calculate $460 multiplied by 7.65 percent, enter 460, press [×], then enter 7.65, and press [%]. The answer 35.19 will appear in the calculator display.

Statement of Employee Earnings and Payroll Deductions

	EARNINGS			DEDUCTIONS							
WEEK ENDED	REGULAR	OVER-TIME	TOTAL	FED. WITH.	SOC SEC.	MEDI CARE.	LIFE INS.	HEALTH INS.	OTHER	TOTAL	NET PAY
2/4	*375.00*		*375.00*	*31.00*	*23.25*	*5.44*	*12.50*	*56.45*	*13.75*	*144.39*	*232.61*

NO. 4798

EXAMPLE 4

Mary Mendosa earned a gross pay of $426 last week. Federal withholding taxes of $39, Social Security taxes of 6.2%, Medicare taxes of 1.45%, health insurance premiums of $45.80, and union dues of $12.56 were deducted from her gross pay. Find Mary's net pay.

SOLUTION

Multiply the gross pay by each tax rate.

$\$426 \times 0.062 = \$26.41$ $\$426 \times 0.0145 = \$6.18$

$\$39 + \$26.41 + \$6.18 + \$45.80 + \$12.56 = \$129.95$ total deductions

$\$426.00 - \$129.95 = \$296.05$ Subtract the total deductions from gross pay.

Mary Mendosa's net pay for the week was $296.05.

✔CHECK YOUR UNDERSTANDING

I. Jay Panetta earned gross pay of $410 last week. From his gross pay the following were subtracted: federal withholding tax, $37; Social Security tax, 6.2%; Medicare tax, 1.45%; health insurance, $34.88; and $40 for his savings plan. Find Jay's net pay.

J. Last week, Rose Petropolis earned gross pay of $820. Her employer deducted $109 in federal withholding taxes, 6.2% in Social Security taxes, 1.45% in Medicare taxes, $74 in health insurance, and $45 in union dues. What was Rose's net pay?

Wrap Up ▶▶▶

Janice earns $520 a week ($13/hr × 40 hr). Using the federal withholding tax table for single persons with 1 withholding allowance, her tax on $520 is $54. In addition, Social Security taxes are $32.24 and Medicare taxes, $7.54. Even if she had no other deductions from her gross pay, her net pay is only $426.22. This amount is not enough to cover the $480 she needs to live alone.

Exercises

Find the sum.

1. $34 + $15.23 + $65.01 + $23.85

2. $87 + $32.71 + $48.14 + $12.09

Find the difference.

3. $523.19 − $106.42

4. $4,456.12 − $98.76

5. $389.28 − $79.52

6. $2,107.88 − $278.43

Rewrite percents as decimals.

7. 8%

8. 2.4%

9. 12.06%

10. 3.67%

11. 89.145%

12. 145%

Find the product.

13. $498 × 5.6%

14. $826 × 3.456%

Find the withholding tax in each exercise using the tables given.

Total Wages	Marital Status	Withholding Allowances	Total Wages	Marital Status	Withholding Allowances
15. $390.00	Single	1	**16.** $487.00	Married	2
17. $411.00	Single	0	**18.** $444.00	Single	4
19. $528.97	Married	5	**20.** $612.81	Married	3
21. $457.07	Single	1	**22.** $438.88	Single	9

Find the Social Security tax and Medicare tax on each weekly wage. Use a 6.2% Social Security tax rate on a maximum of $102,000 gross wages and a 1.45% Medicare tax rate on all wages.

23. $475.00

24. $556.34

25. $249.40

26. $497.45

27. $749.23

28. $180.04

29. $289.48

30. $863.78

Copy and complete the table below. Use a 6.2% Social Security tax rate on a maximum of $102,000 gross wages and a 1.45% Medicare tax rate on all wages.

	Name	Allow-ances	Marital Status	Gross Wages	Income Tax	Social Secur.	Medi-care	Other	Total Deduc.	Net Wages
31.	Ahern	1	Single	$467.29				$82.12		
32.	Brown	0	Single	$399.62				$56.45		
33.	Cali	3	Married	$578.21				$31.51		
34.	Devon	6	Married	$459.65				$75.21		
35.	Ezeka	2	Married	$538.76				$48.22		

Ali Zaheer is married with 3 withholding allowances. Each week his employer deducts federal withholding taxes, Social Security taxes, Medicare taxes, and $38.12 for health insurance from his gross pay. His gross weekly wage is $578.

36. Find the total deductions

37. Find his net pay.

Josh Logan is paid a monthly salary of $8,700.

38. Estimate his Social Security taxes for May.

39. Find the exact Social Security taxes he paid for the month of March.

40. Find the Social Security taxes he paid in December.

41. What Medicare taxes did he pay for the year?

Rachel Radcliff earns an annual salary of $126,000.

42. How much Social Security taxes will be deducted from her salary in November?

43. How much Social Security taxes will be deducted from her salary in May?

44. How much Medicare taxes will be deducted from her salary for the year?

Photodisc/Getty Images

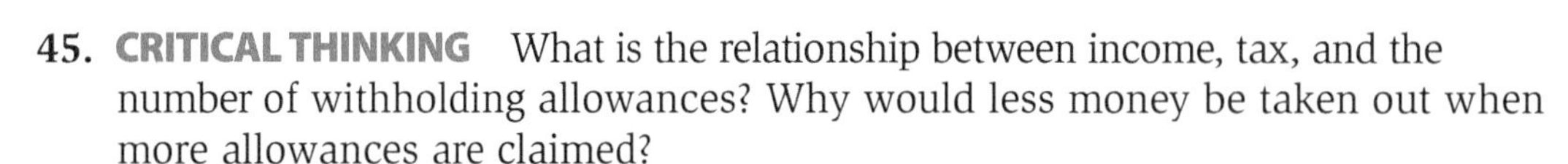

45. **CRITICAL THINKING** What is the relationship between income, tax, and the number of withholding allowances? Why would less money be taken out when more allowances are claimed?

46. **FINANCIAL DECISION MAKING** Your employer allows you to deduct money each week from your gross wages to be placed in a savings plan of your choice. Should you have this deduction taken from your wages each week?

Mixed Review

47. $\$25.60 \times 1\frac{1}{4}$

48. $\frac{1}{4} + 6\frac{3}{4} + 8\frac{1}{2}$

49. $5\frac{1}{4} - 3\frac{1}{2}$

50. $\$8 \times 0.1$

51. $\frac{1}{4}$ more than 40

52. $\$25{,}600 \times 12.5\%$

2-2

Federal Income Taxes

GOALS

- Calculate adjusted gross income and taxable income
- Calculate the income tax due
- Calculate the income tax refund for single dependents

KEY TERMS

- gross income
- adjusted gross income
- taxable income
- deduction
- standard deduction
- exemption

Start Up ▸▸▸

Sven Tole is a high school student who worked during summer vacation. He noticed that federal income taxes were withheld from his paychecks. He didn't think it was fair for him to pay taxes when he made so little during the year. Is he right?

Photodisc/Getty Images

Math Skill Builder

Review these math skills and solve the exercises that follow.

1. **Add** money amounts.
   Find the sum. $2,509 + $1,090 + $56 + $398 = $4,053

   1a. $1,907 + $3,763 + $78 + $189 **1b.** $5,007 + $208 + $976 + $92

2. **Subtract** money amounts.
   Find the difference. $32,459 − $4,108 = $28,351

   2a. $3,766 − $791 **2b.** $28,067 − $1,448 **2c.** $107,390 − $65,288

3. **Multiply** money amounts.
   Find the product. $2,600 × 2 = $5,200

   3a. $2,789 × 4 **3b.** $3,188 × 6 **3c.** $1,572 × 10

Adjusted Gross Income and Taxable Income

Employers deduct money for federal income tax from worker's pay. This is called *a withholding tax*. The amounts withheld are estimates of the tax owed at year's end.

The tax year for individuals ends on December 31. You must calculate and pay any federal income tax due by April 15 of the next calendar year. Income earned and taxes due are reported on *a federal income tax return*.

A completed return shows how much you owe in federal income taxes. If the amount withheld from wages was larger than what was owed, you should claim a refund. If the withholding taxes paid were less than what you owed, you pay the difference.

Gross income is the total income in a year and includes income from wages, salaries, commissions, bonuses, tips, interest, dividends, prizes, pensions, the sale of stock, and profit from a business.

From gross income, you may be eligible to subtract *adjustments to income.* These include business losses, payments to approved retirement plans, alimony, and certain penalties. The amount left is called **adjusted gross income**.

Business *Tip*

Self-employed people must estimate their income taxes for the year. They then pay part of that estimated tax each quarter.

Adjusted Gross Income = Gross Income − Adjustments to Income

From adjusted gross income you subtract the deductions and exemptions for which you qualify. The result is your taxable income. **Taxable income** is the income on which you actually pay tax.

Taxable Income = Adjusted Gross Income − Deductions and Exemptions

Deductions are expenses that reduce the amount of your taxable income. You may deduct interest paid on a home mortgage, property taxes, state and local income taxes, medical and dental expenses, casualty and theft losses, and contributions to charities. You may claim a fixed amount called a **standard deduction**. Or, if your actual deductions are more than the standard deduction, you list all your deductions on your tax return under *itemized deductions.*

An **exemption** is an amount of income per person that is free from tax. You may claim one exemption for yourself unless you are claimed as a dependent on another person's tax return. You can also claim one exemption for a spouse and one exemption for each dependent. For example, a couple with two dependent children can claim four exemptions. A single person with a dependent parent can claim two exemptions.

The amounts allowed for the standard deduction and exemptions change often. In this text, the standard deduction is $5,450 for a person filing an income tax return as an individual and $10,900 for married people filing a return together, *or jointly.* The amount used for each exemption is $3,500.

EXAMPLE 1

Find each person's taxable income.

a. Clara Shane is single and has a gross income of $32,600. She pays $2,600 into an approved retirement plan. Clara has deductions of $6,900. She has one exemption for herself.

b. Andy Cross and his spouse have a gross income of $33,000. They file jointly. They make payments into an approved retirement plan of $3,000. Their itemized deductions were only $4,300. So, they will take the standard deduction of $10,900. They claim two exemptions.

SOLUTION

	a. Clara Shane	b. The Crosses
Gross Income	$32,600	$33,000
Adjustments to Income	− 2,600	− 3,000
Adjusted Gross Income	$30,000	$30,000
Deductions	− 6,900	−10,900
	$23,100	$19,100
Exemptions	− 3,500	− 7,000
Taxable Income	$19,600	$12,100

✔CHECK YOUR UNDERSTANDING

A. Tyronne Gilkey is single and has an adjusted gross income of $65,000. Tyronne has deductions of only $2,900, and so he decides to take the standard deduction. He claims one exemption for himself. What is his taxable income?

B. Alice Greer and her spouse have an adjusted gross income of $50,000. They file jointly. Their itemized deductions are $11,500 and they claim three exemptions. What is their taxable income?

Income Tax Due

Employers withhold money for income taxes from employee paychecks during the year. The amount of tax paid in withholding is an estimate and is probably more or less than the tax the employee actually owes.

To find the tax due, you must complete a tax return. If too much withholding or self-employment tax has been paid, the government will pay back, or *refund* the difference. If too little tax has been paid, you must pay the difference to the government. If taxable income is less than $100,000, a tax table must be used to find the tax. Parts of a recent tax table are shown below.

If line 43 (taxable income) is—		And you are—			
At least	But less than	Single	Married filing jointly *	Married filing sepa-rately	Head of a house-hold
		Your tax is—			
13,000					
13,000	**13,050**	1,563	1,303	1,563	1,394
13,050	**13,100**	1,570	1,308	1,570	1,401
13,100	**13,150**	1,578	1,313	1,578	1,409
13,150	**13,200**	1,585	1,318	1,585	1,416
13,200	**13,250**	1,593	1,323	1,593	1,424
13,250	**13,300**	1,600	1,328	1,600	1,431
13,300	**13,350**	1,608	1,333	1,608	1,439
13,350	**13,400**	1,615	1,338	1,615	1,446
13,400	**13,450**	1,623	1,343	1,623	1,454
13,450	**13,500**	1,630	1,348	1,630	1,461
13,500	**13,550**	1,638	1,353	1,638	1,469
13,550	**13,600**	1,645	1,358	1,645	1,476
13,600	**13,650**	1,653	1,363	1,653	1,484
13,650	**13,700**	1,660	1,368	1,660	1,491
13,700	**13,750**	1,668	1,373	1,668	1,499
13,750	**13,800**	1,675	1,378	1,675	1,506
13,800	**13,850**	1,683	1,383	1,683	1,514
13,850	**13,900**	1,690	1,388	1,690	1,521
13,900	**13,950**	1,698	1,393	1,698	1,529
13,950	**14,000**	1,705	1,398	1,705	1,536

If line 43 (taxable income) is—		And you are—			
At least	But less than	Single	Married filing jointly *	Married filing sepa-rately	Head of a house-hold
		Your tax is—			
23,000					
23,000	**23,050**	3,063	2,671	3,063	2,894
23,050	**23,100**	3,070	2,679	3,070	2,901
23,100	**23,150**	3,078	2,686	3,078	2,909
23,150	**23,200**	3,085	2,694	3,085	2,916
23,200	**23,250**	3,093	2,701	3,093	2,924
23,250	**23,300**	3,100	2,709	3,100	2,931
23,300	**23,350**	3,108	2,716	3,108	2,939
23,350	**23,400**	3,115	2,724	3,115	2,946
23,400	**23,450**	3,123	2,731	3,123	2,954
23,450	**23,500**	3,130	2,739	3,130	2,961
23,500	**23,550**	3,138	2,746	3,138	2,969
23,550	**23,600**	3,145	2,754	3,145	2,976
23,600	**23,650**	3,153	2,761	3,153	2,984
23,650	**23,700**	3,160	2,769	3,160	2,991
23,700	**23,750**	3,168	2,776	3,168	2,999
23,750	**23,800**	3,175	2,784	3,175	3,006
23,800	**23,850**	3,183	2,791	3,183	3,014
23,850	**23,900**	3,190	2,799	3,190	3,021
23,900	**23,950**	3,198	2,806	3,198	3,029
23,950	**24,000**	3,205	2,814	3,205	3,036

If line 43 (taxable income) is—		And you are—			
At least	But less than	Single	Married filing jointly *	Married filing sepa-rately	Head of a house-hold
		Your tax is—			
0	**5**	0	0	0	0
5	**15**	1	1	1	1
15	**25**	2	2	2	2
25	**50**	4	4	4	4
50	**75**	6	6	6	6
75	**100**	9	9	9	9
100	**125**	11	11	11	11
125	**150**	14	14	14	14
150	**175**	16	16	16	16
175	**200**	19	19	19	19
200	**225**	21	21	21	21
225	**250**	24	24	24	24
250	**275**	26	26	26	26
275	**300**	29	29	29	29
300	**325**	31	31	31	31
325	**350**	34	34	34	34
350	**375**	36	36	36	36
375	**400**	39	39	39	39
400	**425**	41	41	41	41
425	**450**	44	44	44	44
450	**475**	46	46	46	46
475	**500**	49	49	49	49
500	**525**	51	51	51	51
525	**550**	54	54	54	54
550	**575**	56	56	56	56
575	**600**	59	59	59	59
600	**625**	61	61	61	61
625	**650**	64	64	64	64
650	**675**	66	66	66	66
675	**700**	69	69	69	69
700	**725**	71	71	71	71

To use a tax table, find your taxable income in the "At least ... but less than" columns. Then read across that line to the column that shows the filing status: single, married filing jointly, etc. The amount where that line and column meet is your tax.

EXAMPLE 2

Use the tax tables to find the income tax due and the amount owed or refunded.

a. Bea O'Shea is single, has taxable income of $13,200, and her employer deducted $2,340 in withholding taxes for the year.

b. Vince Tagliani is married and files a joint return. He and his wife had a taxable income of $23,425 last year. The amount withheld from their wages was $2,669 during the year.

SOLUTION

	a. O'Shea	b. Tagliani
Income tax due from table	$1,593	$2,731
Amounts withheld during year	$2,340	$2,669
Tax owed	—	$62
Refund due	$747	—

Business Tip

A head of household is an unmarried or legally separated person who pays more than half the cost of keeping a home for a dependent father, mother, or child.

✔CHECK YOUR UNDERSTANDING

C. Bill Reston is single and has taxable income of $13,576. His employer deducted $1,827 in withholding taxes for the year. Find Bill's tax due and any refund or amount owed.

D. Vera Yates is married and files a joint return. Vera and her spouse had a taxable income of $23,901 last year. The amount withheld from their wages was $2,509 during the year. Find their tax due and refund or amount owed.

Income Tax Refunds for Single Dependents

Photodisc/Getty Images

Many young, single people, such as students, are listed as dependents on someone else's income tax even though they are employed. They are required to pay income taxes on their earnings, even though the tax they actually owe is usually very low. That means that usually the federal income taxes withheld from their paychecks are greater than the federal income taxes they owe.

To claim a refund on taxes paid, you must file an income tax return. The rules for dependents filing returns are different than for people who are not dependents.

A dependent's income is grouped into two categories: earned income and unearned income. *Earned income* is from the dependent's own labor, such as wages, salaries, and tips. Everything else is *unearned income*, including interest and dividends.

A single dependent who is not blind and under 65 can claim as a standard deduction the higher of these two amounts:

a. $850

b. The amount of earned income, plus $300, up to $5,450 (This is the standard deduction used in this text.)

EXAMPLE 3

Jack Valente is a senior at Bayview High School and his parents claimed him as a dependent on their tax return. Jack worked last summer and earned $2,385. His employer deducted $320 in withholding taxes from his pay. Jack also earned $350 in interest on his savings account and had no adjustments to income. Jack claimed the standard deduction. What was the amount of Jack's refund?

SOLUTION

Find the amount of allowable deduction: $2,385 + $300 = $2,685

Since $2,685 > $850 and $2,685 < $5,450, Jack's allowable deduction is $2,685

Find the amount of adjusted gross income: $2,385 + $350 = $2,735

Find the amount of taxable income: $2,735 − $2,685 = $50

Find the tax due or refund amount: $320 − $6 = $314 Jack's refund

Math *Tip*

The symbol > means greater than. The symbol < means less than.

✔ CHECK YOUR UNDERSTANDING

E. Tina Moore is a junior who earned $3,510 working at a card shop. Her employer deducted $468 in withholding taxes. She has no adjustments to her income or additional income. Her mother claims her as a dependent. How much will Tina receive as a refund?

F. Kim Chung worked for his uncle during the summer. He was paid $2,897 but no withholding tax was deducted. He earned $57 in interest on his savings account. He has no adjustments to his income and his parents claim him on their tax return. How much does he owe in federal income taxes?

Wrap Up ▸▸▸

Young, single, dependent people, such as students, often have federal income taxes withheld from their paychecks, even though they make very little money during a year. Most will pay only a small amount in income taxes and receive a refund when they file their tax return.

Communication

Write a brief paragraph you might include in an e-mail to a friend who works explaining why it is important for the friend to file a tax return.

There are three important guidelines that should be followed when sending e-mail.

1. Your e-mail should cover only one topic.
2. Your message should be brief.
3. Be courteous and professional in your message.

Remember, once the e-mail is sent, you cannot get it back.

Exercises

Find the sum.

1. $2,683 + $5,094 + $94 + $625
2. $8,262 + $853 + $493 + $77

Find the difference.

3. $4,228 − $735
4. $63,163 − $15,926
5. $73,997 − $16,398
6. $125,370 − $73,920

Find the product.

7. $3,034 × 7
8. $6,517 × 23
9. $2,183 × 7
10. $7,525 × 18

Photodisc/Getty Images

Solve.

11. Bo and Dan Brady's income last year was: net income from business, $35,838.67; dividends, $2,312.98; interest, $3,517.45; rental income, $2,672. Adjustments to income totaled $4,628.83. Find their adjusted gross income.
12. In one year, Nestor Ortiz's wages totaled $29,450. His wife, Maria Gomez had a salary of $31,572 and a bonus of $500. The Gomezes also received $2,519 in interest and $953.37 in dividends. They paid $4,850 into a retirement fund and were penalized $52 for removing money from a savings plan early. What was their adjusted gross income that year?

Dee Goer is single and had an adjusted gross income last year of $16,457. Dee's itemized deductions were $5,452. She claimed one exemption of $3,500.

13. What was Dee's estimated taxable income?
14. What was Dee's actual taxable income?

In preparing their tax return, the Rossinis, a married couple, claimed 5 exemptions at $3,500 each, and the standard deduction. Their adjusted gross income was $45,208.

15. Estimate the Rossini's taxable income.
16. What was the Rossini's exact taxable income?

Use the tax tables in this lesson to find the tax.

17. Beth and Ira Stein are married and file a joint tax return. Their taxable income is $23,378. What is their tax?
18. Nicki O'Shea is 25 years old and is single. Her gross income last year was $21,455 and her taxable income was $13,926. Find her tax.
19. How much tax does a head of household with taxable income of $23,798 owe?
20. Jim Bouche is married but is filing a separate tax return. His taxable income is $23,624. Find the amount of his tax.
21. Tien Chou's tax return for last year shows a total tax of $8,278. Tien's employer withheld $8,450 from her wages during the year. What refund should Tien receive?
22. Kumar Panday paid $16,118 in federal withholding. Kumar's total tax shown on his tax return was $17,448. What amount of tax did he owe?

On his federal income tax return, Benito Silva, a single taxpayer, reported income from wages, $27,820; tips, $5,495; and interest earned, $1,466. Benito had these adjustments to income: payments to a retirement plan, $2,000; penalty for early withdrawal of savings, $19. His employer withheld $3,985 from his pay for federal income tax. Benito claims the standard deduction of $5,450 and an exemption of $3,500.

23. What was Benito's gross income?

24. Find the total of his adjustments to income.

25. Find Benito's adjusted gross income.

26. What was his taxable income?

27. What is his amount of tax due?

28. Does he owe or get a refund? How much?

For Exercises 29–35, assume that each person claims the standard deduction, is under age 65, not blind, had no adjustments to income, and is listed on the parents' return as a dependent. Use the tax tables found in this lesson when needed.

Find each dependent's standard deduction.

	Earned Income	Unearned Income	Standard Deduction		Earned Income	Unearned Income	Standard Deduction
29.	$250	$100		**30.**	$1,950	$600	
31.	$4,875	$300		**32.**	$5,575	$0	

Solve.

33. Last summer, Troy Yaeger earned $2,855. His employer withheld $240 of his wages for income taxes. What is the amount of Troy's tax refund?

34. Carmen Reyes worked part-time last year while attending college and earned $4,288. The total withholding taxes she paid were $450. Carmen also earned $78 in interest and $16 in dividends. How much tax refund should she receive?

35. Jon Kent's parents subtracted $3,500 from their adjusted gross income when they listed him as an exemption on their tax return. Jon's taxable income was $148. He paid $277 in withholding taxes. What amount should he expect as a tax refund?

36. **FINANCIAL DECISION MAKING** Molly and Dan Shashack claim one less withholding allowance than they are legally entitled to. They do this so that they always get a large refund when they file their federal income tax return. Is this a good idea? Why or why not?

Mixed Review

37. Find 11% of $250.

38. Find $\frac{1}{2}$% of $4,800.

39. 180 is what percent less than 240?

40. What percent of $90 is $270?

41. 432×0.001

42. $9.053 \div 0.001$

43. Jerry Blanchard sells computers. The average price of the computers is $1,248. His rate of commission is $12\frac{1}{2}$%. How many computers would he have to sell to make $624 a week?

2-3

State and City Income Taxes

GOALS

- Calculate state and city income taxes using a flat tax rate
- Calculate state and city income taxes using a graduated tax rate table

KEY TERM

- flat tax

Start Up ▸▸▸

What are the major uses of city, village, or town taxes in your area? If you live outside a city, village or town, what are the major uses of your county taxes? Make a list of services that your city, village, town, or county provides using tax money.

Photodisc/Getty Images

Math Skill Builder

Review these math skills. Solve the exercises that follow.

1. **Add** money amounts.
   Find the sum. $150 + $280 + $450 + $580 = $1,460

   1a. $45 + $108 + $289 + $310

   1b. $308 + $467 + $589 + $612

2. **Rewrite** percentages as decimals.
   Rewrite 5.6% as a decimal. 5.6% = 0.056

   2a. 4.7% **2b.** 7.14%

   2c. 0.8% **2d.** 14.9%

3. **Multiply** money amounts by decimals and **round** products to the nearest cent.
   Find the product. $54,109 × 0.035 = $1,893.815, or $1,893.82

   3a. $24,780 × 0.07 **3b.** $47,090 × 0.048

   3c. $35,100 × 0.127 **3d.** $249,410 × 0.0345

State and City Flat Income Taxes

Some states and cities tax personal income as a percent of federal taxable income. Some tax personal income as a percent of gross income. Some use a fixed, or **flat tax** rate no matter how much taxable income a person has. That is, the tax rate is the same for every person, regardless of the amount of income they earn in a year.

EXAMPLE 1

Allyson Greve has calculated her federal taxable income to be $45,300. She pays a state income tax rate of 3% on her federal taxable income. Find her state income tax.

Research the list of states with a state income tax. Which state has the highest income tax? Which state has the lowest income tax? Which states have no income tax?

SOLUTION

Rewrite the tax rate as a decimal rate:
3% = 0.03

Multiply the federal taxable income by the decimal tax rate: $45,300 × 0.03 = $1,359
Allyson's state income tax is $1,359.

✔ CHECK YOUR UNDERSTANDING

A. LaDonna Traube has to pay a city income tax of 1.5% of her federal taxable income. Last year her federal taxable income was $34,100. What amount of city income tax did she pay?

B. The state in which Angel Soto lives charges a 3.8% income tax based on a person's federal taxable income. If Angel's federal taxable income last year was $29,900, how much state income tax did he pay?

State and City Graduated Income Taxes

Some states and cities use a *graduated* income tax rate like the federal government. In a graduated tax system, the tax rate gets higher as taxable income gets larger. A portion of a graduated tax rate schedule that might be used by a state is shown below.

For taxable income		
Over —	**But not over —**	**The tax is —**
$ -0-	$8,000	2% of taxable income
8,000	16,000	$160 plus 3% of taxable income over $8,000
16,000	24,000	$400 plus 4% of taxable income over $16,000
24,000	32,000	$720 plus 5% of taxable income over $24,000
32,000	40,000	$1,120 plus 6% of taxable income over $32,000
40,000	48,000	$1,600 plus 7% of taxable income over $40,000
48,000	56,000	$2,160 plus 8% of taxable income over $48,000
56,000	64,000	$2,800 plus 9% of taxable income over $56,000
64,000	72,000	$3,520 plus 10% of taxable income over $64,000

Communication

Which do you favor: A flat tax or a graduated tax? Your local government is considering charging an income tax on the taxable incomes of its citizens. Write a letter to your mayor in which you recommend either a flat income tax or a graduated income tax. Your letter should defend your choice of income tax.

EXAMPLE 2

Your taxable income last year was $25,800. Using the graduated income tax table on the previous page, what was your state income tax?

SOLUTION

Find the income ($25,800) in the table shown on the previous page. It is in the column for over $24,000 but under $32,000.

Find the tax on taxable income up to $24,000: $720

Find the taxable income over $24,000: $25,800 − $24,000 = $1,800

Find the tax on $1,800 at a tax rate of 5%: $1,800 × 0.05 = $90

Find the total state income tax on $25,800: $720 + $90 = $810

✔ CHECK YOUR UNDERSTANDING

C. Jennifer Robler's taxable income last year was $43,600. Use the table given on the previous page to find her state income tax.

D. San-li Pyeon had a taxable income last year of $38,200. Using the table given on the previous page, what is his state income tax?

Wrap Up ▶▶▶

Look back at your list of the uses of local taxes from the beginning of the lesson. Rate the importance of each of the services to you. Place a "1" next to the tax use that is most important to you, a "2" next to the next more important use, and so on.

Exercises

Find the sum.

1. $120 + $270 + $375 + $489

2. $240 + $597 + $812 + $956

Rewrite as decimals.

3. 7.2%

4. 14.7%

5. 24.98%

6. 0.5%

Solve.

7. Renatta Versan pays a city income tax of 3.7% on her taxable income. Her taxable income last year was $58,390. What city income tax amount did she pay?

8. Jason Wiley lives in a state that charges an income tax of 4.52% on all taxable income. Last year Jason's taxable income was $42,189. What state income tax did he pay that year?

Solve.

9. In addition to federal and state income taxes, Rue Lange also has to pay a city income tax. The city income tax rate is $2\frac{1}{2}\%$ of his taxable income. If his taxable income is $2,345, what is his city income tax?

10. The City of Beacon charges its residents an income tax of $\frac{1}{2}\%$ of their taxable income. Dora Feldman lives in Beacon and has taxable income of $34,676. What is her city income tax?

For Exercises 11–18 use the graduated tax tables given.

11. Oren Bradley's taxable income last year was $25,890. What was his state income tax?

12. Lea Kristen's state income tax return shows taxable income of $39,350. What is the state income tax on that amount?

13. Wayne Delvica's income subject to state income tax is $22,690. What is his state income tax?

Bill Stark has taxable income of $26,600. He pays a city income tax of 2% on taxable income in addition to state and federal taxes.

14. What is Bill's city tax?

15. What is his state tax?

16. What is Bill's total city and state tax?

17. Ellen Donald pays city tax of 2.5% on taxable income, in addition to state income tax. Her taxable income last year was $42,870. What was her total state and city income tax?

18. Helmut Schmidt pays a city income tax of $2\frac{1}{4}\%$ on his taxable income of $28,834. In addition, he pays both state and federal income taxes on the same taxable income. If his federal tax last year was $4,538, what was the total of his federal, state, and city income taxes last year?

Photodisc/Getty Images

INTEGRATING YOUR KNOWLEDGE Alma Ruforio is a medical technician. Last year, her employer withheld $520 from her wages for state income tax. When she prepared her tax return, Alma showed gross income of $32,400 less $5,450 as a standard deduction and one exemption for herself.

19. What was Alma's taxable income?

20. What was her state tax for the year?

21. How much is her refund or tax due?

Mixed Review

22. $4\frac{2}{5} + 6\frac{1}{2}$

23. $17\frac{1}{4} - 6\frac{1}{3}$

24. $8\frac{1}{2} \times 5\frac{1}{3}$

25. Valerie Bassett is paid 5% commission on all sales up to and including $25,000, and 7% on all sales over $25,000 in any month. Last month Valerie's sales were $45,250. What was the amount of her total commission?

26. Last year the Pavlos earned $72,800 from salaries. They also earned $1,280 in interest and $2,789 in dividends. The Pavlos paid $6,240 into a retirement fund. What was their adjusted gross income for the year?

2-4

Benefits and Job Expenses

GOALS

- Find total job benefits
- Find net job benefits
- Compare the net job benefits of jobs

KEY TERMS

- employee benefit
- job expense
- net job benefit

Start Up ▸▸▸

Sally's uncle asked her to help with a remodeling job that will take 10 days to complete. He said he would pay Sally $60 a day or start with $1 for the first day and double her pay each day. Which wage do you think will earn Sally the most money?

Photodisc/Getty Images

Math Skill Builder

Review these math skills and solve the exercises that follow.

1. **Add** money amounts.
   Find the sum. $43,112 + $3,078 + $1,087 + $466 = $47,743

   1a. $19,208 + $189 + $2,417 + $25

   1b. $78,297 + $35 + $108 + $3,108

2. **Subtract** money amounts.
   Find the difference. $39,087 − $8,648 = $30,439

   2a. $98,085 − $13,498

   2b. $29,337 − $2,073

   2c. $101,882 − $24,938

   2d. $289,108 − $57,650

3. **Multiply** money amounts.
   Find the product. $8.75 × 40 = $350

   3a. $12.88 × 36 **3b.** $15.39 × 38

   3c. $10.82 × 40 **3d.** $10.10 × 32

Business *Tip*

Health insurance is often too costly for many workers. Some companies provide health insurance at a lower cost as a benefit to their workers.

Total Job Benefits

In addition to wages, many employers provide other things of value called **employee benefits** or *fringe benefits.* For example, employers may provide low-cost health and accident insurance, life insurance, and pensions. They may also provide paid holidays, sick leave, and vacation time, the use of a car, a credit union, uniforms, parking, discounts for purchases of merchandise, recreational facilities, child care, and education or training.

Employee benefits are an important part of a job's total value. Benefits can be worth from 15% to 40% of the amount paid in wages. Benefits may be stated in money amounts or as a benefit rate, which is a percent of gross pay.

Total Employee Benefits = Benefit 1 + Benefit 2

Total Employee Benefits = Benefit Rate × Gross Pay

When you are considering a job offer, the value of employee benefits should be added to the amount of wages to find the *total job benefits.*

Total Job Benefits = Gross Pay + Employee Benefits

> **Calculator Tip**
>
> Add the benefits first to find the total benefits package. Then add the gross pay to the benefits package to get total job benefits.

EXAMPLE 1

Kirby Rosen is a manager with Durable Products, Inc. Last year Kirby earned gross pay of $34,800 and these benefits: paid pension, $2,784; health insurance, $1,892; paid vacation, $1,338; paid holidays, $2,007; and free parking, $425. What total job benefits did Kirby receive last year?

SOLUTION

Gross pay: $34,800

Benefits:	Paid pension	$2,784
	Paid vacation	$1,338
	Free parking	$425
	Health insurance	$1,892
	Paid holidays	$2,007
Total employee benefits:		$8,446

Total job benefits: $34,800 + $8,446 = $43,246

✔ CHECK YOUR UNDERSTANDING

A. Vi Schashack estimated her yearly fringe benefits last year to be: health insurance, $2,580; paid vacations and holidays, $3,133; paid pension, $2,545. Vi's gross pay was $31,807 last year.
(1) What were Vi's total employee benefits for last year?
(2) What were Vi's total job benefits for last year?

B. Lin Ping earned gross pay of $28,089 last year. Her yearly benefits are 33% of her gross pay.
(1) What were Lin's total fringe benefits for last year?
(2) What were her total job benefits for last year?

Net Job Benefits

Almost every job has expenses. Some examples of **job expenses** are union or professional dues, commuting expenses, uniforms, licenses, and tools. To find **net job benefits**, subtract total job expenses from total job benefits.

Net Job Benefits = Total Job Benefits − Job Expenses

EXAMPLE 2

Rita Espinosa had total job benefits of $32,620. Her job expenses were $1,624 for commuting, $135 for a required license, $275 for professional dues, and $75 for the company birthday fund. Find her net job benefits.

SOLUTION

Total job benefits: $32,620

Job expenses: Travel + License + Professional Dues + Birthday Fund
$1,624 + $135 + $275 + $75 = $2,109

Total expenses: $2,109

Net job benefits: $32,620 − $2,109 = $30,511

✔CHECK YOUR UNDERSTANDING

C. Ben Asimov found that his job expenses for last year were: uniforms, $329; licenses, $278; professional dues, $475; commuting costs, $1,077. His total job benefits for the same period were $56,102. Find his net job benefits.

D. Nicki's total job benefits for the previous year were estimated to be $78,299. However, her job expenses for the same job were: licenses, $580; commuting costs, $1,793; technical books, $2,057. What were her net job benefits for the year?

Comparing Net Job Benefits

When you compare jobs you should consider many features about each job, not just the net job benefits offered by each job. For example, you should consider how much you like the job, the chances for raises and promotions, the chances of layoffs, and job security.

EXAMPLE 3

Iko Moro's job pays $33,750 in yearly wages and 26% of her wages in yearly benefits. She estimates that yearly job expenses are $2,354. Another job that she is looking at pays $32,590 in yearly wages and has estimated yearly benefits of 29%, with job expenses of $2,080. Which job offers the greater net job benefits, and how much greater?

SOLUTION

Rewrite Iko's estimated benefit percent as a decimal. Multiply her yearly wages by the decimal rate to get the benefits of the job.

Add the benefits and yearly wage amounts, and subtract the job expenses to find the net job benefits of the job.

Iko's current job	*The other job she is considering*
$33,750 × 0.26 = $8,775	$32,590 × 0.29 = $9,451.10
$33,750 + $8,775 − $2,354 = $40,171	$32,590.00 + $9,451.10 − $2,080.00 = $39,961.10

Subtract the net job benefits of the other job from the net job benefits of Iko's current job.

$40,171.00 − $39,961.10 = $209.90

Iko's current job offers her the greatest net job benefits, by $209.90.

✔ CHECK YOUR UNDERSTANDING

E. Ted Roberts earned a salary of $41,700 last year. His benefits were 32.5% of his salary. His job expenses totaled $3,180. Ted is looking at another job that offers $45,260 in wages and 24% in benefits. His job expenses for the other job total $3,740. Which job offers the greatest net job benefits, and how much greater?

F. Amy Weir had these job expenses last year: union dues, $650; tools, $1,890; uniforms, $375; licenses, $480. She earned $49,500 in wages and received benefits worth 29% of her wages. She has been offered another job at another company that will pay $45,200 in wages and 34% in benefits. Amy's job expenses for the other job are: commuting, $2,560; licenses, $480; tools, $590; parking, $380. Which job offers the greatest net job benefits, and how much greater?

Wrap Up ▶▶▶

Sally would receive $600 ($60 per day × 10 days) in the first offer. The second offer would pay her $1,023. Isn't it amazing how quickly doubling your daily pay starting with only $1 can add up?

Exercises

Find the sum.

1. $1,034 + $215 + $65,901 + $819

2. $82,298 + $3,725 + $406 + $616

Find the difference.

3. $45,823 − $41,637

4. $35,789 − $30,479

5. $52,907 − $48,278

6. $156,980 − $75,345

Rewrite percents as decimals.

7. 56.4%

8. 3.8%

9. 23.6%

10. 13.723%

Find the product.

11. $44,098 × 25.7%

12. $52,491 × 31.8%

13. $29,926 × 23.245%

14. $31,044 × 15.5%

John Bellows earned gross pay of $26,888 last year. He estimates his yearly benefits to be: paid pension, $1,828; health and life insurance, $1,654; paid vacations and holidays, $2,582; free parking, $237.

15. What were John's total estimated benefits for last year?

16. What were his total job benefits for last year?

17. Thomasina Serling had the following job expenses for last year: union dues, $529; licenses, $178; commuting costs, $2,709. Her total job benefits for the same period were $46,192. Find her net job benefits.

BEST BUY Jorge Conseco can work for ABM, Inc. for $437 per week or Zeda, Inc. for $1,408 per month. Benefits average 19% of yearly wages at ABM and 25% at Zeda. Job expenses are estimated to be $1,096 per year at ABM and $636 per year at Zeda.

18. Which job would give Jorge more net job benefits for a year?

19. How much more?

20. **CRITICAL THINKING** The value of some benefits, such as paid holidays, can be figured very accurately. However, the value of other benefits, such as free recreation facilities, can only be estimated. If you were offered a benefit package that included use of a free gymnasium, how would you estimate its dollar value?

INTEGRATING YOUR KNOWLEDGE Nora Bertram works at Radnor Products, Inc. and is paid a salary of $25,000 plus 5% commission on all her sales. Last year her sales were $200,000. Nora's benefits were: paid pension, $3,150; health insurance, $2,400; paid vacations and holidays, $3,365. Her job expenses are $3,007. She is considering a job offer from B-Tree, Inc. that pays a salary of $30,000 plus 6% commission on all sales over $100,000. She estimates her benefits at B-Tree to be $8,489 and her job expenses to be $2,050.

Photodisc/Getty Images

21. If Nora's sales at B-Tree were $200,000, which job would give her more net job benefits?

22. Use your answer from Exercise 22 to determine how much more the net job benefits would be.

Mixed Review

Change to decimals.

23. $\frac{3}{8}$ **24.** 87.6% **25.** 0.5% **26.** $\frac{1}{4}\%$

Change to fractions or mixed numbers and simplify.

27. 25% **28.** 250% **29.** 10%

30. Cromwell, Inc. employs 5 people at a branch office. Their weekly wages are: Fred, $423.34; Erin, $479.14; Bob, $378.98; Susan, $528.20; and James, $462.93. What is the average weekly wage at the branch office?

2-5

Analyze Take-Home Pay

GOALS

- Calculate take-home pay as a percentage of gross pay
- Calculate the impact of a raise on take-home pay
- Calculate potential tax savings of a cafeteria plan

KEY TERM

- cafeteria plan

Start Up

Raul has been working for the same employer for a year. He was given a $75 per month increase to his gross monthly wages. He has been planning how to use an extra $75 per month. What advice would you give Raul?

Image Source/Jupiter Images

Math Skill Builder

Review these math skills, and answer the questions that follow.

1 **Add** money amounts.
Find the sum. $22.85 + $63.22 + $43.18 = $129.25

1a. $23.18 + $14.86 + $55.18

1b. $99.20 + $3.47 + $18.63

2 **Subtract** money amounts.
Find the difference. $675.00 − $83.27 = $591.73

2a. $2,456.00 − $843.67

2b. $458.22 − $86.13

2c. $325.46 − $65.98

3 **Calculate** the percent and round to the nearest percent.
What percent of 160 is 120? 120 ÷ 160 = 0.75 = 75%

3a. What percent of 550 is 485?

3b. 18 is what percent of 45?

3c. What percent of 15 is 12?

3d. 45 is what percent of 60?

4 **Multiply** money amounts by percents, and round the product to the nearest whole cent.
Find the product. $427.25 × 6% = $427.25 × 0.06 = $25.635 or $25.64

4a. $2,842.00 × 3%

4b. $388.49 × 6.5%

4c. $759.32 × 7%

4d. $1,764 × 1.5%

Math *Tip*

To rewrite a decimal as a percent, move the decimal point two places to the right and add a percent sign.

To rewrite a percent as a decimal, move the decimal point two places to the left and drop the percent sign.

Take-Home Pay as a Percentage of Gross Pay

Clearly, there is a difference between gross pay and take-home pay. Taxes, insurance, and other deductions reduce the part of your income that you actually take home. One way to judge the impact that deductions have on your pay is to calculate the percentage of gross pay that you take home.

EXAMPLE 1

Rosalinda's monthly gross pay is $3,000. Out of those wages, $304 in federal withholding, $229.50 in FICA taxes, $60 in state income taxes, and $175 in health insurance premiums were deducted. Find the percentage of gross pay that Rosalinda takes home to the nearest percent.

SOLUTION

Find Rosalinda's take-home pay.

$304 + $229.50 + $60 + $175 = $768.50   Add to get total deductions.

$3,000 − $768.50 = $2,231.50   Subtract the total deductions from gross pay.

$2,231.50 ÷ $3,000 = 0.74 = 74%   Calculate percent.

Rosalinda's take-home pay is 74% of her gross pay.

✔CHECK YOUR UNDERSTANDING

A. Ross earns $500 per week. Each week, $38 is deducted for federal income taxes, $38.25 is deducted for FICA taxes, $30 is deducted for state taxes, $84 is deducted for health insurance, and $2.50 is deducted for union dues. Find the percentage of his gross wages that Ross takes home to the nearest percent.

B. Lisa Dewees earns $15.00 per hour, and she works 40 hours per week. The deductions from her weekly pay include $76 in federal income taxes and $45.90 in FICA taxes. To the nearest percent, find the percentage of her gross wages that Lisa takes home.

Evaluating the Impact of a Raise

Employers may offer employees a raise in pay if they have worked for the employer for a length of time or if they are good workers. In addition, employees can receive a promotion that gives them greater responsibilities and also a raise in pay.

Raises might be expressed as an increase in a dollar amount per hour, per pay period or an increase per year. Other raises may be expressed as a percentage of the current gross pay.

When you are paid a raise, that raise impacts your deductions. Because taxes are a percent of your gross pay, as your income increases, the amount of money you pay in taxes increases.

EXAMPLE 2

Susan earns $450 a week at her job. She is given a 5% raise. She is a single taxpayer who claims one withholding allowance. Federal, FICA, and a 5% state income tax are withheld from her wages. What is the increase in her gross wages? How much does her net pay increase per week with her raise? (Use the withholding table on pp. 41 and 42)

SOLUTION

$450 × 0.05 = $22.50 Find the increase in gross wages.

$450 + $22.50 = $472.50 Add increase to wages.

	Current	With Raise
Gross wages	$450.00	$472.50
Federal income tax	$ 43.00	$ 46.00
FICA	$ 34.43	$ 36.15
State income tax	$ 22.50	$ 23.63
Total deductions	$ 99.93	$105.78
Take-home pay	$450.00	$472.50
	−$ 99.93	−$105.78
	$350.07	$366.72

$366.72 − $350.07 = $16.65

Susan's gross wages increase $22.50 per week. Her net pay increases $16.65.

✔ CHECK YOUR UNDERSTANDING

C. Alejandra makes $500 per week. She is given a 4% raise. She is a single taxpayer who claims zero withholding allowances. Federal income, FICA, and a 3% state income tax are deducted from her pay. What is the increase in her gross wages? How much will her net pay increase?

D. John Daniels earns $11.00 an hour and works 40 hours per week. He is offered a promotion and an 8% raise. He is a married taxpayer who claims 3 withholding allowances. Federal income and FICA taxes are deducted from his wages. How much will his gross wages increase? How much will his net pay increase?

Pre-Tax Deductions

Some employers offer employees the opportunity to participate in a Section 125 plan. Sections 125 plans are nicknamed **cafeteria plans** because they offer a wide variety of options from which employees can choose, much like choosing food in a cafeteria line.

In a cafeteria plan, the employee directs the employer to deduct a certain amount of money from wages before taxes. This money is put into an account that is used to pay for qualified expenses, such as health or life insurance premiums, child care, or healthcare expenses.

By taking the cafeteria plan deductions before taxes, it reduces the amount of wages subject to taxes, thereby reducing the amount paid in taxes.

Consumer Alert

A Section 125, or cafeteria plan, can be a good way to reduce your taxes. However, before participating in a cafeteria plan, workers need to read and understand the terms of the plan, such as which expenses qualify for the plan and how to account for those expenses.

Participants need to understand that under current law, money that is put into a cafeteria plan account must be used for qualified expenses by the end of the calendar year or it is lost. Cafeteria plans are governed by the U.S. Tax Code, and are subject to change.

EXAMPLE 3

Jacob earns $525 a week. He is a married taxpayer who claims two withholding allowances. From his wages, the following amounts are withheld: $65 in health insurance, federal taxes, FICA tax, and a 5% state income tax. If he participates in his employer's cafeteria plan and has the $65 per week for health insurance deducted from his wages before taxes, how much will it reduce his taxes? (Use the withholding table on pp. 41 and 42)

SOLUTION

	Current	**With cafeteria plan**
Gross wages	$525.00	$525.00
Cafeteria plan		$ 65.00
Taxable income	$525.00	$460.00
Federal income tax	$ 24.00	$ 18.00
FICA	$ 40.16	$ 35.19
State income tax	$ 26.25	$ 23.00
Total taxes	$ 90.41	$ 76.19

$90.41 − $76.19 = $14.22.

Jacob will pay $14.22 less in taxes per week if his health insurance is deducted before taxes.

✔CHECK YOUR UNDERSTANDING

E. Tonja is a married taxpayer who claims 2 withholding allowances. She earns $600 per week, and pays $150 per week in childcare expenses. How much will she save in federal income and FICA taxes if she has her childcare expenses deducted from her check before taxes?

F. Enrique makes $11.50 per hour and works 40 hours per week. He is single and claims 0 withholding allowances. He estimates that he will spend $100 per month on qualified expenses. How much will he save per week in taxes if he participates in his employer's cafeteria plan and has $100 per month deducted before taxes? How much will he save in one year, or 50 weeks of work?

Employers benefit from cafeteria plans as well. Because employers are required to pay taxes on each employee's taxable income also, the amount of tax due for an employee that participates in a cafeteria plan is reduced. In addition, by offering cafeteria plans that reduce the amount of taxes employees pay, employers effectively give employees a "raise" without spending more money on wages.

Wrap Up

Although Raul is getting a $75 raise, he will not have an extra $75 per month to spend. Federal, FICA, and possibly state and local income taxes will reduce his raise. Before planning what to do with the extra money, Raul should wait and see how much his take-home pay actually increases.

TEAM Meeting

Form a team with two other students. Each team member should interview one person who receives a paycheck. Each team member should prepare a three-column report of the interview. Make the first column show required deductions; the second, personal or optional deductions; the third, percent of paycheck. Compare lists. Then combine the lists into one, three-column report. When deductions from two or more lists match, show the percents of paycheck as a range of percents using the lowest and highest percents found for that deduction.

PhotoAlto/Getty Images

Exercises

Find the sum.

1. $26 + $18.23 + $6.07 + $14.10
2. $62 + $75.31 + $24.84 + $10.21

Find the difference.

3. $643.81 − $86.22
4. $2,876.42 − $186.40
5. $1,543.28 − $127.33
6. $342.91 − $111.31

Calculate the percent, to the nearest tenth.

7. 63 is what percent of 75?
8. 156 is what percent of 280?
9. What percent of 600 is 120?
10. What percent of 1,500 is 1,226?

Find the product.

11. $629 × 5%

12. $761 × 6.45%

Yolinda earns $4,000 per month. The following amounts are deducted from her gross wages: $584 federal income taxes, $248 Social Security, $58 Medicare, $120 state income taxes, and $120 health insurance.

13. Find the total deductions from Yolinda's check.
14. What is Yolinda's monthly take-home pay?
15. To the nearest percent, what percent of her gross pay does Yolinda take home?
16. Yolinda is offered a 6.5% raise. How much will her gross pay increase each month?
17. With her raise, Yolinda's deductions increase to $1,217.69. What is her new take-home pay?
18. How much did Yolinda's take-home pay increase?
19. After her raise, what percent of her gross pay does Yolinda take home?

Complete the table below for the weekly wages of each employee. Round to the nearest whole percent. Use the withholding tables on pp. 41 and 42.

	Name	Allow-ances	Marital Status	Gross Wages	Income Tax	FICA (7.65%)	Other	Net Wages	% Gross Pay
20.	Jason	1	Single	$463.95			$65.17		
21.	Conn	0	Single	$395.22			$22.80		
22.	Flora	3	Married	$571.14			$73.16		
23.	Marta	6	Married	$452.35			$85.46		
24.	Elena	2	Married	$536.72			$55.21		

John Stone is married with 3 withholding allowances. Each week his employer deducts federal withholding taxes, Social Security taxes, Medicare taxes, and $58.12 for health insurance from his gross pay. His gross weekly wage is $578. His employer gives him a 6% raise.

25. How much will John's gross pay increase?
26. What is John's net pay before his raise?
27. What is John's net pay after his raise?
28. How much did John's net pay increase with his raise?

Sarah Cutler is paid a weekly salary of $525. She is married and claims 3 withholding allowances. Federal withholding taxes, FICA, and a 2% state income tax are deducted from her gross pay. She also has childcare expenses of $75 per week.

29. What amount is deducted each week for taxes?
30. If she had her childcare expenses deducted from her check before taxes, how much would be deducted each week for taxes?
31. How much would she save per week in taxes? How much would she save in a year, or 52 weeks?

Blend Images/Jupiter Images

Raku Tung is a single taxpayer who claims 1 withholding allowance. She makes $485 per week and has federal withholding taxes, FICA taxes, and $43 for health insurance deducted from her check each week. She is planning to get married and will change her withholding allowances to 2. Round any percents to the nearest whole percent.

32. What is her net pay before she gets married?

33. What percent of her gross pay does she take home?

34. What is her net pay after she gets married?

35. What percent of her gross pay will she take home after she gets married?

36. What will her net pay be if she participates in her employer's cafeteria plan and has her health insurance cost deducted before taxes after she is married?

37. What percent of her gross pay will she take home if she participates in the cafeteria plan?

38. **CRITICAL THINKING** Explain why you may take home a lower percentage of your gross pay after you get a raise.

Mixed Review

39. 7.7% of $23,500

40. $35,290 − $1,549

41. $3\frac{1}{8} + 14\frac{2}{3}$

Calli Burns was paid $14.56 an hour for 37.5 hours last week.

42. What was Calli's gross pay for the week?

43. How much was deducted for Social Security taxes at 6.2%?

44. Olaf Svenson worked 8 hours on Monday and Tuesday, 9 hours on Wednesday, and 10 hours on Thursday and Friday. Olaf is paid $12 an hour and time-and-a-half for time past 40 hours in a week. What was Olaf's gross pay for the week?

45. The Carter's taxable income last year was $43,780. They paid a state tax of 3.6% and a city tax of 1.35% on that income. What was the total of the state and city taxes they paid?

46. Terry Jansen works on a piece-rate basis. He completed 70 pieces on Monday, 68 on Tuesday, 74 on Wednesday, and 72 on Thursday. He is paid $1.20 for each piece. How many pieces must he complete on Friday so that his earnings for the 5 days will average $84 a day?

47. Yancy works for $14.35 an hour and gets paid time-and-a-half for overtime hours and double time for weekend hours. This week Yancy worked 40 regular hours and 11 overtime hours. Last week, Yancy worked 35 regular hours and 11 weekend hours. Which week did he earn more in wages and by how much?

Chapter Review

Vocabulary Review

Find the term, from the list at the right, that completes each sentence. Use each term only once.

adjusted gross income
cafeteria plan
deduction
employee benefits
exemption
flat tax
gross income
job expense
net job benefit
net pay
standard deduction
taxable income
withholding allowance
withholding tax

1. An amount subtracted from gross pay is a(n) __?__.
2. Total income in a year that includes income from wages, salaries, commissions, bonuses, tips, interest, dividends, prizes, pensions, the sale of stock and profit from a business is called __?__.
3. Adjusted gross income less deductions and exemptions is __?__.
4. A tax in which the rate does not vary with the amount of income is (a, an) __?__.
5. Gross pay less deductions is __?__.
6. Things of value provided by employers in addition to wages are __?__.
7. The amount left after subtracting adjustments to income from gross income is __?__.
8. An amount of income per person that is free from tax is called a(n) __?__.

2-1 Deductions from Gross Pay

9. Christy Bellows is a married worker earning $612.83 each week. She claims 3 withholding allowances. What amount should be deducted from her weekly earnings for federal withholding taxes?
10. Ron Adams earns $450 in gross wages on January 10. How much is deducted from Ron's gross wages for Social Security at 6.2% and Medicare at 1.45%?
11. Luisa Medina is paid $556 a week. Her employer deducts $56 for federal withholding tax, $54.78 for insurance, 6.2% for Social Security taxes, and 1.45% for Medicare taxes. For the week, what are her Social Security and Medicare taxes, total deductions, and net pay?

2-2 Federal Income Taxes

12. Freida Werner earned gross income of $45,600 last year. She made payments into an approved retirement plan of $3,600. What was her adjusted gross income last year?
13. Jorge Viscano had $38,000 in adjusted gross income last year. He took the standard deduction of $5,450 and one exemption for himself for $3,500. What was his taxable income for the year?

14. Ben Uris is single, has taxable income of $23,500, and his employer deducted $3,640 in withholding taxes. Find Ben's tax due and any refund or amount owed.

15. Elena Alvarez is a high school student and her parents claim her as a dependent on their tax return. Elena works part-time and earned $2,490 last year. Her employer deducted $320 in withholding taxes. Elena also earned $320 in interest on a savings account and had no adjustments to income. Elena wants to file for a refund and claim the standard deduction instead of itemizing deductions. What is the amount of Elena's refund?

2-3 State and City Income Taxes

16. Alan Grey has federal taxable income of $31,500. He pays a state income tax rate of 3.5% on his federal taxable income. Find his state income tax.

17. Haru Umeki's taxable income is $44,200. Find her state income tax if she pays $1,600 plus 7% of her taxable income over $40,000.

2-4 Benefits and Job Expenses

18. Juan Romero earns a gross pay of $44,500. His employee benefits are 18% of his gross pay. What are his total job benefits?

19. Shirley Evans earns $20.50 per hour, working 40 hours a week, 50 weeks a year. Her employee benefits include $1,200 per year for health insurance and $500 per year for free parking. Her expenses include $325 for dues and $1,300 for commuting. What are Shirley's net job benefits?

20. Phyllis Regan's job pays $46,350 plus 24% of wages in benefits. She estimates that her yearly job expenses are $2,256. A job she has been offered pays $49,750 with these estimated benefits: $3,980 in pensions, $450 in free parking, $1,080 in paid vacation, $1,560 in paid holidays, $2,100 in health insurance, and $400 in tools. The job has job expenses of $2,624. Which job offers the greater net job benefits? How much greater?

21. Ursala Thomas had the following job expenses for last year: union dues, $388; licenses, $109; commuting costs, $1,478. Her total job benefits for the same period were $39,256. Find her net job benefits.

2-5 Analyze Take-Home Pay

22. Joe Palucci earns $2,500 per month. Out of those wages, $266 in federal income taxes, $191.25 in FICA taxes, $100 in state income taxes, and $125 in health insurance premiums were deducted. Find the percentage of gross pay that Joe takes home, to the nearest percent.

23. Maggie Ryan earns $350 a week at her job. She is given a 7% raise. She is a single taxpayer who claims one withholding allowance. Federal, FICA, and a 4% state income tax are withheld from her wages. What is the increase in her gross wages? How much does her net pay increase per week with her raise?

24. Joe Marlow's weekly wages are $585. He has federal withholding, FICA, and $125 for health insurance deducted from his check. He is married and claims 2 withholding allowances. How much will he save in taxes if he has his health insurance deducted before taxes?

Technology Workshop

Task 1 Enter Data in a Payroll Sheet Template

Complete a template that calculates the Social Security tax, Medicare tax, and net pay for each employee of the Bainbridge Company.
Open the spreadsheet for Chapter 2 (tech2-1) and enter the data shown in blue (cells E6-F15). Social Security taxes, Medicare taxes, and net pay are calculated for each employee. Your finished spreadsheet should look like the one shown.

	A	B	C	D	E	F	G	H	I	J	K
1	Bainbridge Company										
2	Payroll Sheet for January 15, 20—										
3	Employee No.	Name	Allow-ances	Mar-ried	Gross Wages	Deductions					
4–5						Income Tax	Social Security	Medicare	Other	Total Deductions	Net Pay
6	1	Ajanaku	1	N	421.02	39.00	26.10	6.10	35.45	106.65	314.37
7	2	Bell	1	N	435.89	40.00	27.03	6.32	37.84	111.19	324.70
8	3	Cole	0	Y	502.54	38.00	31.16	7.29	49.75	126.20	376.34
9	4	Dern	1	N	399.50	34.00	24.77	5.79	31.54	96.10	303.40
10	5	Evers	2	Y	575.64	29.00	35.69	8.35	58.97	132.01	443.63
11	6	Ford	5	Y	449.54	0.00	27.87	6.52	50.02	84.41	365.13
12	7	Gomez	0	Y	557.76	45.00	34.58	8.09	57.64	145.31	412.45
13	8	Huang	2	Y	450.89	17.00	27.96	6.54	38.19	89.69	361.20
14	9	Isom	0	N	438.27	50.00	27.17	6.35	37.17	120.69	317.58
15	10	Jackson	3	Y	580.24	23.00	35.97	8.41	61.55	128.93	451.31
16		Totals			4,811.29	315.00	298.30	69.76	458.12	1,141.18	3,670.11
17											
18		Social Security Rate			0.062						
19		Medicare Rate			0.0145						

Task 2 Analyze the Spreadsheet Output

Answer these questions about your completed payroll sheet.

1. Which employee had the largest net pay for the period?
2. Which employee had the largest amount of deductions?
3. Which employee had the greatest number of allowances?
4. Which employees paid more in combined Social Security taxes and Medicare taxes than they paid in income taxes?
5. What was the total amount of income taxes withheld from wages for the week?

Now move the cursor to cell E18, labeled Social Security. Enter the rate 0.065.

Notice how the Social Security tax, total deductions, and net pay amounts all change. These changes show what would happen if the Social Security tax rate was updated from 6.2% to the higher 6.5%.

Answer these questions about your updated payroll sheet.

6. What is the formula used in cell F16? What arithmetic is done in the cell?
7. What is the formula used in cell G6? What arithmetic is done in the cell?
8. What is the formula used in cell H6? What arithmetic is done in the cell?
9. What is the formula used in cell J13? What arithmetic is done in the cell?
10. Why did changing cell E18 change the amounts throughout the rest of the spreadsheet?

Task 3 Design a Job Benefits Spreadsheet

Design a spreadsheet that will allow you to compare net job benefits.
The spreadsheet for Task 1 includes formulas that use subtraction, multiplication, and the SUM function. Create a spreadsheet that will use these same types of formulas for the situation below. The spreadsheet should allow you to compare the jobs based on net benefits. Assume that each job is for a 40-hour week and 52-week year. Your spreadsheet should contain a row for each of the items shown. In addition, you should enter row or column labels and formulas to calculate annual gross pay, pension benefits, total job benefits, total job expenses, and net job benefits.

DATA: You receive two job offers. The expenses and benefits for each job are shown at the right.

	Offer 1	Offer 2
Salary Information		
Hourly Rate	$11.25	$12.05
Annual Benefits		
Health Insurance	$2,500	--
Life Insurance	$250	$325
Health Club Membership	--	$550
Pension*	8%	6%
Free Parking	$650	--
Expense Information		
Commuting Costs	$777	$955
Dues	$98	$150
Tools	--	$380
Uniforms	--	$425

*Stated as a percent of annual salary.

Task 4 Analyze the Spreadsheet Output

Answer these questions about your completed spreadsheet.

11. How did you calculate annual pension benefits?
12. What were the net job benefits of Offer 1? Offer 2?
13. Which job offered the highest total employee benefits? How much higher?
14. Which job benefits package do you think is better?
15. If you were to:
    (a) change the pension percentage rate for Job 1 to 7%,
    (b) change the hourly rate to $12.50 for Job 1, and
    (c) eliminate life insurance as a benefit from both jobs,

    what is the difference in net benefits between the two offers?

Chapter *Assessment*

Chapter Test

Answer each question.

1. Write $\frac{17}{100}$ as a decimal.
2. Rewrite 0.24 as a fraction in lowest terms.
3. $68 is what percent less than $80?
4. Subtract: $34,510.23 − $7,388.04
5. Multiply: 3.85% × $45,076
6. Divide: $563 ÷ 0.1
7. $36 is what percent greater than $20?
8. $\frac{1}{6}$ less than $360 is what number?
9. Add: $506.45 + $108.45 + $78.31 + $1,957.23
10. Rewrite 0.05879 as a percent to the nearest tenth of a percent.

Solve.

11. Phan Am Van earned gross wages of $487.12. Phan's deductions were: $48 in federal withholding taxes, 6.2% in Social Security taxes, 1.45% in Medicare taxes, $28.74 in health insurance, and $12 in union dues. Find the total deductions from Phan's gross pay.
12. Find Phan's net pay.
13. A job you are considering offers these benefits: paid vacations, $2,230; paid holidays, $2,450; paid pension, $4,058.60; paid health insurance, $428.90. What are the total employee benefits of the job?
14. Sally Longfeather earns an annual wage of $43,589. She estimates job benefits at 31.5% of her wages and her job expenses at: insurance, 4% of her wages; transportation, $1,296; dues, $200; and birthday fund, $50. What are her annual net job benefits?
15. A married couple earned a total of $86,340 last year. Their taxable income is $70,000. Their state uses a graduated income tax. They owe $3,520 plus 10% of their taxable income over $64,000. How much do they owe in state income taxes?
16. Evelyn Johnson had gross income of $33,200. She also had adjustments to income of $2,700, itemized deductions of $5,600, and 2 exemptions at $3,500. What was Evelyn's taxable income?
17. Tim O'Leary had a taxable income of $19,600 last year. His employer withheld taxable income, what amount should Tim receive as a refund?
18. Vance Milo had an earned income of $2,250 last year from part-time work. He also earned $24.15 in interest from a savings account. His father claimed him as an exemption on his tax return. If Vance's employer withheld $280 from his wages for withholding taxes, how much should Vance get back as a refund on his federal income taxes?
19. Sonia Ruiz earns $485 per week. The following deductions are taken from her gross pay: $58 federal withholding, $37.10 FICA, $14.55 state income taxes, and $35.00 health insurance. To the nearest percent, what percent of her gross pay does Sonia take home?
20. Genaro Torres makes $400 per week. He receives a 2% raise. He is single and claims zero withholding allowances. Federal income taxes and FICA are withheld from his gross pay. How much does his gross pay increase with his raise? How much does his net pay increase?

Planning a Career in Human Services

David Gilder 2008/Shutterstock.com

There are a wide variety of career choices available in human services. Workers in human services help others to improve their quality of life by providing support, services, or information. Opportunities exist in social work, geriatrics, counseling, nonprofit organizations, and community work. Workers in human services can be found in employment services, nursing homes, governmental social services, hospitals, and charities. If you are patient, optimistic, and enjoy helping people, a career in human services may be a rewarding career pathway for you.

Job Titles

- Social worker
- Case manager
- Gerontology assistant
- Community outreach worker
- Career counselor
- Mental health aide
- Occupation therapist
- counselor
- Human Resources manager

Needed Skills

- excellent communication skills with both clients and coworkers
- strong assessment skills
- outstanding human relation skills
- ability to create or follow treatment plans

- outstanding leadership skills
- ability to work independently, as well as with others

What's it like to work in Social Work?

Social workers assess client need and eligibility, arrange for benefits, and monitor and record client progress. They may work with other professionals like psychiatrists or physical therapists to assist in making and following a client treatment plan to advance life or job skills, improve physical performance, and live more effectively with others. Some social workers are case managers for family services. These case managers work with families in crisis. Families are monitored to ensure participation in treatment plans that bring improvement. These case managers often report their assessments and observations directly to the judicial system for court-ordered treatment plans. Some social workers have a great deal of responsibility and work with little supervision.

What About You?

Can you see yourself working in the field of human services? Which job is most appealing to you?

How Times Have Changed

For Questions 1–2, refer to the timeline on page 39 as needed.

1. In 1990, the government collected $336.3 billion in Social Security and Medicare taxes. In 2000, the government collected $593.3 billion in Social Security and Medicare taxes. If the tax rate was 7.65% for both years, what would account for the increase in collections?
2. If you earn a salary of $35,000, how much more would you pay in Social Security and Medicare taxes at the current rate, versus the previous rate?

Chapters 1–2 *Cumulative Review*

MULTIPLE CHOICE

Select the best choice for each question.

1. Last week Sarah Carver worked 4 overtime hours at time-and-a-half pay. Her regular pay rate is $8.70 per hour. What was her overtime pay for the week?

   A. $13.05 B. $52.20 C. $52.50
   D. $69.60 E. $400.20

2. Last year Jose Inez's gross income was $24,685. She had adjustments to income of $3,640. What was Jose's adjusted gross income last year?

   A. $21,045 B. $22,185 C. $25,825
   D. $28,325 E. $30,825

3. Carmen Rielly is paid piece-rate for each of the 268 items she produces in a week and she receives gross wages of $469. What is Carmen's per piece rate?

   A. $1.50 B. $1.70 C. $1.74
   D. $1.75 E. $2.75

4. Gary Kersting has taxable income of $34,672. He pays a city income tax of 1.5% on taxable income. What is Gary's city tax?

   A. $346.72 B. $490.83 C. $520.08
   D. $5,200.80 E. $34,151.92

5. Maureen Ritter is paid $275 a week and a commission of 7% on all sales. Her sales last week were $3,904. What were her total earnings for the week?

   A. $273.28 B. $548.28 C. $558.28
   D. $2,732.80 E. $3,007.80

6. Jan Morrison's annual salary is $30,605. What is Jan's Social Security tax?

   A. $1,897.51 B. $1,989.32 C. $4,340
   D. $18,975.10 E. $28,707.49

7. Jontay Mays works 8 hours a day Monday through Friday. What is his gross pay for one week if he earns $9.84 per hour?

   A. $78.72 B. $314.88 C. $344.40
   D. $383.60 E. $393.60

8. Morgan Born is paid $11.20 an hour for a 40-hour week. Her estimated benefits are 33% of her wages. What is the value of her total yearly employee benefits?

   A. $165.76 B. $595.84 C. $1,989.12
   D. $7,687.68 E. $7,956.48

9. Dean Stroble is married and claims 2 withholding allowances. His gross weekly wage is $448. His withholding is $16 and FICA is $34.27. What is his net pay?

   A. $397.73 B. $431 C. $389.22
   D. $499.27 E. $465

10. Kim Lui is paid an annual salary of $28,680. She is paid every other week. What is her gross pay for each pay period?

A. $551.54 B. $1,103.08 C. $1,195
D. $1,434 E. $2,390

OPEN ENDED

11. Last week Jason Fields worked: Monday, 7.2 hours; Tuesday, 8.3 hours; Wednesday, 8 hours; Thursday, 8 hours; Friday, 7.4 hours. He is paid $8.90 per hour. What was Jason's gross pay last week?

12. LaDonna Ekwilugo had sales last month of $86,400. Her total earnings for the month were $5,156, which included $1,700 for her monthly salary. What rate of commission was LaDonna paid?

13. A shipping department has five workers: a supervisor who is paid $484 a week, and four other workers who are paid $390, $410, $425, and $430 a week. What is the average weekly pay for shipping department workers?

14. Janice Barton is an assembler in a factory and is paid $1.25 for each hand-held radio she assembles. During one week, Janice assembled these radio: 65 on Monday, 72 on Tuesday, 70 on Wednesday, and 68 on Thursday. How many radios must Janice assemble on Friday to earn $425 for the week?

QUANTITATIVE COMPARISON

Compare the quantity in Column A with the quantity in Column B. Select the letter of the correct answer from these choices:

A if the quantity in Column A is greater;
B if the quantity in Column B is greater;
C if the two quantities are equal;
D if the relationship between the two quantities cannot be determined from the given information.

	Column A	Column B
15. Chris Beltsos is paid a salary of $280 a week and a commission of 5.5% on all sales. His sales last week were $5,025.	Chris' weekly salary	Chris' commission last week
16. Alvin Barr's taxable income last year was $25,800. Jamaal White's taxable income last year was $29,600. Their state income tax rates were 5% for Alvin and 4% for Jamal.	Alvin's state income tax	Jamaal's state income tax
17. Krista Egan worked 4 overtime hours at time-and-a-half pay. Her regular hourly rate is $9.85. Haley Kale worked 2.5 overtime hours at double-time pay. Her regular hourly rate is $11.82 per hour.	Krista's overtime pay	Haley's overtime pay

CONSTRUCTED RESPONSE

18. A friend has just graduated from high school and is looking at two job offers. One pays $12.50 an hour for a 40-hour week. The other pays $14 an hour for a 40-hour week. The friend thinks the choice is a no-brainer. The $14 an hour job pays more and so he should take that job. Write a letter to your friend to explain what other job factors should be examined before making the decision.

Chapter 3

Banking

3-1 Savings Accounts
3-2 Checking Accounts
3-3 Electronic Banking
3-4 Check Register Reconciliation
3-5 Money Market and CD Accounts
3-6 Annuities

Statistical Insights

Leading U.S. Commercial Banks		
Bank	**Headquarters**	**Consolidated Assets (Millions of Dollars)**
JP Morgan Chase Bank	Columbus, OH	1,318,888
Bank of America	Charlotte, NC	1,312,794
Citibank	Las Vegas, NV	1,251,715
Wells Fargo Bank	Sioux Falls, SD	467,861
US Bank	Cincinnati, OH	232,760
HSBC Bank	Wilmington, DE	184,492

Source: FederalReserve.gov

The Federal Reserve System was established in 1913 to serve as the central banking authority of the United States. Among its many responsibilities is to supervise commercial banks and protect consumer credit rights. The banks of the Federal Reserve make loans to commercial banks that make their profits by offering loans to businesses, industries, and individuals.

1. Write the assets of JP Morgan Chase Bank in standard form.
2. Which bank on the list has the least assets?
3. What is the dollar amount difference between the banks with the greatest and least assets?
4. Which two banks have less than a $20,000,000,000 difference in their consolidated assets?
5. **Explain** why a graph or table might use labels such as "Millions of Dollars" or "Amounts in Millions."

How Times Have Changed

When ATMs were first introduced, each machine was installed at a single bank location and customers could only access their accounts at that location. Today, customers can use ATMs all around the world to access their funds.

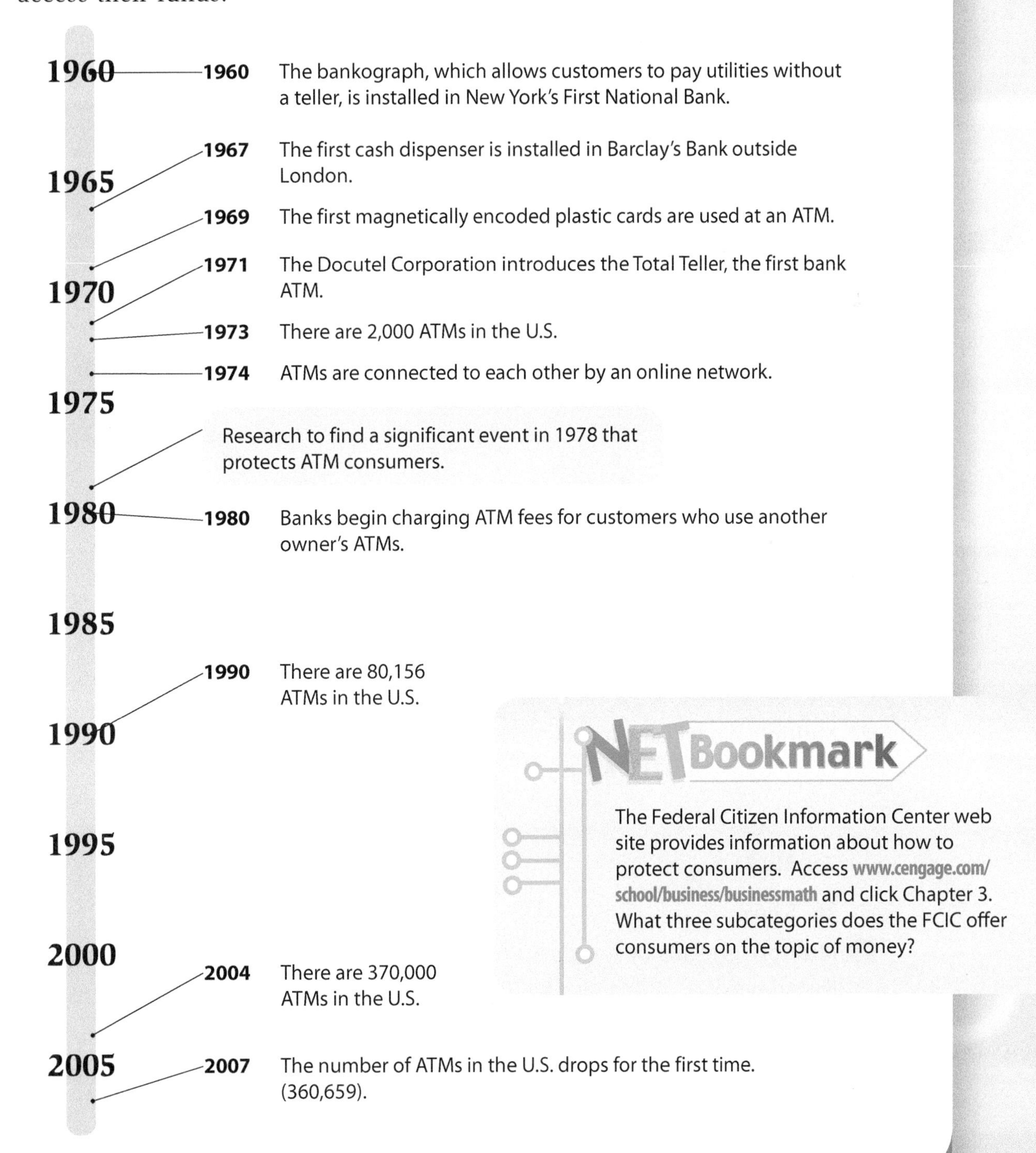

1960
1965
1970
1975
1980
1985
1990
1995
2000
2005
1960 The bankograph, which allows customers to pay utilities without a teller, is installed in New York's First National Bank.
1967 The first cash dispenser is installed in Barclay's Bank outside London.
1969 The first magnetically encoded plastic cards are used at an ATM.
1971 The Docutel Corporation introduces the Total Teller, the first bank ATM.
1973 There are 2,000 ATMs in the U.S.
1974 ATMs are connected to each other by an online network.
Research to find a significant event in 1978 that protects ATM consumers.
1980 Banks begin charging ATM fees for customers who use another owner's ATMs.
1990 There are 80,156 ATMs in the U.S.
2004 There are 370,000 ATMs in the U.S.
2007 The number of ATMs in the U.S. drops for the first time. (360,659).
NET Bookmark
The Federal Citizen Information Center web site provides information about how to protect consumers. Access www.cengage.com/school/business/businessmath and click Chapter 3. What three subcategories does the FCIC offer consumers on the topic of money?

3-1

Savings Accounts

GOALS

- Calculate simple interest on savings deposits
- Calculate compound interest on savings deposits
- Calculate interest using a compound interest table

KEY TERMS

- interest
- transaction
- compound amount
- compound interest

Start Up ▶▶▶

Joanna received her first paycheck. She is trying to decide what to do with the money she has earned. One friend suggests that she cash her paycheck and keep the cash at home. Another friend suggests that she open a savings account. What would you advise Joanna to do?

Photodisc/Getty Images

Math Skill Builder

Review these math skills and solve the exercises.

1 **Multiply** money amounts by decimals and round.
Find the product. $\$1{,}200.56 \times 0.065 = \$78.036$, or $78.04

1a. $\$965 \times 0.043$ **1b.** $\$870 \times 0.0125$

1c. $\$38 \times 0.625$ **1d.** $\$2{,}355 \times 0.82$

2 **Multiply** money amounts by fractions and round.
Find the product. $\$23.876 \times \frac{1}{2} = \$11.938$, or $11.94

2a. $\$12.836 \times \frac{1}{4}$ **2b.** $\$8.3598 \times \frac{1}{2}$

2c. $\$10.75 \times \frac{1}{3}$ **2d.** $\$8.145 \times \frac{1}{5}$

2e. $\$3.65 \times \frac{1}{2}$ **2f.** $\$9.28 \times \frac{1}{4}$

3 **Multiply** percents by fractions.
Find the product. $6\% \times \frac{1}{4} = 1.5\%$

3a. $5\% \times \frac{1}{4}$ **3b.** $8\% \times \frac{1}{2}$

3c. $22\% \times \frac{1}{5}$ **3d.** $25\% \times \frac{1}{8}$

3e. $42\% \times \frac{1}{3}$ **3f.** $108\% \times \frac{1}{4}$

Simple Interest

One reason people open savings accounts is to keep their money safe. Another reason is that they earn interest on their money. **Interest** is money paid to an individual or institution for the privilege of using their money.

When you deposit money into or withdraw money from your account, you should receive a receipt, which is an official record of the transaction. A **transaction** is something that has to be recorded, such as a deposit or withdrawal.

Banks and financial institutions typically have several different kinds of savings accounts. Interest might be figured semiannually, quarterly, monthly, or daily.

To find the simple interest for any period, first find the interest on the deposit for a full year. Then multiply that amount by the fraction of a year, such as $\frac{1}{4}$ or $\frac{1}{2}$, for which you want to find interest.

Interest = Principal × Rate × Time

> **Algebra Tip**
>
> You can calculate the interest using the formula
>
> $I = P \times R \times T$
>
> where I is the interest, P is the principal amount, R is the annual percentage rate, and T is the time in years.

EXAMPLE 1

Find the interest for six months on $400.60 at $1\frac{1}{2}$% annual interest paid semiannually.

SOLUTION

Substitute values into the formula $I = PRT$

$I = \$400.60 \times 0.015 \times \frac{1}{2} = \$3.0045$, or $3.00 interest for 6 months

✔ CHECK YOUR UNDERSTANDING

A. What interest is paid for three months on $860 at $2\frac{1}{4}$% annual interest paid quarterly?

B. Find the interest for one month on $350 at 3.1% annual interest paid monthly.

> **Business Tip**
>
> Interest can be figured on different periods, or fractions, of a year.
>
> Annually: once a year
>
> Semiannually: twice a year, $\frac{1}{2}$
>
> Quarterly: four times a year, $\frac{1}{4}$
>
> Monthly: twelve times a year, $\frac{1}{12}$
>
> Daily: 365 times a year, $\frac{1}{365}$

Compound Interest

At the end of each interest period, the interest due is calculated and added to the previous balance in the savings account. The new balance then becomes the principal on which interest is calculated for the next period, if no deposits or withdrawals are made. When you calculate interest and add it to the old principal to make a new principal on which you calculate interest for the next period, you are *compounding interest*.

Regardless of how often interest is earned, the total money in the savings account at the end of the last interest period is called the **compound amount**, assuming that no deposits or withdrawals have been made. The total interest earned, called **compound interest**, is the difference between the original principal and the compound amount.

EXAMPLE 2

Image Source/Jupiter Images

On January 1, Peter Monroe made an ATM deposit of $800 in a savings account that pays 2% interest, compounded quarterly. He made no other deposits or withdrawals. If interest is calculated and paid on April 1 and July 1, find the account balance (compound amount) and the compound interest on July 1.

SOLUTION

$I = \$800 \times 0.02 \times \frac{1}{4} = \$4$ interest for first quarter

Find the new account balance.

$\$800 + \$4 = \$804$ new balance, or new principal, on April 1

$I = \$804 \times 0.02 \times \frac{1}{4} = \$4.02$ interest for second quarter

Find the new account balance.

$\$804 + \$4.02 = \$808.02$ new balance on July 1

Find the difference between the July 1 balance and the original principal.

$\$808.02 - \$800 = \$8.02$ compound interest for two interest periods

The account balance (compound amount) on July 1 was $808.02. The compound interest for the two quarters is $8.02.

✔CHECK YOUR UNDERSTANDING

C. Your bank pays 3% interest compounded quarterly on October 1 and January 1. You had $700 on deposit on July 1 and made no additional deposits or withdrawals. Find the account balance on January 1.

D. You deposited $400 on July 1 and kept your money on deposit for one year. You made no deposits or withdrawals. If your bank pays 2.5% interest compounded semiannually, what compound interest will you earn in one year?

Compound Interest Tables

When you calculate compound interest for several interest periods, you can use a compound interest table such as one shown below. The table shows the value of one dollar ($1) after it is compounded for various interest rates and periods.

Compound Interest Table								
	Rate Per Period							
Periods	**0.25%**	**0.50%**	**0.75%**	**1%**	**1.25%**	**1.50%**	**2%**	**3%**
4	1.010038	1.020151	1.030339	1.040604	1.050945	1.061364	1.082432	1.125509
6	1.015094	1.030378	1.045852	1.061520	1.077383	1.093443	1.126162	1.194052
7	1.017632	1.035529	1.053696	1.072135	1.090850	1.109845	1.148686	1.229874
8	1.020176	1.040707	1.061599	1.082857	1.104486	1.126493	1.171659	1.266770
20	1.051206	1.104896	1.161184	1.220190	1.282037	1.346855	1.485947	1.806111
30	1.077783	1.161400	1.251272	1.347849	1.451613	1.563080	1.811362	2.427262
40	1.105033	1.220794	1.348349	1.488864	1.643619	1.814018	2.208040	3.262038

To use the table, find the interest rate per period and the total number of interest periods. The number in the table that corresponds to the interest rate per period (column) and the number of periods (row) is the compound interest multiplier. Use the multiplier to calculate the interest.

EXAMPLE 3

Find the compound interest paid on a $400 deposit that earns interest at an annual rate of 2%, compounded quarterly, for 10 years.

SOLUTION

Find the interest rate per period.

$2\% \div 4 = 0.5\%$

Find the number of interest periods.

$4 \text{ quarters} \times 10 \text{ years} = 40 \text{ periods}$

Find the multiplier in the compound interest table: 1.220794

$\$400 \times 1.220794 = 488.3176$ or $488.32 compound amount

Find the difference between the compound amount and the original deposit.

$\$488.32 - \$400 = \$88.32$ compound interest

NET Bookmark

Many web sites have compound interest calculators. You can search to find one and try it using the data from Example 3.

✔CHECK YOUR UNDERSTANDING

E. What compound interest is paid on a $1,100 deposit earning 3% interest, compounded semiannually for 3 years?

F. A deposit of $720 earns 1% annual interest for 7 years. What compound interest will the deposit earn?

If interest is compounded daily, the number of periods for a given time is large. A separate daily interest table gives the multipliers for interest compounded daily. A portion of a daily interest table is given below.

Daily Compound Interest Table						
	Annual Interest Rate					
Years	**1%**	**1.25%**	**1.50%**	**2%**	**3%**	**4%**
1/12	1.000834	1.001042	1.001251	1.001668	1.002503	1.003339
1/4	1.002503	1.003130	1.003757	1.005012	1.007528	1.010050
1/2	1.005012	1.006269	1.007528	1.010050	1.015112	1.020200
1	1.010050	1.012578	1.015113	1.020201	1.030453	1.040808
2	1.020201	1.025315	1.030454	1.040810	1.061834	1.083282
3	1.030454	1.038211	1.046027	1.061835	1.094170	1.127489
4	1.040810	1.051270	1.061835	1.083285	1.127491	1.173501
5	1.051270	1.064493	1.077882	1.105168	1.161827	1.221389
6	1.061836	1.077883	1.094172	1.127493	1.197209	1.271232
10	1.105169	1.133146	1.161831	1.221396	1.349842	1.491792
20	1.221399	1.284020	1.349850	1.491808	1.822074	2.225443

EXAMPLE 4

A $1,200 deposit earns annual interest of 4% compounded daily. Find the compound interest the deposit will earn in 3 years.

SOLUTION

Find the multiplier in the Daily Compound Interest Table that corresponds to 4% and 3 years: 1.127489

$1,200 × 1.127489 = $1,352.99   compound amount

Find the difference between the compound amount and the original deposit.

$1,352.99 − $1,200 = $152.99   compound interest for 3 years

✔ CHECK YOUR UNDERSTANDING

G. An $850 savings deposit is on deposit for 2 years and earns 3% annual interest compounded daily. What will be the compound amount in the account in two years? What interest will have been earned?

H. An account in which interest is compounded daily pays 2% annual interest. What interest will be earned in this account if $1,600 is left on deposit for 20 years?

Algebra *Tip*

There are many ways in which savings institutions calculate compound interest. The most common way uses the formula $B = \pi(1 + \frac{r}{n})^{n}/t$, where *B* is the balance in the account after interest is added, *p* is the principal, *r* is the annual rate, *n* is the number of times a year that interest is compounded, and *t* is the number of years.

Wrap Up ▸▸▸

If Joanna opens a savings account and puts the money in the bank, she will earn interest. In addition, many people are more likely to save, and less likely to spend, money if they have it in the bank, instead of cash in hand.

Financial Responsibility

When selecting a bank in which to deposit your money, you should look for the logo shown. This logo indicates the bank is a member of the Federal Deposit Insurance Corporation (FDIC), which means your deposit is backed by the full faith and credit of the United States government. From the time that the FDIC was established, no depositor has ever lost insured funds. The insurance coverage on qualified accounts is automatic. The consumer does not need to request the coverage or apply for it.

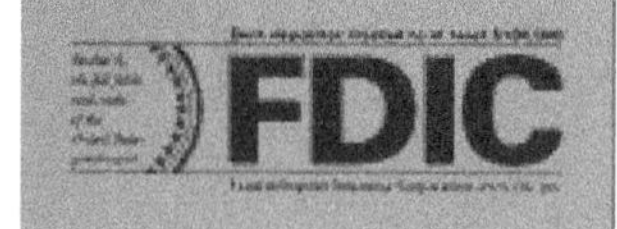

Exercises

Find the sum.

1. $842 + $32.89

2. $430 + $16.13

Find the product and round.

3. $1,286 × 0.0125

4. $984 × 0.026

5. $67.20 × $\frac{1}{12}$

6. $189.78 × $\frac{1}{4}$

Find the simple interest for one quarter.

7. $300 at 5% a year

8. $750 at 3% a year

9. $217.66 at 1.5% a year

10. $1,400 at 0.9% a year

Use the compound interest tables to solve Exercises 11–16.

	Principal	Rate	Time	Compounded	Compound Amount	Compound Interest
11.	$1,000	1.25%	4 years	Annually		
12.	$700	1%	15 years	Semiannually		
13.	$900	3%	2 years	Quarterly		
14.	$1,000	3%	6 months	Monthly		
15.	$600	4%	3 years	Daily		
16.	$800	3%	5 years	Daily		

Your bank pays 4% annual interest compounded quarterly on January 1, April 1, July 1, and October 1. You deposited $840 on April 1 and made no other deposits or withdrawals.

17. Find your savings account balance on January 1 of the next year.

18. How much interest did you earn for these nine months?

Solve.

19. Flora Kenyon made a deposit of $1,600 to her savings account on July 1. For the next year she made no other deposits or withdrawals. Her bank pays annual interest of 3.1% compounded quarterly. Find the estimated and actual interest earned by Flora by July 1 of the next year.

20. Seamus Keane made a $600 deposit on April 1 in the Mac-Oak Bank. The bank pays an annual rate of 2.6% compounded quarterly on the first day of January, April, July, and October. On July 1, Seamus deposited $200 more to his account. He made no other deposits or withdrawals. Find his balance on October 1.

21. Arkin Norris made a $1,200 deposit on June 1 at his bank. The bank pays 2.25% compounded quarterly on July 1, October 1, and January 1, What was his balance on January 1?

22. Sarah made a deposit of $665 into an account that compounds daily and pays 1.5% annual interest. How much interest will she earn if the money in left on deposit for 4 years?

STRETCHING YOUR SKILLS Some banks pay interest only on the minimum or smallest balance on deposit during an interest period. Helen Lamb had a balance of $783 in such an account on July 1. Annual interest is 2.7% compounded quarterly. She withdrew $170 on August 17 and deposited $200 on September 12.

23. What was Helen's minimum balance during the quarter?

24. How much interest was she paid on October 1?

25. How much did Helen have on deposit on October 1?

26. **CRITICAL THINKING** Some people explain compound interest as a way to earn interest on interest. Others say that interest earned is simply added to the previous balance to make a new balance on which interest is computed. Is either view more accurate than the other?

27. **CRITICAL THINKING** Look at the compound interest table. Why are the amounts in the table not rounded to the nearest cent since all money amount answers are rounded to the nearest cent?

Mixed Review

28. $\frac{4}{5} + \frac{1}{4} + \frac{1}{2}$

29. $\frac{2}{5} \times 20$

30. $\frac{8}{9} - \frac{2}{3}$

31. $\frac{4}{7} \div 16$

32. 100 × $26.50

33. 100,000 × $32.18

34. 15 is what percent of 200?

35. 18 is what percent of 240?

36. Find the average of $9.14, $3.83, $1.94, and $4.13.

37. Find the average of $0.90, $0.42, $0.78, $0.13, and $1.25 to the nearest cent.

38. Eva and Trent Blum, a married couple, file a joint tax return for their gross income of $53,000 and claim two exemptions. Their itemized deductions are $3,680. The standard deduction they may use is $10,900. What is their taxable income?

39. Lynn is paid $2.08 for every usable machine part she makes. During one week, she made 220 parts, 14 of which were unusable. What was Lynn's gross pay for the week?

40. Della Lynch is married and earns $470 a week. She claims three withholding allowances. What amount should be deducted from her weekly earnings for federal withholding taxes?

41. Eduardo Rivas works in sales for straight commission. His gross pay of $680 last week was based on his sales of $13,600. What rate of commission was Eduardo paid?

3-2

Checking Accounts

GOALS
- Prepare a deposit slip
- Record entries in a check register

KEY TERMS
- deposit slip
- check register
- balance
- overdrawn

Start Up

A 23-year-old college student who lives at home buys two money orders a month to pay her bills. A 42-year-old single mother gets six bills each month. She also pays by money order. If neither of them have a checking account, should they open one to save the cost of buying money orders?

Mona Makeda/Shutterstock.com

Math Skill Builder

Review these math skills and solve the exercises that follow.

1. **Add** money amounts.
   Find the sum. $17 + $5.25 + $632.19 = $654.44

   1a. $5.60 + $67.49 **1b.** $294.43 + $123

2. **Subtract** money amounts.
   Find the difference. $832.02 − $76.98 = $755.04

   2a. $900.35 − $298.37 **2b.** $1,264.43 − $634.06

   2c. $568.32 − $403.99 **2d.** $1,308.05 − $876.86

3. **Multiply** a money amount.
   Find the product. 5 × $20 = $100

   3a. 15 × $10 **3b.** 13 × $0.50

   3c. 18 × $3.10 **3d.** 32 × $1.50

Deposit Slip

Many people deposit cash in a checking account at a bank and make their payments by check. A checking account is safe and easy to use. The statements provided by the bank give you a record of your payments. A **deposit slip** has been filled in for Sherry and Jamal Taylor to deposit their paychecks and other monies.

Cash deposits of *bills* and *coins* are listed on the line labeled CASH. Each check is listed on a separate line in the space for checks.

If there are more checks than lines available on the front of the deposit slip, list the additional checks separately on the back of the deposit slip.

The total of those checks is entered on the *Total from Other Side* line on the front of the deposit slip. Then, all the amounts are added to find the sum, which is written on the *Subtotal* line.

To receive *cash back* from the bank, write the amount wanted on the *Less Cash Received* line below the subtotal. Cash back is the amount that you want returned to you in cash.

Math Tip

Check all subtraction. The best way to check subtraction is to add the amount subtracted and the difference. The answer should equal the top number.

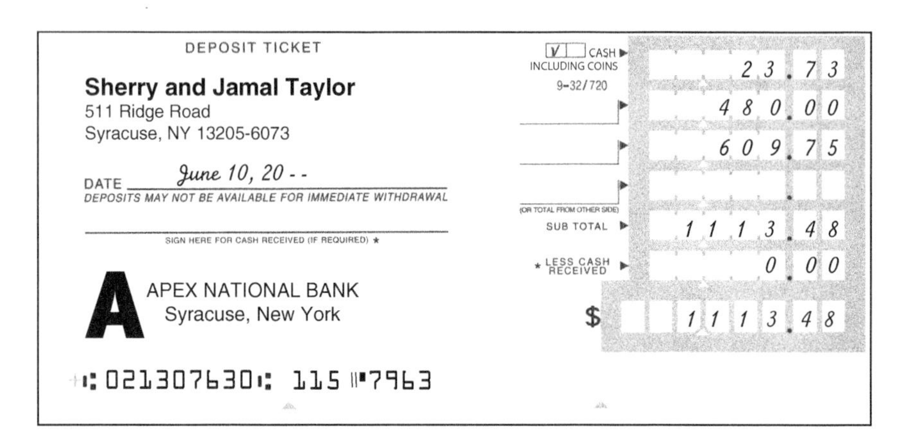
DEPOSIT TICKET

Sherry and Jamal Taylor
511 Ridge Road
Syracuse, NY 13205-6073

DATE *June 10, 20 - -*
DEPOSITS MAY NOT BE AVAILABLE FOR IMMEDIATE WITHDRAWAL

SIGN HERE FOR CASH RECEIVED (IF REQUIRED) ★

APEX NATIONAL BANK
Syracuse, New York

✓ CASH INCLUDING COINS 9-32/720	23.73
	480.00
	609.75
(OR TOTAL FROM OTHER SIDE)	
SUB TOTAL	1113.48
★ LESS CASH RECEIVED	0.00
$	1113.48

⑆021307630⑆ 115 ⑈7963

Find the *Total Deposit* by subtracting the cash received from the subtotal. The total deposit is sometimes called the *net deposit*. You only sign the deposit slip if you receive cash back and the bank requires a signature.

EXAMPLE 1

Alisha Reed made a deposit to her checking account: (bills) 6 twenties, (coins) 40 quarters, (checks) $457 and $18.10. She received 10 one-dollar bills in cash back. Complete a deposit slip.

Photodisc/Getty Images

SOLUTION

Fill in each line with the appropriate amount.

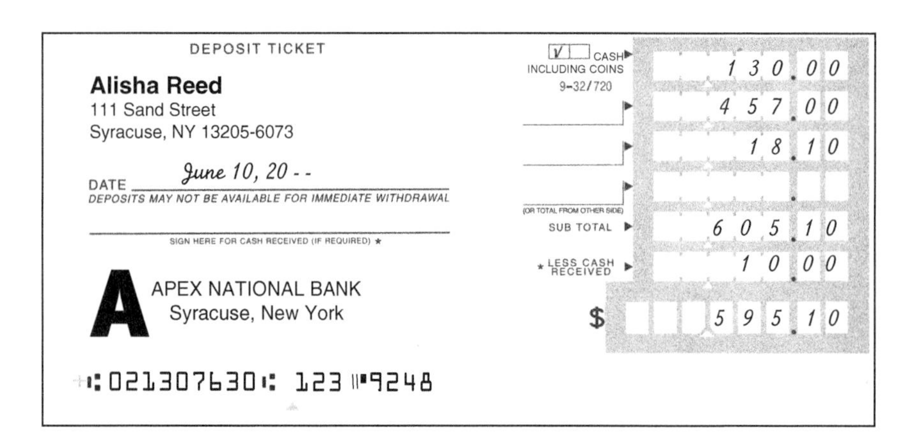
DEPOSIT TICKET

Alisha Reed
111 Sand Street
Syracuse, NY 13205-6073

DATE *June 10, 20 - -*
DEPOSITS MAY NOT BE AVAILABLE FOR IMMEDIATE WITHDRAWAL

SIGN HERE FOR CASH RECEIVED (IF REQUIRED) ★

APEX NATIONAL BANK
Syracuse, New York

✓ CASH INCLUDING COINS 9-32/720	130.00
	457.00
	18.10
(OR TOTAL FROM OTHER SIDE)	
SUB TOTAL	605.10
★ LESS CASH RECEIVED	10.00
$	595.10

⑆021307630⑆ 123 ⑈9248

✔CHECK YOUR UNDERSTANDING

A. Shirley Poe deposited these items into her account at a bank: (bills) 14 twenties; (coins) 21 quarters, 6 dimes; (checks) \$322.94, \$1.45. She received 100 one-dollar bills in cash back. What was the amount of Shirley's deposit?

B. Charles Gray made a deposit for the booster club at his school. He deposited: (bills) 17 twenties, 18 tens, 9 fives, 37 ones; (coins) 49 quarters, 12 dimes; (checks) \$39.86, \$3.83. Charles received no cash back. Find the total deposit.

Check Register

When you write a check you direct the bank to make a payment from your checking account. Checks are numbered to make it easy to keep track of checks.

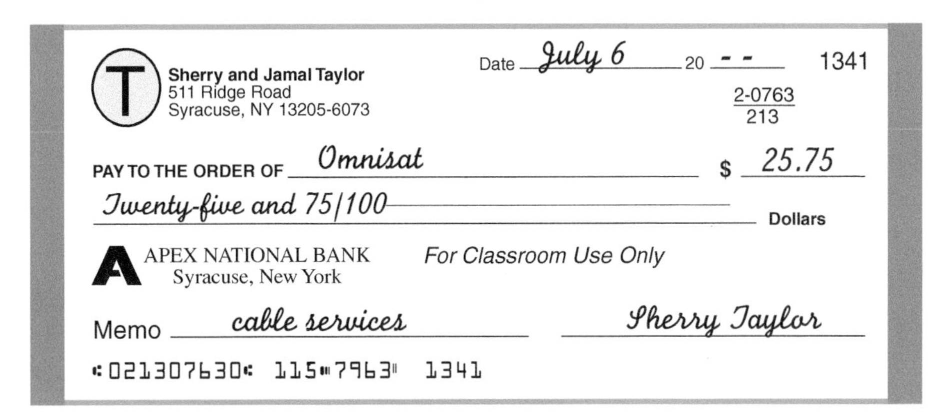
Sherry and Jamal Taylor
511 Ridge Road
Syracuse, NY 13205-6073
Date July 6 20 - -
1341
2-0763
213
PAY TO THE ORDER OF Omnisat
$ 25.75
Twenty-five and 75/100
Dollars
APEX NATIONAL BANK
Syracuse, New York
For Classroom Use Only
Memo cable services
Sherry Taylor
⑆021307630⑆ 115⑈7963⑈ 1341

The Taylors record each deposit made and check written in their **check register**, shown below. The check register is part of the checkbook in which deposits and checks are recorded.

A new balance, called a *running balance*, is calculated after each entry. Each deposit is added to the previous balance. Each check is subtracted from the previous balance. The **balance** is the amount of money in the account.

CHECK REGISTER

NUMBER	DATE	DESCRIPTION OF TRANSACTION	PAYMENT/DEBIT (-)	√ T	FEE (IF ANY) (-)	DEPOSIT/CREDIT (+)	BALANCE
							$ 1500 00
1341	7/6	Omnistat	$ 25 75		$	$	1474 25
	7/6	Sherry's Paycheck				480 00	1954 25
1342	7/7	Syracuse Electric	98 32				1855 93
1343	7/7	APEX Mortgage	772 33				1083 60
1344	7/8	Mid-town Appliance	165 98				917 62
	7/10	Jamal's Pay Check				609 75	1527 37

Business Tip

Check 21 is a federal law that went into effect in 2004. It allows banks to handle many checks electronically by creating a substitute check instead of relying on the movement of paper checks.

Check 21, as well as other changes in technology, means that checks clear banks faster.

EXAMPLE 2

On October 18, Rico Ortez made a deposit of $250 to his checking account that had a previous balance of $1,288.43. He also wrote Check 67 to Mick's Catering for $78.50 and Check 68 to Bev's Tires for $143.78. Record each transaction and a running balance. What was the final balance in his check register?

SOLUTION

CHECK REGISTER									
CHECK NO.	DATE	TRANSACTION	PAYMENT/DEBIT			DEPOSIT/CREDIT		BALANCE	
	10/17	Previous Balance						1288	43
	10/18	Deposit				250	00	1538	43
67	10/18	Mick's Catering	78	50				1459	93
68	10/18	Bev's Tires	143	78				1316	15

The final balance is $1,316.15.

✔CHECK FOR UNDERSTANDING

C. Edwina Moss' check register showed a previous balance of $2,583.45 at the beginning of the week. During the week she made a deposit of $1,220 to her account and wrote checks for $825, $96.40, and $12.78. What final balance did her check register show?

D. Alex Devine made deposits of $500 and $1,236 to his account. He wrote checks for $196, $950, and $87.83. His previous balance before these transactions was $129.74. What was the new balance of his account?

Photodisc/Getty Images

One important reason to record checks and deposits and to keep a running balance in a checkbook register is to keep track of how much money is in the account. If you write a check for more money than is in the account, your account will be **overdrawn**.

Having insufficient funds in your account is called *bouncing* a check and the money will not be paid to the person to whom the check was written. Some banks offer *overdraft protection*, where the bank will transfer money from your savings account into your checking account to cover any check you have written.

Consumer Alert

When a check bounces, the bank returns that check to the person or business to which the check was written. The bank then charges a fee to the person or business who attempted to cash the check.

It is the responsibility of that individual or business to collect the funds and any penalties charged. Businesses usually charge a fee to the person that bounced the check to recover their expenses.

Wrap Up ▶▶▶

Every financial service is provided at a cost. Money orders are available for a fee and may be bought at a variety of places. Banks offer a variety of checking accounts that have some sort of cost or requirement attached to them, so they also are not free. Both women need to compare the monthly cost of buying money orders and using checking accounts. If the decision is made considering only cost, the student and the single mother should choose the plan that is least expensive for them individually.

Exercises

Solve using the indicated operation.

1. $458.59 + $312.03
2. $1,098.45 + $17.31
3. $1,492.49 − $231.56
4. $248.72 − $9.87
5. 14 × $10
6. 12 × $20

Solve.

7. Eva Lanier deposited these checks into her account: $384.39, $12.44, $284.12. She received no cash back. What was her deposit total?
8. June Wilson's account had a beginning balance of $288.43. She made a deposit of $627 and wrote two checks for $35.87 and $263.56. What was the final balance in her check register?
9. The Singer family had a moving sale. The next day they made this deposit with no cash back: 25 twenties, 19 tens, 28 fives, 18 ones; 15 quarters; and checks for $25, $20, and $85. What was the amount of their deposit?
10. On May 1 Rufus Knight's check register showed a balance of $505.23. Rufus had these transactions: May 4, check written for $34.66; May 7, check written for $98.62; May 9, deposit of $259.34; May 10, check written for $112.97. What was his check register's balance on May 10?
11. Yolanda Downey's checking account balance was $182.63. She made a deposit of $218.55 and wrote checks for $25, $30.17, and $46.07. What was the new balance in Yolanda's check register?
12. Donela Boyd's checking account balance was $3,060.65. A deposit of $1,603.48 was made and checks for $1,780.44, $22.74, $9.10, and $125 were written. What was the new balance of her account?

Photodisc/Getty Images

13. **STRETCHING YOUR SKILLS** All-Sports Trophies made this deposit: 6 hundreds, 14 fifties; and checks for $85, $23.50, $32, $45, $17.50, $147, $17.27, $32.25, $65. They got cash back of 50 one-dollar bills, 20 five-dollar bills, 10 ten-dollar bills, and 8 twenty-dollar bills. Find the total deposit.

14. **CRITICAL THINKING** Assume you have a checking account that sends you a monthly statement by mail. If the bank offers you a discount on your checking account charges if you agree to have your statements sent to you via email, would you take advantage of such an offer to save money? Explain your response.

15. **CRITICAL THINKING** Your bank gives you a choice of ordering either 200 or 400 checks at a time. The cost of 400 checks is double the cost of 200 checks, so there is no cost savings. You write about 14 checks a month. How many checks would you order at one time? Why?

16. **STRETCHING YOUR SKILLS** The Floral Place made this deposit: 72 twenties, 126 tens, 57 fives, 235 ones; 287 quarters, 312 dimes, 48 nickels, 347 pennies; and checks for $124.68, $132.08, $1.29, $5.79. They received no cash back. Find the total of the deposit.

INTEGRATING YOUR KNOWLEDGE Stuart Rosenblatt earned a gross salary of $800 in one week. He received a paycheck that showed that taxes and other deductions of $206.75 were subtracted from his gross salary. Stuart also received an $82 purchase refund check and a $200 check from the sale of an old washer and dryer. His previous checking account balance was $502.87.

17. What was the net amount of Stuart's paycheck?

18. Stuart deposited all the checks and got $35 cash back. What was his net deposit?

19. What was Stuart's checking account balance after making the deposit and then writing a check to City Garage for $143.65?

Mixed Review

20. $24,900.53 + $3,102.87

21. $\frac{1}{3} + \frac{1}{2} + \frac{1}{8}$

22. $2,348.81 − $5.99

23. $8\frac{1}{4} - 3\frac{1}{6}$

24. $249.88 \div 1{,}000$

25. $0.14859 \times 100 =$

26. Find the value of N, to the nearest cent: $N = \$352.98 \div 100$.

27. William earns $4,000 per month in salary. If his deductions total $1,350 per month, what percent of his gross pay, rounded to the nearest percent, does William take home?

28. Regina Charren's job pays $45,300 in annual wages and 31% of annual wages in benefits. Her job expenses are estimated to be $3,300 a year. Regina interviewed for another job that pays $47,000 in yearly wages, average benefits of 29.5%, and estimated yearly job expenses of $3,750. Which job offers Regina the greater job benefits, and how much greater?

29. Elmer Hartley is paid a salary of $710 a week. What total salary does Elmer earn in one year?

30. At the end of a craft show, Luma Villa deposited her receipts in a checking account: (bills) 19 twenties, 53 tens, 14 fives, 83 ones; (coins), 17 quarters, 1 nickel; (checks) $64.50 and $49.25. She got back in cash 3 fifties and $5 in dimes. What was her total deposit?

3-3

Electronic Banking

GOAL

- Record electronic banking transactions
- Calculate account balance needed to make online payments

KEY TERMS

- electronic funds transfer (EFT)
- automated teller machine (ATM)
- debit card
- direct deposit
- online banking

Start Up

Your friend, Ronald Billings, received a letter from his bank inviting him to enroll in the bank's online banking and bill paying program. Ronald thinks he would like to use this service. He also wonders whether the monthly charge of $5.95 for online bill paying is worth the cost. What would you say to Ronald that might help him decide what to do?

Photodisc/Getty Images

Math Skill Builder

Review these math skills and solve the exercises that follow.

1 **Add** money amounts.
Find the sum. $873.12 + $12 = $885.12

1a. $18 + $357.38 **1b.** $2,874 + $29.69

2 **Subtract** money amounts.
Find the difference. $1,738 − $622.98 = $1,115.02

2a. $734 − $719.87 **2b.** $287 − $165.45

Electronic Banking

All banks use computers to process transactions electronically. Electronic banking allows bank customers to use telephones, computers, and other technologies in place of paper transactions. Banks use computers to transfer deposits and checks, or "funds," from person to person and bank to bank. This process is called **Electronic Funds Transfer,** or **EFT**.

Individuals can also transfer funds electronically when they use an **ATM** or **Automatic Teller Machine**. By using an ATM card issued by your bank, you can withdraw or deposit money, see account balances, or make transfers between your accounts. A *Personal Identification Number,* or *PIN,* that is known only to you is entered into the ATM before your transaction is processed. The PIN provides protection against unauthorized use of your ATM card.

Your bank's ATM card also allows you to withdraw money at another ATM if it displays the same network logo shown on your card. You may have to pay an ATM fee from your bank and the other bank when you use another bank's ATM.

At many banks, the ATM card is also a **debit card**. Debit cards allow you to pay for your purchases without using cash. When you use your debit card for a purchase, the bank's computer deducts the amount of that purchase automatically from your checking account. You may also use your debit card to receive cash back with a purchase. The amount subtracted from your account is the total of the cash back plus the purchase. Debit cards are sometimes called check cards.

EFTs may also be used to pay monthly bills, such as utility bills. You can instruct your bank to transfer funds automatically each month from your bank account to the account of your utility provider. No checks are written or mailed.

Some companies use EFT to pay their employees by transferring funds directly into their employees' bank accounts without writing any checks to the employees. This is called **direct deposit**.

When you use an ATM or debit card you get a receipt of the transaction. Save the receipt and immediately record the payment or cash withdrawal in your check register.

Many people use the notation "debit" for a debit card purchase and ATM-WD or ATM-DEP for ATM withdrawals or deposits.

Photodisc/Getty Images

EXAMPLE 1

Wan-ying Kuo's checking account had a balance of $512.45 on April 4. Over the next three days she had these electronic transactions: direct deposit of $782.50 on 4/5, ATM cash withdrawal of $100 on 4/6, and debit card clothing purchase of $90.27 on 4/7. What was her final check register balance?

SOLUTION

CHECK REGISTER

CHECK NO.	DATE	DESCRIPTION	PAYMENT/DEBIT		DEPOSIT/CREDIT	BALANCE
	4/4	Previous Balance				512.45
	4/5	Direct Deposit			782.50	1294.95
	4/6	ATM-WD	100.00			1194.95
	4/7	Debit (Clothes)	90.27			1104.68

The final balance is $1,104.68.

✔CHECK YOUR UNDERSTANDING

A. Fred Wilhelm began the day with a $782.88 balance in his checking account. During the day he used his debit card to pay $85 for car repairs and $86.54 for a clothing purchase. Fred also withdrew $50 from his account at an ATM machine. What was the balance of Fred's checking account at the end of the day?

B. In one day Katrina Woll deposited her $890.50 paycheck at an ATM and withdrew $200 in cash. She also made two debit card purchases for $12.87 and $118.94. If she started the day with a bank balance of $1,248.40, what was her balance at the end of the day?

Online-Account Access

Online banking is another form of electronic banking that allows you to do your banking by using your personal computer and the Internet.

Although specific services will vary by bank, the online banking services at most banks allow you to:

- have access to your account 24 hours a day, 7 days a week
- transfer money between your accounts
- have bills sent to you electronically instead of through the mail
- make payments to companies you select
- reorder checks and send messages to bank staff
- look at the history of your accounts

APEX National Bank *online banking*

YOUR ACCOUNTS
- View Accounts
- Download Your Recent Activity
- Transfer Money
- Payees
- Payment
- Electronic Payments

CUSTOMER SERVICE
- Read message
- Send message
- Update personal information
- Reorder Checks

Checking Account: 115 7953

Last Statement Date	11/20/20--	**Balance Last Statement**	$55.71
Current Balance	$216.95	**Interest YTD**	$2.54
Available Balance	$216.95	**Interest Rate**	1.24%

Activity sorted by Date

Date	Description	Amount	Type
11/30	CHECK #1017	-$100.00	CHK
11/29	ONLINE TRANSFER FROM SAVINGS	$800.00	DEP
11/29	ATM WITHDRAWAL	-$20.00	W/D
11/22	TORINO MARKETS	-$16.57	DEB
11/22	ATM DEPOSIT	$20.00	DEP
11/18	METROPAGE, INC	-$22.19	EFT

Many banks offer "free online banking" to their customers. This service allows customers to look at their account balances and make transfers between accounts. In many banks, an online bill payment option is available for a monthly fee. A sample online banking screen is shown above.

The sample screen shows the *available balance* of the checking account. This is the amount that is available to spend. Knowing the balance helps you plan future online and check payments. The *account activity* section of the online screen shows the items that have been processed since the last statement date.

Online banking simplifies the banking process, yet is very similar to using a regular checking account. Both require making decisions to manage the account.

EXAMPLE 2

To help her to plan and make decisions, Annie Clark made a worksheet for her checking account that shows the transactions she expects will take place in March.

DIRECT DEPOSIT: $690 each Fri on dates: 03/01, 03/08, 03/15, 03/22, 03/29	ATM Cash withdrawal of $150 each Saturday	DEBIT CARD Purchases each Fri average $280 weekly	OTHER CHARGES None
EXPECTED PAYMENTS			
Electric bill	$ 53.00	due 03/06	
Natural gas bill	$115.00	due 03/22	
Local phone bill	$35.00	due 03/13	
Long distance phone bill	$17.00	due 03/19	
Medical insurance	$108.00	due 03/24	
Internet service	$18.95	due 03/05	
Car payment	$312.40	due 03/13	
Home improvement loan	$367.00	due 03/06	
House payment	$842.00	due 03/20	

On Tuesday, March 5, Annie decides to pay all the bills listed in the "expected payments" section that are due by March 11. Before she makes any payments on March 5, Annie reviews her account balances: checking, $423.90; savings, $1,513. Her last deposit was the March 1 direct deposit.

Will Annie have enough money in her checking account to make the online payments on March 5 and have a balance of at least $100 in the account? If not, how much money will she have to transfer to checking from savings?

SOLUTION

PAYMENT PLAN WORKSHEET, MARCH 5		
1 Checking Account Balance, March 5		423.90
2 Deposits made after March 1		+ 0.00
3 Subtotal (Line 1 + Line 2)		423.90
4 Online payments, debits, paper checks, ATM _WD		
5 Electric bill	53.00	
6 Internet service	18.95	
7 Home improvement loan	367.00	
8 Subtotal of all payments	− 438.95	
9 Difference (Line 3 − Line 8)		(15.05)

Line 1	Enter last available checking account balance.
Line 2	Enter 0.00 to show that no deposits were made after March 1.
Line 3	Add lines 1 and 2.
Line 4	An instruction line that directs you to list any online payments you wish to make, any checks you will write, and any expected ATM withdrawals.
Lines 5–7	Enter bills due from March 5–11. These include the electric bill, Internet service bill, and home improvement loan. (Add more lines as needed.)
Line 8	Take a subtotal of all payments.
Line 9	When Line 8, \$438.95, is subtracted from Line 3, \$423.90, the result is a negative number, shown in parentheses as (15.05).

There is not enough money to make the payments listed and maintain a minimum balance of \$100 in the account. A total of \$115.05 (\$15.05 + \$100) will have to be transferred to checking from savings.

✔CHECK YOUR UNDERSTANDING

C. The balance of Annie Clark's checking account on March 7 was \$100 and her savings account balance was \$1,397.95. Use Annie's worksheet to determine if on March 12 she can make online payments for all bills due from March 12 through March 18 and still leave a minimum balance of \$100 in the account? If not, how much must she transfer to checking from savings? (Hint: Be sure to include in your calculations any deposits, ATM withdrawals, and debit purchases since the last online payment.)

D. After making his online payments a week ago, Brian Hurley's checking account balance was \$67. In the past seven days, he had these transactions in his checking account: deposit, \$728; debit card purchases of \$36.90 and \$112.85; two checks written for \$270 and \$15. Today Brian is making online payments for boat insurance, \$128; charge card, \$89.23; and medical bill, \$45.50. What will be the balance of his checking account after all the transactions and online payments are entered?

Communication

Banks ask you to select your PIN when they issue an ATM or debit card. Often you don't have much time to think of something that is easily remembered. Assume that your bank asks you to choose a PIN that is six characters long and is alphanumeric, a combination of letters and numbers.

Write guidelines that offer advice to customers that banks should hand out when someone needs to create a personal PIN that cannot easily be guessed by someone who has stolen your card. Be sure to include do and don'ts in your guidelines. Share your guidelines with the rest of the class in an open discussion.

Wrap Up ▶▶▶

Ronald's bank may have a demonstration program of its online service for him to try. He may also talk with users of online banking and bill paying to find out what their experiences have been. Although the bill pay option may have a monthly fee, Ronald should consider how much money he will save in not having to write checks and pay for postage.

TEAM Meeting

As a class, make a list of area banks, credit unions, and savings and loans that offer online banking services. Class members with Internet access should each choose an institution. Research web sites and print screen captures that show how the sites function. Also list the services provided online.

The class should review the printouts and list features common to all institutions. Identify those features that are unique to specific banks, credit unions, or savings and loans.

Exercises

Find the sum or difference.

1. $12,873.29 + $2,498.32
2. $387 + $28.07
3. $1,483.87 − $842.38
4. $248.09 − $74.83

Solve.

5. After work Gladys Schroeder used the ATM to deposit her paycheck for $638.77 and to withdraw $200 cash. If her starting bank balance was $418.03, what is her new balance?

Photodisc/Getty Images

6. Adlise Leiber started the day with a bank balance of $343.64. She used another bank's ATM to withdraw $100 cash. The charge for using the ATM was $2.50. Adlise then used her card to make purchases of $85.10, $23.95, and $8.47. Find the balance in her account after the bank processed these transactions.

7. Toni Nicolet's checking account balance on Monday, April 13, is $540; her savings balance is $980. On Tuesday, April 14, she made an ATM withdrawal from checking of $86. On April 15, Toni plans to make these online payments: income tax bill, $823, utility bill, $98, and charge account bill, $127. How much money, if any, will Toni have to transfer into her checking account from savings to cover the online payments and leave a balance of $50 in the checking account?

8. **FINANCIAL DECISION MAKING** Your bank's online banking web site takes 15 seconds longer to load than that of a competing bank, yet costs $1 less per month than a competitor's system. Will you switch banks and pay $1 more a month to get faster service?

Solve.

9. Kwei-tseng Kuo plans to make these online payments: store charge account, \$160; charitable donation, \$40; rent, \$550; and cable bill, \$42.01. She began the day with a checking account balance of \$16.83. Later that same day she estimated her online payments and transferred \$800 into checking from savings to cover the expected payments. What will be the balance of her checking account after the online payments are made?

10. Andrew Galen had a checking balance of \$1.39 Monday morning. His net wages of \$512.89 were transferred by direct deposit at 11:00 a.m. to his checking account. Later that evening, Andrew made online payments of \$3.24, \$18.30, \$38.96, \$100.34, and \$314.78. His goal is to keep only a \$10 balance in his checking account and have most of his money in savings. How much money was available in Andrew's checking account to be transferred to savings after all the transactions were completed?

11. On Tuesday Henry St. John used a debit card to pay for: garden tools, \$83.12; work clothing, \$46.75; and groceries, \$54.79. If Henry's bank balance was \$437.01 at the start of the day, what is his new balance at the end of the day?

12. Gilbert Conroy withdrew \$200 from his bank's ATM. On a shopping trip he bought an office chair for \$120.87 in cash and paid \$75.11 cash for groceries. He then used his debit card to pay for \$136.50 in painting supplies and \$66.52 for lawn mower repair. What amount was left in Gilbert's account if it had a balance of \$740.12 at the start of the day?

13. **CRITICAL THINKING** Molly Abrams will begin receiving Social Security monthly benefits that the law requires be paid by direct deposit. Molly is used to getting a paper paycheck and going to a bank to cash her check. Molly would prefer to receive her benefits in a paper check. What advantages of direct deposit could you tell Molly about that might convince her that direct deposit is better?

Mixed Review

14. Divide: $8,008 \div 13$

15. Multiply: $2\frac{2}{9} \times 8\frac{3}{5}$

16. Jeanne Williams sells welding equipment and is paid commission every two months. She earns 9.5% on the first \$25,000 of sales and 12.4% of sales from \$25,001 to \$70,000. A commission of 15.1% is paid on all sales over \$70,000 in a pay period. Her sales last month were \$43,000 and \$54,000 this month. What total commission earnings, to the nearest dollar, did Jeanne have for the two months?

Photodisc/Getty Images

17. Justin Niklas made this deposit to the Breakfast Book Club's checking account: (bills) 43 ones, 13 fives; (coins) 9 quarters, 3 half-dollars, 5 dimes, 4 nickels, 12 pennies; (checks) \$397.42, \$192.81. Find the amount of the deposit.

18. The city of Dubline has this income tax schedule for earnings of \$46,000 to \$58,000: \$460 plus 1.25% of earnings over \$46,000. What income tax will Jules Rubin have to pay on earnings of \$53,400?

3-4

Check Register Reconciliation

GOALS

- Reconcile a bank statement
- Reconcile and correct a check register
- Reconcile a checking account with outstanding transactions and other errors

KEY TERMS

- bank statement
- service charge
- outstanding checks
- reconcile

Start Up

Two people are discussing a historical event. The more they talk, the more they disagree about when the event took place, the parties involved, and the outcome. You are asked to settle, or reconcile, the dispute. How would you do this?

Andresr 2008/Shutterstock.com

Math Skill Builder

Review these math skills and solve the exercises that follow.

1. **Add** money amounts.
   Find the sum. $23,487 + $15.34

   1a. $20.08 + $832.58 **1b.** $85.82 + $70.18

2. **Subtract** money amounts.
   Find the difference. $76.28 − $5.39

   2a. $900 − $2.08 **2b.** $175.29 − $38.43

Reconcile the Bank Statement

Banks keep track of checking account transactions and send a monthly report, called a **bank statement**, to depositors. At many banks, you have the option to receive paper statements through the mail or online statements. A sample bank statement for Gerald Booth is shown on the next page.

The bank statement shown lists (1) nine *checks* paid by the bank; (2) four *deposits*, including interest earned, (3) and two *other charges*, an ATM withdrawal and a service charge.

Interest earned is money paid to customers for the use of their money. A **service charge** is a deduction made by the bank for handling the checking account.

Gerald Booth compared the bank statement with his check register. On the bank statement he placed a check mark next to the number of the check when both records agreed.

TN B TRENT NATIONAL BANK
4309 SOUTH BROAD STREET
PHILADELPHIA, PA 19148-3978

09/01	Balance Brought Forward	$ 608.12
	+ Deposits	884.71
	– Checks	1103.85
	– Other Charges	65.00
09/30	Closing Balance	$ 323.98

Gerald Booth
3123 Baltimore Avenue
Philadelphia, PA 19101

Checks

Check	Date	Amount	Check	Date	Amount	Check	Date	Amount
√1072	09/02	34.67	√1075	09/20	7.90	√1078	09/22	61.90
√1073	09/10	8.32	√1076	09/17	311.01	√1079	09/28	450.00
√1074	09/09	125.54	√1077	09/27	26.19	√1082	09/30	78.32
							Total Checks	**1103.85**

Deposits

Date	Explanation	Amount
√09/12	ATM Deposit	481.56
√09/25	Deposit	298.44
√09/28	Deposit	104.55
X09/30	Interest Earned	0.16
	Total Deposits	**884.71**

Other Charges

√09/17	ATM Withdrawal	50.00
X09/30	Service Charge	15.00
	Total Other Charges	**65.00**

Gerald found two checks (numbers 1080 and 1081) that were recorded in his check register, but that were not listed on the statement. These checks are called **outstanding checks**. This means that the checks have not yet been received or paid by the bank.

Gerald also placed a check mark next to each deposit shown on the statement that appeared in his check register. Since all deposits were accounted for, there were no *outstanding deposits.* An outstanding deposit occurs when a deposit is made after the closing date of the bank statement and the deposit is recorded in the check register.

Gerald also placed a check mark next to the ATM withdrawal because it was recorded in his check register. The letter "X" was placed next to the interest earned and the service charge, items that were not recorded in the check register. Gerald's statement shown above is already marked.

When Gerald Booth looked in his check register, he found his last recorded balance for September to be $196.89. The final balance on his bank statement was $323.98. The difference in the balances was the result of the outstanding checks, interest earned, and the service charge.

To bring both balances into agreement and to make sure the bank's records were correct, Gerald Booth has to **reconcile** both records. This is a two-step process. The first step is to reconcile the bank statement. To help him, Gerald can complete the reconciliation form printed on the back of the bank statement.

EXAMPLE 1

Prepare a reconciliation form for Gerald Booth to reconcile the bank statement.

SOLUTION

Complete the reconciliation form.

Reconciliation Form			
Follow these steps:		Outstanding Checks	
1 Enter Closing Balance from Statement	$ 323.98	1080	$48.65
2 Add any deposits outstanding	+ 0.00	1081	$93.28
3 Add lines 1 and 2	= 323.98		
4 Enter total of Checks Outstanding	− 141.93		
5 Subtract line 4 from line 3. This amount should equal your check register balance.	$ 182.05	Total	$141.93

Follow these steps to complete the form:

1. List separately the outstanding checks in the "Outstanding Checks" column.
   Find their total, $141.93.
   Write it in Line 4.
2. Write the closing bank statement balance of $323.98 on Line 1.
3. Write 0.00 on Line 2 to show no "Deposits Outstanding."
4. Do the addition shown on Line 3.
5. Do the subtraction shown on Line 5.
   The result, $182.05, is the "reconciled" bank balance. Since $182.05 does not agree with Gerald Booth's check register balance of $196.89, he will have to also reconcile the check register.

Business Tip

Outstanding items (checks and deposits) are items that you have recorded in your check register but the bank has not yet received. That means they have not subtracted outstanding checks from the bank balance nor added outstanding deposits to the bank balance. These are two reasons why your balance doesn't match the bank's balance.

✔ CHECK YOUR UNDERSTANDING

A. Maria Greeley's bank statement showed a closing balance of $1,383.53, no outstanding deposits, and two outstanding checks for $129.45 and $87.39. Reconcile her bank statement.

B. Xavier Allasandro had a closing balance of $793.57 on his bank statement. The outstanding items were: a deposit of $312.09, Check 278 for $174.85, and Check 280 for $32.78. Reconcile his bank statement.

Reconcile the Check Register

After reconciling his bank statement balance, Gerald Booth found that it still did not agree with the check register balance. The next step is to reconcile his check register.

EXAMPLE 2

Record transactions from the bank statement in Gerald Booth's check register to reconcile the register.

SOLUTION

CHECK REGISTER										
CHECK NO.	DATE	DESCRIPTION	PAYMENT/DEBIT				DEPOSIT/CREDIT		BALANCE	
	9/30	Previous Balance							196	89
	9/30	Interest Earned					0	16	197	05
	9/30	Service Charge	15	00					182	05

The $0.16 interest earned was added to the account by the bank so it also must be added to the balance in the check register. The $15 service charge was deducted from the account by the bank. The service charge must also be deducted from the check register.

The final balance of $182.05 shown in the check register agrees with the final balance of $182.05 shown on the bank's Reconciliation Form. The account is reconciled.

Business *Tip*

Many people use a computer program to record their checkbook transactions. Most programs have a feature that walks the user through the reconciliation process. It is important that if you use a computer program for your banking, that you make backup copies of your data.

✔ CHECK YOUR UNDERSTANDING

C. Mildred Galin's previous check register balance was $727.92. Her bank statement showed a service charge of $18.90 and interest earned of $1.60. Reconcile her check register.

D. Ludwik Sirros had a balance of $457.38 in his check register. His checking account does not pay interest. His bank statement showed a $7.68 service charge. Reconcile his check register.

Photodisc/Getty Images

Reconciliation Problems

Sometimes you must reconcile a bank statement and check register balances when deposits, checks, and EFT transactions are not recorded and when other errors are made.

To understand this process, use Rena Jackson's July bank statement dated August 6. The bank statement is shown on the next page.

Rena compared each item in the check register with each item on the bank statement. On the bank statement she placed a check mark next to the items when the two records agreed.

An "X" was placed next to those items that needed to be investigated. Rena's bank statement is already marked.

NC *North-Central National Bank*

07/01	Balance Brought Forward	$ 822.53
	+ Deposits	1366.70
	– Checks	1012.84
	– Other Charges	544.08
07/31	Closing Balance	$ 632.31

Rena Jackson
111 Central Drive
Indianapolis, IN 46110

Checks

Check	Date	Amount	Check	Date	Amount	Check	Date	Amount
√845	07/01	72.66	√848	07/13	8.90	X853	07/19	72.05
√846	07/05	18.31	√849	07/16	428.00	√854	07/20	215.67
√847	07/08	96.91	X851	07/19	37.16	√855	07/23	63.18
						Total Checks		**$ 1,012.84**

Deposits

Date	Explanation	Amount
√07/10	Direct Deposit	623.47
√07/24	Direct Deposit	623.47
X07/28	ATM Deposit	118.20
X07/31	Interest Earned	1.56
	Total Deposits	**$ 1,366.70**

Other Charges

Date	Explanation		Amount
X07/12	ATM Withdrawal, Trent County Bank		120.00
X07/12	ATM fee, Trent County Bank	1.50	
X07/18	EFT Payment, Truck Loan		416.18
X07/31	DEBIT, All Repair Parts Inc.		6.40
	Total Other Charges		**$ 544.08**

EXAMPLE 3

Reconcile Rena Jackson's bank statement and check register.

SOLUTION

Step 1: Compare the bank statement to the check register and note any differences between them.

Rena's notes about the items where she found a problem are listed.

1. Outstanding checks
   - Check 850 for $8.12
   - Check 852 for $34.28
2. Outstanding deposits
   - Deposit of $85 made on 8/1
3. Transactions not recorded in the check register
   - Check 851 for $37.16 cashed 7/19
   - ATM deposit of $118.20 made on 7/28
   - Interest of $1.56 earned
   - ATM withdrawal of $120 on 7/12
   - ATM fee of $1.50 charged on 7/12
   - EFT loan payment of $416.18
   - Debit charge of $6.40

Photodisc/Getty Images

4. Errors
   - Check 853 for $72.05 was recorded as $27.05 in the checkbook register

Step 2: Prepare a reconciliation form to reconcile the bank statement.

Reconciliation Form		Outstanding Checks	
Follow these steps:			
1 Enter Closing Balance from Statement	$ 632.31	850	$ 8.12
2 Add any deposits outstanding	+ 85.00	852	$34.28
3 Add lines 1 and 2	= 717.31		
4 Enter total of Checks Outstanding	– 42.40		
5 Subtract line 4 from line 3. This amount should equal your check register balance.	$ 674.91	Total	$42.40

The August 1 deposit of $85 not on the July 31 bank statement was added and the two outstanding checks subtracted from the closing balance to reconcile the bank statement.

Step 3: Reconcile the check register.

Check Register

CHECK NO.	DATE	DESCRIPTION OF TRANSACTION	PAYMENT/DEBIT (-)	√ T	FEE (IF ANY) (-)	DEPOSIT/CREDIT (+)	BALANCE
		Previous Balance	$		$	$	1181 39
		Ck 851, July 19	37 16				1144 23
		Ck 853 (wrong amount)				27 05	1171 28
		Ck 853 (correct amount)	72 05				1099 23
		ATM-DEP				118 20	1217 43
		Interest earned				1 56	1218 99
		ATM-WD	120 00				1098 99
		ATM fee	1 50				1097 49
		EFT payment, truck loan	416 18				681 31
		Debit	6 40				674 91

Line 1 The previous balance of $1,181.39 is the last balance in the check register before the reconciliation begins.

Line 2 Record unrecorded Check 851 for $37.16.

Line 3 Check 853 was recorded incorrectly as $27.05. The $27.05 had to be added back into the check register to cancel the error.

Line 4 The correct amount for Check 853 was recorded as $72.05.

Line 5 Record the unrecorded ATM deposit of $118.20.

Line 6 Add interest earned of $1.56 to the register.

Line 7 Record the unrecorded ATM withdrawal of $120 in the register as a payment.

Line 8 Record the unrecorded ATM fee of $1.50 in the register as a payment.

Line 9 Record the unrecorded EFT payment of $416.18 in the check register.

Line 10 Record the debit charge of $6.40 in the register as a payment.

The final balance of $674.91 shown in the Reconciliation Form agrees with the final balance of $674.91 shown in the check register. The checking account is reconciled.

✓ CHECK YOUR UNDERSTANDING

E. On January 30, your check register balance is $107.87 and your bank statement balance is $161.96. Interest earned of $0.43 and an ATM deposit of $56 also appeared on the statement but had not been recorded in the register. You also find that Check 307 for $35.29 had been entered in the register as $32.95. Reconcile the checking account.

F. At the end of October, Allen Springer's check register balance was $812.45. His bank statement balance was $624.77. An examination of his statement and check register showed that an ATM withdrawal of $200 had not been entered in the register, Check 201 for $92.49 was outstanding, and Check 202 for $80.17 was cashed but not recorded in the register. Reconcile the checking account.

Calculator *Tip*

When reconciling your checking account, another error may be created if you incorrectly use a calculator. Be sure to double check all calculations.

Communication

Many employers rely on a Standard Operating Procedure (SOP), which is a guide for all employees to follow in making decisions and completing tasks. The SOP is generally used so that the everyday activities of a company are done consistently even when personnel changes.

The Standard Operating Procedures must be clearly and precisely written. Procedures that are unclear can lead to misunderstandings between the expectations of employers and the performance of employees.

The procedures may also define how employees are to interact with customers. Misunderstandings with customers can lead to a damaged business relationship and a loss of sales.

Prepare a Standard Operating Procedure on how to reconcile a checking account. Assume that you manage a team of bookkeepers and all accounts must be reconciled in the same manner. Your Standard Operating Procedure will serve as guidelines for all who work for you.

Wrap Up ▶▶▶

One way to settle disagreements is to collect and present the facts. Historical records contain information about the event. Presenting the facts to the people who do not agree is a way to begin reconciling the disagreement.

Exercises

Find the sum or difference.

1. $874.20 + $392.29
2. $125.62 + $1.23 + $72.76
3. $17,800.23 − $2,893.98
4. $274.65 − $110.38

Solve.

5. The bank statement of Dottie Weigand showed a closing balance of $3,150.18. The transactions outstanding included three checks for $12.87, $39.47, and $840. Reconcile Dottie's bank statement.
6. Roger Korsak's check register balance was $745.84. His bank statement showed an ATM deposit of $82.67 that was not recorded in his check register, interest earned of $0.87, and an $18 charge for printing new checks. Reconcile Roger's check register.
7. Phoebe Duncan's check register balance on November 30 was $984.09. Her November 30 bank statement showed a balance of $1,462.25. Checks outstanding were 207 for $298.12, 209 for $86.73, and 210 for $105.03. The bank service charge for November was $12.80 and the interest earned was $1.08. Reconcile Phoebe's check register.
8. On May 31, Sue Ware's check register balance was $289.30 and her bank statement balance was $375.37. Checks outstanding were: 543, $86.24; 543, $12.82; and 547, $57.67. A late deposit on June 1 for $68.50 was not on the statement. The statement showed an ATM fee of $2.50 and interest earned of $0.34.

Photodisc/Getty Images

9. On March 31, Goro Hayashi's check register balance was $277.37 and his bank statement balance was $289.23. The statement showed an ATM service charge of $1.25. Checks 85 for $5.37 and 90 for $73.09 were outstanding. Goro also found that he had recorded Check 87 for $20.75 twice in the check register and did not record Check 93 for $86.10.
10. On May 30, Amy Millard's check register showed a balance of $700.61. Her bank statement balance on that date was $1,143.90. The "other charges" part of the statement showed that an EFT car loan payment of $250 was not recorded in the register. Checks outstanding were: 834 for $48.33, 837 for $21.19, 838 for $161.77. A direct deposit of $480 was not recorded in the check register. Check 836 for $86.40 was recorded as $68.40.
11. **CRITICAL THINKING** What would you do if you found that you wrote a check for $78, but it was recorded by the bank as $87?
12. **FINANCIAL DECISION MAKING** Suppose you made a contribution to a charity by check in December. You just received your March bank statement and found the check to be outstanding. If the check is outstanding for several months, can it still be cashed by the charity? Do you need to take any action?

Mixed Review

13. $\frac{3}{4} \times \frac{8}{15} =$

14. $12\frac{3}{4} - 3\frac{1}{8} =$

15. Rewrite 0.8 as a fraction

16. Find 75% of 140

17. Estimate the quotient: 6,132 ÷ 42

18. Divide to the nearest hundredth: 995 ÷ 7.4

19. Round 5,437 to the nearest 10, and then to the nearest 100.

20. Emma's bank statement's closing balance on March 31 is $683. There were three checks outstanding for: $12.33, $8.32; $19, and $274.90. Reconcile Emma's bank statement.

21. Elrod York is paid a 5% commission on all monthly sales. His monthly sales average $82,000. What total commission is he likely to earn for the year?

22. The city in which Leah Zang-Clouse works has a 1.25% tax on all income. What total city income taxes must she pay if her income for a year is $38,200?

23. Kristin Thomas' job expenses last year were: tools, $1,250; work supplies, $180; work clothes, $360; union dues, $480; truck expenses, $3,800. Her total job benefits for the year were $48,000. Find Kristin's net job benefits.

Kasia 2008/Shutterstock.com

24. The owner of a hospital supply company bought four new trucks of different sizes. Their purchase prices were $28,430, $21,692, $22,572, and $25,870. What average price per truck did the owner pay?

25. McCarther Williams, a waiter, wrote $430 in breakfast checks one morning. His tips average 11% of the total checks. What was his tip income for the morning?

26. Elva Rainey is single and has taxable income of $23,810. Her employer deducted $3,718 in withholding taxes for the year. Find Elva's tax due and any refund or amount owed. Use the tax tables in Chapter 2.

27. Walker Levin is paid a 2.75% commission on all sales over $5,500 each week. Last week his sales were $12,650. On how much of his sales did he earn commission and what was the amount paid to him in commission?

28. An assistant chef is paid $120 for every day she works. What is her gross pay for a month in which she works 25 days?

3-5

Money Market and CD Accounts

GOALS

- Calculate interest earned on special savings accounts
- Calculate the penalty for early withdrawals from CD accounts
- Compare the interest earned on savings accounts
- Calculate the effective rate of interest

KEY TERMS

- certificate of deposit (CD)
- term
- maturity date

Start Up

Banks often pay a higher interest rate on savings accounts to customers who keep their money on deposit for a fixed period of time, such as a year, and do not make any withdrawals. Name reasons why banks encourage people to use such savings accounts.

Photodisc/Getty Images

Math Skill Builder

Review these math skills and solve the exercises that follow.

1. **Subtract** money amounts.
   Find the difference. $705.46 − $700 = $5.46

   1a. $870 − $8.78 **1b.** $265.38 − $187.46

2. **Multiply** money amounts by a percent and round.
   Find the product. 2.13% × $673 = 0.0213 × $673 = $14.334, or $14.33

   2a. 4.7% × $547 **2b.** 2.9% × $992

3. **Multiply** money amounts by a fraction and round.
   Find the product. $\frac{1}{4} \times \$27.34 = \$6.835$, or $6.84

   3a. $\frac{1}{2} \times \$68.46$ **3b.** $\frac{1}{12} \times \$56.89$

Special Savings Accounts

In addition to regular savings accounts, many banks also offer special savings accounts for long-term savers or those who keep large savings account balances. The interest rates paid on these special savings accounts are higher than the rates paid on regular savings accounts.

The **certificate of deposit** is widely referred to as a CD. The CD is also known as a *time deposit* or a *savings certificate.* Some government rules apply to certificate of deposit accounts.

In exchange for a fixed higher rate of interest, banks require depositors to:

- Deposit a minimum amount. This may be \$500, \$1,000, \$5,000, or \$10,000.
- Leave the money on deposit for a specified time. The time may be specified in number of days, months, or years. The minimum time is called the **term**. The date that marks the end of the term is the **maturity date**.
- Pay a penalty if money is withdrawn before the end of the term. Most CD accounts can be set up so that the interest is paid out of the account each period with no penalty. This kind of CD is called a simple interest CD.

Like certificates of deposit, money market accounts offer higher interest rates than regular accounts. Special rules apply:

- A minimum balance must be kept in the account. More money may be added to the account at any time.
- The interest rate paid varies with the economy.
- A small number of checks may be written against the account.

NET Bookmark
Research current CD and money market rates.

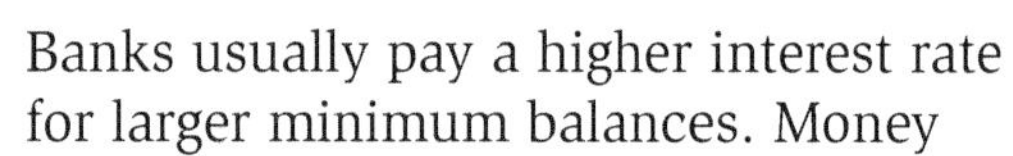
Banks usually pay a higher interest rate for larger minimum balances. Money
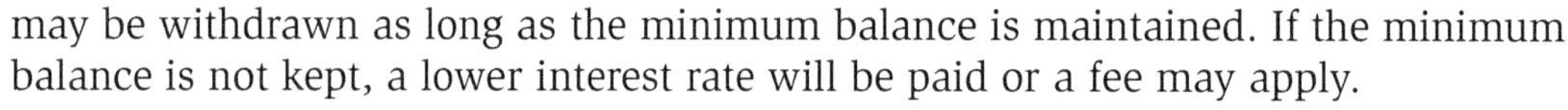
may be withdrawn as long as the minimum balance is maintained. If the minimum balance is not kept, a lower interest rate will be paid or a fee may apply.

EXAMPLE 1

Nick Bolger has a \$1,200 six-month simple interest CD that earns quarterly interest at an annual rate of 4%. How much interest does Nick receive each quarter? How much interest does Nick earn for 6 months?

SOLUTION

Use $I = PRT$ to find interest for each 3-month term.

$I = \$1{,}200 \times 0.04 \times \frac{3}{12} = \$12$ interest for each 3-month period

$\$12 \times 2 = \$24$ total interest earned for six months

✔CHECK YOUR UNDERSTANDING

A. Rose Bannon deposited \$10,000 in a three-year certificate of deposit that pays simple interest at a fixed annual rate of 5.4%. What total interest will Rose have earned at the end of three years?

B. Alex Nugent had \$2,000 on deposit for March and April in a money market account. Interest in the account is paid monthly. In March, the account paid 3.5% annual interest. In April, an annual interest rate of 3.75% was paid. Alex had no other deposits or withdrawals from the account. What total interest did Alex earn for the two months?

Penalties on Certificates of Deposit

By law, banks must charge depositors a penalty for withdrawing money early from a certificate of deposit. Each bank sets its own penalty for early withdrawals. The penalty usually varies with the term of the certificate. For example, a 1-year CD may carry a penalty of 3 months' interest.

If you want to withdraw money from a CD, most banks will require you to cash out the entire CD. You will receive the principal and interest, minus the penalty for early withdrawal.

EXAMPLE 2

Ella Trane invested $5,000 in a 4-year CD that paid 5.2% annual interest. When she cashed out the CD at the end of 3 years, her early withdrawal penalty was 6 months' simple interest. What was the amount of the penalty?

SOLUTION

Use $I = PRT$ to find the penalty.

$I = \$5{,}000 \times 0.052 \times \frac{6}{12} = \$130$ six months' interest penalty for early withdrawal

✔ CHECK YOUR UNDERSTANDING

C. Neil Richards has $2,000 in a one-year time-deposit account that pays an annual interest rate of 2%. Neil cashed out the CD early. The bank charged Neil 3 months' simple interest for the early withdrawal. What penalty did Neil pay?

D. Noelle Hastings' 5-year savings certificate pays an annual interest rate of 4.7%. At the end of the first year she cashed out the $12,000 CD and paid a penalty of 12 months' simple interest. What penalty did she pay?

Compare Savings Accounts

Savings accounts are often compared by the interest earned in each account. To compare, calculate the interest that would be earned by each type of account for the same time period.

Math *Tip*

You can use the Daily Compound Interest Table on p. 83 to find the interest on the savings account.

EXAMPLE 3

In one year, you could earn $18.18 interest on a $900 deposit in a savings account paying 2% daily interest. The $900 could be deposited in a one-year CD paying simple interest at 4.7% annually. How much more interest could you earn in one year by placing your money in a CD?

SOLUTION

Savings account: $18.18 one year's interest

CD: $\$900 \times 0.047 \times 1 = \$42.30$ one year's interest

Find the difference between the interest earned on each account.

$\$42.30 - \$18.18 = \$24.12$ more interest earned by CD

✔ CHECK YOUR UNDERSTANDING

E. A six-month time-deposit account pays 5.35% simple interest. Dora has already calculated that she could earn $10.54 in six months on a $1,400 deposit in a savings account earning 1.5% daily interest. How much more interest could Dora earn if she deposited the $1,400 in the time-deposit account instead of a savings account for 6 months?

Photodisc/Getty Images

F. Jim Russell's $1,500 deposit could earn $18.87 in 12 months in a savings account paying 1.25% daily interest. How much more interest could Jim earn in a 3-month CD that pays 3.26% simple interest every 3 months during the 12-month period?

Effective Rate of Interest

The effective rate of interest is the rate you actually earn by keeping your money on deposit for one year. The annual rate and the effective rate you earn can be different. The effective rate is sometimes referred to as the *annual percentage yield.*

$$\textbf{Effective Rate of Interest} = \frac{\textbf{Interest Earned in One Year}}{\textbf{Principal}}$$

Algebra *Tip*

Effective rate of interest can be calculated using the formula

$ER = \frac{I}{P}$

where ER is the effective rate of interest, I is the interest earned in one year, and P is the principal.

EXAMPLE 4

Find the effective rate of interest to the nearest hundredth percent on $1,000 deposited in an account that pays 5% annual interest, compounded quarterly. Use the compound interest table given in Lesson 3-1.

SOLUTION

5% ÷ 4 = 1.25%

Find the multiplier in the compound interest table: 1.050945

Multiply the deposit amount by the multiplier.

$1,000 × 1.050945 = $1,050.945, or $1,050.95 compound amount

Find the difference between the deposit and the compound amount.

$1,050.95 − $1,000 = $50.95 interest earned in one year

Divide the interest earned for one year by the principal. Round as directed.

$\text{ER} = \frac{\$50.95}{\$1{,}000} = 0.05095$, or 5.1% effective rate of interest

✔ CHECK YOUR UNDERSTANDING

G. A deposit of $2,000 is kept in an account that pays 4% annual interest, compounded quarterly. Find the effective rate of interest to the nearest hundredth percent if the money is on deposit for 1 year. Use the compound interest table.

H. A CD pays 4% yearly interest and compounds interest daily. Find the effective rate of interest to the nearest tenth percent if $6,000 was on deposit for 1 year. Use the daily compound interest table.

Photodisc/Getty Images

Wrap Up ▶▶▶

When an account has a fixed term, such as a year, banks spend less money on personnel and other costs of processing withdrawals or deposits. In turn, the bank can lend the money to a borrower for a longer period of time since they know the money on deposit is not likely to be withdrawn during the term of the deposit.

Communication

Visit a local bank or the Internet site of a bank not in your area to find the types of certificates of deposit offered. Create an overhead transparency that shows the bank name, term of deposit, interest rate, and minimum deposit amount. Write a sentence about the relationship of the length of the term and the interest rate.

When preparing visual presentations, keep in mind the following hints:

- Keep your design simple.
- Include only one bank per visual.
- To maximize effectiveness, be selective in how much information you include on the page.
- Proofread visuals carefully.
- Make sure visuals are large enough to be seen by the entire audience.
- Avoid distorting facts on visuals; be clear and concise.
- When presenting at the overhead, position yourself so the audience may clearly view the visual.
- Make an effort to talk about the data rather than read your visual line by line.

Exercises

Find the sum or difference.

1. $7.58 + $8.34 =

2. $1,430.67 + $78.34

3. $7,757.82 − $257.82

4. $45.63 − $28.08

Find the product and round to the nearest cent.

5. 4.26% × $720
6. 2.78% × $1,570
7. $\frac{1}{2}$ × $27.83
8. $\frac{1}{12}$ × $87.65

Solve.

9. How much interest will you earn per quarter on a 4.5% simple interest CD where interest is paid out quarterly, if you have $1,000 on deposit?

10. A money market account paid annual interest of 4.8% in June and 4.91% in July. A two-month, time-deposit account pays 4.87% annual interest. Which account would have earned more interest if $15,000 were left on deposit in each account for 2 months? How much more?

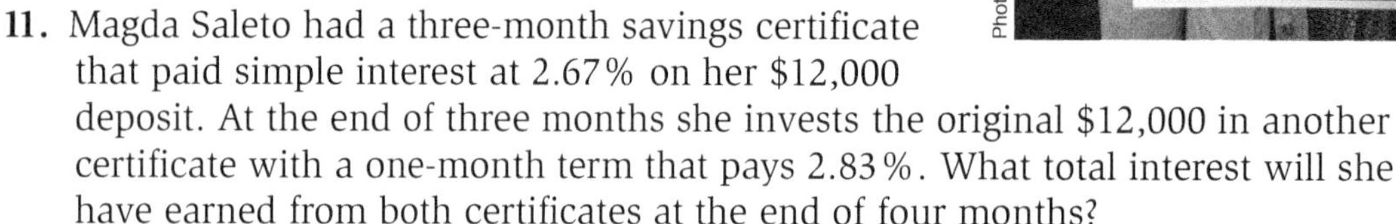

Photodisc/Getty Images

11. Magda Saleto had a three-month savings certificate that paid simple interest at 2.67% on her $12,000 deposit. At the end of three months she invests the original $12,000 in another certificate with a one-month term that pays 2.83%. What total interest will she have earned from both certificates at the end of four months?

12. A $500 investment in a 12-month CD with an interest rate of 4.89% compounded monthly, earns $25.01 interest in one year. What effective rate of interest does this investment earn?

13. **CRITICAL THINKING** A money market account may pay interest of 3.25% provided that a minimum balance of $2,500 is kept on deposit. A deposit that falls below the minimum, to $2,498 perhaps, earns 1% interest. Why is there such a large difference in the rates?

14. **FINANCIAL DECISION MAKING** You sold your house with net cash received of $40,000. You did not need the money for at least two months. You could deposit the $40,000 in a 12-month CD that pays 4.2% simple interest and has an early withdrawal penalty of three-months' interest. You could also deposit the money in a money market account that pays 2% interest, compounded daily. In which account would you deposit your money to earn the most interest if you plan to withdraw the money at the end of the third month?

Mixed Review

15. 907 + 48 + 412 + 62
16. $867.38 − $20.09
17. Write 0.0345 as a percent.
18. $4\frac{2}{3} \div 1\frac{1}{9}$

19. A truck farmer brought 200 pints of berries to sell in a market during each week of the growing season. She sold 80% of the berries the first week and 90% the second week. She sold all the berries in each of the next three weeks. What average number of pints of berries did the farmer sell per week during the five-week period?

20. Jack Nesbitt worked 52.5 hours last week. He is paid $11.29 per hour for regular-time work and time-and-a-half overtime for hours worked past 40 hours in a week. What gross pay did he earn last week?

3-6

Annuities

GOALS

- Calculate the future value of an ordinary annuity
- Calculate the present value of an ordinary annuity

KEY TERMS

- annuity
- future value of an annuity
- present value of an ordinary annuity

Start Up

Manuel and Germaine Montez have a new baby. Each month from the baby's birth, they deposit $50 in an account to pay for the baby's college expenses. If the money earns 5% interest compounded monthly, how much can they expect to have in the college account when the child is 18?

DAJ/Getty Images

Math Skill Builder

Review these math skills and then answer the questions that follow.

1. **Divide** whole numbers and decimals.
Find the quotient. 6 ÷ 4 = 1.5

1a. 3 ÷ 4 **1b.** 2.5 ÷ 4

1c. 1 ÷ 5 **1d.** 3.5 ÷ 8

2. **Multiply** money amounts by decimals and round to the nearest cent.
Find the product. $400 × 5.67234 = $2,268.94

2a. $650 × 1.907534 **2b.** $25 × 3.018432

2c. $325 × 2.046125 **2d.** $105 × 8.07035

2e. $52 × 5.00875 **2f.** $279 × 4.12265

3. **Subtract** money amounts.
Find the difference. $840.27 − $662.32 = $177.95

3a. $2,843.67 − $1,909.14 **3b.** $526.01 − $493.64

3c. $759.45 − $602.89 **3d.** $2,005.75 − $1,945.36

Annuities

An **annuity** is a series of equal payments made at regular intervals of time. Rent, salaries, and loan payments are all examples of annuities. Making regular deposits into a savings account is also an annuity.

An annuity is called an *annuity due* when payments are made at the beginning of each period. An annuity is called an *ordinary annuity* when the payments are made at the end of each period.

In this text, the focus will be ordinary annuities, although with the appropriate tables, the calculations for an annuity due are the same as an ordinary annuity.

Future Value of an Ordinary Annuity

The **future value of an annuity** is the amount of money in an account after a series of equal payments are made to it, including interest the money has earned.

One way to find the future value of an annuity is to use an annuity table. Use the multiplier that corresponds to the interest rate per period and the number of periods of compounding.

Future Value of Ordinary Annuity										
	Rate Per Period									
Periods	0.25%	0.50%	0.75%	1%	1.50%	2%	2.50%	3%	4%	5%
8	8.07035	8.14141	8.21318	8.28567	8.43284	8.58297	8.73612	8.89234	9.21423	9.54911
10	10.11325	10.22803	10.34434	10.46221	10.70272	10.94972	11.20338	11.46388	12.00611	12.57789
12	12.16638	12.33556	12.50759	12.68250	13.04121	13.41209	13.79555	14.19203	15.02581	15.91713
16	16.30353	16.61423	16.93228	17.25786	17.93237	18.63929	19.38022	20.15688	21.8253	23.65749
20	20.48220	20.97912	21.49122	22.01900	23.12367	24.29737	25.54466	26.87037	29.77808	33.06595
24	24.70282	25.43196	26.18847	26.97346	28.63352	30.42186	32.34904	34.42647	39.08260	44.50200
30	31.11331	32.28002	33.5290	34.78489	37.53868	40.56808	43.90270	47.57542	56.08494	66.43885
40	42.01320	44.15885	46.44648	48.88637	54.26789	60.40198	67.40255	75.40126	95.02552	120.7998

EXAMPLE 1

Tan decides to put $100 each quarter into an account earning 4% interest compounded quarterly. How much will be in the account after 5 years? How much of that money will be interest?

SOLUTION

4% ÷ 4 = 1% interest rate per period

4 quarters × 5 years = 20 periods number of periods

Find the multiplier in the Future Value table: 22.01900

$100 × 22.01900 = $2,201.90 amount in the account after 5 years

$100 × 20 = $2,000 amount Tan deposited in 5 years

$2,201.90 − $2,000 = $201.90 amount of interest

Tan will have $2,201.90 in the account after 5 years. He will earn $201.90 in interest.

✔ **CHECK YOUR UNDERSTANDING**

A. Janet Powers saves $75 per month and deposits the money in an account earning 3% annual interest compounded monthly. How much will be in the account after 2 years? How much of that money is interest?

B. Liberty Hodges saves $150 per quarter and deposits the money in an account earning 6% interest compounded quarterly. What is the future value of the annuity after 5 years? How much of that money is interest?

Present Value of an Ordinary Annuity

The **present value of an ordinary annuity** is the balance needed in an account in order to make a series of payments from the account. While the account is being reduced by the payments, interest is still earned on the money that is in the account. In calculating the future value of an ordinary annuity, the payments are coming from the account at the end of the period.

To find the present value of an annuity, use the Present Value of an Ordinary Annuity table in the same way that you used Future Value of an Ordinary Annuity table.

Present Value of Ordinary Annuity										
	Rate Per Period									
Periods	**0.25%**	**0.50%**	**0.75%**	**1%**	**1.50%**	**2%**	**3%**	**4%**	**5%**	**6%**
4	3.97512	3.95050	3.92611	3.90197	3.85438	3.80773	3.71710	3.62990	3.54595	3.46511
5	4.96272	4.92587	4.88944	4.85343	4.78264	4.71346	4.57971	4.45182	4.32948	4.21236
10	9.86386	9.73041	9.59958	9.47130	9.22218	8.98259	8.53020	8.11090	7.72173	7.36009
12	11.80725	11.61893	11.43491	11.25508	10.90751	10.57534	9.95400	9.38507	8.86325	8.38384
14	13.74096	13.48871	13.24302	13.00370	12.54338	12.10625	11.29607	10.56312	9.89864	9.29498
16	15.66504	15.33993	15.02431	14.71787	14.13126	13.57771	12.56110	11.65230	10.83777	10.10590
24	23.26598	22.56287	21.88915	21.24339	20.03041	18.91393	16.93554	15.24696	13.79864	12.55036
30	28.86787	27.79405	26.77508	25.80771	24.01584	22.39646	19.60044	17.29203	15.37245	13.76483

EXAMPLE 2

Rey Garza wants to receive an annuity payment of $200 for each quarter for the four years he will be in college. If his account earns 6% interest, how much money must be in the account by the time he starts college? How much of what he receives will be interest?

SOLUTION

$6\% \div 4 = 1.5\%$ interest rate per period

4 quarters × 4 years = 16 periods number of periods

Find the multiplier in the Present Value table: 14.13126

$200 × 14.13126 = $2,826.25 balance needed

$200 × 16 = $3,200 amount of money received

$3,200 − $2,826.25 = $373.75 amount of interest

Rey needs $2,826.25 in the account. He will receive $373.75 in interest.

✔CHECK YOUR UNDERSTANDING

C. What amount must you invest today at 6% compounded annually so that you can withdraw $5,000 at the end of each year for the next 5 years? How much will you withdraw in interest?

D. Mara Wilmouth wants to take a year off working to travel. She estimates that she will need $2,000 at the end of each month to pay for the following month's expenses. How much will she need to have in an account that pays 3% interest compounded monthly?

Wrap Up ▶▶▶

Manuel and Germaine Montez deposit $50 a month for 18 years in an account earning 5% interest compounded monthly. You can use an online ordinary annuity calculator, or you can use the multiplier 349.20202 to calculate the future value. The account will have about $17,460 after 18 years.

Exercises

Find the quotient.

1. 3 ÷ 12

2. 6 ÷ 12

Find the product.

3. $621.92 × 1.349623

4. $3,500 × 2.36589

5. $400 × 0.99634

6. $2,000 × 1.03987

Find the difference.

7. $733.29 − $548.23

8. $55,296.10 − $36,823.90

9. $1,835 − $1,254.59

10. $852.44 − $729.39

Solve each problem.

Find the future value and the amount of interest for each ordinary annuity.

	Amount Deposited	Frequency	Rate	Time	Future Value	Amount of Interest
11.	$250	Quarterly	4%	4 years		
12.	$2,000	Annually	2.5%	10 years		
13.	$25	Monthly	6%	1 year		
14.	$500	Semiannually	3%	15 years		
15.	$75	Monthly	3%	2 years		

Find the present value and the amount of interest received for each ordinary annuity.

	Amount Received	Frequency	Rate	Time	Present Value	Amount of Interest
16.	$1,500	Quarterly	8%	1 year		
17.	$400	Annually	4%	5 years		
18.	$200	Monthly	6%	2 years		
19.	$5,000	Semiannually	2%	7 years		
20.	$75	Quarterly	3%	3 years		

Chad Holley saves $500 per quarter. He deposits it in an account that earns 4% compounded quarterly.

21. How much will he have in the account after 2 years?
22. How much will he have in the account after 4 years?
23. How much will he have in the account after 10 years?
24. After 10 years, how much interest will he have earned?

Aly Daniels wants to receive an annuity payment of $250 per month for 2 years. Her account earns 6% interest, compounded monthly.

25. How much should be in the account when she wants to start withdrawing?
26. How much will she receive in payments from the annuity?
27. How much of those payments will be interest?

Rachale Martinez is in high school and is saving to buy a car. She estimates that she will spend $5,000 on her car when she graduates high school. For the four years of high school, she plans to save $225 a quarter on an account that earns 4% annually, compounded quarterly.

28. How much will she have after 4 years?
29. How much of the balance after 4 years will be interest?
30. How much more money will she need at the end of the 4 years to buy a car?

31. **STRETCHING YOUR SKILLS** Janell deposits $2,000 and for 5 years leaves it in an account earning 3% annual interest, compounded quarterly. For 5 years, José saves and deposits $150 per quarter in an account earning 3% interest compounded quarterly. Who will have more money in the account after 5 years? How much more?

Mixed Review

32. 15% of $85
33. $485.23 − $16.82
34. John earns 5% commission on sales up to $15,000 and 7% commission on sales over $15,000. If he has $35,000 in sales last month, what was his commission?

Chapter Review

Vocabulary Review

Find the term, from the list at the right, that completes each sentence. Use each term only once.

annuity
Automated Teller Machine, (ATM)
balance
bank statement
certificate of deposit (CD)
check register
compound amount
compound interest
debit card
deposit slip
direct deposit
Electronic Funds Transfer, (EFT)
future value of an annuity
interest
maturity date
online banking
outstanding checks
overdrawn
present value of an annuity
reconcile
service charge
term
transaction

1. A series of regular deposits into or payments from an account is called (a, an) __?__.
2. The amount of money in a checking account is called the __?__.
3. The total in a savings account at the end of a period after interest is added is called the __?__.
4. A printed report of bank transactions given to a depositor is called (a, an) __?__.
5. A savings plan also known as a time deposit or savings certificate is called (a, an) __?__.
6. A record you keep of deposits made and checks written is called (a, an) __?__.
7. The movement of money from one bank's computer to another computer is called __?__.
8. The fixed period of time money is on deposit in a savings certificate is called the __?__.

3-1 Saving Accounts

9. Bob Rowinski made a $4,000 deposit to a savings account paying 1.6% annual interest compounded semiannually. If he kept the money on deposit for 6 months, what would his account balance be after the interest payment is made?
10. On April 1, Preston McCord deposited $1,400 in a savings account that pays annual interest of 3.2% compounded quarterly. If he made no deposits or withdrawals in the account, what interest could he earn by keeping his money on deposit until October 1?
11. Joanna Michael deposits $3,500 in a savings account that earns 2% compounded daily. If no deposits or withdrawals are made, how much will be in the account after 5 years?

3-2 Checking Accounts

12. Justin Nucci listed these items on his deposit slip: (bills) 8 one-hundreds, 27 fifties, 83 fives, 141 ones; (coins) 13 quarters, 117 nickels; (checks) $317.94, $57.89, $527.24 and $77.49. He received cash back of 30 twenties and 11 tens. What total deposit did he make?

13. Zora Omar had a balance of $1,189.17 in her checking account. She wrote checks for $62.41, $224.14, $12.92, and $357.16. Her deposits were $197.34 and $879.13. What was Zora's new bank balance?

3-3 Electronic Banking

14. At the start of the day, Grace Williams' checking account had a $201.87 balance. She used her debit card to pay $56.12 for groceries and $28.45 for cleaning. Grace then transferred $150 to her checking account from savings using her bank's ATM. She also directed her bank to transfer funds electronically from her checking account to pay a charge account bill of $187.12. What was the balance of Grace's checking account at the end of the day?

15. Basil Tomlin's online checking account had a balance of $371.07. The balance did not include a direct deposit of $672.80 that will be transferred to the checking account by Basil's employer at 11:00 a.m. Basil's bank will automatically deduct a house payment of $720 at the end of the business day. Basil plans to make online payments for $34.85, $90.29, and $368.20. How much money does Basil need to transfer to complete these transactions and still have a balance of $100?

3-4 Check Register Reconciliation

16. Giselle Mulroon's bank statement shows a balance of $539.22. Checks outstanding were #841 for $29.67, #843 for $89.02, and #844 for $9.76. A $130 deposit was outstanding. Reconcile Giselle's bank statement.

17. Henry Sokol's bank statement balance on June 30 was $845.43. His check register balance was $247.62. His comparison of the two records showed a check for $85 was recorded twice in the register, a check for $138.11 was recorded as $183.11, and an ATM deposit of $368 and a debit card purchase for $81.10 were not recorded in the register. Also, an ATM withdrawal of $50 and interest earned of $0.82 were not recorded. Three checks were outstanding: $191.22; $23.87; and $15. Reconcile Henry's bank statement and check register.

3-5 Money Market and CD Accounts

18. Sonya Lister's 6-month CD pays 5.3% simple interest. Her deposit to the CD was $4,500. At the end of each 6-month period, Sonya withdraws the interest earned and renews the CD on its original terms. What total interest will Sonya have earned at the end of one year?

19. Dale Lawrence had $80,000 on deposit in a six-month CD paying 2.72% simple interest. At the end of three months, Dale cashed out the CD. The penalty for early withdrawal of money from this CD is one-month's interest. What was the amount of the penalty Dale had to pay?

3-6 Annuities

20. José is planning to save $65 a month for 2 years. The money will earn 3% interest compounded monthly. How much will be in the account at the end of 2 years? How much of that money will be interest?

21. Albert wants to receive a $4,000 annuity payment each year for 10 years. If he will earn 5% interest compounded annually, how much does he need in the account when he starts the withdrawals?

Technology Workshop

Task 1 Comparing the Interest Earned on Savings Accounts

Enter data into a template that calculates the compound interest earned by two savings deposits. Use the results to compare the interest earned by different savings plans.

Open the spreadsheet for Chapter 3 (tech3-1.xls) and enter the data shown in blue (cells B4-7 and C4-7) into the spreadsheet. The compound amount and compound interest of two savings deposits will be calculated and also the difference in the amount of interest earned by the deposits. Your computer screen should look like the one shown below when you are done.

	A	B	C	D
1	**COMPARING INTEREST EARNED IN SAVINGS ACCOUNTS**			
2				
3	**ACCOUNT INFORMATION:**	**ACCOUNT A**	**ACCOUNT B**	**DIFFERENCE**
4	Interest Rate (%)	4.00	4.00	
5	Interest Periods in a Year	12	4	
6	No. of Periods on Deposit	12	4	
7	Amount of Deposit	$1,500.00	$1,500.00	
8	**INTEREST INFORMATION:**			
9	Compound Amount	$1, 561.11	$1,560.91	
10	Less Original Deposit	$1,500.00	$1,500.00	
11	Interest Earned	$61.11	$60.91	
12	**DIFFERENCE: (ACCOUNT A – ACCOUNT B)**			**$0.20**

Task 2 Analyze the Spreadsheet Output

Answer these questions about the interest calculations.

1. What number of years was the money on deposit in both accounts?
2. How much money was on deposit in both accounts?
3. What compounding period was used to compute interest in Account A?
4. What was the total interest earned by the deposit in Account B?
5. Which account earned the most interest, and how much more did it earn?

Now move the cursor to cell C4, which holds the Account B interest rate. Enter a new interest rate of 4.1% without the percent symbol.

6. What is the new difference between Account A and Account B?
7. What does it mean when a number is enclosed in parentheses?
8. If the result in Cell D12 shows a negative number, is there something wrong with the result?

Now use the spreadsheet to calculate the answers to interest problems you have already solved or to compare the terms of savings plans.

Task 3 Design a Bank Reconciliation Spreadsheet

You are to design a spreadsheet that will reconcile the bank statement balance and the check register balance.

The spreadsheet for Task 3 may be solved by using addition and subtraction. The spreadsheet should have two sections, placed side-by-side, one for the bank statement and the other for the check register.

Place all the items to be added or subtracted in one column with the balance labeled at the bottom. You may want to indicate that a number is to be subtracted by placing a minus sign before the number when it is entered, but do so sparingly.

The information shown below lists the usual adjustments necessary to complete the reconciliation of the bank statement and check register. Add any other adjustment items you think are necessary. Be sure to allow enough lines so that all data may be entered.

SITUATION: Your bank statement and check register balances do not agree. The following shows what was found when the bank statement and check register were compared. Prepare a spreadsheet that will result in both the bank statement and check register balances being in agreement.

Bank Statement		Check Register		
Closing Balance	$498.36	Last Balance	$483.41	
Late Deposit	$0.00	Deposit Outstanding	$0.00	
Outstanding Check	$75.30	Other Credit	$20.00	
Outstanding Check	$32.57	Interest Earned	$0.56	
		Check, Debit, or ATM withdrawal		$91.28
		Check, Debit, or ATM withdrawal		$15.00
		Check Error		*
		Service Charge		$4.50
		Other Charge		$2.25

*Check #764 for $17.50 was recorded incorrectly in the check register as $17.05.

Task 4 Analyze the Spreadsheet Output

Answer these questions about your completed spreadsheet:

9. What was the reconciled balance?

10. Why are outstanding checks deducted from the bank statement balance?

11. How was the error made on Check #764 corrected?

12. In what other way could you have corrected the Check #764 error?

Continue testing the spreadsheet by entering the data from reconciliation problems you have already solved.

Chapter *Assessment*

Chapter Test

Answer each question.

1. $237.10 + $76 + $0.56 + $2.48
2. $745.28 − $237.30
3. $850 × 0.00368
4. 6.7% × $\frac{1}{4}$
5. $268.44 × $\frac{1}{12}$
6. 2.37% × $1,200
7. $2\frac{3}{8} - 1\frac{1}{2}$
8. 54 is what percent of 360?
9. 350 increased by $\frac{1}{5}$ of itself is?
10. $288 is $\frac{6}{5}$ of what number

Solve.

11. Stuart Terril listed these items on his checking account deposit slip: (bills) 47 ones, 3 fives; (coins) 129 quarters, 74 dimes; one check for $8. His cash received consisted of 2 twenties and 1 ten. In addition, two other deposits for $200 and $110 were made. A check was written for $650. If the starting balance was $8,411.13, what was the balance after the three deposits and 1 check?
12. On Friday, Carlotta Rowe's checking account balance was $173.56. During the day Carlotta's employer deposited her $516.45 pay directly to her checking account. At lunchtime, she wrote checks for $172, $86.43, and $9.05, and made an ATM withdrawal from checking of $150. What is the balance of Carlotta's checking account at the end of the day?
13. The bank statement sent to Abigail Ochs did not show checks for $158.23, $12.89, and $71.27, and an outstanding deposit of $75 that were listed in the check register. The balance printed on the statement was $289.07. What is the reconciled bank statement balance?
14. The bank statement of Jake Hansen showed a balance of $356.93. His check register showed a balance of $308.34. When Jake compared the two records he found several differences. The bank statement listed these items not recorded in the register: ATM withdrawal, $50; ATM fee, $2.25; direct deposit of paycheck, $624.70; checks for $287.23, $180.11, and $72.89. A check for $85.89 was recorded in the register as $58.89. The items not listed on the bank statement included checks for $72.44, $9.76, and $113.57; and a $50 deposit made after the statement closing date. Debit card purchases of $85.67 and $16.73 appeared on the statement but not in the register. Reconcile Jake's bank statement and check register.
15. Find the interest that a $510 deposit will earn in 3 months at 4.25% simple interest.
16. The daily interest multiplier for a savings account paying 2% annual interest for 180 days is 1.010050. What compound amount will be in a savings account if $5,000 is on deposit in the savings account for 180 days?
17. A 2-month CD pays 2.1% simple interest for the term of the deposit. A savings account pays 1.65% annual interest compounded monthly. In which account will a $3,600 investment earn the most interest for two months, and how much more?
18. What is the future value of $500 invested quarterly at 4% interest compounded quarterly for 5 years?

Planning a Career in Arts, A/V Technology, and Communications

Kasia 2008/Shutterstock.com

There are a wide variety of career choices available in the arts, A/V technology, and communications industries. If you earn a living in the arts, you may find yourself acting on a stage, creating visual arts, or writing songs or music. In the A/V or audio and video technology area, you might choose a career in motion pictures, television, or the video industry. Opportunities in communications exist in journalism, broadcasting, and telecommunications. If you are creative, communicate well, or have musical or artistic talent, a good career choice may be in the arts, A/V technology, and communications industries. Consider your talents and skills, as well as your interest.

Job Titles

- Playwright
- Camera operator
- Graphic designer
- Electric engineer
- Entertainer
- Desktop publisher
- Artist
- Sound technician
- Journalist
- Song Writer

Needed Skills

- natural talent for performing or visual arts
- strong creative writing ability
- computer and technology skills
- good eye for details
- ability to work independently
- able to perform well under pressure

What's it like to work in Photojournalism?

A photojournalist takes pictures of newsworthy events, featured subjects, and people for newspapers, magazines, and television. A photojournalist spends time at the computer editing his or her work. In television, a photojournalist works with a crew that includes a news reporter and sound technician, plus other behind-the-scenes members of a news team. In the printed media, a photojournalist is likely to work independently, receiving assignments and submitting his or her work to an editorial supervisor for publication.

What About You?

What aspect of the fields of arts and communication appeals to you?

How Times Have Changed

For Questions 1-4, refer to the timeline on page 79 as needed. Round answers to the nearest percent.

1. What was the percent increase of ATMs from 1973 to 1990?
2. What was the percent increase of ATMs from 1973 to 2004?
3. What was the percent increase of ATMs from 1990 to 2004?
4. What was the percent decrease of ATMs from 2004 to 2007?

Chapter 4

Credit Cards

Statistical Insights

Percent of Families Holding Specific Debt	
By Age of Head of Family	
Age of Head of Family	**Credit Card Balance**
Under 35 years old	50.7%
35–44 years old	51.3%
45–54 years old	52.5%
55–64 years old	45.7%
65–74 years old	29.2%
75 years old and older	11.2%
By Family Income	
Family Income	**Credit Card Balance**
Less than $10,000	20.6%
$10,000–$24,999	37.9%
$25,000–$49,999	49.9%
$50,000–$99,999	56.7%
$100,000 or more	40.4%

Use the data shown above to answer each question.

1. Using the *By Family Income* data, rank the income levels in order from the group with the greatest percent to the least percent of credit card debt.
2. Which age group is most likely to have a credit card balance?

How Times Have Changed

Credit was used in ancient civilizations more than 3,000 years ago. Ever since then, merchants have been introducing new ways for customers to pay with credit. Some have used the honor system and others have used informal tally systems. In the 19th century, stores issued aluminum charge plates or celluloid charge coins. The term *credit* comes from the Latin word meaning *trust*.

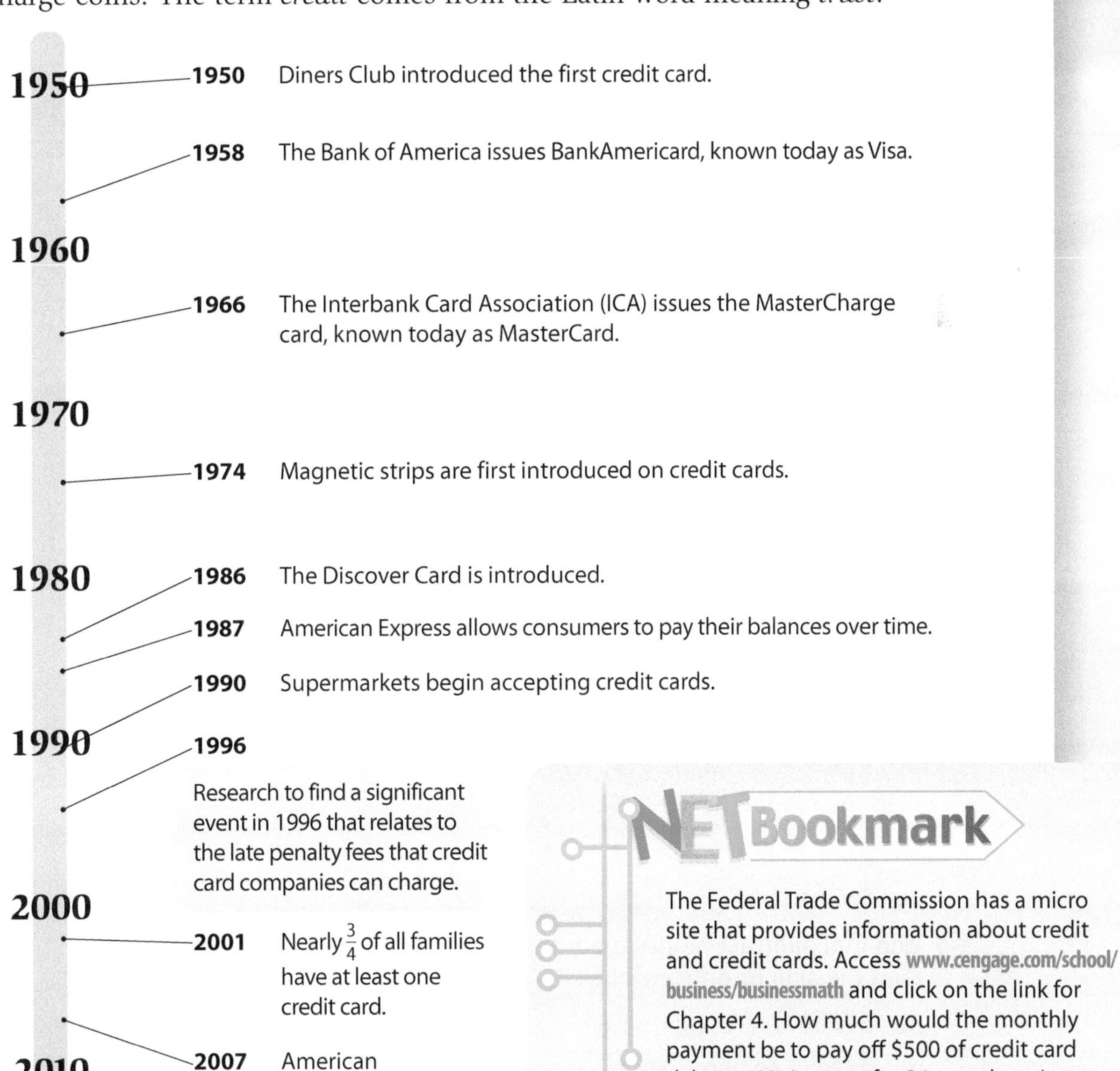

1950
1960
1970
1980
1990
2000
2010
1950 Diners Club introduced the first credit card.
1958 The Bank of America issues BankAmericard, known today as Visa.
1966 The Interbank Card Association (ICA) issues the MasterCharge card, known today as MasterCard.
1974 Magnetic strips are first introduced on credit cards.
1986 The Discover Card is introduced.
1987 American Express allows consumers to pay their balances over time.
1990 Supermarkets begin accepting credit cards.
1996
Research to find a significant event in 1996 that relates to the late penalty fees that credit card companies can charge.
2001 Nearly 3/4 of all families have at least one credit card.
2007 American consumers carry $937 billion dollars in credit card debt.

NET Bookmark

The Federal Trade Commission has a micro site that provides information about credit and credit cards. Access **www.cengage.com/school/business/businessmath** and click on the link for Chapter 4. How much would the monthly payment be to pay off $500 of credit card debt at 18% interest for 24 months using a card without an annual fee? How much will the credit cost?

4-1

Credit Card Costs

GOALS

- Identify important information about credit card terms and conditions
- Calculate the new balance on a credit card
- Verify transactions on credit card statements
- Calculate the cost of using a credit card

KEY TERMS

- periodic finance charges
- credit terms and conditions
- annual percentage rate (APR)
- grace period

Start Up

When Robert receives his credit card statement each month, he simply mails a check for his payment. He decided the time it takes to verify the charges was not worth it. His feeling is that a computer prints out the statement so there will be no errors. What are the dangers in using a credit card as Robert does?

Photodisc/Getty Images

Math Skill Builder

Review these math skills and solve the exercises that follow.

1. **Divide** money amounts and decimals by whole numbers.
   Find the quotients. $366 ÷ 30 = $12.20 $18.25 ÷ 365 = $0.05

   1a. $44.95 ÷ 31   **1b.** $24 ÷ 12   **1c.** 921.25 ÷ 55

2. **Multiply** money amounts by percent.
   Find the product. $500 × 5.5% = $500 × 0.055 = $27.50

   2a. $150 × 20%   **2b.** $868 × 4%

   2c. $650 × 2.8%   **2d.** $1,200 × 6.5%

3. **Add** and **subtract** money amounts.
   Find the sum or difference. $3.45 + $25 = $28.45; $3,298 − $725 = $2,573

   3a. $52.19 + $78.42 + $1.89

   3b. $189 + $92.09 + $3.87

   3c. $4,098.34 − $871.36

   3d. $1,830.43 − $792.19

4. **Rewrite** percents as decimals.
   Rewrite 45.6% as a decimal. 0.456

   4a. 89.08%   **4b.** 0.736%   **4c.** 0.0287%

Using a Credit Card

In a credit card transaction, the issuer of the credit card lends money to the credit card user. That money is paid to a retail merchant on behalf of the credit card user. By signing the merchant's sales slip, the credit card user agrees to pay that money back to the credit card company.

Each month, the credit card company provides a statement to the credit card user that details the purchases, fees, payments and other credits, and the total amount owed.

The total amount may include interest or **periodic finance charges**, as well as other fees. The credit card user makes a payment by a due date for at least the minimum amount given on the statement.

Before applying for a credit card, consumers should read the **credit terms and conditions**. The credit terms and conditions outline the costs associated with using the credit card, and can vary widely among companies that issue credit cards. By law, all credit card solicitations and applications must provide a disclosure box with certain key information.

A sample disclosure box is shown.

Annual percentage rate (APR) for new purchases	2.9% until 11/1/—, after that **14.9%**
Other APRs	Cash advance APR: 15.9%
	Balance transfer APR: 15.9%
	Penalty rate: 23.9% See explanation below.*
Variable-rate information	Your APR for purchase transactions may vary. The rate is determined monthly by adding 5.9% to the Prime Rate.**
Grace period for repayment of balances for purchases	25 days on average
Method of computing the balance for purchases	Average daily balance (excluding new purchases)
Annual fees	$50
Minimum finance charge	$0.50
Transaction fee for cash advances: 3% of the amount received	
Balance transfer fee: 3% of the amount transferred	
Late payment fee: $25	
Over-the-credit-limit fee: $25	
*Explanation of penalty: If your payment arrives more than ten days late two times within a six-month period, the penalty rate will apply.	
**The Prime Rate used to determine your APR is the rate published in the *Wall Street Journal* on the 10th day of the prior month.	

The **annual percentage rate (APR)** is the rate of interest that is charged on a balance that is carried over past the due date, a cash advance, or a balance transfer from another credit card. The APR for cash advances and balance transfers is typically higher than the rate for purchases.

A **grace period** allows you to avoid all finance charges if you pay your balance in full by the due date. Grace periods usually do not apply to cash advances and balance transfers. If you did not pay your balance in full the previous month, you lose your grace period for the next month and finance charges will begin to accrue beginning the date of any new purchases.

There are several accepted methods for computing the balance on purchases. It is this balance that is the amount used to calculate the finance charges. A credit card company may use the average daily balance method (including or excluding new purchases), the adjusted balance method, or the previous balance method. Whichever method is used, most companies have a minimum finance charge.

Credit card companies may charge other fees including annual fees, cash advance and balance transfer fees, late payment fees, and over-the-credit-limit fees. Such fees must be explained in the disclosure box issued by the credit card company.

EXAMPLE 1

Chad Eubanks is applying for the credit card with the terms and conditions in the disclosure box shown on the previous page. After receiving the card, he plans to transfer $2,000 from another credit card because his new card has a lower APR. What fee will he pay to transfer the balance?

Search the Internet to find different types of credit cards such as secured credit cards and prepaid credit cards. Compare and contrast these types of credit cards with a traditional credit card.

SOLUTION

The balance transfer fee is 3% of the amount transferred.

$\$2,000 \times 0.03 = \$60$

The balance transfer fee is $60.

Monkey Business Images 2008/Used under license from Shutterstock.com

✔ CHECK YOUR UNDERSTANDING

A. After Chad received his card, the Prime Rate is published at 9.3%. What is the new APR on purchases? Did the APR increase or decrease?

B. After Chad had been using the credit card for 6 months, he made his second late payment. He received notification that his balance would be subject to the penalty rate. What is his new APR?

Information on a Credit Card Statement

Midori Masami's credit card statement is shown on the next page. Find each italicized word below on the statement:

Transactions are events that need to be recorded on the statement. These events include *purchases, payments* made, and any *fees* Midori has been charged or *credits* made to her account.

Notice that although Midori purchased some clothes on 2/1 from Gale's Fashions, the card company did not record, or *post* the sale to the statement until 2/3.

Midori's *previous balance* from her last statement was $316.15. On 2/25, Midori's payment for her previous balance was received.

The card company allows Midori to carry a maximum balance of $5,000. If she spends more than that *credit limit,* the company will charge her an over-the-credit-limit fee.

OneBank Card

Acct. No. 1200 1200 0000 1200
Statement Closing Date: 02/28/--
Credit Limit: $5,000.00

For customer service call: 1-800-555-5600
Cash Advance Limit: $1,500.00
Payment Must Be Received By: 03/25/--

Transaction Date	Post Date	Reference	Transaction Description		Payments & Credits	Loans, Fees, & Purchases
2/1	2/3	T970R194	Gale's Fashions	Peoria, IL		139.89
2/5	2/7	297B9875	Superfast Gas	Peoria, IL		15.68
2/10	2/12	N978T125	Pasta Garden Restaurant	Peoria, IL		18.25
2/15	2/16	M5980922	Gale's Fashions-Return	Peoria, IL	15.99	
2/17	2/18	9780H142	Boulevard Music Store	Springfield, IL		11.31
2/20	2/20	91089478	Annual membership fee			25.00
2/25	2/25	41978573	Payment Thank You		316.15	

Previous Balance	Purchases & Fees	Payments & Credits	Finance Charge	New Balance	Minimum Payment Due On Due Date	APR 18.9% Monthly Periodic Rate 1.575%
316.15	210.13	332.14	0.00		5.82	

On 2/15, a *credit* occurred when Midori *returned* part of the clothes she had bought from Gale's Fashions. Credits, like payments, are amounts that are subtracted from Midori's card balance.

The statement shows that Midori can borrow up to $1,500 from the company on a *cash advance.* Midori did not take any cash advances during February.

Midori paid no *finance charges* on the previous month's balance. That's because she paid her balance in full during the grace period.

The *statement closing date* is the last date on which transactions are posted to the statement. That date is 2/28 for Midori.

OneBank's finance charge rate is shown in the bottom right corner of the statement. The *annual percentage rate* is 18.9%. The *monthly periodic rate* is the annual percentage rate divided by 12, or 1.575%.

Midori's *new balance* can be found by taking the previous balance, adding purchases and fees plus finance charges, and subtracting payments and credits.

New Balance = Previous Balance + (Purchases + Fees + Finance Charges) − (Payments + Credits)

The *minimum payment due* that Midori must make to the card company is $5.82. Since she doesn't want to pay a finance charge, she plans to pay the new balance in full by the due date, 3/25.

EXAMPLE 2

Midori Masami received the statement shown on the previous page. What is her new balance?

SOLUTION

Use the formula for the new balance.

New Balance = $316.15 + ($210.13 + $0.00) − $332.14 = $194.14

✔CHECK YOUR UNDERSTANDING

C. Enrique Cruz's previous balance on his credit card was $0. He made $25 in purchases, and was charged an annual fee of $75. What should his new balance be?

D. Bev Donnigan's previous credit card balance was $122.36. She charged $114 in new purchases, made a payment of $122.36, and was charged no finance charges. What is her new balance?

Business *Tip*

If your credit card is used by someone else illegally, you can be held responsible for up to $50 per card. If your card is lost or stolen and you notify the card company before any purchases are made, you are not liable for any unauthorized purchases. If someone has stolen your credit card number and not the card, you are not liable for any unauthorized purchases.

Making Payments

Many credit card companies make customers' credit card statements and a record of current transactions available online. This allows customers 24-hour access to account activity. Many people check their account online regularly to check for *unauthorized charges*, which are charges that are on their account but that they didn't make.

Unauthorized charges can happen when a mistake is made or if someone has used your credit card or credit card number illegally.

Credit card users should keep sales receipts for credit card purchases to verify account activity. If any errors are found in a credit card statement or online record of transactions, a customer should contact the credit card company immediately.

Credit card payments can often be made by check or by an online payment from a bank account. If paying by check, you should leave plenty of time for the check to be received before the due date. Any online payments should be scheduled to be transacted on or before the due date.

Sales Receipt

Gale's Fashions, Inc.
2678 Furth St., Peoria, IL 61612
(309) 555-2525
THANK YOU

MERCHANT ID: 346184688345854
One Bank Card Sale/Swiped
Acct: ************1200 Exp: 10/12
Midori Masami

Amount $ 139.89

X *Midori Masami*
Midori Masami

Date: 02/01/-- Day: WED Time: 14:36
Authorized ticket: 564638

TOP COPY-MERCHANT
BOTTOM COPY-CUSTOMER

EXAMPLE 3

Midori compared her sales slips with the statement transactions. She found sales slips for 2/1, 2/5, 2/10, and 2/15 and noted that the amounts were correct. She didn't find a sales slip for the 2/17 transaction and knew she did not buy anything in Springfield during February. The 2/28 payment listed agreed with her checkbook register and she knew that her check was sent late. She also verified that her membership fee was due. What is Midori's correct new balance?

SOLUTION

Subtract the unauthorized purchase from the new balance on the statement.

$194.14 − $11.31 = $182.83   corrected new balance

✔CHECK YOUR UNDERSTANDING

E. When Vondel Bradshaw checked his credit card statement, he found that a sales slip dated 3/2 for $12.49 was posted as $12.99. He also found that a purchase for $56.29 dated 2/19 was unauthorized. If the new balance on the statement was $491.23, what is the correct new balance?

F. Sonja Erickson checked her credit card statement and found a sales slip for $48.99 that was unauthorized. She also found that a sales slip for $17.89 had been listed as $18.79. If the new balance shown on her statement was $208.66, what is her correct new balance?

Cost of Credit Card Use

The finance charges and fees you pay on your credit card can add up. You should calculate the total cost of your credit card to see if using one is of value to you and to compare the cost of your current card to other cards.

EXAMPLE 4

Danny O'Hare switched from the Clarion credit card to the First Bank credit card in April. When he did, he paid an annual membership fee of $50. He also paid a balance transfer fee of 2% of his old card's $420 balance. During the next 12 months, he paid an average monthly finance charge of $33.80 on his unpaid balance. What was Danny's total cost for using his credit card for the year?

SOLUTION

Multiply the monthly finance charge by 12 months.

12 × $33.80 = $405.60   total finance charge for year

Multiply the Clarion card balance by the balance transfer fee rate.

$420 × 0.02 = $8.40   balance transfer fee

Add the total finance charge, balance transfer fee, and membership fee.

$405.60 + $8.40 + $50 = $464   total cost of credit card for year

Business Tip

Stores are charged a fee by the credit card company for accepting credit card purchases. The fee may range between 1.5%–6% of the sale. The store may also be charged a fee for every credit card transaction.

✔CHECK YOUR UNDERSTANDING

G. Lili Favre opened a SkyMail credit card in January. She paid a membership fee of \$45 and a balance transfer fee of \$29 when she moved the balance of her old card to her SkyMail card. During the year, she paid these finance charges: January \$2.68; February \$7.28; June, \$9.22; October \$3.98. What was the total annual cost of the card to Lili?

H. Derwood Kant's credit card statement for May showed a membership fee of \$25, a late fee of \$29, a finance charge of \$3.15, and an over-the-limit fee of \$16. What was the total cost of the card to Derwood in May?

Wrap Up ▶▶▶

Robert may be paying for charges he did not make or fees he should not have to pay. Even though a computer does print the statements, unauthorized use of a credit card does occur. Robert should always check his statements.

Financial Responsibility

Debit cards that carry a credit card logo can be used in two ways. When the card is swiped and a PIN number is requested, the card is being used as a debit card. When the card is swiped and a signature is requested, the card is being processed as a credit card payment. These two types of transactions are processed differently by the bank. If a debit card is used like a credit card, the bank receives fees from the merchant for processing the card.

At many merchants if you use a debit card with a PIN number, you are charged a fee similar to fees charged at ATM machines. When you use a debit card as a credit card, no fee is charged to you.

Exercises

Perform the indicated operation.

1. \$660 ÷ 26 **2.** \$38.41 + \$93.26 + \$9.52 **3.** \$252 + \$63.14 + \$6.62

4. 48 ÷ 8 **5.** \$478 × 30 × 0.000628 **6.** \$350 × 15 × 0.000491

7. \$1,060 ÷ 20 **8.** \$2,617.84 − \$491.53 **9.** \$3,205.33 − \$1,493.59

Rewrite as a decimal.

10. 9.17% **11.** 0.0316%

Solve.

12. Lannie Ickerson checked her credit card statement and found a purchase for \$27.79 that was unauthorized. She also found that a sales slip for \$11.29 had been listed as \$12.19. If the new balance shown on her statement was \$107.09, what is her correct new balance?

13. John Rawlings credit card statement for June 30 showed a previous balance of \$248.67 and new purchases of \$59.89 on 6/10, \$15 on 6/15, and \$28.97 on 6/17. He made a payment of \$248.67 on 6/27. What is John's new balance?

Use the credit card statement for Ana Guzman to answer Exercises 14–16.

UniBank Card

Acct. No. 0200 0200 0000 0200
Statement Closing Date: 11/30/--
Credit Limit: $3,500

Ana Guzman
123 Presidents Place
El Paso, Tx 79915

Transaction Date	Post Date	Reference	Transaction Description		Payments & Credits	Loans, Fees, & Purchases
11/3	11/4	592k9781	Crestwood Gym	El Paso, TX		49.99
11/7	11/9	297B9875	Broadway Videos	El Paso, TX		12.79
11/10	11/12	N978T125	The Corner Gas Station	El Paso, TX		24.59
11/17	11/18	M5980922	Regal Department Store-Return	El Paso, TX	38.29	
11/21	11/23	91089478	Annual membership fee			45.00
11/24	11/24	41978573	Payment Thank You		100.00	
11/30	11/30	41827753	Balance from Transcredit Card			267.88
11/30	11/30	41827756	Balance transfer fee			29.00

For customer service call: 1-800-555-7800
Cash Advance Limit: $1,000.00
Payment Must Be Received By: 12/25/--

Previous Balance	Purchases & Fees	Payments & Credits	Finance Charge	New Balance	Minimum Payment Due On Due Date	APR 18.0% Monthly Periodic Rate 1.5%
249.25	429.25	138.29	3.74		10.88	

14. By what date must she make the minimum payment? What was the date and amount of her last payment? What is her credit limit?

15. How much finance charge does she owe? How much money did she transfer from the Transcredit card to her UniBank card? What fee did she pay for the balance transfer?

16. What should Ana Guzman's new balance be on her credit card?

Solve.

17. Sandra Beal opened a new credit card in January. She paid a membership fee of $15 and a balance transfer fee of 5% when she moved $823.46 from her old card to her new card. During the year, she paid these finance charges: February $3.56; May, $5.82; July, $4.92; and September $2.18. What was the total annual cost of the card to Sandra?

18. Rick Chandler's credit card statements for the year showed a membership fee of $75, two late fees of $25, and an average finance charge of $23.75 a month. What was the total annual cost of the card to Rick?

Solve.

19. Heng-che Pai's credit card statement for May showed a previous balance of \$289.16, new purchases of \$107.99, a membership fee of \$35, a finance charge of \$5.96, and a payment of \$100. What is her new balance?

20. The credit card of Salizar Mendoza for April listed a previous balance of \$419.65, new purchases of \$283.15, a payment of \$300, a finance charge of \$7.23, and a late fee of \$20. What is his new balance?

21. **STRETCHING YOUR SKILLS** BankNote Credit Card Company must pay a store \$409,800 this month for sales the store's customers made using the BankNote credit card. Before paying, BankNote will deduct from the store's check a 3.5% merchant discount fee from the total sales. BankNote will also deduct a \$0.20 transaction processing fee for each of the 21,283 BankNote credit card transactions made at the store during the month. What net amount will the store receive from BankNote?

Photodisc/Getty Images

22. **STRETCHING YOUR SKILLS** Roslynn Rheinhart bought a wood chipper priced at \$575 and received a 4% discount for paying cash instead of using a credit card. What did Roslynn pay for the chipper?

23. **CRITICAL THINKING** Many credit card companies use incentives like offering you rewards or cash back for using their credit card. How can a credit card company afford to give away rewards or give money back?

24. **FINANCIAL DECISION MAKING** You are considering different credit card offers. One card has an APR of 12% on purchases and an annual fee of \$25. The other card has an APR of 15% and no annual fee. Which card would you choose and why?

Mixed Review

25. Write $302\frac{1}{4}\%$ as a decimal.

26. \$4.90 is what percent greater than \$4.20?

27. What amount is $62\frac{1}{2}\%$ smaller than \$88?

28. Gary Feliciano transferred a balance of \$900 to a new credit card with a lower APR. He paid a 3% balance transfer fee. How much was the fee?

29. Carmen Dize used an ATM to deposit a check for \$361.90 and to withdraw \$250 in cash. If her starting bank balance was \$739.18, what is her new balance?

30. On June 30, Tina Nader's check register balance was \$452.88 and her bank statement balance was \$697.55. Checks outstanding were 561, \$39.28; 562, \$121.31; and 564, \$83.19. The statement showed earned interest of \$0.89. Reconcile the check register and bank statement.

31. Ira Morganstein's tax return last year showed gross income of \$56,312 and adjustments to income of \$2,184. What was Ira's adjusted gross income last year?

4-2

Credit Card Finance Charges

GOALS

- Calculate finance charges using a daily or monthly periodic rate
- Calculate finance charges using previous balance method
- Calculate finance charges using adjusted balance method

KEY TERMS

- periodic rate
- previous balance method
- adjusted balance method

Start Up

Holly Winter is comparing two credit card offers. The APR and the annual fees are the same on the two credit cards. What else should Holly consider to evaluate which card is best for her?

Purestock/Jupiter Images

Math Skill Builder

Review these math skills and solve the exercises that follow.

1. **Add** and **subtract** money amounts.
   Find the sum. $209.34 + $345.12 + $16.54 − $516.89 − $28.76 = $25.35

   1a. $507.22 + $397.28 − $44.20 − $579.93

   1b. $183.02 + $97.38 − $88.73 − $38.99

2. **Multiply** money amounts by decimals.
   Multiply: $62.58 × 0.0021 = $0.13

   2a. $398.77 × 0.000673 **2b.** $220.81 × 0.01975

   2c. $710.29 × 0.2978 **2d.** $1,297.55 × 0.008271

Business *Tip*

The monthly or daily finance charge rate on a credit card balance is often called the *periodic rate*. The month for which you are billed is often called the *billing period*.

Monthly and Daily Periodic Rates

If you don't pay your card balance in full by the due date, you will be assessed a finance charge. You also will lose the *grace period* for new purchases. Finance charges will be charged on new purchases from the day they are made.

The finance charge rates on credit cards are advertised by the annual percentage rate (APR), but finance charges are calculated using a monthly or daily **periodic rate**. To find the monthly or daily periodic rate, divide the APR by 12 or 365 and round to the nearest ten-thousandth.

Periodic Finance Charge = Balance Subject to Finance Charge × Periodic Rate × Number of Periods

EXAMPLE 1

Jacquelyn is charged a finance charge on a credit card balance of $500.00. Her card has an APR of 15%. What will her finance charge be if the company uses a monthly periodic rate? What will her finance charges for the month be if the company uses a daily periodic rate for a 31 day billing cycle?

SOLUTION

Find the monthly and daily periodic rates, rounded to the nearest ten-thousandth.

$15\% \div 12 = 1.25\%$; $15\% \div 365 = 0.0411\%$

Rewrite the periodic rates as decimals.

$1.25\% = 0.0125$; $0.0411\% = 0.000411$

Use the finance charge formula for each rate.

Finance Charge = $500 × 0.0125 × 1 = $6.25   using monthly periodic rate

Finance Charge = $500 × 0.000411 × 31 = $6.37   using daily periodic rate

The finance charge with a monthly periodic rate is $6.25. The finance charge with a daily periodic rate is $6.37.

> **Math *Tip***
>
> If a monthly periodic rate is used, divide the APR by 12.
>
> If a daily periodic rate is used, divide the APR by 365.
>
> Round the periodic rate percent to the nearest ten-thousandth.

✔CHECK YOUR UNDERSTANDING

A. The balance of Sue Millis' credit card that is subject to finance charges is $221.68. Her card has an APR of 18% and uses a monthly periodic rate. What are her current month's finance charges?

B. Roland Plewka must pay finance charges on a balance of $1,587. The credit card has an APR of 21%, and uses a daily periodic rate. What will he be charged in finance charges for a 30 day billing cycle?

Previous Balance Method

The amount of the finance charge depends on your periodic rate and how the card company figures the balance subject to finance charges. This balance can be found by several methods.

Laura Solon's credit card statement is shown on the next page. The finance charge, new balance, and minimum payment due boxes in the statement are gray because these amounts will vary with the method used to find the balance on which the finance charge will be applied.

The **previous balance method** charges interest on the balance in the account on the last billing date of the previous month. Any payments, credits, or new purchases in the current month are not included in the previous balance. Use the formula below to find the new balance.

New Balance = Previous Balance + (Finance Charge + New Purchases + Fees) − (Payments + Credits)

Trans-action Date	Post Date	Reference	Transaction Description		Payments & Credits	Loans, Fees, & Purchases
10/3	10/4	3165813T	Cardinal Shoe Stores, Inc.	Detroit, MI		128.99
10/7	10/9	4381R211	Vorax Gas Stations, Inc.	Detroit, MI		21.89
10/10	10/12	4Y659762	The Pasta Barn, Inc.	Detroit, MI		27.79
10/17	10/18	4897W544	Cardinal Shoe Stores-Return	Detroit, MI	35.99	
10/18	10/19	81976534	Annual membership fee			35.00
10/24	10/24	94681322	Payment Thank You		75.00	

Previous Balance	Purchases & Fees	Payments & Credits	Finance Charge	New Balance	Minimum Payment Due On Due Date	APR
225.60	213.67	110.99				18.000%

EXAMPLE 2

Laura Solon's card company uses the previous balance method and a monthly periodic rate to figure the finance charge. Find the finance charge for the month and the new balance.

SOLUTION

Find the monthly periodic rate and rewrite it as a decimal.

$18\% \div 12 = 1.5\%$; $1.5\% = 0.015$

Finance Charge $= \$225.60 \times 0.015 \times 1 = \$3.384$, or $3.38

Add to find the new balance.

$\$225.60 + \$3.38 + \$213.67 - \$110.99 = \$331.66$

Math *Tip*

If a monthly periodic rate is used, the number of periods is 1.

If a daily periodic rate is used, the number of periods is the number of days in the billing cycle.

✔CHECK YOUR UNDERSTANDING

C. John Olden's credit card statement for April showed a previous balance of $309.20, new purchases and fees of $128.45, and payments and credits of $75. The card's annual percentage rate is 24%. What is John's finance charge for April and new balance using the previous balance method and a monthly periodic rate?

D. Sandra Minoro's credit card company uses the previous balance method to calculate finance charges. Its APR is 21% and a daily periodic rate, rounded to the nearest ten-thousandth of a percent, is used. Sandra's credit card statement for June showed a previous balance of $488.32, new purchases and fees of $264.89, and payments and credits of $300. What is Sandra's finance charge for the 30 days in June and her new balance?

Adjusted Balance Method

The **adjusted balance method** subtracts payments and credits during this month from the balance at the end of the previous month. Purchases and fees made during the current month are not included in the adjusted balance.

Adjusted Balance = Previous Balance − (Payments + Credits)

New Balance = Adjusted Balance + Finance Charge + New Purchases + Fees

EXAMPLE 3

Suppose that Laura's card company uses the adjusted balance method and a monthly periodic rate. Find the finance charge for the month and the new balance.

SOLUTION

Find the adjusted balance.

Adjusted Balance = \$225.60 − \$110.99 = \$114.61

Find the periodic finance charge and rewrite it as a decimal. Then find the periodic finance charge.

18% ÷ 12 = 1.5%; 1.5% = 0.015

Periodic Finance Charge = 0.015 × \$114.61 = \$1.719, or \$1.72

Find the new balance.

New Balance = \$114.61 + \$1.72 + \$213.67 = \$330

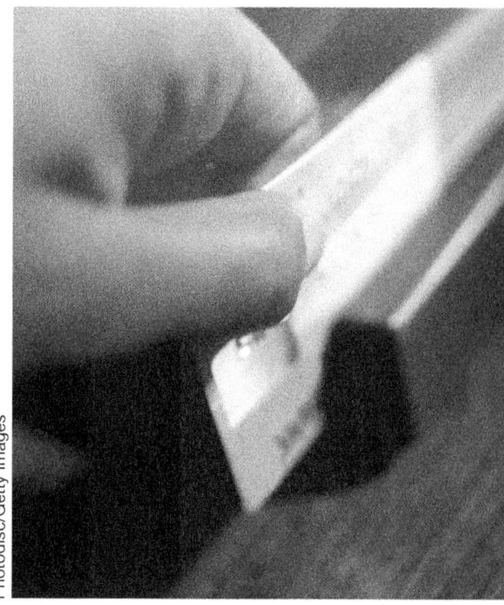
Photodisc/Getty Images

✔ CHECK YOUR UNDERSTANDING

E. Yossi Hussein uses a credit card that carries an 18% APR and uses the adjusted balance method for calculating finance charges. Yossi's statement listed these facts: previous balance, \$310.33; purchases, \$219.67; fees, \$75; payments, \$150; and credits, \$62.69. If the credit card company uses a daily periodic rate, what is Yossi's finance charge and new balance on a 31 day billing cycle?

F. Ricky Luciano's credit card statement showed a previous balance of \$166.98, purchases and fees of \$201.88, and payments and credits of \$75. If his card carried an APR of 21% and used the adjusted balance method with a monthly periodic rate to calculate finance charges, what is Ricky's finance charge and new balance?

Wrap Up ▶▶▶

Although the two credit card offers may look the same since they have the same APR and the same annual fee, they can still vary in many other ways. Two examples are whether the card uses a daily or monthly periodic rate, and how the credit card company finds the balance subject to finance charges.

TEAM Meeting

In most bank lobbies, there is a place where brochures about bank services are available free to the public. Among these brochures is usually a stack of brochures explaining how to apply for a credit card, including the credit card features and terms. With two other students,

- Obtain copies of these brochures from several banks or visit bank web sites for information.
- After reading the brochures, make a chart listing the features common to all credit cards offered by these banks, such as balance transfer fees, late charges, and the method of calculating the balance and finance charge.
- Create columns to let you enter data for each credit card's features.
- Prepare a brief explanation of the chart that includes a discussion of the differences among the cards.

Exercises

Find the result.

1. $98.62 + $978.22 − $34.15 − $98.18

2. $789.23 + $98.21 − $44.63 − $641.09

Find the product.

3. $879.43 × 0.00526

4. $2,097.46 × 0.0002978

Complete the table below.

	Balance Subject to Finance Charge	APR	Period	Periodic Rate	# of Periods	Finance Charge
5.	$213.57	15%	Monthly		1 Month	
6.	$89.22	18%	Daily		30 Days	
7.	$866.73	12%	Monthly		1 Month	
8.	$2,479.01	21%	Daily		31 Days	
9.	$500.00	9%	Monthly		1 Month	
10.	$8,233.86	10%	Daily		30 Days	

Solve.

11. The October credit card statement for Genaro Rios had a previous balance of $175.30, new purchases and fees of $108.85, and payments and credits of $125. The card's annual percentage rate is 21% and the previous balance method is used with a monthly periodic rate. What is Genaro's finance charge for October and new balance?

12. Toni Bando's credit card has an APR of 18% figured on the previous balance. The previous balance on Toni's credit card statement for July was $308.88. During July she had new purchases and fees of $276.49, and payments and credits of $400. If the company uses a daily periodic rate for 31 days, what is her finance charge for July and her new balance?

Solve.

13. Otto Schein's credit card company charges an APR of 21% applied at a monthly periodic rate on the previous balance. Otto's December statement showed: previous balance, $397.90; new purchases, $341.89; fees, $55; payments, $500; and purchase return, $56.99. What is his finance charge for December and new balance?

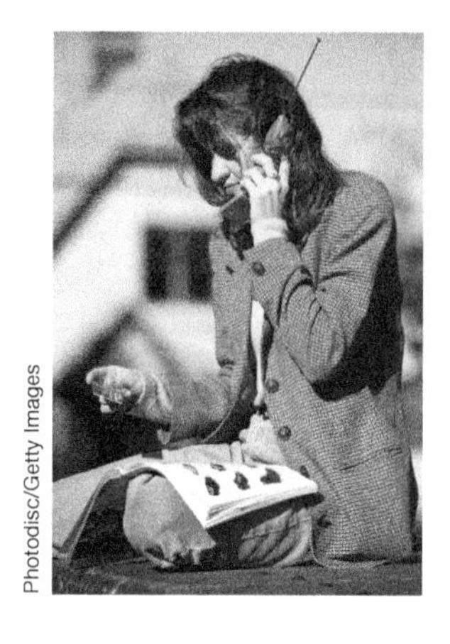
Photodisc/Getty Images

14. A credit card company uses an APR of 15% and the adjusted balance method of computing finance charges using a monthly periodic rate. A credit card statement from the company lists the following: previous balance, $601.87; purchases, $209.88; fees, $75; payments, $400; and credits, $25. What was the finance charge for the month and the new balance?

15. June Christo has a credit card statement that shows an adjusted balance of $598.61, new purchases and fees of $127.88, and payments and credits of $250. Her card company charges an APR of 12% using a monthly periodic rate on the adjusted balance. What is June's finance charge and new balance?

16. **INTEGRATING YOUR KNOWLEDGE** You have two credit cards. The Banker's Card has a previous balance of $301.55, carries an APR of 18% applied using a monthly periodic rate, and uses the previous balance method of figuring finance charges. The MallCard lists a previous balance of $260.61 and payments and credits of $175. It uses the adjusted balance method to find finance charges and carries an APR of 21% with a monthly periodic rate. Find the finance charge on both cards.

17. **FINANCIAL DECISION MAKING** Your credit card statement shows an adjusted balance of $231.86, payments and credits of $125, and purchases and fees of $175.66. Your current card company uses an APR of 15% and the adjusted balance method. Another credit card company that sent you an application in the mail also uses an APR of 15% but uses the previous balance method. If both cards use a monthly periodic rate and have the same annual fee, should you switch companies? Why or why not?

18. **CRITICAL THINKING** Why will the adjusted balance method usually give a lower periodic finance charge than the previous balance method? When would the periodic finance charges be the same for both methods?

Mixed Review

19. Write the ratio of 16 to 236 as a fraction.

20. Write $\frac{8}{24}$ as a percent.

21. Write 175% as a decimal.

22. Rosa Rinaldi's total job benefits for the previous year were estimated to be $62,976. However, her job expenses for the same job were: licenses, $475; commuting costs, $2,108; tools, $197. What were her net job benefits for the year?

23. Maria Hernandez deposited these items on January 11: (bills) 15 twenties, 11 tens, 20 fives, 89 ones, (coins) 35 quarters, 65 dimes, 135 pennies, (checks) $53.69, and $138.98. She received one $100 bill back. What was her net deposit?

4-3

Average Daily Balance Method

GOALS

- Calculate finance charges using average daily balance method

KEY TERM

- average daily balance method

Start Up

Ling Yi is planning on getting her first credit card while she is in college. She does not plan on carrying a balance from month to month. When she looks for a credit card, what should be her priority?

Stephen Coburn 2008/Shutterstock.com

Math Skill Builder

Review these math skills and solve the exercises that follow.

1. **Add** and **subtract** money amounts.
   Find the sum. $65.98 + $486.22 + $684.81 = $1,237.01

   1a. $433.92 + $87.36 + $289.11 **1b.** $582.33 + $86.29 − $46.33

2. **Divide** money amounts by whole numbers.

   2a. $5,832.87 ÷ 31 **2b.** $31,875.32 ÷ 30

3. **Multiply** money amounts by decimals.
   Muliply: $62.58 × 0.0021 = $0.13

   3a. $486.31 × 0.000673 **3b.** $182.65 × 0.01975

   3c. $638.19 × 0.2978 **3d.** $1,856.31 × 0.008271

Average Daily Balance Method

The **average daily balance method** is the most commonly used method for calculating the finance charge on a credit card. When this method is used, the periodic rate is applied to the average daily balance in the account during the billing period. The dates used are the *post* dates. The card company starts with the beginning balance for each day.

The company subtracts any payments or credits posted during that day from the beginning balance. Fees posted for that day are added to the balance. If the credit card company is using the *average daily balance method including new purchases* new purchases will be added as well. If the credit card company is using the *average daily balance method excluding new purchases* then new purchases will not be added to the daily balance. The ending balances for every day are then totaled and divided by the number of days in the billing period to get the *average daily balance.*

The following formulas are used for finding the average daily balance when new purchases are included in the calculations.

$$\textbf{Daily Balance with New Purchases} = \textbf{Beginning Balance} - (\textbf{Payments} + \textbf{Credits}) + (\textbf{Purchases} + \textbf{Fees})$$

$$\textbf{Average Daily Balance} = \frac{\textbf{Sum of Daily Balances}}{\textbf{Number of Days in the Billing Cycle}}$$

$$\textbf{New Balance} = \textbf{Previous Balance} + (\textbf{Finance Charges} + \textbf{New Purchases} + \textbf{Fees}) - (\textbf{Payments} + \textbf{Credits})$$

EXAMPLE 1

Jake Ramiro's credit card company uses the average daily balance method including new purchases to figure the finance charge. The APR is 18% on new purchases, using a monthly periodic rate. Find the finance charge for the month and the new balance.

SOLUTION

Create a chart like the one below. A transaction is either a payment or credit and subtracted from the balance, or a purchase or fee and added to the balance. The balance at the end of a day is the previous balance plus or minus any additions or deductions. The Number of Days is the number of days that the balance is in effect. The Sum of Daily Balances is the balance multiplied by the number of days.

Post Date	Transactions	Balance at End of Day	Number of Days	Sum of Daily Balances
10/1 (Bal.)	0.00	225.60	1	225.60
10/2–10/3	0.00	225.60	2	451.20
10/4	+128.99	354.59	1	354.59
10/5–10/8	0.00	354.59	4	1,418.36
10/9	+21.89	376.48	1	376.48
10/10–10/11	0.00	376.48	2	752.96
10/12	+27.79	404.27	1	404.27
10/13–10/17	0.00	404.27	5	2,021.35
10/18	−35.99	368.28	1	368.28
10/19	+35.00	403.28	1	403.28
10/20–10/23	0.00	403.28	4	1,613.12
10/24	−75.00	328.28	1	328.28
10/25–10/31	0.00	328.28	7	2,297.96

Problem Solving *Tip*

Finding the average daily balance is another use of weighted averages.

Add the column, "Sum of Daily Balances," and divide by 31, the number of days the statement covers.

$225.60 + $451.20 + $354.59 + $1,418.36 + $376.48 + $752.96 + $404.27 + $2,021.35 + $368.28 + $403.28 + $1,613.12 + $328.28 + $2,297.96 = $11,015.73

$11,015.73 ÷ 31 = $355.346, or $355.35 average daily balance

Find the monthly periodic rate and use the periodic finance charge formula to find the finance charge.

18% ÷ 12 = 1.5% = 0.015 monthly periodic rate

Finance charge = $355.35 × 0.015 = $5.33

Find the new balance.

New Balance = $225.60 + ($5.33 + $213.67) − ($35.99 + $75) = $333.61

✔ CHECK YOUR UNDERSTANDING

A. Jade Hameed's credit card statement for August showed these items: 8/1, previous balance, $108.15; 8/5, purchase, $56.89; 8/10, purchase, $61.88; 8/14, purchase, $190.23; and 8/25, payment, $150. Jade's card company uses a 1.6% monthly periodic rate and the average daily balance method including new purchases. What is Jade's finance charge for August and the new balance?

B. The credit card statement of Gloria Herrera for January listed these items: 1/1, previous balance, $89.27; 1/5, purchase, $159.34; 1/9, purchase, $108.45; 1/24, payment, $150; and 1/28, fee, $25. The card company uses the average daily balance method including new purchases and a daily periodic rate of 0.000575. What is Gloria's finance charge for January and what is her new balance?

Problem Solving *Tip*

The periodic finance charge formula is the same no matter which method is used to find the balance subject to finance charge.

Periodic Finance Charge = Balance Subject to Finance Charge × Periodic Rate × Number of Periods

When a credit card company uses the average daily balance method excluding new purchases, the same method is used except daily balances do not include any new purchases.

The following formulas are used with finding the average daily balance when new purchases are not included in the calculations.

Daily Balance Excluding New Purchases = Beginning Balance − (Payments and Credits) + Fees

$$\textbf{Average Daily Balance} = \frac{\textbf{Sum of Daily Balances}}{\textbf{Number of Days in the Billing Cycle}}$$

New Balance = Previous Balance + (Finance Charges + New Purchases + Fees) − (Payments + Credits)

EXAMPLE 2

Jake Ramiro's credit card company uses the average daily balance method excluding new purchases to figure the finance charge. Find the finance charge for the month and the new balance. The transactions on 10/4, 10/9, 10/12 are purchases. The transaction on 10/19 is a fee. (Refer to Example 1.)

SOLUTION

Create a chart like Example 1, excluding the transactions on 10/4, 10/9, and 10/12.

Post Date	Transactions	Balance at End of Day	Number of Days	Sum of Daily Balances
10/1 (Bal.)	0.00	225.60	1	225.60
10/2–10/17	0.00	225.60	16	3,609.60
10/18	−35.99	189.61	1	189.61
10/19	+35.00	$224.61	1	$224.61
10/20–10/23	0.00	$224.61	4	$898.44
10/24	−75.00	149.61	1	149.61
10/25–10/31	0.00	149.61	7	1,047.27

Add the column, "Sum of Daily Balances" and divide by 31, the number of days in the billing cycle.

$\$225.60 + \$3{,}609.60 + \$189.61 + \$224.61 + \$898.44 + \$149.61 + \$1{,}047.27 = \$6{,}344.74$

$\$6{,}344.74 \div 31 = \$204.67$ average balance excluding new purchases

Find the daily periodic rate and rewrite as a decimal.

$15\% \div 365 = 0.0411\%;\ 0.0411\% = 0.000411$

Periodic Finance Charge $= \$204.67 \times 0.000411 \times 31 = \$2.61$

Find the new balance.

New Balance $= \$225.60 + (\$2.61 + \$213.67) - (\$35.99 + \$75) = \$330.89$

✔CHECK YOUR UNDERSTANDING

C. In January, Lincoln Hurt's credit card statement has a beginning balance of $498.67. He made a payment of $200 on the 25th of the month. If the credit card company uses an average daily balance method excluding new purchases with a monthly periodic rate of 1.5%, what are the finance charges?

D. Jorge Itarra has a credit card with a beginning balance on 3/1 of $221.48. He made a payment of $150 on 3/15. He made other purchases during the month totaling $183.46. If the credit card has a daily periodic rate of 0.0575% applied to the average daily balance excluding new purchases, what are the finance charges for March? What is the new balance?

Problem Solving *Tip*

Remember that finance charges include interest plus fees.

Wrap Up ▸▸▸

If she plans to pay off her credit card each month, then a low annual fee is more important than the APR. However, the APR should be a secondary consideration in case she has some months where she is unable to pay the entire balance.

Exercises

Find the result

1. $1,668.43 ÷ 30
2. $2,721.24 ÷ 31
3. $984.56 × 0.000575
4. $221.67 × 0.15

When Glorica Batic received her May credit card statement she found these items listed: 5/1, previous balance, $281.59; 5/7, purchase, $168.99; 5/10, purchase, and $57.98; 5/25, payment, $200. Glorica's card company uses a 1.8% monthly periodic rate.

5. Find the finance charge if Glorica's card company uses the average daily balance including new purchases.
6. Find Glorica's new balance if the card company uses the average daily balance including new purchases.
7. Find the finance charge if Glorica's card company uses the average daily balance excluding new purchases.
8. Find Glorica's new balance if the card company uses the average daily balance excluding new purchases.

The credit card statement of Luiz Lopea for June listed these items: 6/1, previous balance, $193.39; 6/11, purchase, $175.39; 6/15, purchase, $71.84; and 6/24, payment, $75. The card company uses a daily periodic rate of 0.056%.

9. What is Luiz's finance charge for June and what is his new balance if the card company uses the average daily balance including new purchases?
10. What is Luiz's finance charge for June and what is his new balance if the card company uses the average daily balance excluding new purchases?
11. **FINANCIAL DECISION MAKING** Your credit card company uses the average daily balance method excluding new purchases. Your credit card bill is due on the 15th day of the month, and you get paid on the first of the month. How will it impact your periodic finance charges if you make an online payment on the 2nd day of the month instead of the 15th?

Mixed Review

12. Write 0.0293% as a decimal.
13. Round to the nearest ten-thousandth: 0.0583569
14. Find the average: 85, 73, 92, 77
15. Julia Hazlett earns a yearly salary of $54,000. What are her gross monthly earnings?

4-4

Cash Advances

GOALS
- Calculate total finance charges on cash advances
- Calculate credit card balances that include cash advances

KEY TERMS
- cash advance
- total finance charge

Start Up

Joaquim received cash convenience checks with his credit card statement. Printed with the checks was the statement: "GET CASH WHEN YOU NEED IT! Use the attached checks for your Cash Advance needs." Joaquim is considering using the checks to pay for a stereo that he has had a hard time saving the money for. How would you advise Joaquim?

Losevsky Pavel 2008/Used under license from Shutterstock.com

Math Skill Builder

Review these math skills and then answer the questions that follow.

1. **Divide** numbers and round.
Find the quotient and round to the nearest ten-thousandth.
$22 \div 12 = 1.8333$

1a. $14 \div 365$ **1b.** $13 \div 12$
1c. $16 \div 365$ **1d.** $18 \div 12$
1e. $22 \div 365$ **1f.** $5 \div 12$

2. **Rewrite** percents as decimals.
Rewrite 1.0835% as a decimal. 0.010835

2a. 0.0575% **2b.** 1.86% **2c.** 0.0423%
2d. 3.65% **2e.** 0.0683% **2f.** 0.235%

3. **Multiply** money amounts by percents.
Find the product. $\$404 \times 7\% = \$404 \times 0.07 = \$28.28$

3a. $\$249.00 \times 4\%$ **3b.** $\$478.83 \times 7.2\%$ **3c.** $\$219.32 \times 1.45\%$
3d. $\$875.49 \times 3.8\%$ **3e.** $\$1,500 \times 6\%$ **3f.** $\$346.29 \times 2.85\%$

Math *Tip*

To rewrite a percent as a decimal, move the decimal point two places to the left and drop the percent sign.

Finance Charges on Cash Advances

When you can get cash from an ATM using your credit card, or by using checks provided by the credit card company, you are borrowing money from the credit card company. Transactions of this type are called **cash advances**.

The charges associated with cash advances are high. The periodic finance charges begin on the day you withdraw the money and there is typically no grace period for cash advances.

The APR for cash advances is higher than the APR for purchases. In addition, credit card companies often charge a fee for the cash advance.

To find the periodic finance charge, use the same formula as periodic finance charges on a regular balance. A daily periodic rate is usually applied for the number of days in the billing cycle since the withdrawal was made. The **total finance charges** for the cash advance is the sum of the periodic finance charges and any fee charged for the cash advance.

Periodic Finance Charge = Balance Subject to Finance Charge × Periodic Rate × Number of Periods

Total Finance Charges = Periodic Finance Charges + Fees

EXAMPLE 1

Benita Moya borrowed $500 for 20 days on his credit card using a cash advance. His card company charged a cash advance fee of 4% of the cash advance and a daily periodic interest rate of 0.0573%. What was the total finance charge on the cash advance?

SOLUTION

Rewrite the daily interest rate as a decimal: 0.0573% = 0.000573

Find the periodic finance charge and the fee for the cash advance.

Periodic Finance Charge = $500 × 0.000573 × 20 = $5.73

Fee = $500 × 0.04 = $20

Total Finance Charges = $5.73 + $20 = $25.73

> **Problem Solving *Tip***
>
> Remember that total finance charges include the periodic finance charges plus fees.

✔CHECK YOUR UNDERSTANDING

A. Vera Millay used her credit card in an ATM to borrow $200 on a cash advance. Her card company charged a cash advance fee of $5 and a daily periodic interest rate of 0.0487%. If Vera paid the cash advance and finance charges back at the end of 25 days, what was the total finance charge on the cash advance?

B. Akvar Assam borrowed $150 on a cash advance from his credit card company. The card company charged a cash advance fee of 3% and a daily periodic interest rate of 0.058% for the 35 days he had the cash advance. What total amount did Akbar need to pay off the cash advance and finance charges?

Credit Card Balances with Cash Advances

Finance charges for credit card accounts that have a balance with both purchases and cash advances must be figured separately because the rate for cash advances is higher than the rate for purchases.

In addition, most credit card companies apply payments to the purchase balance first, before the cash advance balance. This means that in order to pay off a cash advance, you must pay the entire balance of the credit card, or your payments will not be applied toward the cash advance and you will continue to pay the higher rate of interest on the cash advance balance.

EXAMPLE 2

Jolinda's credit card has an APR of 14% for purchases and 21% for cash advances. They use an average daily balance method with a daily periodic rate for purchases. They use a daily periodic rate for cash advances and a cash advance fee of 3%. In a 31 day billing cycle, Jolinda has an average daily balance for purchases of $543.26. She took a cash advance of $200 during the billing cycle and must pay finance charges for 26 days. What are her total finance charges?

Simone van den Berg 2008/Shutterstock.com

SOLUTION

Find the daily periodic rates for purchases and for cash advances. Rewrite as decimals.

Purchases: $14\% \div 365 = 0.0384\% = 0.000384$ periodic rate for purchases

Cash advances: $21\% \div 365 = 0.0575\% = 0.000575$ periodic rate for cash advances

Find the periodic finance charges and fees for the purchases and for the cash advance.

Purchases: $\$543.26 \times 0.000384 \times 31 = \$6.47$ periodic finance charges for purchases

Cash advance: $\$200 \times 0.000575 \times 26 = \$2.99$ periodic finance charge for cash advance

Cash advance: Fee $= \$200 \times 0.03 = \$6$ cash advance fee

Total Finance Charges $= \$6.47 + \$2.99 + \$6 = \$15.46$

✔ CHECK YOUR UNDERSTANDING

C. Jolinda made a payment on her credit card, but she did not pay off the balance. The next month, she has an average daily balance for purchases of $263.98. What are her total finance charges if it is a 30 day billing cycle?

D. It takes Jolinda a total of 15 months to pay off her credit card. On the cash advance, she paid the first month's fee and periodic finance charge plus an average of $3.50 per month for the remaining months. How much did she pay in finance charges on the $200 cash advance?

Business *Tip*

Credit card companies apply your payments to your purchase balance before your cash advance balance.

Wrap Up ▸▸▸

Joaquim can use the cash convenience checks, but they will most likely come with a hefty price in finance charges. In fact, he will probably pay more in finance charges than if he charged the price of the stereo on his credit card and made payments.

Exercises

Divide and round to the nearest ten-thousandth.

1. 19 ÷ 12
2. 16 ÷ 365

Rewrite the percent as a decimal.

3. 1.8745%
4. 0.0683%

Find the product.

5. $550 × 6%
6. $685 × 2.4%
7. $485 × 4.6%
8. $856 × 2.5%
9. Jareen Knabe borrowed $800 for 30 days from her credit card company using a cash advance. The daily finance charge was 0.0543%. What was the periodic finance charge on her cash advance?

Yvonne Clark used her credit card in an ATM for a $325 cash advance. Her card company charges a daily periodic interest rate of 0.06% and a cash advance fee of $19. Yvonne repaid the cash advance in 15 days.

10. What total finance charge did Yvonne pay for the cash advance?
11. How much did she pay back to the credit card company in total?

Isabel Aponte borrowed $150 on a cash advance from her credit card company. The company charges a 4% fee and an APR of 21%, calculated using a daily periodic rate, on cash advances. Isabel paid the money back in 26 days.

12. What was the total finance charge on the cash advance?
13. What total amount did Isabel pay back?

Bryan's credit card has an APR of 12% for purchases and 19% for cash advances. They use a daily periodic rate for purchases and cash advances. There is a cash advance fee of 3%. In a 31 day billing cycle, Bryan's purchase balance that is subject to finance charges is $286.26. In addition, he took an advance of $500 during the billing cycle and must pay finance charges on it for 15 days.

14. What are the periodic finance charges for the purchases?
15. What are the total finance charges for the cash advance?
16. What are the total finance charges?
17. Bryan did not pay off his credit card, and the next month the purchase balance subject to finance charges was $622.86. What are the total finance charges for a 31 day billing cycle?
18. **CRITICAL THINKING** Why is it advantageous for a credit card company to apply your payments to the purchase balance before the cash advance balance?

19. **FINANCIAL DECISION MAKING** Write guidelines for how you would use a cash advance. Include when you might take out a cash advance and how you will handle paying it back.

20. **STRETCHING YOUR SKILLS** Lupe's credit card statement for October listed these items: 10/1, previous balance, \$243.86; 10/5, purchase, \$45.21; 10/8, cash advance, \$350; 10/12, purchase, \$129; and 10/25, payment, \$75. The credit card company uses the average daily balance method including new purchases. The card carries an APR of 15% on purchases and 20% on cash advances. There is a 4% fee for cash advances. If the company uses a daily periodic rate to calculate finance charges, what are the finance charges for October and what is Lupe's new balance?

Mixed Review

21. \$25 × 100
22. \$8.25 × 1,000
23. \$5.43 × 10
24. \$43.86 + \$22.57
25. \$500 × 20%
26. 15% of \$250

Sharon Donald has a credit card with an APR on purchases of 16%. The balance transfer APR is 8% and there is a 2% balance transfer fee.

27. How much will she pay in fees to transfer \$500 to the card?

28. What is the daily periodic rate on purchases, rounded to the nearest ten-thousandth?

29. If Sharon has a balance of \$227.83 that is subject to daily periodic finance charges for a 31-day billing cycle, how much will she pay in periodic finance charges?

30. Lung Shen used his credit card in an ATM to borrow \$300 on a cash advance. His card company charged a cash advance fee of \$15 and a daily periodic interest rate of 0.049%. Lung paid the cash advance back at the end of 15 days. What was the total finance charge on the cash advance?

31. Johanna has a 3-year CD for \$5,000 that pays 5.5% annual interest. She cashes the CD after 1 year and pays a penalty of 6 months' interest. How much is the penalty?

Stockbyte/Getty Images

32. Alfred's paycheck has 3% withheld for state taxes. If his gross yearly wages are \$35,000, how much will be withheld for state taxes?

33. Dahlia earns a salary of \$50,000. What are her gross bi-monthly wages?

34. Tom Ito works 10 hours on Monday, 5 hours on Tuesday, 9 hours on Wednesday, 8 hours on Thursday, and 12 hours on Friday. His employer pays \$10.50 per hour for a 40 hour work week. Any overtime hours are paid at time-and-a-half. What are Tom's gross wages for the week?

4-5

Debt Management

GOALS

- Analyze a credit card account when minimum payments are made
- Calculate debt-to-income ratio

KEY TERMS

- debt
- debt-to-income ratio
- credit score
- consolidation loan

Start Up ▶▶▶

In her first month of having a credit card, Julianna charged $500. She was pleased to see on her statement that the minimum payment was only $7.50. She plans to make minimum payments each month until she pays the credit card off. What advice would you give Julianna?

Purestock/Jupiter Images

Math Skill Builder

Review these math skills and then answer the questions that follow.

1. **Add** and **subtract** money amounts.
   Find the total. $582.41 + $12.65 + $187.15 − $27.59 = $754.62

   1a. $432.91 + $65.81 + $4.55 − $200 **1b.** $103.11 + $26.43 + $1.15 − $75

2. **Multiply** money amounts by percents and round the product to the nearest dollar.
   Find the product. $221.67 × 3% = $221.67 × 0.03 = $6.65, or $7.00

   2a. $1,369.00 × 1.5% **2b.** $249.53 × 5% **2c.** $223.81 × 2.5%

3. **Divide** money amounts and round to the nearest percent.
   Divide. $650 ÷ $1,400 = 0.464 = 46%

   3a. $800 ÷ $1,375 **3b.** $375 ÷ $900 **3c.** $1,250 ÷ $4,000

Debt

Credit cards are a form of **debt** because you are using someone else's money with the promise to pay it back. Credit cards are easy to use, but that convenience can result in people spending more money than they can repay, leaving them with a situation that is difficult to get out of.

EXAMPLE 1

On January 1, Sonia has a $2,000 beginning balance on her credit card. She charges an average of $100 per month on her credit card, and each month she makes a minimum payment that is 1.5% of her current balance, rounded to the nearest dollar. Her credit card has an APR of 15%. Monthly periodic finance charges are calculated using the previous balance method. How much will she owe in 6 months? How much will she have paid in periodic finance charges?

SOLUTION

Find the monthly periodic rate.

$15\% \div 12 = 1.25\% = 0.0125$

Make a table to calculate each month's balance, finance charges, and payments.

Calculations for January:

$\$2{,}000 \times 0.0125 = \$25$ finance charge

$\$2{,}000 + \$25 + \$100 = \$2{,}125$ current balance

$\$2{,}125 \times 0.015 = \$31.88$ minimum payment

$31.88 rounded to the nearest dollar is $32.00.

$\$2{,}125 - \$32 = \$2{,}093$

Month	Previous Balance	Finance Charge	New Purchase	Current Balance	Payment	Final Balance
Jan.	$2,000	$25	$100	$2,125	$32	$2,093
Feb.	$2,093	$26.16	$100	$2,219.16	$33	$2,186.16
March	$2,186.16	$27.33	$100	$2,313.49	$35	$2,278.49
April	$2,278.49	$28.48	$100	$2,406.97	$36	$2,370.97
May	$2,370.97	$29.64	$100	$2,500.61	$38	$2,462.61
June	$2,462.61	$30.78	$100	$2,593.39	$39	$2,554.39

Total finance charges $= \$25 + \$26.16 + \$27.33 + \$28.48 + \$29.64 + \$30.78 = \$167.39$

At the end of six months, Sonia will owe $2,554.39.

She will have paid $167.39 in finance charges.

EXAMPLE 2

Sonia (in Example 1) decides to stop making any charges on her credit card and continue to make the minimum payment each month. To the nearest percent, what percent of her payments went towards paying off the balance of the card?

SOLUTION

Continue the table from Example 1 to calculate each month's balance, finance charges, and payments.

To find the percentage of her payments that went towards paying off the balance, subtract the final December balance from the July previous balance and divide by the total payments she made.

Month	Previous Balance	Finance Charge	New Purchase	Current Balance	Payment	Final Balance
July	$2,554.39	$31.93	$0.00	$2,586.32	$39	$2,547.32
Aug.	$2,547.32	$31.84	$0.00	$2,579.16	$39	$2,540.16
Sept.	$2,540.16	$31.75	$0.00	$2,571.91	$39	$2,532.91
Oct.	$2,532.91	$31.66	$0.00	$2,564.57	$38	$2,526.57
Nov.	$2,526.57	$31.58	$0.00	$2,558.15	$38	$2,520.15
Dec.	$2,520.15	$31.50	$0.00	$2,551.65	$38	$2,513.65

$2,554.39 − $2,513.65 = $40.74   Amount of balance paid off in 6 months

$39 + $39 + $39 + $38 + $38 + $38 = $231   Amount paid in 6 months

$40.74 ÷ $231 = 0.176 or 18%

Only 18% of Sonia's payments went towards paying off the balance.

✔ CHECK YOUR UNDERSTANDING

A. Beginning in January, Sonia decides to pay $200 each month and make no more charges on her credit card. What is her balance at the end of 6 months?

B. What percent of Sonia's payments went towards paying off the balance of the card?

Even without making new purchases, paying only the minimum balance on a credit card will extend the payments on a card for a long time. The table below shows how long it will take to pay off a balance if a minimum payment of 1.5% is made on a credit card with an APR of 15%.

Time Required to Pay off a Credit Card with a 15% APR (making a minimum payment of 1.5%, and no new purchases)

Beginning Balance	Time to Pay Off	Interest Paid	Total Paid
$500	6 years, 7 months	$289.59	$789.59
$1,000	25 years, 7 months	$2,442.31	$3,442.31
$2,000	48 years, 8 months	$7,442.42	$9,442.42
$3,000	62 years, 2 months	$12,442.54	$15,442.54
$5,000	79 years, 2 months	$22,442.25	$27,442.25
$8,000	94 years, 9 months	$37,442.81	$45,442.81

Consumer Alert

You are entitled to receive a free credit report every 12 months from the three major credit bureaus. Obtaining your reports annually to verify the information is a good idea.

Assessing Debt

A person's **debt-to-income ratio** is a ratio, usually expressed as a percent, that indicates the percent of one's income is spent on housing and other debts.

To find the debt-to-income ratio, add up monthly debt payments, including credit card minimum payments, loan payments, and money spent on housing. Then divide that amount by the monthly gross income.

$$\textbf{Debt-to-income Ratio} = \frac{\textbf{Debt Payments}}{\textbf{Gross Income}}$$

The following guidelines are used in the financial industry to classify a person's debt load. The debt-to-income ratio is used by lenders to determine who qualifies for a loan and how much that loan should be.

- 36% or less: A healthy debt load for the majority of people.
- 37%–42%: Begin reducing debts.
- 43%–49%: Likely in financial trouble.
- 50% or more: Dangerous financial position

EXAMPLE 3

Sam earns $1,500 each month. He pays $400 per month for housing, $250 per month for car loan, and $25 per month on his credit card. Find Sam's debt-to-income ratio and evaluate his financial health based on the guidelines above.

SOLUTION

Find the total debts. Divide by the monthly income.

$\$400 + \$250 + \$25 = \$675$

Debt-to-income ratio $= \$675 \div \$1{,}500 = 0.45$ or 45%

Sam's debt-to-income ratio is 45%. Sam is likely in financial trouble.

© AVAVA 2008/Shutterstock.com

✔ CHECK YOUR UNDERSTANDING

C. Hollister earns $3,000 each month. He pays $500 per month for housing, and has $50 per month in other debt payments. What is his debt-to-income ratio? Evaluate his financial health.

D. Lorna earns $2,050 each month. She pays $800 per month for housing and has $250 per month in other debt payments. What is her debt-to-income ratio? Evaluate her financial health.

Credit bureaus gather information on your credit accounts and generate a **credit score,** generally between 300 and 850. The higher your score, the less risk you pose to a creditor.

In addition to your debt-to-income ratio, credit scores may be used to evaluate your financial health. Many banks and credit card companies use your credit score to determine if you qualify for a credit card or a loan, as well as the interest rate for which you qualify.

Managing and Reducing Debt

Many people struggle with debt. Below are guidelines to reduce or manage debt.

- Use cash, not credit cards.
- Transfer high interest credit card balances to lower interest accounts.
- Pay more than the minimum payment on credit cards.
- Pay off credit cards with the highest interest rates first.
- Consider debt-management counseling.
- Consider a consolidation loan.

A **consolidation loan** is money that is borrowed to pay off all of your debts. Instead of several creditors to pay each month, your debts are combined into one loan payment. This can be a viable option if you can secure a loan at a better interest rate. One pitfall, however, is that many people do not stop using their credit cards and so in time they have the consolidation loan payment and additional balances on credit cards.

Factors that Influence Credit Scores

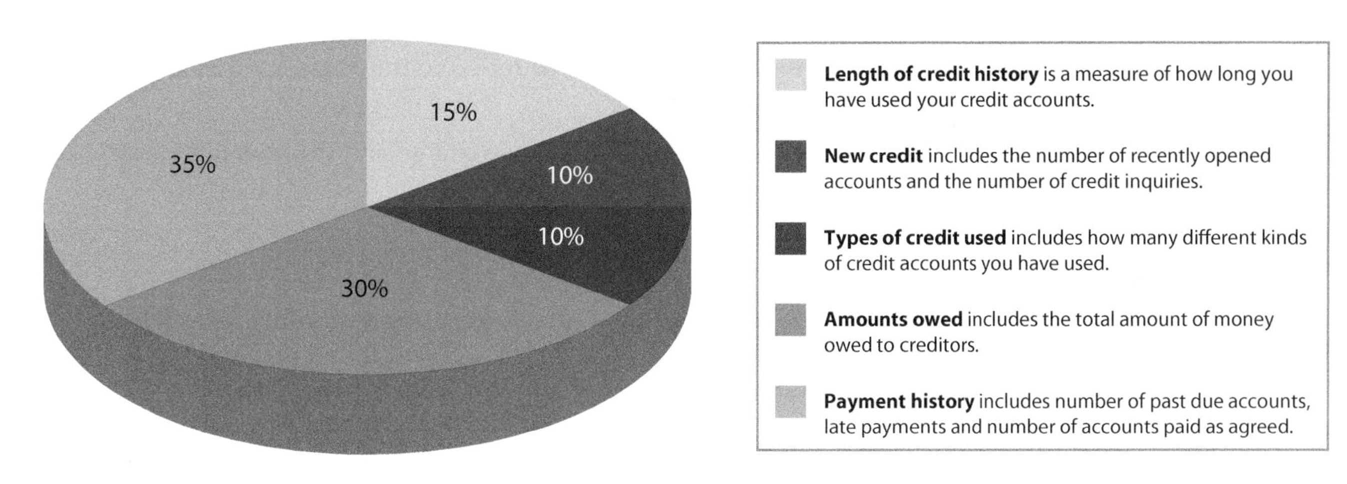

15%
10%
10%
35%
30%
Length of credit history is a measure of how long you have used your credit accounts.
New credit includes the number of recently opened accounts and the number of credit inquiries.
Types of credit used includes how many different kinds of credit accounts you have used.
Amounts owed includes the total amount of money owed to creditors.
Payment history includes number of past due accounts, late payments and number of accounts paid as agreed.

Financial Responsibility

There are warning signs that you have too much debt. These include:

- You only make the minimum payments on your credit cards.
- You continue to use your credit cards while you are trying to pay a balance off.
- You have at least one credit card that is near, at, or over its credit limit.
- You use cash advances to pay other bills.
- You have been denied credit.

Wrap Up ▶▶▶

If Julianna continues to pay only the minimum balance on her credit card, she will be paying on that balance for a very long time, even if she stops using her credit card. If she makes new purchases, she may put herself in the situation that the card will never be paid off. Julianna should make higher payments on the credit card than the minimum amount due.

Exercises

Find the sum.

1. $99.62 + $6.81 + $24.55 − $50
2. $280 + $22.76 + $8.14 − $22

Find the product and round to the nearest whole dollar.

3. $368.00 × 1.5%
4. $145.62 × 5%
5. $529.38 × 2%
6. $2,107.88 × 2.5%

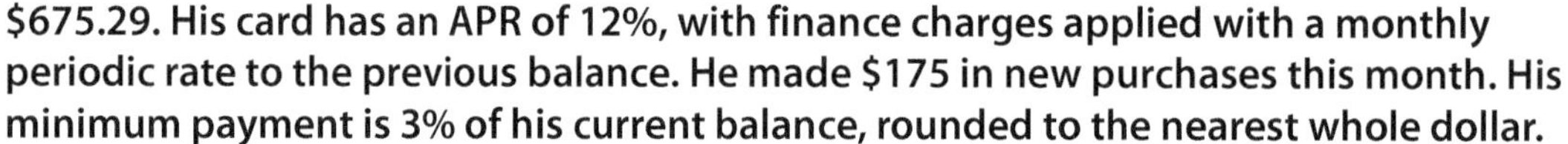

Photodisc/Getty Images

The previous balance on Josiah Evanston's credit card is $675.29. His card has an APR of 12%, with finance charges applied with a monthly periodic rate to the previous balance. He made $175 in new purchases this month. His minimum payment is 3% of his current balance, rounded to the nearest whole dollar.

7. What are his periodic finance charges?
8. What is his current balance, including the finance charge and the new purchases?
9. What is his minimum payment?
10. What will his final balance be after he makes a minimum payment?

Anya Mueller's credit card has an APR of 10%, calculated using a monthly periodic rate on the previous balance. Her minimum payment is 2%, rounded to the nearest whole dollar. Complete the table below, assuming Anya makes the minimum payment each month.

	Month	Previous Balance	Finance Charge	New Purchase	Current Balance	Payment	Final Balance
11.	July	$500		$50			
12.	Aug.			$50			
13.	Sept.			$50			
14.	Oct.			$50			
15.	Nov.			$50			
16.	Dec.			$50			

17. How much has Anya paid in periodic finance charges in 6 months?

Solve.

18. Enrique has the following monthly housing and debt payments: rent, $350; credit card minimum payment, $15; student loan payment, $60; and monthly income, $1,800. What is Enrique's debt-to-income ratio, to the nearest percent?

19. Shawna has the following monthly housing and debt payments: house payment, $600; credit card minimum payment, $50; car loan, $125; monthly income, $2,000. What is Shawna's debt-to-income ratio, to the nearest percent?

LaToya Thompson has a credit card with an APR of 15%, calculated using a monthly periodic rate on the previous balance. Her minimum payment is 4%, rounded to the nearest whole dollar. Complete the table below, assuming LaToya makes the minimum payment each month.

	Month	Previous Balance	Finance Charge	New Purchase	Current Balance	Payment	Final Balance
20.	July	$500		$0.00			
21.	Aug.			$0.00			
22.	Sept.			$0.00			
23.	Oct.			$0.00			
24.	Nov.			$0.00			
25.	Dec.			$0.00			

26. How much money has LaToya made in payments in 6 months?

27. How much has LaToya paid in periodic finance charges in 6 months?

28. How much of the balance has she paid off in 6 months?

29. To the nearest percent, what percent of her payments went towards paying off the balance?

30. **FINANCIAL DECISION MAKING** Tristen has the following monthly housing and debt payments: rent, $385; credit card minimum payment, $46; car loan, $200; monthly income, $1,500. What advice would you give Tristen?

Alistair Cotton 2008/Shutterstock.com

31. **CRITICAL THINKING** A graph showing different factors that influence credit scores appears on page 157. Based on those factors, what are things you can do to have a better credit score?

32. **STRETCHING YOUR SKILLS** Janica Johannsen has just graduated from college. She has a job that pays $2,200 per month. She has monthly student loan payment of $125 and no other debts. What is the most she can spend on housing without exceeding a debt-to-income ratio of 36%?

Mixed Review

33. What is 29% of $360?

34. $\frac{1}{2} + \frac{3}{4} + 5\frac{1}{2}$

35. Write $\frac{5}{20}$ as a percent

36. John earns a 6% commission. If his sales for the first quarter were $43,000, how much did he earn in commission?

37. Susanna Ray's gross monthly income is $4,000. She has 3% withheld from her check for state income taxes. How much is withheld per month for state income taxes?

38. Christian Rogers earns 5% annual interest on a 5-year $10,000 CD. How much will he earn in interest the first year?

Chapter *Review*

Vocabulary Review

Find the term, from the list at the right, that completes each sentence. Use each term only once.

- adjusted balance method
- annual percentage rate
- average daily balance method
- cash advance
- consolidation loan
- credit terms and conditions
- credit score
- debt
- debt-to-income ratio
- grace period
- periodic finance charge
- periodic rate
- previous balance method
- total finance charges

1. The percent of your income that is going towards housing and other debts is called _?_
2. The sum of periodic finance charges and fees is called _?_.
3. The method that calculates the balance subject to finance charges by subtracting payments and credits from the previous balance is called the _?_.
4. The time for which you are not charged finance charges on a credit card balance is called a _?_.
5. The monthly or daily rate charged on a credit card balance is called the _?_.
6. Money borrowed on a credit card and received in cash is called a _?_.
7. The yearly interest rate charged on a credit card is called the _?_.
8. Money borrowed to pay off all of your debt is a _?_.
9. A _?_ is a number between 300 and 850 generated by credit bureaus regarding your credit accounts.
10. The document issued by the credit card company that outlines the costs associated with using their credit card is the _?_.

4-1 Credit Card Costs

11. Cierro McClendon is going to transfer a $3,500 balance from one credit card to a new one. The credit card charges a 3% fee for balance transfers. What fee will he pay to transfer the balance?
12. Derek Wilson checked his credit card statement and found a purchase for $26.99 that was unauthorized. He also found that a sales slip for $35.89 had been listed as $38.59. If the new balance on his statement was $140.68, what is his correct new balance?
13. Loni Dramin's credit card statement for April showed a membership fee of $55, a late fee of $25, a finance charge of $6.45, and an over-the-limit-fee of $12. What was the total cost of the card to Loni in April?
14. Jose DeLeon's previous credit card balance was $356.83. He made $259.01 in new purchases, and received a credit of $56.83 for a return. He made a payment of $400 and was charged $4.46 in periodic finance charges. What is Jose's new credit card balance?

4-2 Credit Card Finance Charges

15. Lacy Clure is charged a finance charge on a credit card balance of $800.00. Her card has an APR of 15%. What is her monthly finance charge if the company uses a monthly periodic rate?

16. Juan Mendoza's credit card statement for October showed a previous balance of $239.80, new purchases of $174.50 and payments and credits of $95. The card's annual percentage rate is 24%. The company uses a daily periodic rate. What is Juan's finance charge and new balance for October using the previous balance method?

17. What is Juan's finance charge and new balance for October using the adjusted balance method?

4-3 Average Daily Balance Method

18. Madison Andrickson's credit card statement for August showed these items: 8/1, previous balance, $220.56; 8/5, purchase, $56.89; 8/14, purchase, $190.23; and 8/25, payment, $150. Madison's card company uses a 1.6% monthly periodic rate. What is Madison's finance charge for August and the new balance using the average daily balance method including new purchases?

19. What is Madison's finance charge for August and the new balance using the average daily balance method excluding new purchases?

4-4 Cash Advances

20. Ula Johan borrowed $250 on a cash advance from her credit card company. She was charged a daily periodic rate of 0.053% for the 25 days she had the cash advance. She was also charged a 4% cash advance fee. What was the total amount Ula had to pay back?

21. Javier's credit card has an APR of 14% for purchases and 21% for cash advances. They use an average daily balance method with a daily periodic rate for purchases. They use a daily periodic rate for cash advances and a cash advance fee of 3%. In a 31 day billing cycle, Javier has an average daily balance for purchases of $143.66. He took a cash advance of $300 during the billing cycle and must pay finance charges on the cash advance for 23 days. What are his total finance charges?

4-5 Debt Management

22. The previous balance on Delaney Shuba's credit card is $354.87. He made $210 in new purchases this month, and his periodic finance charges are $5.32. His minimum payment is 3% of his current balance, rounded to the nearest whole dollar. What is the minimum payment?

23. Raven Ingram's credit card has a previous balance of $125.62. She made $175 in new purchases during the billing cycle and made a payment of $50. She was charged $1.88 in finance charges. She wants to pay off the entire balance of the card. How much should she pay?

24. Veronica Cole earns $2,500 per month. She has a monthly student loan payment of $120 per month, and a car payment of $200 per month. She and her roommate share an apartment and split the $800 per month rent. What is Veronica's debt-to-income ratio, rounded to the nearest percent?

Technology Workshop

Task 1 Enter Data Into An Average Daily Balance Finance Charge Template

Complete a template that calculates the average daily balance, the finance charge, and the new balance for a credit card statement.

Open the spreadsheet for Chapter 4 (tech4-1.xls) and enter the data shown in blue (cells A5-B12, D19 and E19) into the spreadsheet. The values you enter are from Example 1 in Lesson 4-3. Your computer screen should look like the one shown when you are done.

The spreadsheet will calculate the:

1. Balance at the end of each day
2. Number of days the balance is in effect
3. Sum of the daily balances
4. Total of the sum of the daily balances
5. Number of days in the month
6. Average daily balance
7. Finance charge
8. New balance

	A	B	C	D	E
1	Average Daily Balance				
2	Finance Charge Calculator				
3	Date	Transaction	Balance at	Number	Sum of Daily
4			End of Day	of Days	Balances
5	10/1	225.60	225.60	3	676.80
6	10/4	128.99	354.59	5	1,772.95
7	10/9	21.89	376.48	3	1,129.44
8	10/12	27.79	404.27	6	2,425.62
9	10/18	-35.99	368.28	1	368.28
10	10/19	35.00	403.28	5	2,016.40
11	10/24	-75.00	328.28	8	2,626.24
12	11/1	0.00	328.28	0	0.00
13			Sums	31	11,015.73
14					
15	*When the last transaction is entered, enter the first day of the*				
16	*next month in the Date .*				
17					
18				Monthly	Daily
19			Periodic Rate	0.015	0.00000000
20		Number of Days in Month		31	31
21			Average Daily Balance	355.35	355.35
22			Finance Charge	5.33	0.00a
23			New Balance	333.61	328.28

The template can be used for either a monthly periodic rate or a daily periodic rate. To calculate values for a daily periodic rate, enter the daily periodic rate in cell E19.

After you enter the last transaction for the month, enter the first day of the next month in the next blank Date cell. In Task 1, this required you to enter the first day of the next month in cell A12.

Task 2 Analyze The Spreadsheet Output

Move the cursor to row 19 and column D, the cell for Monthly Periodic Rate. Enter the rate 0.02. Notice how the finance charge changed. This change shows what would happen if the monthly periodic rate were raised from 1.5% to 2%.

Move to cell A11 and change the date to 10/25. Notice how the sums of the daily balances, total sums of daily balances, the finance charge, and the new balance have changed.

Answer these questions about your updated spreadsheet.

1. What function is used in cell E13?
2. What arithmetic is done in cell E13?
3. What is the formula used in D21?
4. What arithmetic is done in cell D21?
5. What formula is used in cell D22?
6. What arithmetic is done in cell D22?
7. What arithmetic is done in cell D23?

Task 3 Design a Cash Advance Finance Charge Spreadsheet

You are to design a spreadsheet that will calculate the interest, total finance charge, and total amount needed to pay off a cash advance.

SITUATION: You want to borrow $500 on a cash advance for 20 days using your credit card. Your card company charges a cash advance fee of $15 and a daily periodic interest rate of 0.045%. You want to know how much interest and finance charges you must pay and what amount will be needed to pay off the loan.

Task 4 Analyze the Spreadsheet Output

Answer these questions about your completed spreadsheet:

8. How did you calculate the interest on the cash advance?
9. What would the interest be on the loan?
10. What would the finance charge be on the loan?
11. What amount is needed to pay off the loan in 20 days?
12. If you were to (a) change the interest rate to 0.07%, (b) change the cash advance fee to $20, and (c) extend the loan to 30 days, what would be the interest, finance charge, and payoff amount?

Chapter *Assessment*

Chapter Test

Answer each question.

1. Rewrite 1.5% as a decimal.
2. Rewrite 0.67 as a percent.
3. Find what percent $42 is of $120
4. What is 3% of $500?
5. Multiply: $242.12 × 0.000575
6. Divide: $400 ÷ $1,600
7. Multiply: $583 × 0.0493% × 30
8. Divide: $4,852.67 ÷ 31
9. Add and subtract: $442.87 + $281.10 + $15.42 − $250
10. Divide and round to the nearest ten-thousandth of a percent: 14.9% ÷ 12

Solve.

11. Jamika Hurty found an unauthorized charge of $76.03 on her credit card statement. If the new balance on her statement was $281.48, what is her correct new balance?
12. Jaime Escobido switched to a new credit card. He paid an annual membership fee of $40. He also paid a balance transfer fee of 3% of his old card's $335.81 balance. During the next year, he paid finance charges of $13.85, $5.92, $2.63, and $1.81. What was the total cost for using his credit card for the year?
13. Davida Thompson has the following monthly debts: house payment, $500; car payment, $225; credit card minimum due, $52. Her monthly income is $1,800. To the nearest percent, what is her debt-to-income ratio?
14. Joe Oyoumick's credit card requires a minimum payment of 3% of the current balance each billing cycle. If Joe's current balance is $125.40, what is his minimum payment due?
15. Julie Myrick has a credit card that has an APR of 15%, applied at a daily periodic rate to the previous balance. Her previous balance is $567.31. She made new purchases totaling $128.92, and a payment of $125 in the month of June. What is her new balance?

Jacob Snap has a credit card with a previous balance of $678.32. The APR on the card is 12% and periodic finance charges are applied with a monthly rate. During the month of June, he made a $175 purchase on 6/15 and made a payment of $240 on 6/20.

16. If the card company uses the previous balance method, what is his finance charge?
17. If the card company uses the adjusted balance method, what is his finance charge?
18. If the credit card company uses the average daily balance method including new purchases, what is his finance charge for June?
19. If the credit card company uses the average daily balance method excluding new purchases, what is his finance charge for June?

Jim Settler takes a cash advance of $400 for 22 days. The credit card company charges a 5% cash advance fee and an APR of 21%, applied daily.

20. What are the total finance charges for the cash advance?
21. How much will he have to pay back to the credit card company?

Planning a Career in Education

Careers in education and training entail teaching children and adults. Teachers explain concepts and how to perform particular skills to their students. Principals, secretaries, counselors, and others, support the educational environment. Workers in the field of education and training can be found in schools, colleges, training centers, child care centers, and in businesses. If you are an excellent communicator and are effective at giving instructions to others, a career in education and training may be for you.

Jaimie Duplass 2008/Shutterstock.com

Job Titles

- Teacher
- Reading specialist
- School counselor
- Child care provider
- Principal
- College advisor
- Speech pathologist
- School superintendent
- College professor
- Teacher's aid

Needed Skills

- outstanding communication skills
- ability to work and make decisions independently

- content area knowledge and skills
- computer and technology skills
- strong leadership skills
- genuinely like people and want to see them succeed

What's it like to work in Education?

Teacher assistants, also called teacher aids or instructional aids, support classroom teachers by supervising students, reinforcing concepts taught in class, performing clerical duties, such as keeping records or grading papers, and managing special projects with groups of students. Teacher assistants frequently work alone or in small groups with students to give them individualized instruction. Because special education students are included in the regular classroom environment whenever possible, teacher assistants often support special needs students, those with disabilities or language barriers.

What About You?

What aspect of education appeals to you? What age level are you most interested in teaching?

How Times Have Changed

Refer to the timeline on page 127 to answer the following question.
In 2007, the average American family carried about $9,000 of credit card debt. How many American households contributed to the total amount of credit card debt?

Chapters 3-4 *Cumulative Review*

MULTIPLE CHOICE

Select the best choice for each question.

1. Lu Yang paid an average of $32 per month in finance charges on his credit card. His credit card has an annual fee of $75. What was Lu's total cost of credit for the year?

   A. $107 B. $252 C. $267
   D. $459 E. $1,284

2. Olga Kirosky deposited these items in a bank: (bills) 12 twenties, (coins) 11 quarters, and 16 dimes; (checks) $125.99, and $41.65. She received 100 one-dollar bills in cash back. What was the net amount of her deposit?

   A. $311.99 B. $319.99 C. $399.99
   D. $411.99 E. $419.99

3. Bella Tuller has a credit card with an APR of 15% and a monthly periodic rate. The credit card company uses the previous balance method to calculate finance charges. If Bella's previous balance is $900, what are the finance charges for the month?

   A. $0.37 B. $11.10 C. $11.25
   D. $32.67 E. $135

4. The bank statement sent to Abigail Ochs did not show checks for $158.23, $12.89, and $71.27, and an outstanding deposit of $75 that were listed in the check register. The balance printed on the statement was $289.07. What is the reconciled bank statement balance?

   A. $121.68 B. $167.39 C. $28.32
   D. $456.46 E. $606.46

OPEN ENDED

5. The bank statement of Jake Hansen showed a balance of $356.93. His check register showed a balance of $308.34. When Jake compared the two records he found several differences. The bank statement listed these items not recorded in the register: ATM withdrawal, $50; ATM user fee, $2.25; direct deposit of paycheck, $624.70; checks for $287.23, $180.11, and $72.89. A check for $85.89 was recorded in the register as $58.89. The items not listed on the bank statement included checks for $72.44, $9.76, and $113.57; and a $50 deposit made after the statement closing date. Debit card purchases of $85.67 and $16.73 appeared on the statement but not in the register. Reconcile Jake's bank statement and check register.

6. Justin Niklas made this deposit to the Breakfast Book Club's checking account: (bills) 43 ones, and 13 fives; (coins) 9 quarters, 3 half-dollars, 5 dimes, 4 nickels, and 12 pennies; (checks) $397.42, and $192.81. Find the amount of the deposit.

7. After work Gerry Mathews used the ATM to deposit her paycheck for $868.39 and to withdraw $400 in cash. If her starting bank balance was $1,528.93, what is her new balance?

8. Drew Moro's check register balance was $349.64. His bank statement showed an ATM deposit of $59.39 that was not recorded in his check register, interest earned of $1.27, and a $15 charge for printing new checks. Reconcile Drew's check register.

9. Etta Warnicke deposited $20,000 in a three-year certificate of deposit that pays simple interest at a fixed annual rate of 4.9%. What total interest will Etta have earned at the end of three years?

10. Lung Shen used his credit card in an ATM to borrow $300 on a cash advance. His card company charged a cash advance fee of $15 and a daily periodic interest rate of 0.049%. Lung paid the loan back at the end of 15 days. What was the total finance charge on the cash advance?

QUANTITATIVE COMPARISON

Compare the quantity in Column A with the quantity in Column B. Select the letter of the correct answer from these choices:

A if the quantity in Column A is greater;

B if the quantity in Column B is greater;

C if the two quantities are equal;

D if the relationship between the two quantities cannot be determined from the given information.

	Column A	Column B
11.	New balance on an account that had a beginning Balance of $43.65 with checks written for $2.98, $78.23 and deposits of $211.86, $12.45	New balance on an account that had a beginning balance of $215.44 with checks written for $56.89, $112.90 and a deposit of $85
12.	1 twenty dollar bill, 6 fives, 12 dimes, and 25 pennies	4 ten dollar bills, 8 ones, 12 quarters, and 9 nickels
13.	simple interest on a $1,250 CD for 2 years at 12.5% interest rate	simple interest on a $3,200 CD for 2.5 years at 4.5% interest rate

CONSTRUCTED RESPONSE

14. A friend has a $5,000 balance on her credit card. The card has a 15% APR applied at a monthly periodic rate. Your friend charges an average of $175 per month, and makes a $150 payment each month. Find the balance on the credit card after 3 months. What advice would you give your friend?

Chapter 5

Loans

Statistical Insights

Mortgage Loans

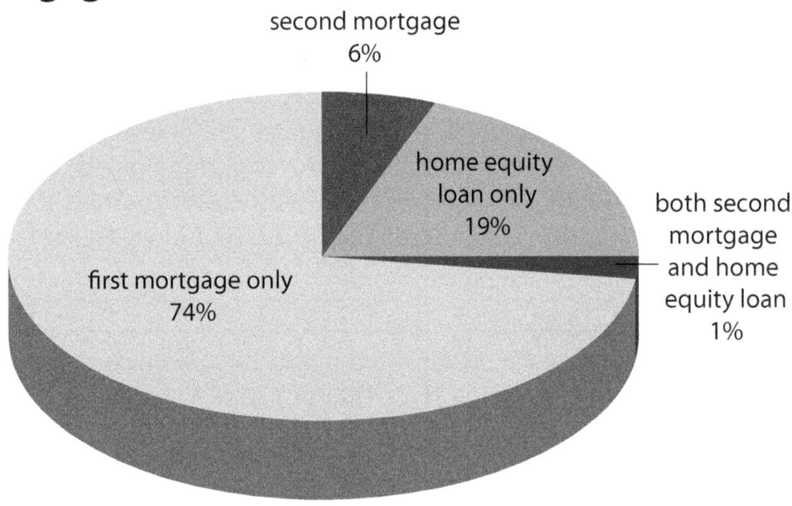

Source: 2006 Census Reports

When a person borrows money to buy a house, the secured loan the bank issues is a mortgage. The house serves as the security the bank needs to justify the risk of loaning money. If a homeowner takes out another loan on the same house, the secured loan is a second mortgage. Some homeowners use their homes as collateral to get a home equity loan. In 2006, there were 51,234,170 houses used as collateral for mortgages or home equity. Use the pie chart to answer Questions 1–5.

1. How many homeowners have only taken a first mortgage?
2. How many homeowners have only one loan against their home?
3. How many homeowners have three loans against their home?
4. How many loans do the data in the pie chart represent?
5. **Explain** why the answer to Question 4 is greater than the number of houses in the report.

How Times Have Changed

The Federal Reserve System, created in 1913, is the organization under which the Federal Open Market Committee (FMOC) operates. Since 1933 FMOC meets at least eight times a year to set the Federal Funds Target Rate (FFTR). The FFTR, the most influential benchmark in regulating the U.S. economy, is the interest rate that banks charge each other for overnight loans. From this rate, the U. S. Prime Rate is determined. The U.S. Prime Rate is the interest rate that banks typically charge customers to borrow money. Since the second quarter of 1994, the U.S. Prime Rate has averaged 3 percentage points greater than the FFTR.

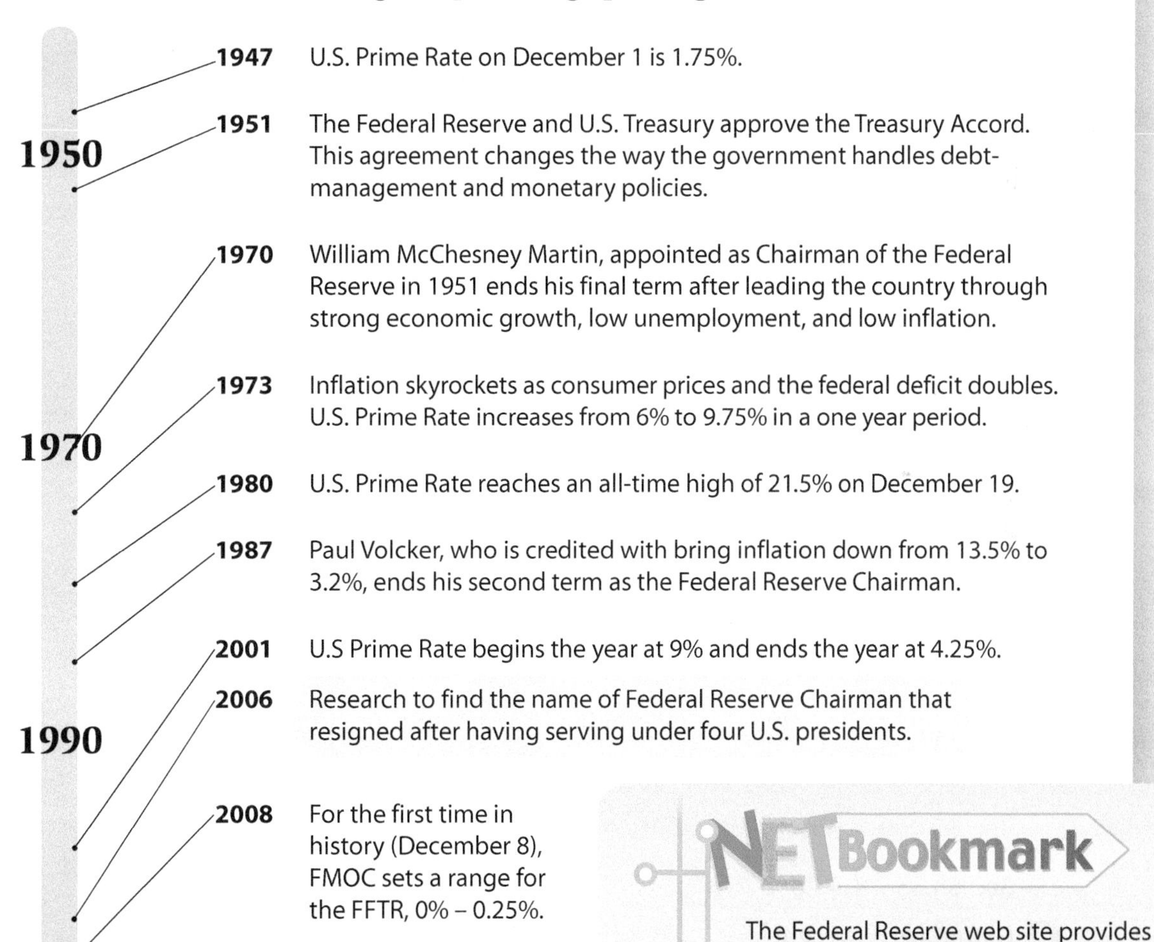

1950
1970
1990
2010
1947 U.S. Prime Rate on December 1 is 1.75%.
1951 The Federal Reserve and U.S. Treasury approve the Treasury Accord. This agreement changes the way the government handles debt-management and monetary policies.
1970 William McChesney Martin, appointed as Chairman of the Federal Reserve in 1951 ends his final term after leading the country through strong economic growth, low unemployment, and low inflation.
1973 Inflation skyrockets as consumer prices and the federal deficit doubles. U.S. Prime Rate increases from 6% to 9.75% in a one year period.
1980 U.S. Prime Rate reaches an all-time high of 21.5% on December 19.
1987 Paul Volcker, who is credited with bring inflation down from 13.5% to 3.2%, ends his second term as the Federal Reserve Chairman.
2001 U.S Prime Rate begins the year at 9% and ends the year at 4.25%.
2006 Research to find the name of Federal Reserve Chairman that resigned after having serving under four U.S. presidents.
2008 For the first time in history (December 8), FMOC sets a range for the FFTR, 0% – 0.25%.
2009 President Obama appoints Paul Volcker to lead White House economic advisory committee.

NET Bookmark

The Federal Reserve web site provides information about the responsibilities of Federal Reserve System. Access **www.cengage.com/school/business/businessmath** and click on the link for Chapter 5. Other than setting the Federal Funds Target Rate, what is the primary object of the committee?

5-1

Promissory Notes

GOALS

- Calculate interest on interest-bearing promissory notes
- Calculate interest using the exact interest method
- Calculate interest using the ordinary interest method
- Calculate the rate of interest

KEY TERMS

- promissory note
- interest
- principal
- rate of interest
- exact interest method
- ordinary interest method

Start Up

Make a list of the advantages and the disadvantages of borrowing money. If you borrowed $1,000 from a bank for a year, how much extra do you think you should pay them back for the benefit of using their money for a year?

Photodisc/Getty Images

Math Skill Builder

Review these math skills and solve the exercises that follow.

1 **Write** percents as decimals.
Write 67% as a decimal. 67% = 0.67

1a. 45% **1b.** 150% **1c.** 0.5%

2 **Multiply** money amounts by percents.
Find 4% of $3,500. 4% = 0.04; 0.04 × $3,500 = $140

2a. 3% of $2,580 **2b.** 6.5% of $6,340 **2c.** 5.07% of $855

3 **Multiply** money amounts by whole numbers and fractions.
Find the product: $\$3{,}000 \times 2\% \times \frac{1}{2}$; $\$3{,}000 \times 0.02 \times \frac{1}{2} = \$30$

3a. $\$5{,}000 \times 4\% \times \frac{1}{4}$ **3b.** $\$7{,}520 \times 6.2\% \times 1\frac{1}{2}$

3c. $\$8{,}356 \times 4.25\% \times 3$ **3d.** $\$2{,}698 \times 1.42\% \times 3\frac{1}{3}$

4 **Calculate** a percent.
What percent of $350 is $14? $14 ÷ $350 = 0.04, or 4%

4a. $15 is what percent of $750? **4b.** $43 is what percent of $860?

5 **Simplify** fractions.
Simplify $\frac{24}{48} \cdot \frac{24 \div 24}{48 \div 24} = \frac{1}{2}$

5a. $\frac{180}{360}$ **5b.** $\frac{260}{360}$ **5c.** $\frac{126}{360}$ **5d.** $\frac{240}{365}$

Interest-Bearing Promissory Notes

When you borrow money, you usually sign a promissory note. A **promissory note** is your written promise, or IOU, that you will repay the money to the lender on a certain date. Usually you also have to pay for using the lender's money.

That cost is called **interest**. A note that requires you to pay interest is called an *interest-bearing note*.

Lenders may require a borrower to deposit or pledge property as security for a loan. This property is called *collateral*. Types of collateral that are often used to secure loans are cars, stocks, bonds, and life insurance. If the loan is not repaid, the lender can seize the collateral and sell it to recover the borrowed money.

Many lenders offer *home equity loans* to home owners. *Home equity* is the difference between what the home could be sold for and what is owed on it. To get a home equity loan, the borrower pledges the equity in the home as collateral for the loan.

Most promissory notes today are lengthier documents than the one shown below. In whatever form the note is, it will contain the same basic information.

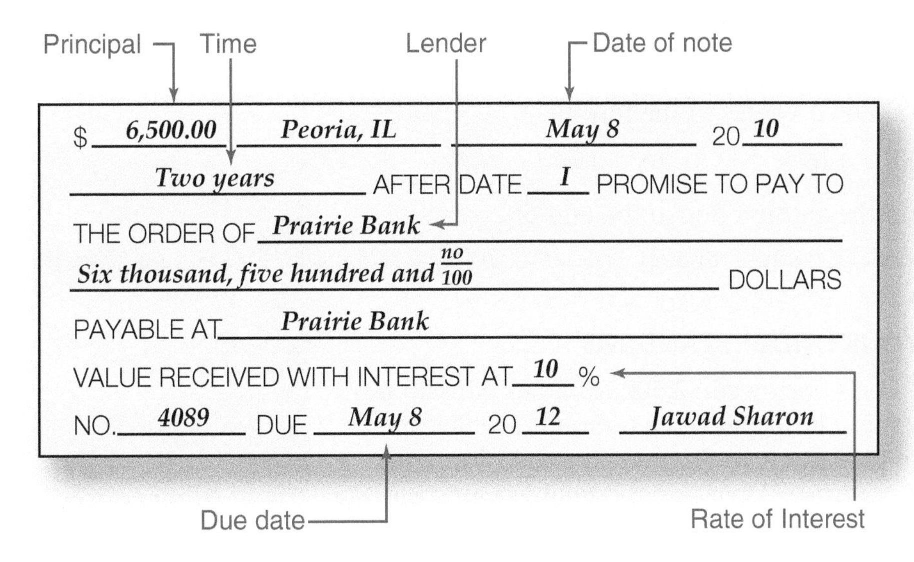
$ 6,500.00 Peoria, IL May 8 20 10

Two years AFTER DATE I PROMISE TO PAY TO

THE ORDER OF Prairie Bank

Six thousand, five hundred and no/100 DOLLARS

PAYABLE AT Prairie Bank

VALUE RECEIVED WITH INTEREST AT 10 %

NO. 4089 DUE May 8 20 12 Jawad Sharon

The amount of money borrowed on a promissory note is the *face*, or **principal**.

The date the note was signed is called the *date of the note*. The date on which the money must be repaid is the *due date*, or *maturity date*. The amount of time between the date of a note and the date that the note is due is the time of the note. Time should be expressed in years.

The rate of interest to be paid is the **rate of interest**. The amount of money that must be paid on the due date is the *maturity value* or the total amount due.

Interest rates are stated as rates *per year*. The interest you pay on a loan is proportional to the time for which you borrow the money. For a loan of three months, or $\frac{1}{4}$ of a year, the amount of interest is one-fourth of the interest for a full year. For a loan of 2 years, the amount of interest is double the interest of one year.

To calculate interest on a note, you use the same formula as you used to calculate simple interest paid on savings.

Interest = Principal × Rate × Time

$$I = P \times R \times T$$

To find the amount due on the due date, you add the interest to the principal.

Amount Due = Principal + Interest

$$A = P + I$$

Math Tip

Interest formula:

$\boldsymbol{I = P \times R \times T}$

P = *principal*

R = *rate*

T = *time*

I = *interest*

EXAMPLE 1

Jawad Sharon borrowed $6,500 from his bank to buy a boat, which he used as collateral for the loan. Jawad signed a 2-year promissory note at a 10% interest rate. Find the amount of interest Jawad must pay. Then find the total amount he must repay when the note is due.

SOLUTION

Rewrite the interest rate as a decimal: 10% = 0.10

Substitute known values in the formula.

$I = \$6,500 \times 0.10 \times 2 = \$1,300$ Use $I = PRT$.

Add to find the amount due at the end of 2 years.

$A = \$6,500 + \$1,300 = \$7,800$ Use $A = P + I$

Photodisc/Getty Images

✔ CHECK YOUR UNDERSTANDING

A. Leslie Regis borrowed $2,500 from her bank to pay for a cruise. Leslie signed a 6-month promissory note at 11% interest. Find the amount of interest Leslie must pay. Then find the total amount she must repay to her bank when the note is due.

B. Raanan Beilin borrowed $3,500 for 18 months from his bank to have his house repainted. Raanan signed a promissory note that carried 12% interest. Find the amount of interest Raanan must pay. Then find the total amount he must repay to his bank on the due date.

Exact Interest Method

When the time of a note is shown in days, interest may be calculated by the **exact interest method**. Exact interest uses a 365-day year. The exact interest method is used by the United States government and by many banks and other businesses.

To find exact interest, you show the time as a fraction with 365 as the denominator. For example, you would show 79 days as $\frac{79}{365}$.

EXAMPLE 2

Rosa Chavez borrows $1,000 at 6% exact interest for 85 days.

SOLUTION

Rewrite the interest rate as a decimal: 6% = 0.06

$I = \$1{,}000 \times 0.06 \times \frac{85}{365} = \$13.97$ Use $I = P \times R \times T$

You can also use and simplify fractions to find the solution.
Write the percent as a fraction. $6\% = \frac{6}{100}$

Write and simplify the time fraction: $\frac{85}{365} = \frac{17}{73}$

$\overset{10}{\cancel{\$1{,}000}} \times \frac{6}{\underset{1}{\cancel{100}}} \times \frac{17}{73} = \frac{1020}{73} = \$13.97$ Use $I = P \times R \times T$ with fractions

Calculator *Tip*

To use a calculator:
Enter 1000
Press ×
Enter .06
Press =
Press ×
Enter 85
Press =
Press ÷
Enter 365
Press =

✔ **CHECK YOUR UNDERSTANDING**

C. Ana Lopez borrows $5,000 for 75 days at 8% exact interest. Find how much interest she must pay on the loan and how much will be due at maturity.

D. Albert O'Malley signs a promissory note for $3,500 for 150 days at 9% exact interest. Find the interest he must pay and the total amount due on the due date.

Ordinary Interest Method

The **ordinary interest method**, or *banker's interest method* is used in place of the exact interest method by some businesses. With this method of finding interest, a year has only 360 days. The 360-day year has 12 months of 30 days each and is known as the *banker's year*. Of course, there really is no such year. It is used because it is easier to calculate with than a 365-day year.

EXAMPLE 3

Rosa Chavez borrows $1,000 at 6% ordinary interest for 85 days.

SOLUTION

$I = \$1{,}000 \times 0.06 \times \frac{85}{360} = \$14.17$ Use $I = P \times R \times T$

$I = \overset{1}{\cancel{\overset{10}{\cancel{\$1{,}000}}}} \times \frac{\overset{1}{\cancel{6}}}{\underset{1}{\cancel{100}}} \times \frac{85}{\underset{6}{\cancel{\underset{60}{\cancel{360}}}}} = \frac{85}{6} = \$14.17$ Use $I = P \times R \times T$ with fractions

✔ **CHECK YOUR UNDERSTANDING**

E. Ikuko Kimura signed a promissory note for $5,900 at 12% ordinary interest for 180 days. Find the interest and amount due she will pay when the note is due.

F. On May 6, Solomon Kaufman borrowed $4,000 signing a promissory note at his bank. The note carries 9% ordinary interest and is due in 4 months. Find the interest and amount due that Solomon must pay at maturity.

Rate of Interest

If you know the principal and the amount of interest for one year, you can find the rate of interest by dividing the interest by the principal.

Rate of Interest = Interest for One Year ÷ Principal

If the interest given in the problem is not for a year, you must first find how much the interest would be for one year.

Algebra Tip

Use the rate of Interest formula:

$R = \frac{I}{P}$

where R is rate of interest, I is interest for one year, and P is the principal

EXAMPLE 4

Ella Stein paid $30 interest on a loan of $1,000 for 3 months. Find the rate of interest she paid.

SOLUTION

Find the amount of interest for one year by finding the number of 3-month periods in one year.

12 months ÷ 3 months = 4 number of 3-month periods in one year

$30 × 4 = $120 interest for one year

$R = \$120 \div \$1{,}000 = 0.12$, or 12% Use $R = \frac{I}{P}$

✔ CHECK YOUR UNDERSTANDING

G. Trish Newcomb must pay $320 in interest on a promissory note for $8,000 due 4 months from the date of the note. Find the rate of interest she will pay.

H. Susilo Wahyudi paid $450 in interest on a 3-month note for $12,000. Find the rate of interest he paid.

Photodisc/Getty Images

Consumer Alert

Payday Loans = HIGH Interest

Payday loans are usually short term loans designed to provide money until the next payday. While payday loans might seem like a convenient way to get out of a jam, the interest rate for these loans can range from 400% – 800%.

A common payday loan is for $500 with $25 per $100 due for interest. If the loan is to be repaid in two weeks, what is the interest rate for this loan?

Companies offering payday loans may ask borrowers to leave a check for the principal and the interest that will be cashed on the day the loan is due. Online companies may ask for electronic access to the borrower's bank account to electronically withdraw funds on the due date. The Consumer Federation of America recommends never transmitting bank account numbers, Social Security numbers, or other personal financial information over the Internet or by fax to unknown companies.

Wrap Up ▶▶▶

Look back at the list of the advantages and disadvantages of borrowing money. What advantages or disadvantages would you add to the list? Calculate the rate of interest for the amount of interest you identified in the Start Up on a one-year, $1,000 loan.

TEAM Meeting

The rate of interest on a loan is directly tied to the risk to the lender. Lenders typically give the lowest interest rates to people and businesses that they believe will be responsible to pay back the loan. Organize a team and brainstorm a list of things that a bank might consider when evaluating whether a person or a business is a low or high credit risk. Contact several banks to evaluate and edit your list.

Exercises

Find the product.

1. 2% of $4,689
2. 150% of $84
3. $2,200 × 6% × 5

Find the percent.

4. 75 as a percent of 3,000
5. $150 as a percent of $1,200
6. Write these as a decimal: 2.3%, 230%, 23%

Find the interest to be paid for each promissory note.

	Principal	Rate	Time in Years	Interest
7.	$2,500	15%	2	
8.	$12,500	12%	$3\frac{1}{2}$	
9.	$500	8%	$\frac{1}{2}$	

Find the interest and the amount due at maturity for each note.

	Face of Note	Time	Rate	Interest	Amount Due at Maturity
10.	$500	3 yr	12%		
11.	$150	3 mo	18%		
12.	$920	$2\frac{1}{4}$ yr	$5\frac{1}{2}$%		

Solve.

13. To finance the remodeling of her kitchen, Rosa borrowed \$26,400 on an 18-month home equity loan. She signed a promissory note bearing interest at $7\frac{1}{2}\%$. What total amount did Rosa pay on the due date?

Photodisc/Getty Images

14. Rondel Wilson borrowed \$2,000 to replace the furnace in his house. The promissory note he signed was for 3 months at $15\frac{1}{4}\%$ interest. How much did Rondel have to pay when the note came due?

15. Khalil Hamid Ali borrowed \$12,000 and paid \$1,890 in exact interest when the loan came due $1\frac{1}{2}$ years later. What rate of interest did Khalil pay?

16. Lynn Wessel borrowed \$2,500 for 18 months. The total interest she paid was \$315. What rate of interest did Lynn pay?

Find the exact interest to the nearest cent. Then find the ordinary interest to the nearest cent.

17. \$360 @ 14% for 210 days
18. \$1,500 @ 15% for 36 days
19. \$1,200 @ 6% for 240 days
20. \$2,400 @ 9% for 60 days
21. \$450 @ 12% for 146 days
22. \$1,450 @ 7% for 100 days

Bill Rich signed a 180-day note for \$1,250. He repaid the loan when due with interest at an annual rate of 12% using a banker's year.

23. How much interest did Bill pay?
24. What total amount did he pay?

Find each interest rate.

25. Kelly Bullock borrowed \$4,800 for 6 months and paid \$264 interest. What rate of interest did she pay?
26. Tony Colito paid \$19.50 in interest on a loan of \$2,600 for 1 month. What rate of interest did he pay?

Tara Long borrowed \$10,000 for 180 days. She paid exact interest at an annual rate of 12%.

27. Estimate the interest Tara owed.
28. What is the exact amount of interest she had to pay?
29. What total amount did she have to repay?

30. **FINANCIAL DECISION MAKING** Maria is shopping for a $2,000 loan for 1 year. One lender offers her a loan with 9% interest. The other lender offers her a loan with 7.5% interest, but she must use her car as collateral for the loan. What would you advise Maria to do?

FINANCIAL DECISION MAKING You can borrow $5,600 at 12% interest for 90 days from a lender that uses the exact interest method. You can borrow $5,600 at 12% for 90 days from a lender that uses the ordinary interest method.

31. Which lender offers the loan with the lowest interest?

32. How much less interest will you pay?

33. **STRETCHING YOUR SKILLS** A family sells their home for $150,000 through a real estate agent who deducts $9,000 commission. Other costs they were charged to complete the sale totaled $1,875. What percent of the sale price did the family receive, to the nearest whole percent?

Mixed Review

34. $632.7 + 25.23 + 0.17$

35. 4.15×0.822

36. Write $5\frac{7}{8}\%$ as a decimal.

37. $80 is $6\frac{1}{2}\%$ of what amount?

38. $5.32 is 5% less than what amount?

39. Ellen Carson's sales for 5 months were $26,908, $28,386, $28,730, $27,290, and $29,009. What must be her sales next month if she wants her monthly sales average to be $28,000 for the 6 months?

40. Klaus Reinhardt, a secretary, is paid a yearly salary of $24,960. This is equal to how much a week?

41. On April 31, Steve Daley's balances were checkbook, $339.11, and bank statement, $394.62. A service charge of $1.74 had not been deducted in the checkbook. Checks outstanding were 134, $41.32; 135, $3.18; 137, $12.75. Prepare a reconciliation statement for Steve.

42. Shannon Burke is a substitute teacher. On the days she works she is paid $85 a day. What is Shannon's pay for 3 days of work?

Anthony Delgado has a credit card with an APR of 15%. His balance that is subject to finance charges is $285.92

43. What is his monthly periodic rate?

44. What are the periodic finance charges using a monthly periodic rate?

Photodisc/Getty Images

5-2

Calculating Interest

GOALS

- Calculate interest using simple interest tables
- Calculate interest using the daily interest factor

KEY TERM

- daily interest factor

Start Up ▶▶▶

If today is November 2nd and you have a project due on December 16th, how many days do you have to complete the project?

Math Skill Builder

Review these math skills and solve the exercises.

1 Find the number of days between the two given dates.
Find the number of days between January 3 and January 15: 15 − 3 = 12

1a. March 11 and March 31 **1b.** September 5 and September 10

2 **Divide** by 100.
Find the quotient. $7,200 ÷ $100 = 72

1a. $890 ÷ $100 **1b.** $12,089 ÷ $100 **1c.** $103,278 ÷ $100

3 **Multiply** money amounts by decimals and round to nearest cent.
Find the product. $475 × 0.6283 = $298.442, or $298.44

2a. $1,800 × 0.4186 **2b.** $943 × 0.5289 **2c.** $11,093 × 0.3752

Simple Interest Tables

Most banks use computers or specially programmed calculators to find interest and calculate payment amounts on loans. However, some lenders use interest tables (like the one shown on the next page) as a quick reference chart.

The table shows the interest on $100 for a 365-day year for up to a 31-day loan. To find the interest on any amount of money using the table, follow these steps:

1. To find the number of hundreds of dollars in the principal, divide the principal by $100. Simply move the decimal in the principal two places to the left.
2. Use the number from the chart that matches your interest rate and time and multiply it by the number of hundreds in the principal.

SIMPLE INTEREST TABLE

Interest on $100 for a 365-Day Year

Time (Days)	8%	$8\frac{1}{2}\%$	9%	$9\frac{1}{2}\%$	10%	$10\frac{1}{2}\%$	11%	$11\frac{1}{2}\%$	12%	$12\frac{1}{2}\%$
1	0.0219	0.0233	0.0247	0.0260	0.0274	0.0288	0.0301	0.0315	0.0329	0.0342
2	0.0438	0.0466	0.0493	0.0521	0.0548	0.0575	0.0603	0.0630	0.0658	0.0685
3	0.0658	0.0699	0.0740	0.0781	0.0822	0.0863	0.0904	0.0945	0.0986	0.1027
4	0.0877	0.0932	0.0986	0.1041	0.1096	0.1151	0.1205	0.1260	0.1315	0.1370
5	0.1096	0.1164	0.1233	0.1301	0.1370	0.1438	0.1507	0.1575	0.1644	0.1712
6	0.1315	0.1397	0.1479	0.1562	0.1644	0.1726	0.1808	0.1890	0.1973	0.2055
7	0.1534	0.1630	0.1726	0.1822	0.1918	0.2014	0.2110	0.2205	0.2301	0.2397
8	0.1753	0.1863	0.1973	0.2082	0.2192	0.2301	0.2411	0.2521	0.2630	0.2740
9	0.1973	0.2096	0.2219	0.2342	0.2466	0.2589	0.2712	0.2836	0.2959	0.3082
10	0.2192	0.2329	0.2466	0.2603	0.2740	0.2877	0.3014	0.3151	0.3288	0.3425
11	0.2411	0.2562	0.2712	0.2863	0.3014	0.3164	0.3315	0.3466	0.3616	0.3767
12	0.2630	0.2795	0.2959	0.3123	0.3288	0.3452	0.3616	0.3781	0.3945	0.4110
13	0.2849	0.3027	0.3205	0.3384	0.3562	0.3740	0.3918	0.4096	0.4274	0.4452
14	0.3068	0.3260	0.3452	0.3644	0.3836	0.4027	0.4219	0.4411	0.4603	0.4795
15	0.3288	0.3493	0.3699	0.3904	0.4110	0.4315	0.4521	0.4726	0.4932	0.5137
16	0.3507	0.3726	0.3945	0.4164	0.4384	0.4603	0.4822	0.5041	0.5260	0.5479
17	0.3726	0.3959	0.4192	0.4425	0.4658	0.4890	0.5123	0.5356	0.5589	0.5822
18	0.3945	0.4192	0.4438	0.4685	0.4932	0.5178	0.5425	0.5671	0.5918	0.6164
19	0.4164	0.4425	0.4685	0.4945	0.5205	0.5466	0.5726	0.5986	0.6247	0.6507
20	0.4384	0.4658	0.4932	0.5205	0.5479	0.5753	0.6027	0.6301	0.6575	0.6849
21	0.4603	0.4890	0.5178	0.5466	0.5753	0.6041	0.6329	0.6616	0.6904	0.7192
22	0.4822	0.5123	0.5425	0.5726	0.6027	0.6329	0.6630	0.6932	0.7233	0.7534
23	0.5041	0.5356	0.5671	0.5986	0.6301	0.6616	0.6932	0.7247	0.7562	0.7877
24	0.5260	0.5589	0.5918	0.6247	0.6575	0.6904	0.7233	0.7562	0.7890	0.8219
25	0.5479	0.5822	0.6164	0.6507	0.6849	0.7192	0.7534	0.7877	0.8219	0.8562
26	0.5699	0.6055	0.6411	0.6767	0.7123	0.7479	0.7836	0.8192	0.8548	0.8904
27	0.5918	0.6288	0.6658	0.7027	0.7397	0.7767	0.8137	0.8507	0.8877	0.9247
28	0.6137	0.6521	0.6904	0.7288	0.7671	0.8055	0.8438	0.8822	0.9206	0.9589
29	0.6356	0.6753	0.7151	0.7548	0.7945	0.8342	0.8740	0.9137	0.9534	0.9932
30	0.6575	0.6986	0.7397	0.7808	0.8219	0.8630	0.9041	0.9452	0.9863	1.0274
31	0.6795	0.7219	0.7644	0.8068	0.8493	0.8918	0.9342	0.9767	1.0192	1.0616

EXAMPLE 1

Find the interest from July 8 to July 28 on $850 at 12%.

SOLUTION

Find the number of days from July 8 to July 28. $28 - 8 = 20$ days

Find the interest on $100 for 20 days at 12%

$0.6575   interest from the table

$850 ÷ $100 = 8.5   the number of $100s in the principal

Multiply the interest by the number of 100s.

$0.6575 × 8.5 = $5.588, or $5.59

Problem Solving *Tip*

To avoid reading the wrong table amount, place a ruler or piece of paper under the line you need to read.

✔ CHECK YOUR UNDERSTANDING

A. Find the interest from April 6 to April 18 on $620 at 10%.

B. Find the interest from December 7 to December 18 on $550 at $9\frac{1}{2}\%$

When the number of days you want is not shown in the table, you must combine multipliers to get the number.

EXAMPLE 2

Find the interest from May 20 to June 25 on $450 at 8%.

SOLUTION

May 20 − May 30 = 10 days

June 1 − June 25 = 25 days

Total = 35 days

$0.6575 interest on $100 for 30 days at 8%

$0.1096 interest on $100 for 5 days at 8%

Add the interest for 30 days and 5 days

$0.6575 + $0.1096 = $0.7671 interest on $100 for 35 days

$450 ÷ $100 = 4.5 the number of $100s in the principal

Multiply the interest for 35 days by the number of 100s in the principal.

$0.7671 × 4.5 = $3.452, or $3.45

Interest for a rate not shown on the table can be found in much the same way. For example, the interest on $100 @ $18\frac{1}{2}$% for 20 days is the sum of the amount for 9% ($0.4932) and the amount for $9\frac{1}{2}$% ($0.5205).

✔ CHECK YOUR UNDERSTANDING

C. Find the interest from May 1 to June 9 on $1,320 at 10%.

D. Find the interest from August 4 to October 5 on $740 at 12%.

Problem Solving Tip

This old poem has been used by many to remember the number of days in the months:

Thirty days hath September, April, June and November.

All the rest have thirty-one, except February, which has 28.

Photodisc/Getty Images

Daily Interest Factor

The **daily interest factor** tells you how much interest a note is accumulating per day. You can use the daily interest factor to calculate the interest on a note.

Before you to substitute a number for the rate into the daily interest factor formula, change the percent to a decimal.

$$\textbf{Daily Interest Factor} = \textbf{Principal} \times \frac{\textbf{Rate}}{\textbf{Number of Days in Year}}$$

Business Tip

Exact interest method uses 365 days in a year.

Ordinary interest method uses 360 days in a year.

EXAMPLE 3

Find the daily interest factor for $100 borrowed at 9% exact interest.

SOLUTION

Use the daily interest factor formula.

$100 \times \frac{0.09}{365} = \$0.0247$

✔ CHECK YOUR UNDERSTANDING

E. What is the daily interest factor for $600 borrowed at 18% exact interest?

F. What is the daily interest factor for $1,200 borrowed at 6% ordinary interest?

You can use the daily interest factor to calculate the interest due over a period of days.

EXAMPLE 4

Find the ordinary interest from March 3 to March 15 on $1,000 at 15% interest.

SOLUTION

Find the daily interest factor and multiply it by the number of days

$1{,}000 \times \frac{0.15}{360} = \$0.4167$ daily interest factor

$15 - 3 = 12$ days number of days from March 3 to March 15

$\$0.4167 \times 12 = \$5.00$

✔ CHECK YOUR UNDERSTANDING

G. Find the ordinary interest from November 8 to November 22 on $750 at 9% interest.

H. Find the ordinary interest from August 5 to September 4 on $2,500 at 10% interest.

Math *Tip*

Round the daily interest factor to the nearest ten-thousandth. If you round to the nearest cent, there will be too great of an error when you use the daily interest factor in other calculations.

Spreadsheet *Tip*

Spreadsheets can be used to find the days between dates. Enter the first date in B1 and the other date in B2. Then in B3 enter: = B2 − B1. Next format B3 to display a *number* with *no decimal places.*

Communication

Search the Web to find at least 4 calculators that you can use online. Include simple interest and compound interest calculators, Then, test each calculator by solving these two problems with them:

1. Find the simple interest on $1,200 for $\frac{1}{2}$ year at 12%.
2. Find the compound interest on $1,200 for 1 year at 5% compounded monthly.

Choose 1 of the calculators and write a review containing the name of each calculator, the major types of calculations it can handle, its Web address, the ease with which you can understand how to use it, and whether it answered the problems above correctly.

Wrap Up ▸▸▸

The elapsed time you found should have been 45 days. What does "elapsed time" mean? Explain the steps you took to determine elapsed time. Why do you need to find the elapsed time? Work with a partner. Have one person name a starting date and an ending date and the other find the elapsed time. Exchange roles and repeat.

Exercises

Find the number of days between the two given dates.

1. April 5 to April 27

2. September 18 to September 30

3. January 19 to February 16

4. July 6 to September 12

Find the product or quotient.

5. $418 ÷ 100

6. $297 × 0.2851

7. $310,790 ÷ 100

8. $5,300 × 0.9184

9. $51,280 ÷ 100

10. $31,627 × 0.5028

Use the Simple Interest Table to find the interest to the nearest cent.

11. $500 @ 8% for 20 days

12. $380 @ 12% for 12 days

Use the Simple Interest Table to find the interest for the given dates to the nearest cent.

13. $6,150 @ 9% for 12/5 to 12/30

14. $275 @ $12\frac{1}{2}$% for 3/8 to 3/28

15. $400 @ $9\frac{1}{2}$ %for 5/10 to 6/24

16. $725 @ 11% for 7/12 to 9/10

17. $270 @ 18% for 4/22 to 8/20

18. $540 @ 21% 1/28 to 2/17

Lucia Flores borrowed $1,700 on a note for 60 days with interest at 12%.

19. Using the Simple Interest Table, what interest did she pay?

20. What total amount did she owe when the note was due?

Find the daily interest factor. Round to the nearest ten-thousandth.

21. $825 @ 17% exact interest

22. $975 @ 15% exact interest

23. $450 @ 11% ordinary interest

24. $625 @ 12% ordinary interest

25. $2,250 @ 9% exact interest

26. $1,250 @ 6% exact interest

27. $1,500 @ 10% ordinary interest

28. $1,000 @ 18% ordinary interest

Use the daily interest factor to find the interest to the nearest cent.

29. \$725 from April 12 to April 28 @ 11% exact interest
30. \$1,200 from October 2 to October 21 @ 9% ordinary interest
31. \$550 from January 19 to February 15 @ 6% exact interest
32. \$950 from March 29 to April 10 @ 15% ordinary interest
33. \$2,150 from December 15 to February 9 @ 7% ordinary interest
34. \$1,650 from September 23 to November 2 @18% exact interest

INTEGRATING YOUR KNOWLEDGE On March 5, Jake Lowry signed a note for \$15,000 at 15% exact interest. He paid the note on June 3.

35. What amount of interest did Jake owe?
36. What was the total amount due on June 3?
37. On July 13, Rosa D'Lario borrowed \$9,000 to buy a new car for \$14,500. She signed a note with exact interest at 8%. If she paid 100% of the note off on September 24, what total amount did she owe?
38. **FINANCIAL DECISION MAKING** You can borrow \$2,500 at 15% ordinary interest for 6 months from one lender. Another lender offers you a loan of \$2,500 at 14% ordinary interest for 8 months. Which loan is the better deal? Justify your decision.

Mixed Review

39. $\frac{1}{8} + \frac{1}{6}$
40. 2,078.01 − 43.098
41. $12\frac{1}{4} - 3\frac{1}{2}$
42. What amount is $2\frac{1}{4}$% of \$900?
43. What amount is $5\frac{3}{5}$% of \$1,200?
44. The interest on a loan of \$8,600 for 3 months is \$258. What is the rate of interest?
45. Jeanne Dixon is paid \$2.25 for each fan she assembles. During the five days of last week, she assembled these fans: 27, 28, 33, 30, 31. What was Jeanne's gross pay for the week?
46. Find the interest on \$745 for 25 days at 8% using the Simple Interest Table.
47. Residents that live in Devon have an average annual taxable income of \$38,445. The city's income tax rate is $\frac{7}{8}$%. How much does the average resident pay in city tax each year?

Photocisc/Getty Images

48. Use the Simple Interest Table in Lesson 3-2 to find the interest on a \$1,400 loan at $12\frac{1}{2}$% for 29 days.

5-3

Installment Loans

GOALS

- Calculate the installment price and finance charge on an installment plan purchase
- Calculate the number and amount of monthly payments
- Calculate the interest, principal payment, and new balance on an installment loan

KEY TERMS

- down payment
- finance charge
- installment loan

Start Up

Darrel just graduated from high school. He needs a car to get a job but he also needs a job to afford a car. What choices does Darrel have to solve his problem?

Photodisc/Getty Images

Math Skill Builder

Review these math skills and solve the exercises that follow.

1. **Add** and **subtract** dollar amounts.
Find the sum. $109.45 + $29.01 = $138.46
Find the difference. $244.21 − $62.09 = $182.12

1a. $1,905.34 + $804.23 **1b.** $15,378.82 + $598.38

1c. $5,074 − $2,985 **1d.** $56.89 − $12.98

1e. $5,083.20 − $799.43 **1f.** $8,001.35 + $56.08

2. **Multiply** dollar amounts by whole numbers and decimals.
Find the products. $34.19 × 6 = $205.14 and $106.80 × 0.015 = $1.602, or $1.60

2a. $537 × 8 **2b.** $1,790 × 0.2

2c. $450 × 0.045 **2d.** $0.95 × 120

3. **Divide** dollar amounts by whole numbers and dollar amounts.
Find the quotients. $340 ÷ 8 = $42.50 and $56.30 ÷ $500 = 0.1126

3a. $1,080 ÷ 12 **3b.** $36 ÷ $1,200

3c. $171 ÷ $3,800 **3d.** $54 ÷ 36

Installment Price and Finance Charge

Beneda Miroslav 2009/Shutterstock.com

Sound systems, boats, cars, furniture, and many other items can be bought on an installment plan, also called a time payment plan. When you buy on an *installment plan,* you are borrowing money and paying it back in part payments.

You may have to make a **down payment**, or part of the purchase price. An *installment contract* will outline the responsibility for paying the unpaid balance.

The installment price is higher than the cash price because the seller adds a **finance charge** to the cash price. This charge pays the seller interest on the money and covers the extra cost of doing business on the installment plan. The finance charge is the difference between the installment price and the cash price.

EXAMPLE 1

A desktop computer system has a cash price of $1,200. To buy it on an installment plan, you pay $100 down and $38 a month for 36 months. Find the finance charge. By what percent is the installment price greater than the cash price?

SOLUTION

Add the total of the monthly payments and the down payment to find the installment price.

$\$38 \times 36 = \$1{,}368$ total monthly payment

$\$1{,}368 + \$100 = \$1{,}468$ installment price

Subtract the cash price from the installment price to find the finance charge.

$\$1{,}468 - \$1{,}200 = \$268$ finance charge

Divide the finance charge by the cash price to find the percent that the installment price is greater than the cash price.

$\$268 \div \$1{,}200 = 0.22\overline{3}$, or $22\frac{1}{3}\%$ percent greater

✔CHECK YOUR UNDERSTANDING

A. You can buy a watch for $125 cash or pay $25 down and the balance in 12 monthly payments of $9. What is the installment price? By what percent would your installment price be greater than the cash price?

B. A digital audio player that sells for $169.95 can be bought for $20 down and $26.17 a month for 6 months. What is the installment price? By what percent, to the nearest tenth, does the installment price exceed the cash price?

Monthly Installment Payments

Sometimes you may know the installment price and down payment and need to find the amount of the monthly payment or the number of months to pay.

EXAMPLE 2

The installment price of a set of water skis is \$190. You must pay \$50 down and make payments for 16 months. What will be your monthly payments?

SOLUTION

Subtract the down payment from the installment price to find the remaining amount to pay.

$\$190 - \$50 = \$140$ remainder to pay

Divide the remaining amount to pay by the number of months to pay to find the monthly payment.

$\$140 \div 16 = \$8.75$ monthly payment

Photodisc/Getty Images

✔ CHECK YOUR UNDERSTANDING

C. A scuba diver's wetsuit costs \$175 on the installment plan. You must make a down payment of \$25 and make payments for 15 months. What will be your monthly payments?

D. A refrigerator sells for \$1,044 on the installment plan. After making a down payment of \$100, you pay \$59 a month. How many months will it take to pay for the refrigerator?

Installment Loans

Rather than pay a retail store a down payment and monthly payments, you can obtain an **installment loan** from a bank or credit union. You repay the principal and interest in installments, usually monthly. Typically, the interest rate on an installment loan from a bank will be less than the rate offered for a retail store installment plan.

Many lenders calculate payments so that each payment is the same amount. This payment method is called the *level payment plan*. From each payment, the interest due for that month is deducted. The payment amount remaining after deducting the interest is applied to the principal.

If the interest rate on the loan is 1.5% a month, this rate is equal to 18% a year $(1.5 \times 12 = 18)$. Sometimes a service charge is added to the cost of the loan.

On the next page is a schedule of payments on a one-year, \$500 loan at 18%. The loan was repaid in 12 equal monthly payments of \$45.84.

Spreadsheet *Tip*

The spreadsheet function *PMT* lets you compute the monthly payment needed in a level payment loan. In Excel, the function for a \$500 loan for 1 year at 18% is =PMT(0.18/12,12,500,0) 0.18/12 is the monthly interest rate. The number of months is 12. The principal is 500 and 0 means the monthly payment is made at the end of the month. A 1 would mean the monthly payment is made at the beginning of the month.

Loan Repayment Schedule				
Month	Monthly Payment	Interest Payment	Applied to Principal	Balance
1	45.84	7.50	38.34	461.66
2	45.84	6.92	38.92	422.74
3	45.84	6.34	39.50	383.25
4	45.84	5.75	40.09	343.15
5	45.84	5.15	40.69	302.46
6	45.84	4.54	41.30	261.16
7	45.84	3.92	41.92	219.24
8	45.84	3.29	42.55	176.68
9	45.84	2.65	43.19	133.50
10	45.84	2.00	43.84	89.66
11	45.84	1.34	44.50	45.16
12	45.84	0.68	45.16	0.00
Totals	550.08	50.08	500.00	

Notice that the interest paid in any month is equal to the unpaid balance multiplied by the monthly interest rate. A loan that uses this method of allocating interest is called a *simple interest installment loan.*

EXAMPLE 3

The Winstons borrowed \$500 on a one-year simple interest installment loan at 18% interest. The monthly payments were \$45.84. Find the amount of interest, amount applied to the principal, and the new balance for the first monthly payment.

SOLUTION

Calculate the monthly interest rate: $18\% \div 12 = 1.5\%$

Use $I = P \times R \times T$: $I = \$500 \times 0.015 \times 1 = \$7.50$

Subtract the interest from the monthly payment: $\$45.84 - \$7.50 = \$38.34$

Subtract the amount applied to principal from the previous balance.

$\$500.00 - \$38.34 = \$461.66$ new balance

Math *Tip*

Interest Formula:

$I = P \times R \times T$

I = *interest*

P = *principal*

R = *rate*

T = *time*

✔ CHECK YOUR UNDERSTANDING

E. Benito Diaz borrowed \$1,000 on a one-year simple interest installment loan at 15% interest. The monthly payments were \$90.26. Find the amount of interest, amount applied to the principal, and the new balance for the first monthly payment.

F. Lillian Dish signed a \$2,500, 6-month simple interest installment loan at 18% interest. The monthly payments were \$438.81. Find the amount of interest, amount applied to the principal, and the new balance for the first two monthly payments.

Many lenders advertise their rates on the Internet. Access www.cengage.com/school/business/businessmath and click on the link for Chapter 5. Compare the interest rates to local lenders.

Wrap Up ▸▸▸

Darrel is going to have to borrow money to purchase a car. He can borrow the money from the auto dealer, buying the car on the installment plan. Or, he can borrow the money through an installment loan from another lender, such as a bank, credit union, or finance company. Since Darrel probably doesn't have other collateral for the loan, he will have to use the car he is buying as collateral. That means that if Darrel fails to make his loan payments, the lender can take back, or *repossess* the car.

TEAM Meeting

With two other students, investigate the borrowing terms of lenders for car loans. Check out the interest rates and monthly payments they would charge for a four-year, $15,000 auto loan. The lenders should include (a) an auto dealer, (b) a bank, (c) a credit union, and (d) a finance company.

Prepare a comparison chart that includes the following for each lender.

- interest rate
- monthly payments
- total payments
- finance charge

Exercises

Find the sum or difference.

1. $2,500 + $89.15
2. $159.95 + $12.28
3. $650 − $75
4. $1,079.34 − $418.73

Find the product or quotient.

5. $2,450 × 0.18
6. $389.67 × 2.5
7. $157 × 9
8. $4,584 ÷ 6
9. $81.60 ÷ $4,800
10. $1,000 ÷ 10

Find the installment price and finance charge for each item.

	Item	Cash Price	Installment Terms
11.	Digital Camera	$1,500	$100 down; $70.57 a mo. for 24 mos.
12.	Sleeper Sofa	$950	$50 down; $51.99 a mo. for 20 mos.
13.	Computer System	$2,150	$225 down; $228.40 a mo. for 9 mos.

Solve.

14. You buy a CD changer for a car's audio system for $25 down and a total installment price of $297.34. You pay $30.26 per month. For how many months will you have to make payments?

© Ramzi Hachicho 2008/Shutterstock.com

15. A novelty watch that sells for $60 cash may be bought for $6 down and $5.76 a month for 10 months. By what percent is the installment price greater than the cash price?

16. You can buy aluminum louvers for the rear window of your car for $180 cash or pay $45 down and the balance in 12 monthly payments of $13.50. By what percent would your installment price be greater than the cash price?

Find the monthly interest payment, amount applied to principal, and new balance for the first month for each simple interest installment loan.

	Amount Financed	Number of Payments	Monthly Payment	Annual Interest Rate
17.	$500	6	$87.02	15%
18.	$1,200	6	$207.06	12%
19.	$800	12	$73.34	18%

20. **CRITICAL THINKING** Look at the Loan Repayment Schedule chart. If the interest rate is 18% per year, why is the total of the interest payments less than $90, or $500 × 0.18?

A member of a credit union borrows $920 on a simple interest installment loan at 12% agreeing to repay it in 12 equal monthly payments of $81.74.

21. What was the total finance charge on the loan?

22. What was the interest paid for the first month?

23. What was the amount applied to principal at the end of the first month?

24. What was the new balance at the end of the first month?

Kay borrowed $400 from a finance company on a simple interest installment loan and repaid it in 6 monthly payments of $70.81. The finance charge rate was 21%.

25. What was the total finance charge on the loan?

26. What was the interest paid for the first month?

27. What was the amount applied to principal at the end of the first month?

28. What was the new balance at the end of the first month?

29. **FINANCIAL DECISION MAKING** You can buy a flat panel computer screen for $720 in cash or $50 down and 12 monthly payments of $63.47 on the installment plan from the dealer. You can also obtain a simple interest installment loan from another lender by signing a promissory note and using your car as collateral. The face of the note would be for $720 and would be payable, along with interest at 12%, one year later. If you do not have the cash but want the screen now, which loan would be the best for you? Why?

Mixed Review

30. $43{,}109 \times 150$

31. $7{,}082 \div 1{,}000$

32. $2\frac{2}{5} \times 5\frac{1}{8}$

33. $\$24 \times \frac{3}{4}$

34. $\frac{2}{5} + \frac{2}{3}$

35. What is $36\frac{1}{2}\%$ as a decimal?

36. Find the simple interest on $2,812 for 1 month at 9% annually.

37. The town of Glen Gary charges its residents an income tax of $\frac{1}{2}\%$ of their taxable income. Julio Gonzalez lives in Glen Gary and has a taxable income of $120,560. What is his income tax?

38. Sallie Woo worked these hours last week: Monday, 10; Tuesday, $8\frac{1}{4}$; Wednesday, $9\frac{1}{2}$; Thursday, 6; Friday, 8; Saturday, 5. Sallie is paid $11.40 per hour for regular hours, time-and-a-half for overtime during the week, and double time for weekend hours. If Sallie works on an 8-hour day basis, what was her pay for the week?

39. Tischa Fogoros' credit card statement for June showed no previous balance on June 1. A purchase of $263.18 was posted on June 9, and a $15.85 credit for returned merchandise has posted on June 17. Tischa's credit card company uses the average daily balance method to compute finance charges based on a 1.52% monthly APR. Find the finance charge for June and the new balance.

40. Luann Weber is paid a salary of $520 a week and commission of 4.5% on all sales. Her sales last week were $8,600. Find her total earnings for the week.

41. Tom Ridley invested $10,000 in a certificate of deposit. He cashed the CD before the end of the certificate's term. The penalty for early withdrawal was 1 month's interest at 6% annual percentage rate. What was the amount of the penalty?

42. You have $500 on deposit in a one-year certificate of deposit that pays 6% annual interest compounded quarterly. What is the effective annual interest rate, to the nearest tenth percent?

43. Ann Quinland bought a set of golf clubs on the installment plan for $395. She paid $45 down and the balance in equal monthly installments of $25 each. How many months did it take Ann to pay for the set?

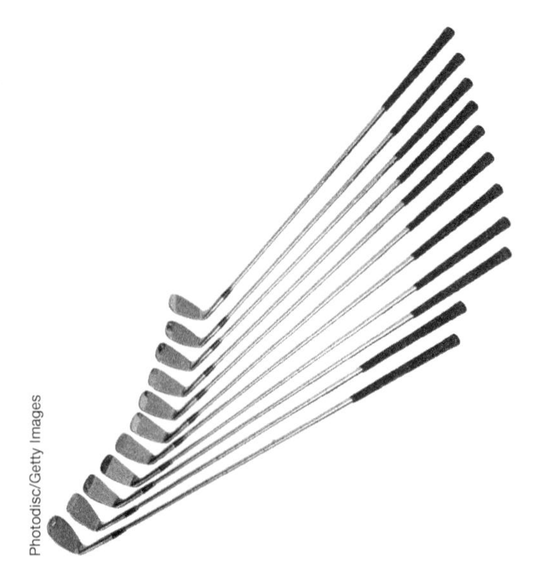
Photodisc/Getty Images

5-4

Early Loan Repayments

GOALS

- Calculate the final payment to pay an installment loan off early
- Calculate the savings in interest to pay an installment loan off early

KEY TERM

- prepayment penalty

Start Up ▶▶▶

Eva Lewis borrowed $500 on a 12-month installment loan at 18% interest. After she made her first payment, she received a $600 bonus from work. She is trying to decide whether to use the bonus to pay off the loan early or to keep making the monthly payments. What should Eva consider to help her make her decision?

Digital Vision/Getty Images

Math Skill Builder

Review these math skills and solve the exercises that follow.

1 **Rewrite** percentages as decimals.
Rewrite 84% as a decimal. 84% = 0.84

1a. 52.9% | **1b.** 253%

1c. 0.6% | **1d.** 7.76%

2 **Subtract** dollar amounts.
Find the difference. $42,648.29 − $35,763.11 = $6,885.18

2a. $25,754 − $22,904

2b. $169.29 − $75.98

2c. $132.05 − $98.99

2d. $70,000 − $37,549

3 **Multiply** dollar amounts by whole numbers and decimals.
Find the product. $390.30 × 0.027 = $10.538, or $10.54

3a. $6,572 × 3 | **3b.** $31,290 × 0.3

3c. $145 × 0.074 | **3d.** $331 × 2.45

3e. $515 × 9 | **3f.** $2.65 × 425

Early Loan Repayment

For simple interest installment loans, the monthly interest rate is applied to the unpaid balance of the loan. If you pay the loan off early, you simply pay the unpaid balance plus the current month's interest as the final payment. In some cases, there may be a **prepayment penalty**, which is a fee charged if you pay the loan off early. A prepayment penalty must be disclosed in the original terms of the loan.

EXAMPLE 1

Vern Goode took out a $5,000 simple interest loan at 6% interest for 24 months to buy a car. His monthly payment is $221.60. After making payments for 12 months, his balance is $2,574.79. He decides to pay the loan off with his next payment. How much will his final payment be?

© Stepanov 2008/Shutterstock.com

SOLUTION

Find the interest due for the next month and add it to the balance. Calculate the monthly interest rate.

$6\% \div 12 = 0.5\% = 0.005$ monthly interest rate

Substitute known values in the interest formula

$I = \$2{,}574.79 \times 0.005 \times 1 = \$12.87$ $I = P \times R \times T;\ T = 1$ mo

Add the balance to the current month's interest:

$\$2{,}574.79 + 12.87 = \$2{,}587.66$ final payment

✔ CHECK YOUR UNDERSTANDING

A. Mario Mineto had a 12-month, $2,000 simple interest loan at 9% interest. He repaid the loan in full with the sixth payment when his balance was $1,188.40. How much was his final payment?

B. Emily Polinkski repaid a 9-month, $3,000 installment loan at the end of 6 months. Her interest rate was 15%, and her balance was $1,374.82. How much was her final payment?

One reason to pay off an installment loan early is to pay less interest.

EXAMPLE 2

How much interest did Vern Goode (from Example 1) save by paying off his loan early?

SOLUTION

Calculate how much Vern would have paid if he had paid it on the payment schedule for all 24 months. Subtract the amount that Vern did pay.

Multiply the monthly payment amount by 24.

$\$221.60 \times 24 = \$5{,}318.40$ total that would be paid in 24 payments

Multiply the monthly payment amount by the number of payments Vern paid.

Business Tip

One way to evaluate whether to pay off a loan early is to consider the costs and benefits of paying off the loan versus investing the money and continuing to pay on the loan.

$221.60 × 12 = $2,659.20   amount paid to date before final payment

Add the amount paid to the final payment.

$2,659.20 + $2,587.66 = $5,246.86   total amount paid with early payoff

Subtract the amount Vern paid from the amount of scheduled payments.

$5,318.40 − $5,246.86 = $71.54   amount of interest saved

✔ CHECK YOUR UNDERSTANDING

C. In Exercise A, Mario's monthly payment was $174.90. How much interest will Mario save by paying off his loan with the sixth payment?

D. In Exercise B, Emily's monthly payment was $354.51. How much interest will Emily save by paying off her loan in the sixth month?

Consumer Alert

Penalty for Prepayment

Before signing a loan document, you should read the terms of the loan and also ask questions. An important question to ask is if there is a prepayment penalty or if the loan is a Rule of 78 loan.

A prepayment penalty is a financial penalty for paying off the loan early. Prepayment penalties may be a set amount, a percentage of the loan or a certain number of months' interest. In a Rule of 78 loan, you pay more interest in the early months of the loan, so if you pay off the loan early you will have paid more in interest than with a simple interest loan. Rule of 78 loans are illegal in many states.

If you do not pay off the loan early, prepayment penalties and Rule of 78 loans will not increase the total amount that you pay. In some cases, you can get a better interest rate if you will agree to certain prepayment penalties. Before taking out a loan with prepayment penalties, consider your future plans and financial situation carefully.

Wrap Up ▶▶▶

Eva should consider how much she could save by paying off the loan early. She should find out if the loan has any prepayment penalties. She should consider the value of not having a monthly debt payment.

Exercises

Rewrite as a decimal.

1. 245.6%

2. 0.35%

Perform the indicated operation.

3. $4,208 − $3,489

4. $63,418.36 − $48,897.57

5. $189.87 × 12

6. $45.63 × 3.148

Solve.

7. Rachel Carr has a 2-year, $4,000 car loan at 9%. After paying on the loan for 18 months, Rachel has a balance of $1,068.22. She decides to pay off the loan with the next payment. How much will she have to pay for the final payment?

8. Terry O'Doole has a 4-year, $6,000 loan at 12%. After paying on the loan for 24 months, Terry has a balance of $3,356.52. He decides to pay off the loan with the next payment. How much will he have to pay for the final payment?

Lucy Smola borrowed $2,500 at 18% for 12 months. Her monthly payment is $229.20.

9. If Lucy makes the monthly payment for 12 months, how much will she pay back to the lender?

10. Lucy has the opportunity to pay off the loan with her fourth payment. Her current balance is $1,916.23. How much will she have to pay to pay off the loan?

11. How much did Lucy pay in total for the loan if she pays off the loan with her fourth payment?

12. How much will Lucy save if she pays off the loan early?

Levi Stein's 18-month, $6,500 loan has a 12% interest rate. His monthly payment is $396.38. After making payments for 9 months, the loan's balance is $3,395.43.

13. How much will Levi have to pay in order to pay off the loan with the next payment?

14. How much will Levi save if he pays off the loan early?

15. **FINANCIAL DECISION MAKING** Romero has a simple interest installment plan to pay for his computer. For his birthday, he receives a gift of money that is enough to pay off his installment loan. His friend encourages Romero to spend the money on other things and keep the loan on the computer. What would you advise Romero? Why?

16. **CRITICAL THINKING** Some lenders charge a prepayment penalty if you pay off a loan early. Why do you think some lenders do this?

Mixed Review

17. Round 29,458 to the nearest hundred.

18. Estimate the product of $34.56 × 24.8.

19. What is the average of $35.12, $36.20, $32.98, and $33.56?

20. The cash price of a large color TV was $2,400. Dee Hart bought it for $240 down and 12 monthly payments of $196. By what percent did the installment price exceed the cash price?

21. Sheila Wiggins' regular-time pay rate is $12.60 an hour, time-an-a-half for overtime, and double time for work on Saturdays and Sundays. What is her overtime rate? What is her double-time rate?

5-5

Annual Percentage Rates

GOAL
- Calculate the APR on a loan

KEY TERM
- annual percentage rate

Start Up

José is planning to get an installment loan for $2,000 for 12 months. One lender offers him a 13% interest rate. Another lender offers him a 10% interest rate with additional finance charges of $50. Which loan is the better deal?

Photodisc/Getty Images

Math Skill Builder

Review these math skills and solve the exercises that follow.

1. **Round** dollar amounts to the nearest cent.
   Round $0.2978 to the nearest cent. $0.2978 = $0.30

   1a. $3.985 **1b.** $2.0793
   1c. $8.0049 **1d.** $0.455

2. **Multiply** dollar amounts by 10, 100, and 1,000.
   Find the product. $489.43 × 100 = $48,943

   2a. $189 × 100 **2b.** $208.97 × 1,000 **2c.** $790.72 × 10

3. **Divide** dollar amounts by dollar amounts to the nearest thousandth.
   Find the quotient to the nearest thousandth. $500 ÷ $3,000 = 0.1666, or 0.167

   3a. $420 ÷ $1,600 **3b.** $1,241 ÷ $8,560 **3c.** $2,445 ÷ $16,308

Annual Percentage Rate (APR)

Loans from different lenders can have very different rates, terms and fees. To evaluate and compare loans, you can use the **annual percentage rate**, the APR, which is the cost of credit for one year, expressed as a percentage.

To find the rate of interest on a single-payment loan for one year, you divide the interest paid in a year by the principal. Finding the rate of finance charges on an installment loan is not as easy. The cost of borrowing money may also include service charges. Also, since you make payments on the loan each month, you are not borrowing the whole principal for the full time of the loan.

The *Truth in Lending Act* makes the lender tell the borrower what annual percentage rate (APR) is charged on the loan. The APR is usually higher than the interest rate of your loan.

The easiest way to find the annual percentage rate is to use tables like the ones shown below. To use the tables, you need to know the number of monthly payments for the loan and the finance charge per $100 of the amount financed.

To find the *finance charge per $100* of the amount financed, divide the finance charge by the amount financed. Then multiply the quotient by 100.

$$\textbf{Finance Charge per \$100 of Amount Financed} = \frac{\textbf{Finance Charge}}{\textbf{Amount Financed}} \times \$100$$

After you have found the finance charge per $100, you can use the tables to find the annual percentage rate

Number of Payments	Annual Percentage Rate										
	$12\frac{3}{4}\%$	13%	$13\frac{1}{4}\%$	$13\frac{1}{2}\%$	$13\frac{3}{4}\%$	14%	$14\frac{1}{4}\%$	$14\frac{1}{2}\%$	$14\frac{3}{4}\%$	15%	$15\frac{1}{4}\%$
	Finance Charge per $100 of Amount Financed										
3	2.13	2.17	2.22	2.26	2.30	2.34	2.38	2.43	2.47	2.51	2.55
6	3.75	3.83	3.90	3.97	4.05	4.12	4.20	4.27	4.35	4.42	4.49
9	5.39	5.49	5.60	5.71	5.82	5.92	6.03	6.14	6.25	6.35	6.46
12	7.04	7.18	7.32	7.46	7.60	7.74	7.89	8.03	8.17	8.31	8.45
15	8.71	8.88	9.06	9.23	9.41	9.59	9.76	9.94	10.11	10.29	10.47
Number of Payments	Annual Percentage Rate										
	$15\frac{1}{2}\%$	$15\frac{3}{4}\%$	16%	$16\frac{1}{4}\%$	$16\frac{1}{2}\%$	$16\frac{3}{4}\%$	17%	$17\frac{1}{4}\%$	$17\frac{1}{2}\%$	$17\frac{3}{4}\%$	18%
	Finance Charge per $100 of Amount Financed										
6	4.57	4.64	4.72	4.79	4.87	4.94	5.02	5.09	5.17	5.24	5.32
12	8.59	8.74	8.88	9.02	9.16	9.30	9.45	9.59	9.73	9.87	10.02
Number of Payments	Annual Percentage Rate										
	$18\frac{1}{4}\%$	$18\frac{1}{2}\%$	$18\frac{3}{4}\%$	19%	$19\frac{1}{4}\%$	$19\frac{1}{2}\%$	$19\frac{3}{4}\%$	20%	$20\frac{1}{4}\%$	$20\frac{1}{2}\%$	$20\frac{3}{4}\%$
	Finance Charge per $100 of Amount Financed										
6	5.39	5.46	5.54	5.61	5.69	5.76	5.84	5.91	5.99	6.06	6.14
12	10.16	10.30	10.44	10.59	10.73	10.87	11.02	11.16	11.31	11.45	11.59

EXAMPLE 1

The finance charge for a 6-month, $1,200 installment loan is $72. Find the annual percentage rate on the loan.

Math *Tip*

When necessary, round the finance charge to the nearest cent.

SOLUTION

Divide the finance charge by the amount financed.

$72 ÷ $1,200 = 0.06

0.06 × $100 = $6 Multiply the result by $100.

The finance charge per $100 of amount financed is $6.

Use the Annual Percentage Rate Tables. Read across the rows for 6 payments until you come to the amount closest to $6. Since $5.99 is the closest amount, use the rate, $20\frac{1}{4}\%$.

The annual percentage rate is $20\frac{1}{4}\%$.

✔CHECK YOUR UNDERSTANDING

A. Melina Cavaletti borrowed $800 on a loan with a finance charge of $78. Find the finance charge per $100 of the amount financed.

B. Chris Mathers borrowed $250 on a 12-month loan that had a finance charge of $20. Find the finance charge per $100 of the amount financed and the annual percentage rate.

Wrap Up ▶▶▶

One way to evaluate the loans is to consider the APR. The monthly payment for the loan at 13% is $178.63. He will pay back $2,143.56, or $143.56 in finance charges. The finance charge per $100 of amount financed is $7.18, or an APR of 13%. The monthly payment for the loan at 10% is $175.83. Including the $50 finance charge, he will pay back $2,159.96, or $159.96 in finance charges. The finance charge per $100 of amount financed is $8.00, or an APR of about $14\frac{1}{2}$%. Although the interest rate on the 13% loan is higher, the APR is lower. This means the cost of the credit per year is less than the loan with the 10% interest rate.

Financial Responsibility

Your Good Credit

Using credit cards and obtaining loans are very common practices to satisfy the desire to have something now. That debt does not come without risk or cost, however.

Consider some of the risks of buying on credit or with a loan:

- The item costs more because of the cost of interest.
- Payments may last longer than the pleasure you get from owning the item.
- Changes in your income can impact your ability to make timely payments.
- A cycle of debt can take months or years from which to recover.
- Falling behind on payments or defaulting on a loan will have a long-term impact on your ability to secure credit or a good interest rate in the future.

Chris is researching computer installment plans. The interest rate for the plans varies from 10%–29% depending on the credit score of the borrower. Below are three interest rates and the corresponding monthly payments for a 48-month installment plan to finance a $1,500 computer.

10%: $38.04 per month

19%: $44.85 per month

29%: $53.14 per month

Calculate the amount of interest paid over the life of the loan for each credit standing.

1. If Chris has good credit, he can get a 10% interest rate.
2. If Chris has average credit, he can get a 19% interest rate.
3. If Chris has poor credit, he can get a 29% interest rate.

Exercises

Round to the nearest cent.

1. $4.5627
2. $105.3978

Find the product.

3. $98,208 × 100
4. $389.74 × 1,000
5. $1,078.43 × 10
6. $38.95 × 100

Find the quotient to the nearest thousandth.

7. $186 ÷ $828
8. $482 ÷ $4,298.18
9. $384 ÷ $605
10. $78 ÷ $152.98

Find each finance charge per $100.

11. A loan of $3,200 with a finance charge of $480
12. A loan of $12,600 with a finance charge of $1,638

Find the annual percentage rate for the following loans.

13. Finance charge of $5.75 per $100 for 9 payments.
14. Finance charge of $2.50 per $100 for 3 payments.
15. Finance charge of $8.85 per $100 for 15 payments.
16. Finance charge of $23.55 per $500 for 6 payments.
17. Finance charges of $19.13 per $750 for 3 payments.
18. **CRITICAL THINKING** What specific pieces of information should you look for in any loan contract? Why is this information important?

You borrow $2,600 and repay the loan in 12 monthly installments of $232.

19. What was the finance charge on your loan?
20. What was the finance charge per $100 of the amount financed?
21. What was the annual percentage rate?

Olga Pozinski borrowed $1,300 and repaid it in 12 monthly payments of $116.

22. What was the finance charge on the loan?
23. What was the finance charge per $100 of amount financed?
24. What was the annual percentage rate?

Find the APR on each loan.

25. Charles repaid a loan of $1,600 in 15 monthly installments of $116.40 each.
26. Kieran repaid a loan of $900 in 6 monthly installments of $157.98 each.
27. Nyla repaid a loan of $2,500 in 12 monthly installments of $224.77.
28. Sule repaid a loan of $1,750 in 12 monthly payments of $160.85.
29. Vega repaid a loan $4,250 in 15 monthly installments of $348.8 each.
30. Thai repaid a loan of $1,125 in 9 monthly payments of $131.90 each.

INTEGRATING YOUR KNOWLEDGE Violeta Ramos is considering financing $4,000 of electronic equipment on a simple interest installment loan. She would make 12 monthly payments of $360.

31. Find the total amount Violeta will pay over the life of the loan.

32. How much interest will Violeta pay?

33. What is the finance charge per $100 of amount financed?

34. What is the annual percentage rate?

35. Violeta goes to a bank and is offered one year promissory note for 8%. How much interest will she pay on the note? How does this compare to the interest charges on the installment loan?

Mixed Review

36. $788 × 100 =

37. $9.75 × 1,000 =

38. $10.78 × 10 =

39. $3.97 × 100 =

40. Write $\frac{1}{4}$ as a ratio.

41. Rewrite $12\frac{5}{8}$ as a percent.

42. On January 31, Edith Nagel's bank statement balance was $516.24. Her check register showed that a deposit for $382.10 was outstanding. The following checks were also outstanding: #108, $45.93; #109, $108.12; #111, $6.87. What was the corrected bank statement balance?

43. Rhonda Peterson pays a city tax of 1.5% on her taxable income. She also pays a state tax of 3.25% on her taxable income. What total city and state income taxes does she pay if her taxable income is $38,109?

Photodisc/Getty Images

44. Hector Vadillo is a waiter and had these total checks for the week: Tuesday, $245.19; Wednesday, $299.74; Thursday, $349.73; Friday, $612.50; and Saturday, $575.33. If Hector received an average of 15% for tips during the week, how much money did he receive in tips?

45. Find the exact interest on $1,050 at 14% for 126 days.

Cory Mathis is married with 4 withholding allowances. Each week his employer deducts $18 in federal withholding taxes, 6.2% in Social Security taxes, 1.45% in Medicare taxes, and $58.77 for health insurance from his gross pay. His gross weekly wage is $609.

46. Find the total deductions.

47. Find his net pay.

48. Faye Rivera worked these hours last week: Monday, 8 hours; Tuesday, 6 hours; Wednesday, 7 hours; Thursday, 8 hours; and Friday, 7 hours. If she is paid $12 an hour, what was Faye's gross pay for the week?

49. Tyrone Wilson's yearly pay for the last three years was: $24,800, $25,900, and $29,760. What was Tyrone's average yearly pay?

Chapter *Review*

Vocabulary Review

Find the term, from the list at the right, that completes each sentence. Use each term only once.

annual percentage rate
daily interest factor
down payment
exact interest method
finance charge
interest
ordinary interest method
prepayment penalty
principal
promissory note
rate of interest
total amount due

1. A way to find interest that uses a 365-day year is the _?_.
2. The _?_ tells you how much interest a note is accumulating per day.
3. Interest, fees, and other charges paid on an installment loan or purchase is the _?_.
4. The true rate of interest on an installment loan is called _?_.
5. The principal of a loan and the interest on the loan is the _?_.
6. A financial penalty for paying a loan off early is a(n) _?_.
7. The amount charged for the use of money is called _?_.
8. The amount that is borrowed is called the _?_.
9. The _?_ is the part of the purchase price of an item that is paid at the time you take out a loan.
10. When you sign a(n) _?_, you are giving a lender a written promise you will repay the money that you have borrowed.

5-1 Promissory Notes

11. How much interest will you for a 18-month promissory note that has a principal of $3,200 and a rate of 7.875%?
12. What is the amount due at maturity for the loan in Question 11?
13. Phyllis Snow borrowed $3,200 to pay for a new roof. She signed a 6-month promissory note at 12% interest. Find the amount of interest Phyllis must pay. Then find the amount she must repay to her bank when the note comes due.
14. Oki Saga signed a promissory note for $1,500 at 8% interest for 90 days. Find the interest and amount due she will pay when the note is due using a) ordinary interest and b) exact interest.
15. Mohamed Jatmiko paid $420 in interest on a 6-month note for $5,600. Find the rate of interest he paid.
16. Brandon Jones paid $305.25 in interest on a 2-year note for $1,850. Find the rate of interest paid.

5-2 Calculating Interest

17. Use the simple interest table to find the exact interest on: a) $470 for 10 days at 9%; b) $470 for 40 days at 10%.
18. What is the daily interest factor for $1,250 borrowed at 13.5% exact interest?
19. Find the ordinary interest from March 5 to May 12 on $5,500 at 8% interest.
20. Find the exact interest from June 13 to August 15 on $6,200 at 9%.
21. Find the interest from March 15 to May 31 on $1,900 at 10%.

5-3 Installment Loans

22. You can buy a DVD player for $250 cash or pay $50 down and the balance in 12 monthly payments of $18. What is the installment price? By what percent would your installment price be greater than the cash price?
23. A stove costs $525 on the installment plan. You must make a down payment of $75 and make payments for 15 months. What will be your monthly payments?
24. Hector Morales borrowed $2,400 on a one-year simple interest installment loan at 12% interest. The monthly payments were $213.24. Find the amount of interest, amount applied to the principal, and the new balance for the first monthly payment.
25. Find the installment price and the finance charge that Harold paid for a $1,200 television because he paid $95 down and $78.85 a month for 18 months.

5-4 Early Loan Repayments

26. Terrence Moore has a 4-year, $9,500 car loan at 8%. After paying on the loan for 36 months, he has a balance of $2,666.13. He decides to pay off the loan with the next payment. How much will he have to pay?
27. Gaby Frazier has a 1-year, $2,200 loan at 10%. After paying on the loan for 4 months, Gaby has a balance of $1,490.87. She decides to pay off the loan with the next payment. How much will she have to pay?
28. Penn Peloko borrowed $10,500 at 8.5% for 5 years. His monthly payment was $215.42. Penn did not pay off the loan early. How much did he pay back to the lender?

5-5 Annual Percentage Rates

29. Maria Medina borrowed $400 on a 12-month loan with a finance charge of $39. Find the finance charge per $100 of the amount financed and the annual percentage rate.
30. Find the annual percentage rate for a loan with a finance charge of $6.25 per $100 for 9 payments.
31. Calida repaid a loan of $2,100 in 15 monthly installments of $152.40 each. Find the APR on her loan.
32. Find the annual percentage rate for a loan with a finance charge of $40.20 per $500 for 12 payments.
33. Pedro repaid a loan $3,900 in 15 monthly installments of $284 each. Find the APR on his loan.

Technology Workshop

Task 1 Payments of a Loan

Complete a template that calculates balances after each loan payment to show interest paid and remaining balance due. Then analyze the impact of applying additional money to the principal each month.

Open the spreadsheet for Chapter 5 (tech5-1.xls) and enter the data shown in blue (cells C1, C3 and B7) into the spreadsheet. The value you enter in cell C3 is the interest rate on the loan. The amount of the loan is the value you enter in cell C1. The monthly loan payment is the amount you enter in cell B7. Note that only 30 of the 60 months needed to pay off the loan is shown in the spreadsheet.

The spreadsheet will calculate the:

1. amount of interest paid for the month
2. amount of the payment applied to the principal each payment
3. remaining balance after each payment
4. total interest paid
5. total amount of interest paid
6. total amount of principal paid

	A	B	C	D	E	F
1	Loan amount		$17,500.00			
2	Length of loan		5 years			
3	Interest rate		5.50%			
4	Month	Loan Payment	Interest	Principal	Additional	Loan Balance
6						$ 17,500.00
7	1	$ 334.27	$ 80.21	$ 254.06	$ –	$ 17,245.94
8	2	$ 334.27	$ 79.04	$ 255.23	$ –	$ 16,990.71
9	3	$ 334.27	$ 77.87	$ 256.40	$ –	$ 16,734.32
10	4	$ 334.27	$ 76.70	$ 257.57	$ –	$ 16,476.75
11	5	$ 334.27	$ 75.52	$ 258.75	$ –	$ 16,217.99
12	6	$ 334.27	$ 74.33	$ 259.94	$ –	$ 15,958.06
13	7	$ 334.27	$ 73.14	$ 261.13	$ –	$ 15,696.93
14	8	$ 334.27	$ 71.94	$ 262.33	$ –	$ 15,434.60
15	9	$ 334.27	$ 70.74	$ 263.53	$ –	$ 15,171.07
16	10	$ 334.27	$ 69.53	$ 264.74	$ –	$ 14,906.34
17	11	$ 334.27	$ 68.32	$ 265.95	$ –	$ 14,640.39
18	12	$ 334.27	$ 67.10	$ 267.17	$ –	$ 14,373.22
19	13	$ 334.27	$ 65.88	$ 268.39	$ –	$ 14,104.83
20	14	$ 334.27	$ 64.65	$ 269.62	$ –	$ 13,835.20
21	15	$ 334.27	$ 63.41	$ 270.86	$ –	$ 13,564.35
22	16	$ 334.27	$ 62.17	$ 272.10	$ –	$ 13,292.25
23	17	$ 334.27	$ 60.92	$ 273.35	$ –	$ 13,018.90
24	18	$ 334.27	$ 59.67	$ 274.60	$ –	$ 12,744.30
25	19	$ 334.27	$ 58.41	$ 275.86	$ –	$ 12,468.44
26	20	$ 334.27	$ 57.15	$ 277.12	$ –	$ 12,191.32
27	21	$ 334.27	$ 55.88	$ 278.39	$ –	$ 11,912.92
28	22	$ 334.27	$ 54.60	$ 279.67	$ –	$ 11,633.25
29	23	$ 334.27	$ 53.32	$ 280.95	$ –	$ 11,352.30
30	24	$ 334.27	$ 52.03	$ 282.24	$ –	$ 11,070.07
31	25	$ 334.27	$ 50.74	$ 283.53	$ –	$ 10,786.53
32	26	$ 334.27	$ 49.44	$ 284.83	$ –	$ 10,501.70
33	27	$ 334.27	$ 48.13	$ 286.14	$ –	$ 10,215.56
34	28	$ 334.27	$ 46.82	$ 287.45	$ –	$ 9,928.12
35	29	$ 334.27	$ 45.50	$ 288.77	$ –	$ 9,639.35
36	30	$ 334.27	$ 44.18	$ 290.09	$ –	$ 9,349.26
38	Totals		$ 1,877.36	$ 8,150.74	$ –	

Task 2 Analyze the Spreadsheet for Early Loan Pay Off

If you wanted to pay off the loan early and save money paid as interest, you can pay additional money each month that will be applied directly to the principal.

Enter an amount in cell E7. Use the Fill Down feature to duplicate that amount in each row. Try entering different amounts in that cell to see how much interest and time you can save paying on this loan.

Answer these questions about your spreadsheet.

1. What is being multiplied to find the interest in Column C?
2. How is the amount in the Principal column calculated?
3. After the 30th loan payment how much of the principal is still due without paying any additional money each month?
4. How much interest has already been paid to the bank?
5. If $100 additional money was paid each month, what would be the loan balance after the 30th payment?
6. How much less interest would have been paid?
7. About how much additional money (rounded to the nearest $10) would you have to pay each month to get the loan paid off in 30 months instead of 60 months? How much interest will have been saved?

Task 3 Design a Spreadsheet to Pay Off a Loan Early

You are to design a spreadsheet that will calculate the amount of your monthly payment that is interest and the amount applied to the principal, so that you can determine the additional amount needed to pay off a loan early.

SITUATION: Your monthly payment is $198.85 for a $6,500 loan you have taken out at a rate of 6.375% for 3 years. You want to know the additional amount you can pay each month so that you can pay the loan off in 2 years.

Task 4 Analyze the Spreadsheet for Early Loan Pay Off

Enter an amount in cell E7. Use the Fill Down feature to duplicate that amount in each row. Try entering different amount in that cell to see how much interest and time you can save paying on this loan.

Answer these questions about your spreadsheet.

1. Enter an amount in your additional column cells. Based on the balance of the loan in the row for the 24th month, should increase or decrease the additional amount.
2. What additional amount (rounded to the nearest $10) will allow you to pay off the loan in two years?
3. How much interest will you save?

Chapter *Assessment*

Chapter Test

Answer each question.

1. Rewrite 0.06% as a decimal.
2. Rewrite $7\frac{1}{4}\%$ as a decimal.
3. Multiply: $540 × 3.5%
4. Multiply: $398.77 × 0.000673.
5. Multiply: $426 × 0.057%
6. Find what percent $35 is of $140.
7. Divide: 720.15 ÷ 100.
8. Divide: $2,400 ÷ 48.
9. Add: $209.34 + $345.12
10. Find the number of days between January 4 and January 23.

Solve.

Use the Simple Interest Table on page 179 to find the interest to the nearest cent.

11. $5,600 at 10.5% from 2/10 to 2/28
12. $950 at 12% from 11/1 to 12/31
13. Cora Fear borrowed $1,500 for 18 months from her bank. Cora signed a promissory note that carried 15% interest. Find the amount of interest Cora must pay. Then find the amount she must repay to her bank on the due date.
14. Ted Nash must pay $1,600 in interest on a promissory note for $40,000 due 4 months from the date of the note. Find the rate of interest he will pay.
15. You can buy a product for $750 cash or pay $150 down and the balance in 12 monthly payments of $61.25. What is the installment price? By what percent would your installment price be greater than the cash price?
16. Don Crawitz borrowed $780 on a loan with a finance charge of $68. Find the finance charge per $100 of the amount financed.
17. What is the daily interest factor for $3,450 borrowed at 9.5% exact interest?
18. Brenda Vowlski has a 2-year, $2,500 car loan at 9%. After paying on the loan for 3 months, Brenda has a balance of $2,211.46. She decides to pay off the loan with the next payment. How much will she have to pay?
19. Thane Davis has a 3-year, $5,200 loan at 7.5%. After paying on the loan for 15 months, he has a balance of $1,411.31. He decides to pay off the loan with the next payment. How much will he have to pay?
20. Revi repaid a loan $1,450 in 12 monthly installments of $132.75 each. Find the APR on his loan.

Use the APR table for Questions 21 and 22.

21. Find the interest from June 14 to August 10 on $975 at 8.5%.
22. Zoe Lendon borrowed $3,800 and repaid it in 15 monthly payments of $275.50. How much did she pay in finance charges on the loan? What was the annual percentage rate?

Planning a Career in Finance

Supri Suharjoto 2008/Shutterstock.com

Careers in finance involve working with numbers. A loan officer, financial advisor, or insurance claims adjuster works closely with the public. Other careers in finance, such as insurance underwriter, some types of accountants and budget analysts have little contact with the public. Workers who have a strong aptitude for working with numbers thrive in the banking, accounting, financial, and business industries. If you enjoy math and want to turn it into a successful career, a job in finance might be the career path for you.

Job Titles

- Financial clerk
- Accountant
- Tax auditor
- Technical writer
- Loan underwriter
- Insurance claims adjuster
- Economist
- Bank manager
- Financial advisor

Needed Skills

- very detail oriented
- excellent mathematical ability
- computer and technology skills
- honesty

- ability to work and make decisions independently
- excellent communication skills with clients and coworkers
- outstanding analytical and problem solving skills
- understand and maintain confidentially

What's it like to work in Finance

Loan officers generally work in financial institutions like banks, credit unions, or savings and loans. Loan officers often specialize in the types of loans they service; consumer loans, mortgages, or commercial loans. A loan officer will guide the client through the loan process, explain the process and types of loans available, request appropriate information, answer questions, and assist with applications. A loan officer can analyze a prospective client's credit worthiness and can sometimes assist to improve chances of being approved for a loan.

What About You?

In the finance industry, where can you see yourself in the future? Would you like a job with customer contact, or a job behind the scenes?

How Times Have Changed

For the following questions, refer to the timeline on page 169 as needed.

1. What was the difference in U.S. prime interest rate at the end of 1980 to the end of 2001?
2. What was the range of rates U.S. consumers likely paid for loans in December 2008?
3. In December of 2008, on a simple interest loan of $5,000 for one year, what was likely the least interest amount paid by a typical U.S. bank customer? the greatest interest amount paid? How does that compare to the interest that would have been paid in 1980?

Chapter 6

Own a Home or Car

6-1 Borrowing to Buy a Home
6-2 Renting or Owning a Home
6-3 Property Taxes
6-4 Property Insurance
6-5 Buying a Car
6-6 Car Purchases and Leases
6-7 Depreciating a Car
6-8 Cost of Owning a Car

Statistical Insights

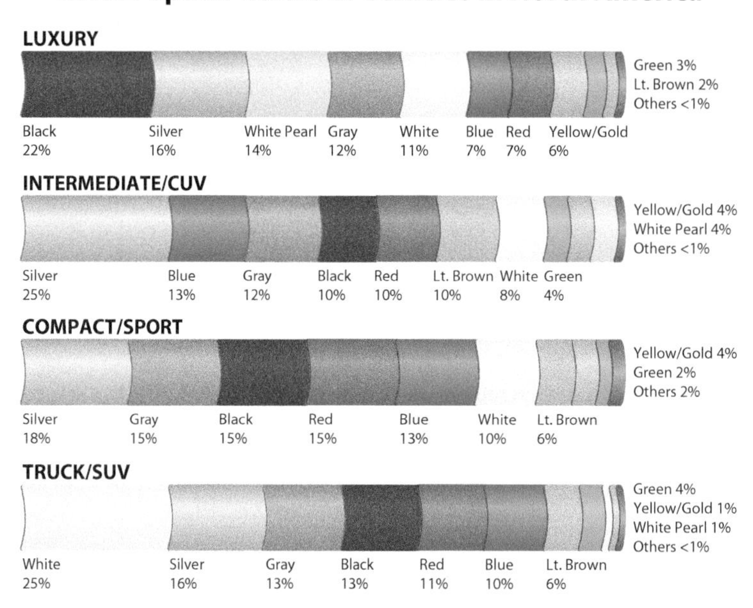

Source: www2.dupont.com

Use the data shown above to answer each question.

1. What color is favored among truck and SUV owners?
2. If you were the person responsible for placing the order for the compact and sports cars that are to be sold from a dealer's lot, what three car colors would you order most?
3. What color is among the top three favored colors for all vehicles?
4. Fifteen black vehicles in each category are ordered and arrive at a dealer's lot on the same day. Assuming each type of vehicle sells at about the same rate, which type of black vehicle would sell out first? Which type of black vehicle would sell out last?

How Times Have Changed

The percentage of American households that own homes is the homeownership rate. At the beginning of the 20th century, this rate was below 50%. By the end of the century, the homeownership rate was above 67%. The greatest annual drop of the homeownership rate ever recorded was in the year 2007.

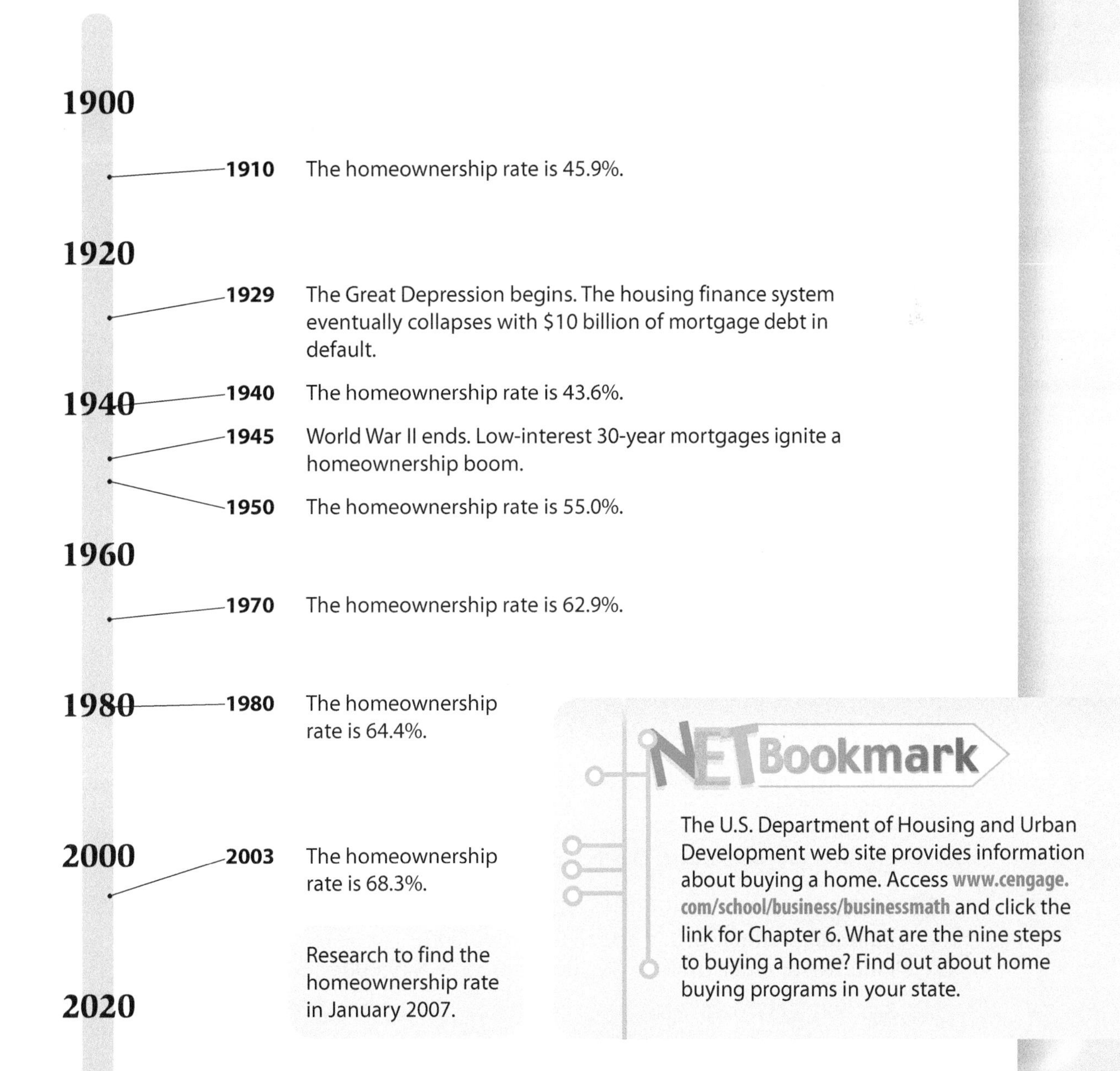

1900
1910 The homeownership rate is 45.9%.
1920
1929 The Great Depression begins. The housing finance system eventually collapses with $10 billion of mortgage debt in default.
1940
1940 The homeownership rate is 43.6%.
1945 World War II ends. Low-interest 30-year mortgages ignite a homeownership boom.
1950 The homeownership rate is 55.0%.
1960
1970 The homeownership rate is 62.9%.
1980
1980 The homeownership rate is 64.4%.
2000
2003 The homeownership rate is 68.3%.
2020
Research to find the homeownership rate in January 2007.

NET Bookmark

The U.S. Department of Housing and Urban Development web site provides information about buying a home. Access www.cengage.com/school/business/businessmath and click the link for Chapter 6. What are the nine steps to buying a home? Find out about home buying programs in your state.

6-1

Borrowing to Buy a Home

GOALS

- Calculate the down payment, closing costs, and mortgage loan amount
- Calculate the total interest cost of a mortgage loan
- Calculate the savings from refinancing mortgages

KEY TERMS

- down payment
- mortgage loan
- principal
- closing costs

Start Up ▸▸▸

A home located near a school bus parking lot may cost less to buy than a similar home located several blocks away. Why might this be true?

Photodisc/Getty Images

Math Skill Builder

Review these math skills. Solve the exercises.

1 **Add** money amounts.
Find the sum. $127 + $258 + $98 = $483

1a. $956 + $32 + $128

1b. $125 + $85 + $275

2 **Multiply** whole numbers.
Find the product. 20 × 12 = 240

2a. 30 × 12

2b. 25 × 12

2c. 120 × 12

2d. 60 × 12

3 **Multiply** money amounts by whole numbers and percents.
Find the product. 240 × $416.70 = $100,008
Find the product. 15% × $49,000 = 0.15 × $49,000 = $7,350

3a. 300 × $316.98

3b. 360 × $501.05

3c. 20% × $115,800

3d. 25% × $78,200

4 **Subtract** money amounts.
Find the difference. $95,600 − $42,000 = $53,600

4a. $106,892 − $41,985

4b. $3,582 − $1,284.34

Down Payments and Closing Costs

The total cost of buying a home includes the purchase price, the cost of borrowing money for the purchase, and closing costs.

Feverpitch/Shutterstock.com

Most people make a cash **down payment**, or a percentage of the total cost of the house paid at the time of purchase, to their lender. Some lenders require no down payment; others ask for as much as 30%. The more money you can put as a down payment, the less you need to borrow.

The balance of the purchase price (after the down payment) is usually borrowed through a **mortgage loan** taken with a bank or other lender. The money borrowed is called the **principal**. Interest must be paid on the mortgage loan. A mortgage gives the lender the right to take the property if the loan is not repaid as agreed. The length and terms of mortgages vary; 15-, 20-, and 30-year mortgages are common.

Closing costs are fees and expenses paid to complete the transfer of ownership of a home. Closing costs may range from 3% to 6% of the purchase price of the home. Typical closing costs include legal fees, recording fees, title insurance, loan application fees, appraisal and inspection fees, land surveys, prepaid taxes, and prepaid interest charges known as *points*. Interest rates and closing costs vary among lenders, so it pays to compare when you are looking for a lender.

To calculate the amount of the loan you need, subtract the down payment from the purchase price.

Mortgage Loan = Purchase Price − Down Payment

To calculate the amount of money you need to buy a home, add the down payment and the closing costs.

Cash Needed to Buy a Home = Down Payment + Closing Costs

Business *Tip*

At the time you pay the down payment and closing costs, you must also provide proof that you have insured the property, which is another cost of buying and owning a home.

Consumer Alert

Always Get a Good Faith Estimate

Within three days of applying for a mortgage, the lender must provide a Good Faith Estimate, which is an itemized list of the fees and costs associated with the loan. The Good Faith Estimate document is a way you can compare mortgages from different lenders. Be aware, though, that the Good Faith Estimate is an estimate, and actual costs can vary.

EXAMPLE 1

Hilda Mikon is buying a home for $74,000. She will make a 20% down payment and estimates closing costs as:

legal fees	$950
title insurance	$140
property survey	$250
inspection	$175
loan processing fee	$ 84
recording fee	$740

What amount of money will she need to borrow for her mortgage loan? What amount of cash will she need when she buys the house?

SOLUTION

Multiply the purchase price by the down payment percent.

20% × $74,000 = 0.2 × $74,000 = $14,800   amount of down payment

Subtract the down payment from the purchase price.

$74,000 − $14,800 = $59,200   amount of mortgage loan

Add the closing costs to find their total.

$950 + $140 + $250 + $175 + $84 + $740 = $2,339   total closing costs

Add the down payment and the total closing costs.

$14,800 + $2,339 = $17,139   cash needed to buy house

✔ CHECK YOUR UNDERSTANDING

A. Ricky Alberts' lender requires him to make a 25% down payment to get a mortgage on a home that costs $86,000. What amount will Ricky have to borrow to purchase the home?

B. Terri Wilburn will be able to purchase a condominium by making a 5% down payment on its $64,000 purchase price. She estimates her closing costs to be 3.5% of the purchase price. What amount of money will Terri need to pay the down payment and closing costs?

Math *Tip*

When you multiply a number by two different percents, you may add the percents and do one multiplication to find the answer.

Mortgage Loan Interest Costs

There are many different types of mortgages. Two of the most common types are fixed rate mortgages and variable rate mortgages. With *a fixed rate mortgage*, the same rate of interest is paid for the life of the loan. With a *variable rate mortgage*, the rate of interest is not guaranteed and may be increased or decreased.

Most mortgages are repaid gradually, or *amortized*, over the life of the mortgage in equal monthly payments. Each payment pays off part of the principal plus the interest due each month.

At first, most of the monthly payment goes to pay interest. As time passes, the amount that goes to repay the principal increases. The following table shows the amounts of interest and principal paid in different months on a 30-year, $70,000 loan at 9.75% interest. The monthly payment is $601.41.

	Month In Which Payment Is Made		
Payment Breakdown	**No. 1**	**No. 180**	**No. 358**
Interest	$568.75	$462.39	$14.42
Principal	$ 32.66	$139.02	$586.98
Total Payment	$601.41	$601.41	$601.41

Most lenders allow customers to make additional payments toward the principal so the mortgage can be paid off earlier. These added payments reduce the total interest paid.

The amortization table below shows the monthly payments needed to amortize mortgage loans over different periods of time using interest rates of 6%, 7%, and 8%.

To use the table to find a monthly payment, locate the box where the interest rate, the term of the loan, and the amount of the loan cross, or intersect.

Photodisc/Getty Images

AMORTIZATION TABLE
MONTHLY PAYMENTS NEEDED TO PAY A LOAN

Dollar Amount of Loan	Interest Rate								
	6.00%			7.00%			8.00%		
	Term of Loan								
	20 yrs	25 yrs	30 yrs	20 yrs	25 yrs	30 yrs	20 yrs	25 yrs	30 yrs
$ 40,000	$286.57	$257.72	$239.82	$310.12	$282.71	$266.12	$334.58	$308.73	$293.51
45,000	322.39	289.94	269.80	348.88	318.05	299.39	376.40	347.32	330.19
50,000	358.22	322.15	299.78	387.65	353.39	332.65	418.22	385.91	366.88
60,000	429.86	386.58	359.73	465.18	424.07	399.18	501.86	463.09	440.26
70,000	501.50	451.01	419.69	542.71	494.75	465.71	585.51	540.27	513.64
80,000	573.14	515.44	479.64	620.24	565.42	532.24	669.15	617.45	587.01
90,000	644.79	579.87	539.60	697.77	636.10	598.77	752.80	694.63	660.39
100,000	716.43	644.30	599.55	775.30	706.78	665.30	836.44	771.82	733.76
110,000	788.07	708.73	659.51	852.83	777.46	731.83	920.08	849.00	807.14

EXAMPLE 2

Amira Okano wants to buy a home that costs $83,000. She has $13,000 for the down payment, and her bank will lend her $70,000 on a 25-year, 8% mortgage. Find Amira's monthly payments and the total amount of interest she would pay over the term of the mortgage.

SOLUTION

On the $70,000 line of the amortization table in the 25-year column under 8% is the amount $540.27, the monthly payment.

Multiply the number of months in a year by the number of years in the loan to find the total number of months the loan will last.

$25 \times 12 = 300$ months

Multiply the number of months in the loan by the monthly payment to find the total amount needed to pay off the loan over 25 years.

$300 \times \$540.27 = \$162,081$ total payments

Subtract the amount of the mortgage from the total monthly payments to find the total interest paid over 25 years.

$162,081	total payments
−$ 70,000	amount of mortgage
$ 92,081	interest paid over the 25-year period

✔CHECK YOUR UNDERSTANDING

C. Marvin Zack bought a home for $95,000 with a $15,000 down payment. His mortgage is for 20 years at 7%. Find Marvin's monthly payments and the total amount he will pay in interest over the 20-year loan period.

D. Joseph and Rhoda Flynn bought a modular home as a future retirement home. The Flynns made a $25,000 down payment and got a $40,000 loan to pay for the home. If the loan is for 25 years at 6%, what monthly payment will they make? What total interest will they pay on the loan over the 25 years?

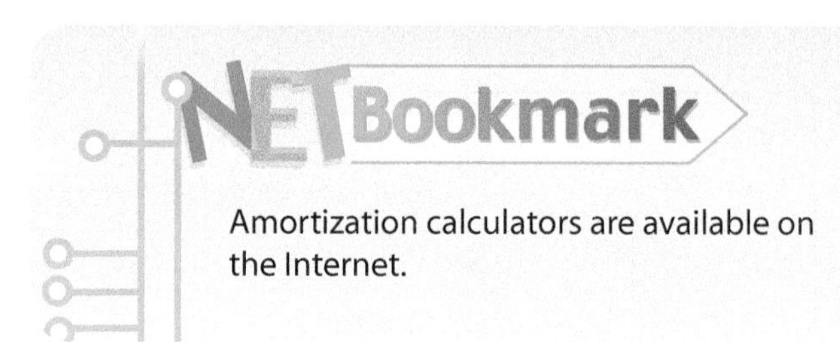

Amortization calculators are available on the Internet.

Refinancing a Mortgage

When interest rates go down, business firms and property owners may refinance or replace their fixed or variable rate mortgages with another mortgage at a lower interest rate.

When you refinance a mortgage, you take out a new mortgage and use that money to pay off the old mortgage. When you refinance your mortgage, you also pay closing costs on the new loan. There may also be other fees, such as the application costs for a loan or a prepayment penalty charged for paying off the first mortgage before it is due.

Business *Tip*

Due to loan costs, the savings in the first year is small. Finding the difference between loans over several years will show the total savings of refinancing.

EXAMPLE 3

The Rowes had a fixed rate mortgage at 9.65% with an unpaid balance of $40,000. The monthly payment on the old mortgage was $511.09. They got a new mortgage at 7.98% for the amount of the unpaid balance from another lender. Their new monthly payment is $340.73.

To get the new mortgage, they had to pay closing costs of $935. To pay off the old mortgage before it was due, they had to pay a prepayment penalty of $500. How much did they save during the first year by getting the new mortgage?

SOLUTION

Multiply the old monthly payment by the number of months in a year.

$12 \times \$511.09 = \$6,133.08$ one year's payment under old mortgage

Multiply the new monthly payment by the number of months in a year.

$12 \times \$340.73 = \$4,088.76$ one year's payment under new mortgage

$\$6,133.08 - \$4,088.76 = \$2,044.32$ difference in yearly payments

$\$935 + \$500 = \$1,435$ total of closing costs and prepayment penalty

$\$2,044.32 - \$1,435 = \$609.32$ amount saved in first year

✔ CHECK YOUR UNDERSTANDING

E. Nancy Ouimet's monthly mortgage payment is $597. She can refinance her loan with a new mortgage with monthly payments of $465. The total cost of getting a new mortgage is $836. What net first-year savings will she have with the new mortgage?

F. Will Ryan can refinance a mortgage by paying $716 in closing costs and $485 for a prepayment charge. His current monthly mortgage payment is $982. The monthly payment for the refinanced mortgage will be $876. If he refinances the mortgage, how much will he save in the first year?

Wrap Up ▶▶▶

The location of property is usually the most important factor in determining its value. The increased traffic, the exhaust odors, and the noise may make the home near the parking lot less desirable.

Exercises

Find the sum.

1. $128 + $65 + $902

2. $3,404 + $1,783

Find the product.

3. 12 × 15

4. 12 × 40

5. 12 × $653.87

6. 180 × $326.85

7. 20% × $184,600

8. 2.5% × $78,400

Find the difference.

9. $145,874 − $60,000

10. $4,329.34 − $2,508.85

The Colburns want to buy a condominium priced at $135,700. They will need to make a down payment of 15% and pay closing costs of 3% of the purchase price.

11. How much cash will they need for the down payment?

12. How much of the purchase price will they have to borrow?

13. How much cash will they need for the closing costs?

Photodisc/Getty Images

Ethel and Hector Ward bought a house at its market value of $82,000. They made a 5% down payment and paid these closing costs: legal fees, $550; property survey, $310; title insurance, $275; inspection fees, $240; points, $1,558.

14. What was the amount of the down payment?

15. What was the total of the closing costs?

16. How much of the purchase price will they have to borrow?

17. How much cash will they need to pay the down payment and closing costs?

Use the loan amortization table to help solve problems 18–23.

	Loan Amount	Interest Rate	Loan Term	Monthly Payment	Total Interest Paid Over Loan Term
18.	$45,000	6%	30 yrs		
19.	$60,000	8%	20 yrs		
20.	$110,000	7%	25 yrs		
21.	$90,000	7%	30 yrs		
22.	$50,000	6%	20 yrs		
23.	$100,000	8%	30 yrs		

Your old $47,000, 30-year, 12.8% mortgage has a monthly payment of $512.58. Over the 6 years since you took out the loan, mortgage rates have dropped. You can now get a mortgage at 9.05%, which will result in a new monthly payment of $369.36. To refinance, you must pay $1,020 in closing costs and a $480 prepayment penalty.

24. Find the net amount you will save in the first year.

25. Find the net amount you will save in the second year.

26. After being in effect for 4 years, the rate of interest on Syd Mutin's variable rate mortgage increased to 6.48% from 4.7%. Syd's old monthly payment was $259.32. His new monthly payment is $327.18. How much more will Syd pay in one year at the new mortgage rate?

27. **FINANCIAL DECISION MAKING** What do you think are the advantages and disadvantages of low down payments and long-term loans?

28. **CRITICAL THINKING** When do you think that refinancing a mortgage may not be a good idea?

Mixed Review

29.
$$\begin{array}{r} \$\ 78{,}412.23 \\ \$129{,}808.49 \\ +\ \$\ \ \ 3{,}642.01 \\ \hline \end{array}$$

30. $283,371 − $127,319

31. Round to the nearest hundred: 5,565.17

32. Write $12\frac{1}{2}\%$ as a decimal.

33. $2\frac{1}{8} + 3\frac{3}{8}$

34. A laptop computer has a cash price of $1,800. To buy it on an installment plan, you pay $100 down and $60 per month for 36 months. Find the finance charge.

35. Benjamin Toomey borrowed $4,200 from his bank for two years at 10.5% annual interest. What total amount must he pay his bank at the end of two years?

36. The finance charge-on an $800 loan for 6 months was $46. Use the Annual Percentage Rate table to find the annual percentage rate of the loan.

37. Lee Moore's monthly gross pay is $2,900. Out of those wages, $304 in federal withholding, $221.85 in FICA taxes, $58 in state income taxes, and $175 in health insurance premiums were deducted. Find the percentage of gross pay that Lee takes home, to the nearest percent.

Supri Suharjoto/Shutterstock.com

38. You take a cash advance on your credit card for $450. There is a $20 cash advance fee, plus daily periodic interest at an APR of 23%. You pay off the cash advance in 25 days. What is the total finance charge for the cash advance?

39. Felicia Delaney has a credit card that charges an APR of 12% applied using a monthly periodic rate on the average daily balance including new purchases. If Felicia's previous balance on 4/1 was $572.01 and she made a purchase of $125 on 4/3 and a payment of $400 on 4/30, what is her finance charge for April?

40. How much is a 3% balance transfer fee on a transfer of $2,000?

6-2

Renting or Owning a Home

GOALS

- Calculate the costs of home ownership
- Calculate the cost of renting a home or apartment
- Compare the costs of renting vs. owning

KEY TERMS

- depreciation
- security deposit

Start Up

Several students who will attend the same college next fall plan to rent apartments. They have been talking about pooling the money they will be paying for rent and using it instead to buy an old, large house in which all of them could live. They believe they could save money by making monthly mortgage payments instead of paying rent. One of the students thinks his father would be willing to sign for the loan. What advice would you give the students?

Photodisc/Getty Images

Math Skill Builder

Review these math skills and solve the exercises that follow.

1. **Add** money amounts.
Find the sum: $3,600 + $180 + $1,267 + $683 = $5,730

1a. $5,120 + $1,568 + $462
1b. $12,874 + $1,236 + $658
1c. $2,806 + $4,902 + $765
1d. $16,234 + $6,498 + $2,555

2. **Subtract** money amounts.
Find the difference: $11,005 − $1,643

2a. $12,905 − $1,238
2b. $12,858 − $11,983
2c. $18,600 − $8,932
2d. $6,788 − $4,076

3. **Multiply** money amounts by percents and whole numbers.
Find the product: 2% × $78,400 = 0.02 × $78,400 = $1,568
Find the product: 12 × $764 = $9,168

3a. 3.5% × $126,900
3b. 1.75% × $56,513
3c. 12 × $45.60
3d. 12 × $567 =

Costs of Home Ownership

After the home is bought, a homeowner has many ongoing expenses. Cash has to be paid out for property taxes, repairs, maintenance, utilities, insurance, mortgage interest, and special services such as trash pickup. Two other less obvious expenses are depreciation and the loss of income on the money invested in the home.

Depreciation is the loss in value of property caused by aging and use. The loss in value may be caused by the wearing out of parts of the home, such as the roof. It may also occur as home styles change or if the home becomes too expensive to heat and cool as energy costs rise. Most housing depreciates slowly at about 1% to 4% of its original value per year.

The amount of depreciation cannot be calculated until a house is sold. Until that time, depreciation must be estimated. Estimates of depreciation are often shown as a percent of the original purchase price.

Loss of income occurs because the money initially invested in buying the property (down payment and closing costs) could have been deposited in a savings account or other investment and earned interest.

One financial benefit homeowners have is that they may include the interest they pay on their home mortgage and the property taxes they pay on their property as itemized deductions on their income tax return. This reduces the income tax they pay. Homeowners also build equity in their homes. *Equity* is the difference between what is owed on a home and its value.

EXAMPLE 1

The Hansens want to buy a home. The interest they will pay on their mortgage in the first year will be $5,244. The annual property taxes on the home are $2,350, and an insurance policy on the home will cost $360 a year. They estimate that the home will depreciate $1,240 the first year, utilities will cost $1,710, and maintenance and repairs will cost $1,535. They lose $1,270 interest on their down payment.

They estimate that they will save $1,428 in income taxes in the first year because the mortgage interest and property taxes will raise their itemized deductions above the standard deduction allowed. What will be the net cost of owning the home in the first year for the Hansens?

SOLUTION

Add all the expense items.

$5,244	mortgage interest
2,350	property taxes
360	insurance
1,240	depreciation
1,710	utilities
1,535	maintenance, repairs
+ 1,270	lost interest
$13,709	total expenses

Subtract the tax savings.

$13,709	total expenses
− 1,428	tax reductions
$12,281	net cost for the first year

The Hansen's net cost of owning the home in the first year is $12,281.

✔CHECK YOUR UNDERSTANDING

A. The Krafts want to buy a home. Their estimated first-year expenses are: mortgage interest, $6,848; property taxes, $3,782; insurance, $560; depreciation, $1,790; utilities, $1,300; and maintenance and repairs, $2,050. They estimate lost interest income on savings to be $1,562. Income tax savings are estimated to be $1,320. Find their net cost of home ownership for the first year.

B. The Khurana family is building a home for $87,000 on a lot they own. They estimate their expenses for the first year to be: mortgage interest, $5,788; property taxes, $1,904; insurance, $347; lost interest income, $1,140; depreciation, 2% of the home's cost; maintenance and repairs, $900. The cost of heating, electricity, and water is estimated to be $1,860. The Khurana's expect to save $1,050 in income taxes as a result of owning the home. What will be the net cost of the home the first year?

Math *Tip*

When depreciation is shown as a percent, multiply the depreciation rate times the home's value to find the amount of depreciation.

Costs of Property Rental

Some people rent until they can afford to buy a home. Others prefer to rent. Many people prefer not to worry about the expense and effort of maintaining a property.

Renters usually pay a one-time **security deposit** in addition to their first month's rent when they sign a *lease,* or rental agreement. Return of the security deposit at the end of the lease is not guaranteed. The property owner may keep the security deposit to pay for repairs to the rental property or to clean the property, if the renters have not done so.

EXAMPLE 2

Mary Beth Berkovic plans to rent an apartment for $710 a month, including heat and water. The security deposit will be one month's rent. She estimates that other annual costs connected with the apartment will be: insurance, $110; utilities, $600; covered parking space, $240. What is Mary Beth's first-year cost of renting the apartment?

SOLUTION

$12 \times \$710 = \$8,520$ yearly rent

$\$8,520 + \$710 + \$110 + \$600 + \$240 = \$10,180$ total first-year cost

✔CHECK YOUR UNDERSTANDING

C. Rick Cassell rented an apartment for one year and paid $625 monthly rent. His other apartment-related costs for the year were: security deposit of $625; insurance, $85; utilities, $1,210; replacement of lost mailbox key, $10. What was the cost of renting the apartment for the one year?

D. Belinda Pryor's monthly rent on a house she is leasing is $1,250. The security deposit is one month's rent. Belinda is responsible for mowing the lawn and clearing snow and estimates she will spend $100 a month to have this done. Her other annual costs include $136 for insurance and $1,700 for utilities. What will be her first-year costs of renting this home?

Compare Renting and Owning Homes

When you buy or rent property you have some expenses that are similar, such as insurance and utilities. Most other expenses are quite different as are the sizes of the properties and their locations. For these reasons, it is very difficult to compare the purchase of a home to the rental of an apartment. However, the costs of owning a home and renting a similar home can be compared.

EXAMPLE 3

Cedric Thorne has $17,000 he can use as a down payment on a house that sells for $85,000. The interest for a year on his mortgage would be $5,184. He estimates his property taxes to be $1,720. Other costs of home ownership would total $3,270. He would lose $680 interest on his down payment and closing costs, but would save $895 on income taxes. Cedric can rent a similar home for $850 a month with a $1,200 security deposit. His other annual expenses of renting would be $130 for insurance and $1,400 for utilities. For the first year, is it less expensive for Cedric to buy or rent a home, and what is the difference?

SOLUTION

Find the net cost of home ownership.

($5,184 + $1,720 + $3,270 + $680) − $895 = $9,959   cost of home ownership

Find the cost of renting.

12 × $850 = $10,200   annual rent

$10,200 + $1,200 + $130 + $1,400 = $12,930   cost of renting

Find the difference between owning and renting.

$12,930 − $9,959 = $2,971

Buying is $2,971 less expensive than renting for the first year.

✔ CHECK YOUR UNDERSTANDING

E. Lynette Wolfe estimates her loan interest, taxes, insurance, and maintenance to be $9,100 on a house she bought for $83,600. She estimates her house would depreciate $2,926 a year. Lynette would lose $560 interest on the cash invested in her home, but would pay $1,600 less in income taxes. She could have rented the house for $785 a month, paid insurance of $115, utilities of $1,150, and not had any maintenance expenses except for the cost of routine cleaning. She also would have paid a security deposit of $200. For the first year, would it have been cheaper to rent or buy the house?

F. If Roscoe Tippin bought a manufactured home instead of continuing to rent an apartment, these expenses would increase by the amount shown: insurance, $276; utilities, $980. His annual interest on the mortgage would be $4,060, and he would have to rent a lot on which to place the home for $240 a month. The current monthly rent on his apartment is $510. Depreciation on the home is estimated to be $764. If he bought the home he would lose $57 interest on the cash invested, pay yearly property taxes of $945, and have income tax savings of $274. Is it more expensive for Roscoe to rent the apartment or buy the home, and how much more?

Financial Responsibility

When considering whether to buy a home or rent a place to live, making a list of the advantages and disadvantages for each option is one way to make an informed decision. A person's actual list can look similar to the list below, but will also likely include specific amounts for the costs associated with both.

Buy or Rent?	
Advantages of Renting • *No down payment* You can earn interest on the money you would have used for a down payment. • *More predictable housing costs* No responsibility for repair and upkeep beyond normal cleaning. • *Mobility* You can move whenever you want to after giving the required notice.	**Disadvantages of Renting** • *No ownership* Your monthly payments do not give you ownership of anything. • *Rent increases* Over time, you can expect the cost to rent the same place will increase. • *Restrictions* The property owner sets the rules for what you can do in and to your living space. • *No tax deduction* Rent payments are not deductible on federal income taxes.
Advantages of Buying • *Ownership* Monthly payments lead to ownership. • *Potential value* Some dwellings will increase in value over time. • *Tax deduction* Money paid in interest and property taxes can decrease your taxes. • *Fewer restrictions* Since you own the property, you can do what you want on your property within the law.	**Disadvantages of Buying** • *Down payment and closing costs* You need money up front to obtain a mortgage. • *Property taxes* Taxes add to the cost of owning a home. • *Restricted mobility* You may have to sell the house to relocate, and there are no assurances that the value of your home at that time will have increased from when you purchased. • *More responsibility* You must handle all maintenance and repairs.

Wrap Up ▸▸▸

Buying a house instead of just renting an apartment presents several problems when expenses are shared. There is no guarantee that all the students will remain in school, and some will be in school longer than others. All would have to cooperate in maintaining the property to keep up its value.

Also, the student group may not be able to afford to pay the cost of major, unexpected repairs. The purchase of a home also requires money up front to make the down payment and pay closing costs. Finding a parent willing to make such a financial investment may be a possibility, but not a certainty.

Communication

The monthly rent for a large luxury apartment in an exclusive high-rise building in New York City is $7,000 a month. An apartment of the same size located in a mid western city costs only $2,000 a month. Write the reasons you think could account for the difference in rents. Share your thoughts with other class members in class discussion.

Exercises

Find the sum.

1. $12,876 + $1,003
2. $284 + $128 +$265

Find the difference.

3. $12,736 − $9,375
4. $1,680 − $735

Find the product.

5. 2.5% of $140,300
6. 1.5% of $67,200
7. 12 × $578
8. 12 × $67.50

Solve.

9. Mort Silver owns a home. He estimates his expenses as mortgage interest, $10,180; property taxes, $3,690; insurance, $833; depreciation, $3,800; maintenance and repairs, $900; lost interest income, $2,375; and utilities, $2,450. He expects to save $3,700 in income taxes. What is his net cost of home ownership?
10. Joyce Navarro-Martin wants to buy a home for $73,200. She estimates her first-year expenses to be mortgage interest, $6,810; lost interest of $971; property taxes, $1,585; insurance, $395; depreciation at $1,098; maintenance and repairs, $1,800; and utilities, $1,250. She expects to save $1,847 in taxes. What will be the net cost of owning the home in the first year?
11. Patricia McCarthy lives in a subsidized apartment complex for low-income senior citizens. She pays monthly rent of $297 for a unit that could rent for $650 a month elsewhere. The monthly rent provides heat and water, but does not cover the $42 average monthly cost of electricity. She carries no insurance on the contents of the apartment. What are her annual costs of renting the apartment?

Photodisc/Getty Images

12. Tyrone Northrup pays monthly rent of $1,240 for a one-bedroom apartment in a large city. He pays an extra monthly charge of $160 to the rental company for a parking space in an attended lot. His other costs are: electricity, $90 a month; other utilities, $860 a year; and insurance, $170 a year. What is his annual cost of renting?
13. Sybil Kline rents a home for $970 a month. Her other annual expenses of renting total $1,305. If she buys the home, her estimated yearly expenses would be mortgage interest, $5,100; property taxes, $1,890; depreciation, $2,400; maintenance, $1,100; insurance, $479; utilities, $1,470; and lost interest income, $1,020. She would save $1,536 in income taxes. Are her net annual costs of housing lower by renting or buying? How much is saved?

Carole Finney rents an apartment for $640 a month and pays $120 for insurance and $820 for utilities yearly. She can buy a home with about the same space for $52,000. If she buys the home, she must withdraw $10,400 from her savings account and lose $624 interest. Her other home ownership expenses are estimated to be $9,300. She also estimates that owning a home will save her $1,428 in income taxes.

14. What is her total cost of renting for the year?

15. Which costs more, renting or owning? How much more?

16. **INTEGRATING YOUR KNOWLEDGE** Stephanie Larken presently rents an apartment for $580 a month. She estimates that her rent will increase 4% each year and that she will make total rent payments of $207,592.68 over 20 years. Her insurance and utility bills will average $800 a year over the 20 years. Stephanie has savings that she could use to buy a home with the same space as her apartment for $62,000. If she makes a 5% down payment and pays closing costs of $1,900, she can get a 20-year mortgage with a monthly payment of $465.53. Stephanie estimates her average annual ownership expenses as: depreciation, $1,150; maintenance, $1,400; insurance, $280; property taxes, $1,970, utilities, $1,390. Her income tax savings would average $965 a year. Interest of $310 a year could be earned on the money used for the down payment and closing costs. What is the total amount Stephanie might pay out over 20 years for renting the apartment and buying the home? Which plan do you think would work best for Stephanie?

17. **CRITICAL THINKING** Some homes are worth more now than when they were purchased. These homes are said to have appreciated, or increased in value. In this lesson you calculated the estimated depreciation on homes. Can homes appreciate and depreciate at the same time?

Mixed Review

18. $92.75 + $11.49 + $102 + $42.11 + $905.72

19. Round to the nearest 10,000: 129,817

20. Round to the nearest 10,000: 487,299

21. Lou Miller paid $39 interest on a promissory note of $1,200 for 3 months. Find the interest rate he paid.

22. Stella Sabo worked these weekly hours in February: $27\frac{1}{2}$, $30\frac{3}{4}$, $38\frac{1}{2}$, $31\frac{1}{2}$. What average number of hours did she work per week for these four weeks, to the nearest tenth hour?

23. What amount of tax will be owed on an income of $63,280 if the city income tax rate is 1.5% of all income?

24. Find the finance charge per $100 on a loan of $8,260 with a finance charge of $1,230.

25. Felicia Delaney uses her credit card to take a cash advance on September 6. She pays off the loan on November 3. For how many days did she borrow money from her credit card company?

6-3

Property Taxes

GOALS

- Calculate the decimal tax rate
- Calculate property taxes for tax rates per $100 or $1,000
- Calculate property taxes for tax rates in mills or cents per $1

KEY TERMS

- property tax
- assessed value

Start Up ▸▸▸

Renters of apartments or homes do not own the property they rent. Do the renters have to pay any property taxes?

Mirenska Olga/Shutterstock.com

Math Skill Builder

Review these math skills and solve the exercises that follow.

1. **Multiply** money amounts by decimals.
   Find the product. $72,000 × 0.0587 = $4,226.40

   1a. $1,000 × 0.0564 **1b.** $84,000 × 0.0642

2. **Divide** money amounts by money amounts.
   Find the quotient. $75,126 ÷ $1,000 = 75.126

   2a. $46,300 ÷ $100 **2b.** $59,317 ÷ $1,000

3. **Divide** money amounts and round to the nearest ten-thousandth.
   Find the quotient. $507,000 ÷ $8,000,000 = 0.06338, or 0.0634

   3a. $483,200 ÷ $6,000,000 **3b.** $889,600 ÷ $9,800,000

Decimal Tax Rate

Property taxes are taxes on the value of real estate such as homes, business property, or farm land. Taxes are collected annually or semiannually by the tax departments of local tax districts such as cities and towns in which the property is located.

Services that are often supported by taxes include schools, government operations, fire and police protection, and parks and road maintenance.

The amount of property tax paid is based on the **assessed value** of a property. Local tax assessors calculate this value. For example, the Watson's tax bill below shows their property has a fair market value of $150,000. It is assessed at 40% of its market value, or $60,000. The assessed values of properties are usually less than their market values. Similar properties in the same community should have similar assessed values.

Local tax districts determine the tax rate needed to pay for the services they provide. They estimate their expenses for the coming year and prepare an expense budget. They also estimate income from sources other than the property tax, such as licenses, fees, fines, rents, state aid, and so on. The difference between the total budget and the income from other sources is the amount that must be raised by the property tax.

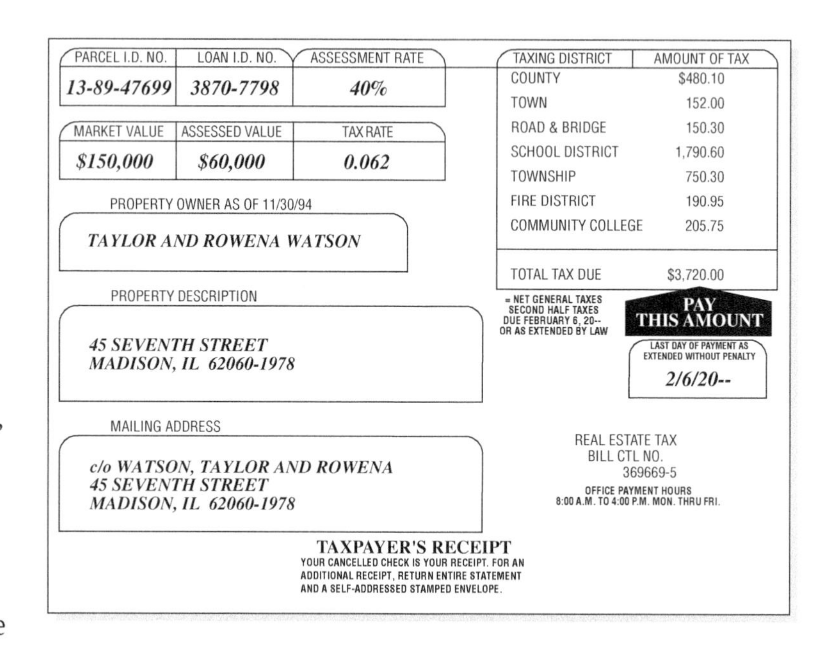

PARCEL I.D. NO.	LOAN I.D. NO.	ASSESSMENT RATE
13-89-47699	*3870-7798*	*40%*

MARKET VALUE	ASSESSED VALUE	TAX RATE
$150,000	*$60,000*	*0.062*

PROPERTY OWNER AS OF 11/30/94

TAYLOR AND ROWENA WATSON

PROPERTY DESCRIPTION

45 SEVENTH STREET
MADISON, IL 62060-1978

MAILING ADDRESS

c/o WATSON, TAYLOR AND ROWENA
45 SEVENTH STREET
MADISON, IL 62060-1978

TAXING DISTRICT	AMOUNT OF TAX
COUNTY	$480.10
TOWN	152.00
ROAD & BRIDGE	150.30
SCHOOL DISTRICT	1,790.60
TOWNSHIP	750.30
FIRE DISTRICT	190.95
COMMUNITY COLLEGE	205.75
TOTAL TAX DUE	$3,720.00

= NET GENERAL TAXES SECOND HALF TAXES DUE FEBRUARY 6, 20-- OR AS EXTENDED BY LAW

PAY THIS AMOUNT

LAST DAY OF PAYMENT AS EXTENDED WITHOUT PENALTY

2/6/20--

REAL ESTATE TAX BILL CTL NO. 369669-5

OFFICE PAYMENT HOURS 8:00 A.M. TO 4:00 P.M. MON. THRU FRI.

TAXPAYER'S RECEIPT
YOUR CANCELLED CHECK IS YOUR RECEIPT. FOR AN ADDITIONAL RECEIPT, RETURN ENTIRE STATEMENT AND A SELF-ADDRESSED STAMPED ENVELOPE.

Local tax districts then determine the *decimal tax rate*, which is the tax rate at which property is to be taxed. They find the decimal tax rate by dividing the amount to be raised by the property tax by the total assessed value of all property in the district.

$$\textbf{Decimal Tax Rate} = \frac{\textbf{Amount to be Raised by Property Tax}}{\textbf{Total Assessed Value}}$$

EXAMPLE 1

The Columbia School District's total budgeted expenses last year were $6,000,000. Estimated income from other sources was $1,800,000. The total assessed value of all taxable property in Columbia last year was $39,000,000. Find the decimal tax rate needed to meet expenses, rounded to the nearest hundred thousandth.

Photodisc/Getty Images

SOLUTION

Subtract the income from other sources from the total budgeted expenses to find the income to be raised from property taxes.

$6,000,000 − $1,800,000 = $4,200,000   property tax income needed

Divide the income to be raised from property taxes by the total assessed value of all property.

Decimal tax rate = $4,200,000 ÷ $39,000,000 = 0.107692 = 0.10769

✔ **CHECK YOUR UNDERSTANDING**

A. Filber County's budget for a year is $6,750,000. Of that, $650,000 is raised from other income, and the rest from property taxes. The total assessed value of the county's property is $80,000,000. What is the decimal tax rate, rounded to the nearest thousandth?

B. The Gayle Fire District must raise $1,950,000 from property taxes. The assessed value of property in the district is $48,200,000. What is the decimal tax rate needed, to the nearest ten thousandth?

In some communities the decimal tax rate is shown as a rate per $1,000 or $100, cents per $1, or mills per $1. So, the Watson's tax rate of 0.062 could also be stated as:

$62 per $1,000 (0.062 × $1,000 = $62)

$6.20 per $100 (0.062 × $100 = $6.20)

6.2 cents per $1 (0.062 × 100 cents = 6.2 cents)

62 mills per $1 (0.062 × 1,000 mills = 62 mills)

All the tax rates are equivalent, and the tax due on the Watson's property in all cases is $3,720.

Math *Tip*

A mill is one-tenth of a cent and one thousandth of a dollar.

Tax Rates per $100 or $1,000

To find the tax due on property when the rate is in $100 or $1,000, first find the number of $100 units or $1,000 units in the assessed value. Then, multiply the numbers of units by the tax rate to find the tax due.

EXAMPLE 2

Calculate the property tax due on the Watson's property if their tax rate is stated as $6.20 per $100.

SOLUTION

Divide the assessed value by $100.

$60,000 ÷ $100 = 600   number of $100 units in the assessed value

Multiply the tax rate per $100 by the number of $100 units.

600 × $6.20 = $3,720   property tax due

✔ **CHECK YOUR UNDERSTANDING**

C. The tax rate for the town of Beal is $3.736 per $100. Find Rita's tax bill if her property in Beal is assessed at $42,000.

D. Find the tax on property assessed at $120,000 if the tax rate is $4.128 per $100.

Math *Tip*

To quickly divide by 100, move the decimal point two places to the left; move the decimal point three places to the left to divide by 1,000.

EXAMPLE 3

Calculate the property tax due on the Watson's property if their tax rate is stated as \$62 per \$1,000.

SOLUTION

$\$60{,}000 \div \$1{,}000 = 60$ number of \$1,000 units in the assessed value

$60 \times \$62 = \$3{,}720$ property tax due

✔ CHECK YOUR UNDERSTANDING

E. What tax must Art pay on his home, assessed for \$67,500 if his tax rate is \$50.08 per \$1,000?

F. The Gilbey family owns a cabin and land with an assessed value of \$13,500. What property tax do they pay if the tax rate on the property is \$25.83 per \$1,000?

Tax Rates in Mills or Cents per Dollar

Some communities show the tax rate in mills. A mill is one tenth of a cent, and one thousandth of a dollar. There are ten mills in one cent and 1,000 mills in one dollar.

To find the tax due when the rate is in mills or cents per \$1 of assessed value, change the rate to a rate in dollars. Then multiply that rate by the assessed value.

To change mills to dollars, divide the number of mills by 1,000. To change cents to dollars, divide the number of cents by 100.

EXAMPLE 4

Calculate the tax due on the Watson's property if their tax rate is stated as either 62 mills or 6.2 cents per \$1 of assessed value.

SOLUTION

$62 \text{ mills} \div 1{,}000 = \$0.062$ mills rate changed to rate in dollars

$6.2 \text{ cents} \div 100 = \$0.062$ cents rate changed to rate in dollars

$\$60{,}000 \times \$0.062 = \$3{,}720$ total tax amount

✔ CHECK YOUR UNDERSTANDING

G. The city tax rate in Milser is 52 mills per \$1 of assessed value. Find the tax to be paid on property assessed at \$38,400.

H. What tax must Michelle Nolan pay on a condominium assessed at \$32,100 if her tax rate is 3.8 cents per \$1?

Wrap Up ▶▶▶

Renters pay property taxes indirectly through the rent they are charged. The owners of the real estate being rented must pay property taxes on its assessed value. The rents the owners charge usually cover their expenses, including property taxes.

Communication

Contact your city or town tax office and find out what taxes are charged property owners to support local government services and schools. Also find the rates at which taxes are levied. Write a short report summarizing your findings.

Exercises

Find the product.

1. 900 × $6.86
2. 1,370 × $5.07

Find the quotient.

3. $125,300 ÷ $100
4. $467,890 ÷ $1,000

Find the quotient, correct to the nearest ten thousandth.

5. $264,000 ÷ $9,300,000
6. $1,315,000 ÷ $24,100,000

Solve.

7. Property in the Bello School District has a total assessed value of $98,500,000. The district's budget for next year shows expenses totaling $5,000,000. The district expects to receive $3,200,000 from sources other than property tax. What decimal property tax rate, to the nearest ten thousandth, will the district use to raise enough money to meet budgeted expenses?
8. Hazel Forest City plans to spend $4,470,000 next year. Income from sources other than property tax will be $1,430,000. The taxable property in the city has an assessed value of $78,000,000. Find the tax rate, correct to the nearest hundred thousandth.
9. Carlos' property is assessed at $92,700. The school tax rate in his district is 1.45 cents per $1. What is Carlos' school tax?
10. Redfield Township levied a property tax of 1.5 mills per $1 to pay for new equipment for its fire department. If the total value of all property in the township is $790 million, what amount will be raised by this tax to pay for new equipment?

Photodisc/Getty Images

11. The town of Chester has a tax rate of 47.079 mills per $1. Find the tax on property in Chester worth $350,000, assessed at 60% of its market value.
12. Voters in Harmon approved a library tax of 0.75 mills per $1 of assessed value. What amount of library tax will a business owner pay for property assessed at $240,000?

Find the amount to be raised by property tax and the tax rate. Show the rate as a decimal, correct to the nearest thousandth.

	Assessed Value	Total Expenses	Other Income	Raised by Property Tax	Tax Rate
13.	$27,000,000	$989,000	$87,000		
14.	$36,000,000	$878,000	$97,500		
15.	$22,750,000	$382,700	$68,400		
16.	$ 7,900,000	$396,300	$45,600		

Find the tax due for Exercises 17–24.

	Assessed Value	Tax Rate	Tax Due
17.	$18,000	$5.20 per $100	
18.	$48,500	$71.10 per $1,000	
19.	$37,000	3.5 cents per $1	
20.	$25,300	77.3 mills per $1	
21.	$59,100	$4.747 per $100	
22.	$89,000	$87.45 per $1,000	
23.	$60,200	4.28 cents per $1	
24.	$19,800	56.82 mills per $1	

25. **CRITICAL THINKING** Business firms are often given a tax abatement to encourage them to build or expand their operations in a community. The tax abatement usually reduces the amount of tax paid by the business over several years. For example, the abatement could be a 50% reduction in taxes for 20 years. Are such tax abatements to businesses fair to homeowners who do not get such tax reductions?

26. **INTEGRATING YOUR KNOWLEDGE** Maxine Campau lives in a city that has a flat tax rate of 0.5% on all income. The city property tax rate is 3.43 cents per $1. The county in which Maxine lives also charges tax on property at a rate of 2.3 mills per $1 to pay for county operations. Maxine expects her income this year to be $52,000. Her home has an assessed value of 50% of its market value of $110,000. What total amount will Maxine expect to pay this year in city income tax and property taxes?

Mixed Review

27. 0.004×84.27

28. $1\frac{7}{8} \times 6$

29. Write $\frac{9}{20}$ as a decimal

30. $4,228 ÷ 7

31. Estimate. Then, find the exact product: 80.35×29

32. Round the result of $135.7 \div 5.2$ to the nearest hundredth.

33. Winston Chambers borrows $3,000 at 7% exact interest for 43 days. What interest will he pay on the loan?

34. A futon has a cash price of $650. To buy it on an installment plan, you pay $100 down and $40 a month for 18 months. What finance charge will you pay for this purchase?

35. You borrowed $300 for 15 days using your credit card's cash advance feature. The credit card company charges a $5 fee and a daily periodic interest rate of 0.0482% on the cash advance. What was the total finance charge on the cash advance?

6-4

Property Insurance

GOALS

- Calculate property insurance premiums for homeowners
- Calculate property insurance premiums for renters
- Calculate how much can be collected on insurance claims

KEY TERMS

- homeowners insurance
- premium
- renters policy

Start Up

Some people insure their homes for less than they are worth to save money on their home insurance policy. Is this a wise decision?

Michael Shake/Shutterstock.com

Math Skill Builder

Review these math skills and solve the exercises that follow.

1. **Add** money amounts.
   Find the sum. $462 + $89 = $551

   1a. $728 + $127 **1b.** $327 + $46

2. **Subtract** money amounts.
   Find the difference. $58,000 − $750 = $57,250

   2a. $1,284 − $500 **2b.** $6,738 − $250

3. **Multiply** money amounts by percents.
   Find the product. $86,000 × 80% = $86,000 × 0.8 = $68,800

   3a. $75,000 × 90% **3b.** $42,500 × 80%

4. **Multiply** money amounts by whole numbers and **round** to the nearest dollar.
   Find the product. 762 × $0.83 = $632.46 = $632

   4a. 1,051 × $0.54 **4b.** 797 × $0.67

   4c. 2,020 × $0.34 **4d.** 1,565 × $0.88

5. **Divide** money amounts by money amounts.
   Find the quotient, to the nearest thousandth. $38,000 ÷ $56,000 =

   5a. $43,000 ÷ $60,000 **5b.** $85,000 ÷ $120,000

   5c. $117,000 ÷ $155,000 **5d.** $99,000 ÷ $135,000

Property Owners Insurance Premiums

A policy that covers your home and protects you against other risks is called **homeowners insurance**. Basic homeowners insurance covers:

- *Dwelling*, the home in which you live
- *Other structures*, such as a garage
- *Personal property*, includes the contents of a home
- *Additional living expense*, which pays for the extra costs of living when you cannot use your own home because of damage
- *Personal liability*, which protects you in case of lawsuits by persons injured on your property
- *Medical payments to others*, but not to you or your family, for medical expenses in case of injury on your property

Business *Tip*

Basic policies often do not cover the full value of jewelry, cameras, computers, furs, and valuable collections. Special insurance called a personal articles floater is required to insure such items to full value.

The amount for which your home is insured is called the *face value* of the policy. That amount determines the amount of insurance you have in other categories.

For example, if your home is insured for a face value of $60,000, personal property is usually covered for 50% of that amount, or $30,000. Additional living expense coverage is typically 20% of the face value, or $12,000.

Photodisc/Getty Images

Homeowners policies may provide other options as well. For example, a policy may insure personal property when you are away from home. This coverage is called *off premises* and is usually for 10% of the amount of the policy.

For example, if the luggage and clothes you take on vacation are stolen, their loss would be covered under the off premises policy feature.

REPLACEMENT COST POLICIES Under replacement cost policies, the insurance company will pay the cost of replacing your property at current prices. If a leather chair that costs $600 is destroyed by fire, the insurer will pay for a replacement chair that now costs $900 even though the cost is higher than the original purchase price.

Before issuing this type of policy, insurers usually require a survey and inspection of the property. Also, the property must be insured for 100% of its current replacement value with automatic annual adjustments for inflation.

Premiums for this type of policy are 10–15% higher than a standard policy because of the extra protection it offers.

Business *Tip*

Most homeowners policies do not cover damage caused by earthquakes, floods, termite and other pest damage. In addition, policies will typically not cover losses if the house is vacant 60 days or more.

INSURANCE PREMIUMS The money paid to an insurance company for property insurance is the **premium**. The premiums you pay depend on many things, such as how much and what kind of coverage you buy, how your house or apartment is built, and where it is located. For example, the premium rates for a brick house near a fire department will be less than for a house made of wood that is far from a fire department.

Some items such as computer systems, jewelry, and expensive entertainment systems may not be covered by a basic policy. You will have to buy additional insurance, called a *rider*, or *floater* to cover possible loss.

Property insurance rates are usually based on $100 units of insurance.

NOTE: Homeowners insurance premium charges are rounded to the nearest dollar.

EXAMPLE 1

Marion Duval insured his house for $89,000 at an annual rate of $0.51 per $100. Find his premium.

SOLUTION

Find the number of $100 units in the insured amount.

$89,000 ÷ $100 = 890 number of $100 units

Multiply the rate per $100 by the number of $100 units.

890 × $0.51 = $453.90, or $454 premium rounded to the nearest dollar

✔ CHECK YOUR UNDERSTANDING

A. Nolan Harwood insured his home for $61,000. Find the annual premium, to the nearest dollar, he will pay for a policy that costs $0.46 per $100.

B. Mandy Wisko insures her home for $43,000. What annual premium will she pay if the policy cost is $0.74 per $100?

Renters Insurance Premiums

If you rent a house or an apartment, you can buy a **renters policy** that provides nearly the same coverage as a homeowners policy except for loss of the dwelling and other structures. Annual premiums for a renters policy are based on the amount of insurance on the contents of your apartment or rental home. The table below shows the annual premium charged by one company for a basic renters policy.

Maximum Amount of Coverage on Contents	Distance From Fire Station	
	Less Than 5 Miles	5 Miles or More
$ 5,000	$120	$138
$10,000	$129	$148
$15,000	$140	$161
$20,000	$152	$175
$25,000	$165	$190
$30,000	$177	$204

EXAMPLE 2

Myron Segal rents an apartment that is 4.1 miles from a fire station. He insures its contents for $10,000. A computer system Myron owns is also insured, but at an extra cost of $27 per year. What total annual premium will Myron pay for this coverage?

SOLUTION

Locate the correct insurance amount row and distance from fire station column to find the basic premium.

The basic premium is $129.

Add to the basic premium the cost of additional insurance, if any.

$129 + $27 = $156 total annual premium

✔CHECK YOUR UNDERSTANDING

C. Ed and Kathryn Bosh want to insure their apartment's contents for $25,000. In addition, they decide to insure jewelry appraised at $3,000 for an additional premium of $31. They live one block from the fire station. Find their total premium for one year.

D. Samantha Hilliard rents a home that is located 12 miles from the nearest fire station. She insures the home's contents for $5,000. What annual premium will she pay?

Collecting on Insurance Claims

Photodisc/Getty Images

If your property is damaged by fire or a theft occurs, you have to file a claim with your insurance company. The company
will send an adjuster to look at the property and decide on the amount of loss. The amount of the loss your insurance company pays depends on the type of coverage you have.

If you have a basic policy, the company will pay the full amount of the loss up to the face value of the policy. It will not pay more than the amount of your policy.

Your basic policy usually contains a *deductible*. With a $100 deductible, you are responsible for the first $100 of loss. The insurance company pays the full amount less the deductible up to the face value of the policy. The higher the deductible the lower the premium.

Business Tip

When a loss occurs, coinsurance policies pay only the depreciated value of personal property, not its replacement cost.

EXAMPLE 3

Your policy has a face value of $30,000 with a $1,000 deductible. How much will the insurance company pay if your loss is $7,800?

SOLUTION

Find the amount paid by the insurance company by subtracting the deductible from the loss amount.

$7,800 − $1,000 = $6,800 amount insurance company pays

✔CHECK YOUR UNDERSTANDING

E. How much will an insurance company pay for a loss of $10,200 if property is insured for $18,000 with a $250 deductible?

F. Property insured for $70,000 with a $500 deductible suffers a loss of $65,000. How much will the insurance company pay?

If you have a *coinsurance policy*, you purchase insurance up to a stated percent of the value of the property. This is usually 80% of the property's value. If you have a property valued at $50,000 and insure it at 80%, the coinsurance coverage is $40,000. Because the property is insured for less, the annual premium will be less also.

If you have a *coinsurance policy* and carry the required insurance, the insurance will pay for losses up to the face value of the coinsurance policy. If the coinsurance carried is less than 80% (or the agreed upon coinsurance percent), an insurance company will pay only a fractional part of the damages. The formula used is:

$$\text{Amount Paid by Insurance Company} = \frac{\text{Face Value of Policy}}{\text{Required Amount of Coinsurance}} \times \text{Amount of Loss}$$

EXAMPLE 4

A building with a value of $50,000 is insured for $24,000 under an 80% coinsurance policy. The building had fire damage of $7,200. What amount did the insurance company pay?

SOLUTION

The face value of the policy is $24,000.

$\$50{,}000 \times 80\% = \$50{,}000 \times 0.8 = \$40{,}000$ required coinsurance

Divide the face value of the policy by the required amount of coinsurance. Then multiply the result times the loss amount.

$\frac{\$24{,}000}{\$40{,}000} \times \$7{,}200 = 0.6 \times \$7{,}200 = \$4{,}320$ amount insurance pays

✔CHECK YOUR UNDERSTANDING

G. Betsy Rowan has an 80% coinsurance policy $41,600. Her home is worth $65,000. What insurance on a $4,000 loss?

H. Casey Maynard's home is worth $90,000. He on an 80% coinsurance policy. What amount pay on a $42,000 loss?

yurok/Shutterstock.com

Wrap Up ▶▶▶

Carrying less insurance is one way to save money, but the property owner carries a great risk to save a relatively small amount of money. Since a home is the greatest investment that most people will make in their lifetime, it is good practice to protect the investment and insure the property for full value.

TEAM Meeting

A homeowners insurance policy that provides additional coverage is known as a Broad Form, HO2 policy. Find out what the HO2 policy covers. Also find out what other forms of homeowners insurance exist and what they cover. Summarize in a paragraph what the general differences are, if any, between the policies. Suggested sources include insurance company sites on the Internet insurance agents, or a search of the Internet using the term Broad Form, HO2 policy.

Exercises

Find the sum.

1. $820 + $79
2. $455 + $66

Find the difference.

3. $26,870 − $250
4. $1,025 − $750

Find the product.

5. $67,000 × 85%
6. $137,100 × 90%
7. $78,000 × 75%
8. $326,400 × 65%

Find the quotient to the nearest thousandth.

9. $36,000 ÷ $51,000
10. $74,000 ÷ $78,000

Find the quotient.

11. $150,500 ÷ $100
12. $36,440 ÷ $100
13. $189,350 ÷ $100
14. $24,600 ÷ $100

Solve.

15. Find the premium, to the nearest dollar, for one year for a $61,000 policy at $0.47 per $100.
16. A $31,000 policy costs $0.68 per $100. Find the premium for one year, to the nearest dollar.
17. A home valued at $200,000 is insured for $0.56 per $100 on a basic policy. Earthquake insurance costs $850 extra a year. What total premium must be paid to insure this home?

The Woodman family moved to a new home in the same town. Their new house had the same value as their old home, $129,500. Because the new home is located more than 1,000 feet from a fire hydrant, their homeowners insurance rate increased by $0.11 per $100.

18. Estimate the amount their annual premium will increase?

19. Calculate their actual premium increase for a year.

20. Glenda Hope insures the basic contents of her apartment for $25,000 at the rate shown in the Renter's Insurance Table. She lives in the city a few blocks from the fire station. Glenda's depression glass collection is not covered by her basic policy, and she gets extra coverage at $0.35 per $100 of its $12,500 value. What annual premium does she pay?

Photodisc/Getty Images

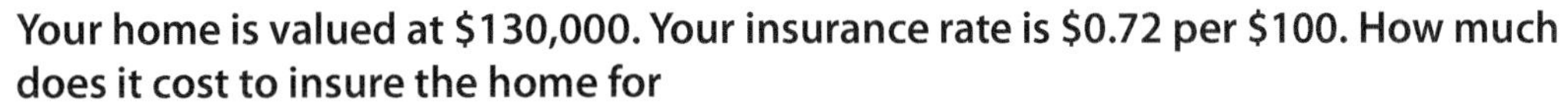

Your home is valued at $130,000. Your insurance rate is $0.72 per $100. How much does it cost to insure the home for

21. 100% of its value?

22. an 80% coinsurance rate of its value?

Find the amount paid by an insurer for each loss in problems 23–27.

	Face of Policy	Amount of Loss	Value of Property	Coinsurance Percent	Amount Paid
23.	$45,000	$ 8,000	$75,000	80%	
24.	$63,000	$ 5,600	$90,000	80%	
25.	$40,000	$13,200	$60,000	80%	
26.	$45,000	$ 7,000	$62,500	80%	
27.	$49,000	$22,500	$70,000	90%	

28. Allison Renfrew had a $750 deductible, replacement cost policy. Her laser printer and fax machine were stolen. Their value was $824. How much did Allison collect from the insurance company?

29. Your policy has a face value of $20,000 with a $1,000 deductible. How much will your insurance company pay if your loss is $7,400? $12,000? $19,000?

30. **CRITICAL THINKING** A flood plain is an area where floods could possibly occur even though it may be some distance from a river. Why does the insurance on homes within the flood plain cost more even if the area has not been flooded for many years?

31. **CRITICAL THINKING** Homes located in certain coastal areas of the United States may be destroyed by hurricanes. Is an insurance company obligated by law or a principle of good business practice to provide insurance even for high risks?

32. **FINANCIAL DECISION MAKING** Benita Lahr insures her home for $60,000 and pays insurance at a rate of $0.50 per $100. She will receive a 2% discount if she installs smoke detectors and dead-bolt locks and has a fire extinguisher. These items will cost $250 to purchase and install. Should she install the safety devices to save 2%?

Mixed Review

33. $1,429 ÷ 100

34. $\frac{1}{5} \times \frac{1}{7}$

35. $5\frac{3}{4} - 3\frac{1}{4}$

36. 15 − 12.78

37. 8% of $1,675.40

38. Find $\frac{3}{10}$ of 28.

39. Rewrite 0.44 as a fraction in lowest terms.

40. In the number 1,427.589, what is the place value of the 4? the 9?

41. The previous balance on Nina Garcia's credit card is $225.83. Her credit card company charges an APR of 11%, applied at a monthly periodic rate. If the card company uses the previous balance method to calculate finance charges, how much will Nina be charged in finance charges if she has lost the grace period?

42. The installment price of a set of golf clubs is $947. You must pay $200 down and make payments for 18 months. What will be your monthly payments?

43. When Wallace Figueroa checked his March 8 credit card statement, he found a sales charge of $128.30 that was unauthorized. He also found that a sales slip for $36.25 dated March 11 did not appear on the statement. What is Wallace's correct new balance if the new balance listed on the statement is $301.68?

44. Find the ordinary interest from June 4 to June 16 on $1,000 at 15% interest.

45. The income tax rate for the city of Allenby was 1% last year, and is 1.25% this year. Syd Johnson expects his income this year to be the same as last year, $48,400. What city income tax will Syd owe this year? By what percent did Syd's income taxes increase this year?

June Marie Sobrito/Shutterstock.com

46. JP Grimshaw earns $3,750 each month. He pays $1,200 per month for housing and has a $400 per month car payment, $120 in minimum credit card payments, and a personal loan with a $420 monthly payment. To the nearest percent, what is JP's debt-to-income ratio? If you were a lender and JP came to you for a loan, what would you advise him?

47. Zechariah decides to put $200 each quarter into an account earning 4% interest compounded quarterly. How much will be in the account after 5 years? How much of that money will be interest?

48. Find the interest for six months on $400.60 at 1.25% annual interest paid semiannually.

6-5

Buying a Car

GOALS

- Calculate the MSRP for a new car, including optional equipment
- Calculate the delivered price and the balance due for a new car
- Calculate the delivered price and the balance due for a used car

KEY TERMS

- MSRP (Manufacturer's Suggested Retail Price)
- negotiation

Start Up

Most people who buy a new car make a down payment, get a loan, and then make regular monthly payments until the loan is repaid. Others make regular payments into a savings account until they have enough money to pay the entire purchase price in cash. Which way is better?

Photodisc/Getty Images

Math Skill Builder

Review these math skills and solve the exercises that follow.

1 **Add** money amounts.
Find the sum. $23,456 + $938.24 + $198 = $24,592.24

1a. $19,276 + $235 **1b.** $187 + $12,390.04

1c. $57,143 + $789 **1d.** $90,872 + $13,776

2 **Subtract** money amounts.
Find the difference. $18,629 − $1,500 = $17,129

2a. $14,592 − $500 **2b.** $27,582 − $3,985

2c. $52,005 − $32,864 **2d.** $16,936 − $12,999

3 **Multiply** money amounts by percents.
Find the product. 5.5% × $26,780 = 0.055 × $26,780 = $1,472.90

3a. 3% × $12,833 **3b.** 20% × $9,384

3c. 5% × 24,556 **3d.** 40% × 16,789

3e. 10% × 46,882 **3f.** 12% × $894

Manufacturer's Suggested Retail Price

Most car buyers are familiar with a car's **MSRP (Manufacturer's Suggested Retail Price)**, or *sticker price*. This is the price printed on a sticker pasted on the window of a new car. The sticker also lists the equipment on the car and mileage information.

Car buyers do not usually pay the full MSRP for their car because of discounts given by the car dealer or manufacturer. A car in very high demand may sell for a price higher than the MSRP.

Buyer's Guide to Features

THOMSON MOTOR CAR COMPANY
Product Features: MR35 Sports Sedan, 2-door

Features	Model			Features	Model		
	STD	CTM	PRM		STD	CTM	PRM
Air bags	S	S	S	Power windows	NA	0	S
Air conditioning	S	S	S	Seats, cloth	S	S	S
Alarm system	0	0	0	Seats, leather	NA	NA	0
Cruise control	S	S	S	Side mirror, manual	S	NA	NA
Defogger, rear	S	S	S	Side mirror, electric	0	S	S
Engine, 4 cylinder	S	NA	NA	Ski rack	0	0	0
Engine, 6 cylinder	0	S	S	Sound, standard	S	NA	NA
Light package	S	S	S	Sound, deluxe	0	S	NA
Message center	NA	0	S	Sound, premium	0	0	S
Power brakes	S	S	S	Tilt steering	S	S	S
Power seats	0	0	S	Trim, bright	S	S	S
Power steering	S	S	S	Trim, color matched	NA	NA	0

Code: S–Standard; 0–Optional; NA–Not Available

New Car Warranty On All Models: 12 Months or 12,000 Miles

In many car purchases, an important part of the process is **negotiation**, where the buyer and seller of the vehicle come to an agreed price for the vehicle. Other sellers may offer a "no-haggle" price, which indicates that the vehicle is priced as low as it will be sold and negotiation is not generally utilized.

Now assume that you are interested in buying the Thomson MR35, a small sports car. The features found on three MR35 models are shown above. By looking to the right of each feature, you can determine if the feature is standard (S), optional (O), or not available (NA) on each model.

The features listed in the table above are available on many cars. The buyer's guide you get from a new car dealer would list more options for you to consider.

Business *Tip*

New car warranties pay for the cost of correcting defects and making most repairs. The buyer is reponsible for routine care such as oil changes and replacing parts, such as brakes, that may wear.

The table below shows the MSRP base price for each MR35 model and the prices for longer warranties and optional features. The *base price* is the price paid for a model equipped with all the standard features shown in the buyer's guide. For example, the base price for the STD model is $23,208.

The basic new car warranty, abbreviated as 12/12,000, covers the car "bumper to bumper" for 12 months, or 12,000 miles, whichever comes first. The *extended warranty* provides extra coverage for the number of months and miles shown.

EXAMPLE 1

A customer wants to buy the MR35 Sports Sedan, STD model with the optional six-cylinder engine, the power seats, and the 36/36,000 extended warranty. Using the information from the MSRP list, find the price of this car.

Business Tip

When shopping for a new car, read reports found in libraries or through an Internet search that compare the repair and insurance costs and the safety features of various makes of cars. Also ask owners for their opinion about the car you may want and take a test drive to make sure the car suits your needs.

Manufacturer's Suggested Retail Price List

THOMSON MOTOR CAR COMPANY MSRP: MR35, Sports Sedan, 2-door			
Car Model	**Base Price**	**Optional Features**	**Price**
Model STD	$23,208	Alarm system	$481
Model CTM	$24,185	Engine, 6 cylinder	$796
Model PRM	$25,096	Message center	$348
Extended Warranty	**Price**	Power seats	$410
		Power windows	$306
24 months; 24,000 miles	$240	Seats, leather	$680
36 months, 36,000 miles	$375	Side mirror, electric	$197
48 months, 48,000 miles	$575	Ski rack	$365
24 months, 30,000 miles	$400	Sound, deluxe	$187
36 months, 45,000 miles	$585	Sound, premium	$246
48 months, 60,000 miles	$750	Trim, color matched	$248

SOLUTION

Find the base price of the STD model. Then, add the base price and the cost of the optional features and extended warranty, if any.

$23,208 base price

$23,208
$ 796
$ 410
$ 375

$24,789 MSRP

✔CHECK YOUR UNDERSTANDING

A. Loretta wants the MR35 PRM model with an alarm system and color matched trim. What is the MSRP of this car?

B. Jack is interested in the MR35, Model CTM, with the message center and ski rack. What is the car's MSRP?

Cost of New Car Purchases

The key items in a new car purchase are listed below.

PURCHASE PRICE The price negotiated by the dealer and the buyer. The price includes the car and any options installed by the dealer.

SALES TAX Tax computed on the purchase price.

REGISTRATION FEES License and title transfer fees.

NON-TAXABLE ITEMS Any items such as extended warranties that may be exempt from sales tax depending on state tax laws.

REBATES Discounts, if any, given by the manufacturer or car dealer.

DELIVERED PRICE Often called the "out-the-door" price. It is the total of the purchase price, sales tax, registration fees, and non-taxable items, less any rebates.

DOWN PAYMENT A cash payment made by the customer or the value of a vehicle given as a trade-in.

BALANCE DUE The amount the customer has left to pay. This amount is usually borrowed from the car dealer or other lender.

Use the following formulas, find the delivered price and balance due on a new car purchase.

Delivered Price = Purchase Price + Sales Tax + Registration Fees + Non-taxable Items − Rebates

Balance Due = Delivered Price − Down Payment

EXAMPLE 2

The purchase price on a new car bought by Gretchen Cerna is $23,340. She is charged a 5% sales tax on the purchase price. She received a manufacturer's rebate of $1,250. Registration costs were $128. Gretchen's down payment was a trade-in of $4,300 given for her old car. Find the delivered price and the balance due.

Photodisc/Getty Images

SOLUTION

Find the sales tax on the purchase price.

$5\% \times \$23{,}340 = 0.05 \times \$23{,}340 = \$1{,}167$ sales tax

Add the purchase price, sales tax, and registration costs.

$\$23{,}340 + \$1{,}167 + \$128 = \$24{,}635$

Deduct the rebate.

$\$24{,}635 - 1{,}250 = \$23{,}385$ delivered price

Subtract the down payment from the delivered price.

$\$23{,}385 - \$4{,}300 = \$19{,}085$ balance due

✔CHECK YOUR UNDERSTANDING

C. Tim Garner and a car dealer agreed on a $20,067 price for a car. Tim then decided to buy an extended service warranty for $250 extra. He used his old license plates, but still had to pay registration and title fees of $134.85. A 4.5% state sales tax is charged on all purchases, except warranties. Tim made a cash down payment of $6,000. Find the delivered price and the balance due for this purchase.

D. Lillian Weinstein's purchase price for an economy car is $16,238. Sales tax is charged at 6% in her state. Plates, title transfer, and other fees totaled $186. Lillian made a down payment of 10% of the purchase price of the car. Find the delivered price and the balance due.

Cost of Used Car Purchases

Used cars may be purchased from new car dealers who resell trade-ins, used car dealers, car rental agencies, and individual car owners. The used cars are generally sold "as is," without warranty.

The *purchase price* for a used car refers to the price on which the buyer and the seller agree and is the price on which sales tax is figured.

The *delivered price* of used cars is equal to the sum of the purchase price, sales tax, and registration fees. The balance due is the delivered price less the down payment.

Business *Tip*

A used car seller may give a limited warranty to the buyer to make the sale. Buyers of used cars with part of the manufacturer's warranty still in effect are often allowed to transfer the warranty to their name.

EXAMPLE 3

The purchase price of a 3-year-old used car is $12,450. Other costs include registration fees of $128 and sales tax of 4%. The buyer made a down payment of $3,800. What is the delivered price of the car and the balance due?

SOLUTION

Find the sales tax on the purchase price.

$4\% \times \$12,450 = 0.04 \times \$12,450 = \$498$ sales tax

Add the purchase price, sales tax, and registration fees.

$\$12,450 + \$498 + \$128 = \$13,076$ delivered price

Subtract the down payment from the delivered price to find the balance due.

$\$13,076 - \$3,800 = \$9,276$ balance due

✔CHECK YOUR UNDERSTANDING

E. Arnold Knapp agreed to buy a 1-year-old car for $16,500 cash. Sales tax of 7% is charged on the sale. Other costs included license plates, $85; title transfer, $47. What was the total cost of the car?

F. The purchase price of a used van bought by Frances Sauger was $11,370. She paid $200 extra for a 2-year warranty on the transmission. In her state, warranties are exempt from the 5% sales tax charged on merchandise. Registration costs were $129. What is the balance due that Frances needs to finance if she makes a down payment of 25% of the purchase price?

Wrap Up ▶▶▶

People who need a car now and have little money must buy a car on the monthly payment plan. Those who are able to save money and pay cash for a car earn interest on their savings and do not pay interest on their loan. The best way is the one that fits the circumstances.

TEAM Meeting

Visit the "make your own car" portion of the web site of one of the major car manufacturers. Select three different car lines and find the price of the most basic model within each car line. Then find the price for the same basic model with every option. Prepare a chart showing the prices you found and calculate the amount and percent of increase from the lowest to the highest prices in a car line. Share your findings with the rest of the class.

Exercises

Find the sum or difference.

1. $18,458 + $239
2. $456 + $318 + $148
3. $28,346 − $7,086.50
4. $34,829 − $14,340

Find the product.

5. 4% × $36,784
6. 30% × $29,006

Use the buyer's guide and MSRP table given in the lesson to solve Exercises 7–11. (Hint: For Exercises 7–10, if the features the customer wants are either standard features or are not available for the model, do not include their cost in the MSRP you calculate.)

	Car Model	Features Wanted	Warranty Wanted	MSRP
7.	STD	alarm system, air conditioning, tilt steering	24/24,000	
8.	PRM	color matched trim, leather seats	48/60,000	
9.	CTM	premium sound, alarm system, leather seats	12/12,000	
10.	PRM	power package: brakes, seats, steering, windows	48/48,000	

11. A customer wants to buy the STD model with the three features that are standard on the CTM model but optional on the STD model. Would the customer save money by buying the base CTM model instead? If so, how much would be saved?

Solve.

12. The purchase price of a new car bought by Kay Terchek is $30,875. The car's MSRP was $32,560. Her other costs were: sales tax at 7%, non-taxable extended warranty at $390, registration fees at $128. Kay got a $250 rebate and made a $4,300 down payment. What was the delivered price of the car and the amount due?

13. Herman Ollender's purchase price for a new car was 95% of the $21,400 MSRP. Sales tax was figured at 5%. Registration fees totaled $284. He received a customer loyalty rebate of $500. A trade-in value of $3,170 for Herman's old car was used for the down payment. Find the car's delivered price and the amount due.

14. A new car's usual purchase price of $21,480 was reduced by $1,007 because it had been used as a demonstrator car. Eugene Basanese paid a 3.5% sales tax, registration fees of $172, and made a $3,200 down payment. What is the balance due on the car?

15. A used car's price is $4,850. The buyer pays a combined city/state sales tax of 4.6%. Registration fees are $85 for license plates and $31 for title transfer. If the buyer pays for the car in cash, how much will be due?

16. A car that was bought for $23,700 nine years ago was offered for sale at $4,125. Because the car had some body rust, the buyer asked the seller to reduce the price by $350. The seller agreed. The buyer paid $167 in registration fees and a 6% sales tax. What was the total cost of this car to the buyer?

17. **CRITICAL THINKING** Tom believes that most new cars are very reliable and should run with no problems for 125,000 miles. He also believes that buying extended warranties is a waste of money since they will never be needed. What do you think?

18. **FINANCIAL DECISION MAKING** Betty is offered $7,400 for her old car as a trade-in on a new car. She looks in the want ads and sees that many cars the same age as hers sell for up to $3,000 more. Should Betty trade the car or try to sell it herself for a higher price?

Mixed Review

19. Round 10.09 to the nearest tenth and to the nearest unit.

20. Multiply $5\frac{1}{4}$ by $2\frac{1}{3}$

21. $2{,}142 \div 9$

22. Rewrite $1\frac{1}{8}$ as a decimal.

23. 0.17×0.34

24. Colin Marshall's credit card has an APR of 15% and uses a monthly periodic rate. His balance subject to finance charge is $542.75. How much is the finance charge for one billing cycle?

25. Frank Rozier worked five days last week and was paid a per diem rate of $109.50. If his workday is 7.5 hours, how much does Frank earn per hour of work?

6-6

Car Purchases and Leases

GOALS

- Calculate the total amount paid and the finance charge for installment loan car purchases
- Calculate the cost of leasing cars
- Compare the costs of leasing and buying cars

KEY TERMS

- lease

Start Up

Bobbi tells you that she doesn't like her family's cars because they are kept too long, for 8–10 years. She claims that she will always lease cars so she can drive a new car all the time. She asks you whether you agree with her. What would you say?

Diego Cervo/Shutterstock.com

Math Skill Builder

Review these math skills and solve the exercises that follow.

1 **Add** money amounts.
Find the sum. $24,875 + $3,100 + $450 = $28,425

1a. $180 + $570 + $43

1b. $24,765 + $1,250

1c. $562 + $19 + $1,235

2 **Subtract** money amounts.
Find the difference. $34,279 − $33,892 = $387

2a. $18,367 − $17,907

2b. $38,431 − $8,400

2c. $64,395 − $16,338

3 **Multiply** money amounts by whole numbers and percents.
Find the product. 24 × $582 = $13,968
Find the product. 13% × $16,300 = 0.13 × $16,300 = $2,119

3a. 48 × $648

3b. 60 × $610.45

3c. 21% × $19,250

3d. 17% × $9,670

3e. 64 × $19.95

3f. 35% × $13,842

Financing Car Purchases

The delivered price of a car purchase may be paid in cash. Most buyers, however, make a down payment and take out an installment loan.

EXAMPLE 1

The delivered price of Lydia Zollner's new car is $23,560. She makes a $2,000 down payment and pays the balance in 48 monthly payments of $560. What total amount did Lydia pay for the car? What was the finance charge?

SOLUTION

Add the total of the monthly payments and the down payment.

$48 \times \$560 = \$26{,}880 \qquad \$26{,}880 + \$2{,}000 = \$28{,}880$ total paid

Subtract the cash price from the total paid.

$\$28{,}880 - \$23{,}560 = \$5{,}320$ finance charge

✔CHECK YOUR UNDERSTANDING

A. Steve Ruhlin bought a used truck for $9,650. He paid for the truck with a $2,650 down payment and 36 monthly payments of $234.30. What total amount did the truck cost?

B. Iris DiNeise bought a luxury car for $47,851 and made a $4,500 down payment. She got a special loan rate of 2.1% for 60 months. Iris' monthly payments were $761.74. What was her finance charge on this car?

Costs of Leasing

People who lease cars sign a lease. A **lease** is a contract made between the company that owns the car (the lessor) and the person who will be given the right to use the car (the lessee). Leasing a car is similar to renting a car. You use the car for a time and once the lease period is over you turn in the car and walk away.

Before signing a lease, be sure you understand the lease contract, including how many miles you are allowed to drive the car each year.

A LEASE CONTRACT Leasing is based on the idea that you agree to make a monthly payment that covers the depreciation, finance charges, prepaid mileage, and other fees. An additional payment may also be required at the time the lease is signed. The typical lease contract includes these items:

LEASE PRICE The price negotiated by you and the dealer. It is the price on which the monthly lease payments are usually figured.

DOWN PAYMENT This is an amount that may be required by the lease contract or voluntarily paid by a buyer. A down payment reduces the lease price and results in smaller monthly lease payments.

RESIDUAL VALUE The expected value of the car at the end of the lease period. This may also be thought of as the depreciated value of the car.

Business *Tip*

Most leases are closed-end leases that have a fixed residual value regardless of what the market price of the car may be at the lease end.

INTEREST RATE The rate used to compute the finance charge.

LEASE TERM The length of the lease, usually stated in months.

SECURITY DEPOSIT Money held by the dealer to pay for any possible damage to the leased car. The security deposit is refundable.

LOAN FEE A charge for processing the lease contract and making credit checks.

REGISTRATION FEES The cost of license plates and title registration.

MILEAGE ALLOWED The number of miles the car may be driven each year for the term of the lease.

When you lease a car you must buy insurance and pay for gas, oil, and other routine maintenance expenses.

EXAMPLE 2

Jay Sluman leased a car at $307 a month for 48 months with a $995 down payment. At the end of the lease he was charged $0.22 a mile for the 2,800 miles he drove over his lease mileage allowance. What were his total lease costs?

SOLUTION

Find the total of the monthly lease payments and the excess miles charge. Add the two answers to the down payment.

48 × $307 = $14,736 total lease payments

$0.22 × 2,800 = $616 excess miles charge

$14,736 + $ 616 + $995 = $16,347 total lease costs

✔ CHECK YOUR UNDERSTANDING

C. Candace Ortisi had a two-year car lease with $528 monthly payments. Her lease had a mileage limit of 12,000 miles a year and charged 20 cents a mile for each mile over the limit. Her total mileage for the two years was 30,850 miles. What was the total cost of the lease over its term?

D. Terrance Duggan's 4-year lease cost $332 a month and allowed him to drive 12,000 miles a year. He bought 3,000 extra miles each year for an additional charge of 7 cents a mile. Lease processing fees and a down payment totaled $418. What was the four-year cost of the lease?

Compare Leasing and Buying

When you buy a car you pay for its total cost and end up owning a car that still has some value. When you lease the car you pay for only part of its cost, so your monthly payments are lower, but you have no ownership claim on the car.

When you lease a car you have the option to buy the car at the end of its lease. The price you would pay is the *residual value,* which is the estimated value of the car. The method used in this book to compare leasing and buying will be to calculate the total cost of buying the car under both plans.

EXAMPLE 3

Photodisc/Getty Images

Wilbur Frye and his dealer negotiated a price of $27,400 for a new car. Wilbur can lease the car for $496 a month for 36 months and buy it at the end of the lease for its residual value of $16,200. If he buys the car now his monthly loan payment will be $822 for three years after making a $2,000 down payment. What is the total cost of purchasing the car under each plan? Which plan is less expensive?

SOLUTION

Find the total of the monthly lease payments. Then, add the residual value and the down payment, if any.

$36 \times \$496 = \$17,856$ total lease payments

$\$17,856 + \$16,200 = \$34,056$ total cost to purchase leased car

Find the total of the monthly loan payments, add the down payment amount, and find the difference.

$36 \times \$822 = \$29,592$ total monthly loan payments

$\$29,592 + \$2,000 = \$31,592$ total cost to purchase outright

$\$34,056 - \$31,592 = \$2,464$ difference between two costs

The outright purchase cost is $2,464 less expensive.

✔ CHECK YOUR UNDERSTANDING

E. A car that costs $18,500 can be leased for $436 monthly with a $500 down payment and an $11,000 residual value. The monthly loan price cost is $728 with a $3,100 down payment. The lease and loan terms are 24 months. Which is less expensive, leasing or buying, and how much less?

F. A car that sells for $37,960 today is expected to have a residual value of $19,050 in four years. With a $3,800 down payment the loan payments for four years will be $873 monthly. The monthly lease price for four years with a $995 down payment is $623. Does leasing or buying cost more? How much more?

Select the car you would most like to own. You can find web sites on the Internet that rank cars by the number of miles per gallon they get. List the names and mileage information for four cars: the car you selected, the most efficient car, the least efficient car, and the average car. Assume that all the cars will be driven 15,000 miles in a year and their owners will buy fuel at your local gas station. Calculate how much would be spent per year on fuel for each car listed. Write a one-page report, including a chart, about your findings.

Wrap Up ▶▶▶

Having a fairly new car all the time is one of the appealing features of leasing if you are willing to pay the cost. Bobbi should realize that her family has to make a choice on how they spend their money. Perhaps they know that buying and keeping a car for 10 years costs less than leasing 5 different cars for 2 years at a time.

Exercises

Find the sum.

1. $47 + $167 + $476
2. $12,450 + $1,250 + $1,800

Find the difference.

3. $31,374 − $29,857
4. $26,473 − $23,826

Find the product.

5. 72 × $457
6. 24 × $518
7. 14% × $21,560
8. 11% × $19,870

Solve.

9. Maggie Holden's 2-year car lease costs $597 a month. She had to pay a $600 security deposit that she got back at the end of the lease. She made a down payment of $500 and paid fees of $125 to get the lease. What was the net cost of the lease to Maggie at the end of the lease?
10. A 3-year lease costs $324 a month and a 2-year lease costs $392 a month for the same car. On the lease signing date Matt Grove makes a $990 down payment and pays lease application costs of $175. How much more would Matt pay for leasing during the first year under the more expensive plan?
11. A car that costs $20,990 can be bought for $1,800 down and 48 monthly loan payments of $488. It can be leased for $354 a month for 48 months with $760 down and a residual value at lease end of $9,445. Looking at the total purchase costs, will buying or leasing cost more? How much more?
12. The monthly payment is $356 for a 36-month lease with a $250 down payment for a car that sells for $18,042. Its residual value is $10,284. With a $1,700 down payment, the monthly payments on a 36-month loan would be $546. Compare the total purchase costs of leasing and buying. Which costs more, and how much more?

Amy Stiles bought a new car for $24,300. She made a $1,800 down payment. Her monthly loan payments for four years are $569. Had she bought the car two months later and made the same down payment, she would have received a $1,250 rebate that would have lowered her monthly loan payments to $538.

13. What total amount will she pay for the car?
14. What total amount would she have paid if she could have received the rebate and the lower loan payment rate?

Mike Huber is buying a car for $24,300. He wants to make a maximum down payment of $3,000. Two deals are available, both for 48-month loans. In the first deal, with a factory rebate his down payment will be $1,000. His monthly loan payments would be $573.21 at 8.4% interest. In the second deal, he gets no rebate and makes a $3,000 down payment, but his interest rate is 0.9% resulting in a $451.95 loan payment.

15. What total amount will Mike pay with the first deal?

16. What total amount will Mike pay with the second deal?

17. **FINANCIAL DECISION MAKING** A dealer offers you a 2-year lease with no money down at $675 a month for a car that costs $29,100. The estimated residual value of the car is $18,915 at the end of the 2-year lease. You want to buy the car over four years with a $1,200 down payment. The monthly loan payment would be $712. What is the difference between the total cost of buying the car under both plans? Would you choose the less expensive plan?

18. **CRITICAL THINKING** When you lease a car you are responsible for returning the car in good condition with only the normal amount of wear and tear. How would you define what is normal "wear and tear?" Do you think your lease contract defines it in the same way?

Mixed Review

19. Estimate. Then, find the actual quotient of 27,414 ÷ 9

20. Divide 4.37 by 0.023.

21. $7\frac{1}{2} + 1\frac{3}{5}$

22. Divide 5,639 by 0.1 and by 1,000.

23. $48.32 decreased by $\frac{1}{8}$ of itself

24. $52.44 increased by $\frac{1}{3}$ of itself

25. $19.45 decreased by $\frac{2}{5}$ of itself

26. The exact interest charge per $100 is $0.8630 at $10\frac{1}{2}$% interest for 30 days and $0.4315 per $100 for 15 days. What interest will be paid on a $700 loan for 45 days at $10\frac{1}{2}$%?

27. A home entertainment center sells for $3,200 cash. It may be purchased for $400 down and monthly payments of $89 for 48 months. By what percent is the installment price greater than the cash price? Round to the nearest tenth percent.

28. Mike Ruskey's charge statement showed these figures: previous balance, $346.90; new balance, $684.70; cash advance limit, $750; credit limit, $1,250; minimum payment due, $19.12. If Mike makes the minimum payment, what is the most he can charge next month and not exceed his credit limit?

Photodisc/Getty Images

6-7

Depreciating a Car

GOALS

- Calculate average annual depreciation on a car
- Calculate the rate of depreciation

KEY TERMS

- resale value
- trade-in value

Start Up

What do you think has more value: a 3-year-old car with 30,000 miles or a 1-year-old car of the same make and model with 80,000 miles? List at least one source where you could verify your answer.

Photodisc/Getty Images

Math Skill Builder

Review these math skills and solve the exercises that follow.

1. **Divide** money amounts by whole numbers.
   Find the quotient. $14,240 ÷ 8 = $1,780

   1a. $23,580 ÷ 9 **1b.** $12,438 ÷ 3

2. **Divide** money amounts by money amounts to find a percent.
   Find the percent, to the nearest tenth. $540 ÷ $10,500 = 0.0514, or 5.1%

   2a. $1,260 ÷ $9,400 **2b.** $5,200 ÷ $28,600

Average Annual Depreciation

A car loses value as it grows older. This loss of value is called *depreciation.* The total depreciation on a car is the difference between its original cost and its resale, or trade-in, value. **Resale value** is the market value, or the amount you get when you sell the car to someone else. The **trade-in value** is the amount you get for your old car when you trade it in to buy a new car.

Depreciation = Original Cost − Trade-in or Resale Value

When you buy a car, you can only estimate what the depreciation will be. The actual amount of depreciation will be known only when the car is sold or traded in. However, by making some good guesses about your car's future value, you can calculate the estimated *average annual depreciation.*

To calculate the *estimated* average annual depreciation on a car or other motor vehicle follow these steps:

1. Estimate the number of years the car will be kept.
2. Estimate the value of the car when it is resold or traded in.
3. Subtract trade-in or resale value from the original cost to estimate total depreciation.
4. Divide the total depreciation by the number of years the car will be kept.

To calculate the *actual* average annual depreciation, also follow these four steps, keeping in mind that you are using actual, not estimated, amounts. For example, you do not have to estimate the length of time you keep the car or its resale or trade-in value. You use the actual time and dollar amounts.

Business Tip

Cars depreciate much more quickly in their first few years of use than in their last years of use. Cars with many defects or those of poor design will depreciate even more quickly. Cars of high quality depreciate less.

You can use the following formula to find estimated and actual average annual depreciation.

$$\textbf{Average Annual Depreciation} = \frac{\textbf{Original Cost} - \textbf{Trade-In or Resale Value}}{\textbf{Number of Years}}$$

EXAMPLE 1

LaWanda Turgill bought a car for $14,800. She estimates its trade-in value will be $5,800 at the end of 4 years. Find the estimated total and the estimated average annual depreciation of the car.

SOLUTION

Subtract the estimated trade-in value from the original cost.

$14,800 − $5,800 = $9,000 estimated total depreciation

Divide the estimated total depreciation by the number of years.

$9,000 ÷ 4 = $2,250 estimated average annual depreciation

Business Tip

Go to used car web sites to get data that will help you find the estimated depreciation for a car.

✔ CHECK YOUR UNDERSTANDING

A. Roland Corbett bought a new car for $19,500. He has been told that his car will probably be worth $9,800 at the end of two years. What will be his estimated total and average annual depreciation for two years?

B. Genevieve Prekova bought a car 9 years ago for $14,130. She sold it recently for $1,800. What was the total and average annual depreciation on the car?

Monkey Business images/Shutterstock.com

When the average annual depreciation is calculated as it was in Example 1, this is called the *straight-line method.* It assumes the car depreciates the same amount each year.

Rate of Depreciation

When the straight-line method of finding depreciation is used, the average annual depreciation may be shown as a percent of the original cost. The percent is called the *rate of depreciation.*

Rate of Depreciation = Average Annual Depreciation ÷ Original Cost

EXAMPLE 2

A $12,000 car is sold 3 years later for $6,960. What is the rate of depreciation?

SOLUTION

Find the total depreciation.

$12,000 − $6,960 = $5,040

Find the average annual depreciation.

$5,040 ÷ 3 = $1,680

Divide the annual depreciation by the original cost to find the rate of depreciation.

$1,680 ÷ $12,000 = 0.14, or 14%   rate of depreciation

✔ CHECK YOUR UNDERSTANDING

C. A new car that cost $23,000 is worth $16,100 a year later. What was the rate of depreciation for the one year?

D. Billy Macon sold his car for $368. He paid $9,200 for the car when he bought it 12 years ago. What was the annual rate of depreciation?

Wrap Up ▶▶▶

The value of a used car depends not only on its age and the miles it has been driven, but also on its overall condition and how it has been maintained. After the two cars are seen and their service records examined, their true value is determined by what price they will bring in the market. The Kelley Blue Book is one source that provides used car pricing information. You may also want to look at the want ad prices for similar cars or check the prices posted on used car web sites or the Blue Book site.

TEAM Meeting

Meet with three to four members of your class and select one specific make and model that would fit into each of these vehicle categories: luxury car, luxury SUV, economy car, economy SUV. Find the approximate selling price of each vehicle a year ago and what each vehicle would sell for today as a one-year-old vehicle. Calculate the percent of depreciation for each vehicle. Study the depreciation percents and write a paragraph about any differences you find.

Exercises

Find the difference or quotient.

1. $28,459 − $14,286
2. $34,120 − $20,875
3. $16,457 ÷ 7
4. $18,372 ÷ 12

Find the percent, to the nearest percent.

5. $2,368 ÷ $18,300
6. $1,280 ÷ $15,340

Find the total depreciation.

	Type of Vehicle	Original Cost	Resale or Trade-in Value	Total Depreciation
7.	Mid-size Car	$21,606	$9,375	
8.	Sports Utility	$28,461	$12,225	
9.	Pickup Truck	$14,187	$1,560	
10.	Luxury Car	$43,597	$20,150	
11.	Mini Van	$20,500	$12,800	

Solve.

12. Brandon Merritt paid $11,300 for a car 7 years ago. He bought a new car recently at a total cost of $11,100 after deducting the $950 he got as a trade-in for his old car. What was the total depreciation on the 7-year-old car?
13. A bakery bought a truck for $21,088. After four years a new truck was bought that cost $23,420. A trade-in value of $4,790 was given for the old truck. To the nearest dollar, find the average annual depreciation of the old truck.
14. A van that costs $24,444 is estimated to be worth $6,300 after four years. Find the rate of depreciation on the van, to the nearest percent.

Find the average annual depreciation for each.

	Original Cost	Resale or Trade-in Value		Average Annual Depreciation
		At end of	Amount	
15.	$12,800	3 years	$6,710	
16.	$23,100	7 years	$7,210	
17.	$19,750	2 years	$10,120	
18.	$28,980	5 years	$10,800	

Find the rate of depreciation. Round answers to the nearest percent.

	Original Cost	Resale or Trade-in Value		Rate of Depreciation
		At end of	Amount	
19.	$14,500	4 years	$5,600	
20.	$28,350	3 years	$14,700	
21.	$9,450	6 years	$1,800	
22.	$12,680	2 years	$7,700	

Trudy Winslow bought a car for $8,850 three years ago. A car dealer offered Trudy $3,825 as the trade-in value, but she feels she can sell the car for $4,500.

23. What will be the average annual depreciation if she takes the trade-in offer?

24. What will be the average annual depreciation if she is able to sell the car at the price she wants?

25. **STRETCHING YOUR SKILLS** Nora sold her car for $3,100 after owning it for five years. She found the average annual depreciation was $1,860 and the rate was 15%. What did Nora pay for the car when it was new?

26. **CRITICAL THINKING** A new car may depreciate as much as 35% of its original cost in the first year of use. Would it make more sense not to buy the car new but to wait one year and buy it used for less money?

Mixed Review

27. Find the average of 8, 11, 15, 3, and 27.

28. $\frac{5}{6} - \frac{2}{3}$

29. $\frac{2}{9} + \frac{11}{18}$

30. $\frac{7}{12} \div \frac{5}{6}$

31. Nell Burton borrows $4,500 at 6% banker's interest for 24 days. What amount of interest does she pay?

32. Alex Sims borrowed $750 on a one-year simple interest installment loan at 18%. His monthly payment on the loan was $68.76. Find the amount of interest, amount applied to the principal, and new balance for the first monthly payment.

33. Ella Rankin sells printer paper to retail stores. She is paid a commission of 1.25% on all sales for a month. What are her commission earnings for a month where her sales are $382,000?

34. You have a monthly income of $2,400. Your monthly mortgage payment is $550. Your credit card minimum payment is $55, and you have other monthly loan obligations of $385. To the nearest percent, what is your debt-to-income ratio?

Last year, Chandna Venkatraman was paid a salary of $700 a week for 52 weeks. Her federal taxable income for the same period was $21,200. She paid to the city of Carthage in which she worked an income tax of 3.5% on her taxable income. She could have worked at another firm in the city of Hamden. Her annual salary at the other firm would have been $39,000. The city in which the other firm is located charges an income tax of 1.8% on the gross salary.

35. In which city would Chandna have paid more city income taxes?

36. How much more money would she have paid in city income taxes?

37. Lori Schneider can buy a large TV for $700 cash. On the installment plan, Lori must make a down payment of $100 and pay $54.50 for 12 months. How much more is the installment price than the cash price?

38. Gilbert Conroy withdrew $200 from his bank's ATM. On a shopping trip he bought an office chair for $120.87 in cash and paid $75.11 cash for groceries. He then used his ATM card to pay for: $136.50 in painting supplies; $66.52 for lawn mower repair. What amount was left in Gilbert's account if it had a balance of $740.12 at the start of the day?

6-8

Cost of Owning a Car

GOALS
- Calculate car insurance premiums
- Find the cost of operating cars

KEY TERMS
- bodily injury
- property damage
- collision
- comprehensive damage

Start Up

Insurance companies charge young drivers below age 25 much higher premiums for insuring their cars than they would older drivers. What factors do you think they especially consider in setting their rates for young people?

Photodisc/Getty Images

Math Skill Builder

Review these math skills and solve the exercises.

1. **Rewrite** a percent as a decimal. Rewrite as a decimal. 80% = 0.8

   1a. 25% **1b.** 5%

2. **Multiply** money amounts by percents.
   Find the product. 20% × $387 = 0.20 × $387 = $77.40

   2a. 5% × $468 **2b.** 10% × $763

Car Insurance

There are four basic types of insurance or coverage for motor vehicles that protect you against the risk of financial loss:

Bodily injury Covers your liability for injury to others.

Property damage Covers damage to other people's property, including their vehicles.

Collision Covers damage to your own motor vehicle.

Comprehensive damage Covers damage or loss to your vehicle from fire, theft, vandalism, hail, and other causes.

Business Tip

The minimum car insurance you must carry by law for bodily injury and property damage may not offer the financial protection you need.

States require car owners to carry minimum amounts of car insurance. Some states combine bodily injury and property damage coverage into one minimum amount of insurance required. The insurance applies regardless of whether there is injury to one or more persons or whether there is damage to property of others in a single accident.

In addition to requiring minimum amounts of the basic types of insurance coverage, states may require car owners to carry additional coverage.

A car owner may be required to buy *uninsured* and *underinsured motorists insurance,* which protects against damage to the car or injury to persons in the car caused by a driver who carries no or insufficient insurance.

Premiums for automobile insurance may vary from state to state and within a state. Each insurance company sets its own rates following state regulations. Premiums may be higher in large cities than in small cities and rural areas. Premiums may also be higher on cars used for business than those used for pleasure driving. Premiums are usually higher for drivers under 25 years of age than for those over 25.

Business Tip

A bodily injury coverage limit of $25/50,000 means a maximum of $25,000 per person, and $50,000 per accident will be paid by the insurance company.

Sample car insurance annual premiums are found in the table below. As you study the table, notice how the premium changes depending on the use of the car, the coverage limits, and the deductible amount.

Sample Annual Car Insurance Premiums

Type of Insurance Coverage	Coverage Limits	Annual Premiums for: Pleasure Use Only	Driving to Work	Business
Bodily Injury	$25/50,000	$ 20.58	$ 22.84	$ 29.71
	50/100,000	30.88	34.27	44.68
	100/300,000	53.95	59.35	79.74
Property Damage	$25,000	$ 135.80	$ 150.74	$ 196.50
	50,000	161.67	179.44	233.92
	100,000	190.19	211.11	274.44
Collision	$100 deductible	$ 466.53	$ 517.84	$ 574.70
	250 deductible	324.03	358.24	461.81
	500 deductible	261.95	290.77	378.01
Comprehensive	$50 deductible	$ 125.32	$ 137.85	$ 179.21
	100 deductible	93.99	104.33	135.62

EXAMPLE 1

Emma Jane Cooke wants a basic insurance policy for her car that she uses only for pleasure driving. She chooses this coverage: bodily injury, $25/50,000; property damage, $25,000; collision, $500 deductible; comprehensive, $100 deductible. Using the rates in the premiums table above, what annual premium will Emma Jane pay for car insurance?

SOLUTION

Find the premiums in the pleasure use only column.

$20.58 + $135.80 + $261.95 + $93.99 = $512.32 annual premium

✔CHECK YOUR UNDERSTANDING

A. Oliver Trainor insures the car that he drives to work. His coverage is $25/50,000 bodily injury, $50,000 property damage, $250 deductible collision, and $100 deductible comprehensive. Using the premiums shown in the premiums table, what will be Oliver's annual car insurance premium?

B. Harriet Driscoll's car is insured for business use. She chooses the highest coverage limits for bodily injury and property damage and $100 deductibles for both collision and comprehensive. Using the rates shown in the premiums table, what is her annual premium?

Costs of Operating Cars

The total operating cost for a car is the sum of all the annual expenses of using the car. These expenses may include insurance, gas, oil, license and inspection fees, tires, repairs, garage rent, parking fees, taxes, and general upkeep. They also include depreciation and interest lost on a down payment.

EXAMPLE 2

Gerri Forbes paid $18,700 for her car. Her annual payments for insurance, gas, oil, repairs, and other expenses total $2,300. The car depreciates 16% a year. Gerri could have earned $145 interest on her investment in the car. What was her total annual cost of operating the car?

SOLUTION

Find the annual depreciation on the car. Then add to that amount the annual expenses and lost interest.

16% × $18,700 = 0.16 × $18,700 = $2,992 annual deprecation

$2,992 + $2,300 + $145 = $5,437 annual cost of car operation

✔CHECK YOUR UNDERSTANDING

C. Conrad bought a used car for $12,480. His expenses for the first year were gas and oil, $1,070; repairs, $512; insurance, $981; license plates, $83; loss of interest on his car investment, $561; and depreciation, 12%. Find the total operating cost for the year, rounded to the nearest dollar.

D. Ester McHugh, a high school student, bought an old, used car for $600. Her expenses for the year were gas, $610; maintenance and repairs, $780; property damage and liability insurance, $630; license plates, $63; depreciation, 10%; and lost interest, $15. What was Ester's annual cost of operating the car?

Wrap Up ▸▸▸

Insurance companies base their insurance rates on statistics that show that young drivers in the 16–25 years age group are more likely to be involved in accidents, especially fatal accidents. Higher rates are charged to cover the increased damage payments insurance companies will have to make for the young driver group as a whole compared to the general population. Some companies decrease their rates for good students and for those who have taken approved driver's education training.

TEAM Meeting

Form a small group and find out what types and amounts of car insurance are required in the state in which you live. Contact a local insurance agent or your state's car registration office, or search insurance web sites to access the information you need. Present your findings to the class.

Exercises

Write as a decimal.

1. 5%
2. 2.5%
3. 10.35%
4. 0.7%
5. 105%
6. $6\frac{1}{2}\%$

Find the product. Round to the nearest cent.

7. $2\% \times \$563.98$
8. $3\frac{1}{4}\% \times \$762.37$

For the car insurance problems in this textbook, use the premiums table to find the cost of insurance. If the insurance coverage is not given, assume it is one of these standard coverages: bodily injury, $25/50,000; property damage, $25,000; collision, $100 deductible; comprehensive, $50 deductible. The same rate will apply to all types of motor vehicles unless otherwise indicated.

9. What is the total premium for standard insurance coverage on a car driven to work?
10. On a truck he drives to work, Norbert carries bodily injury insurance of $50/100,000 and $250 deductible on collision. Other coverage is standard. Find his annual premium.

June Driscoll uses her truck for business and insures the truck with standard coverage.

11. What annual premium does she pay?
12. If June took the highest deductibles, what amount would she save annually on her total car insurance bill?

Solve.

13. Wasaburo Sumida owns two cars. One car, used for pleasure only, is insured at standard coverage. His business car is insured for the greatest amount of bodily injury and property damage coverage and the highest deductibles. Because he insures both cars with the same company, he gets a 10% discount on his total premium. Find the premium for insuring both cars for one year.

14. Because of her three speeding tickets, Louella Burchette cannot get car insurance unless she pays a premium of 2.2 times the rate for standard coverage. She uses her car to drive to work. What is her premium for one year?

15. Owners of cars with antilock brakes and an alarm system get a 3.5% discount on their total insurance premium. What premium would they pay for standard coverage if they used their car to drive to work?

16. A truck used on a farm is insured at the same rate as if it were being driven to work. It has standard coverage with the highest deductibles. Since the truck is seldom driven outside the farm, it can be insured for 70% of the usual rate. Find the annual premium.

Nancy Tripp/Shutterstock.com

17. Vicki Dirkel paid $5,700 for her car. Her annual payments for insurance, gas, oil, repairs, and other expenses total $1,600. The car depreciates 8% a year. Gerri could have earned $57 interest on her investment in the car. What was her total annual cost of operating the car?

18. After buying a car for $12,230, JoAnn Zimmer estimates her first-year car expenses as: gas, $1,458; maintenance and repairs, $312; license plates, $64; insurance, $511; depreciation, 10% of the car's purchase price; lost interest, $305. Find JoAnn's total cost of operating the car for the first year.

Karl Trattner is 16 years old and owns his car. Because of his age and the type of car he owns, Karl must pay four times the usual rate for his insurance.

19. What annual premium must he pay for pleasure driving with standard coverage?

20. To reduce the amount he must pay, Karl is considering not covering his car for collision and comprehensive damage. What would be his annual premium with this reduced coverage?

21. **CRITICAL THINKING** Insurance companies offer discounts to owners of cars that have features such as air bags, anti-lock brakes, and alarm systems. How can they justify giving such discounts?

22. **CRITICAL THINKING** Insurance companies may charge higher rates to very old drivers or give them insurance with restrictions, such as allowing them to drive only in daylight hours. Is this a form of age discrimination?

23. **FINANCIAL DECISION MAKING** A car that you drive to work is worth about $1,200. If you did not insure your car for comprehensive and collision coverage, you would save about $500 a year in insurance costs. Should you drop these two coverages to save money?

24. **STRETCHING YOUR SKILLS** The IRS allows taxpayers to deduct the expense of operating a vehicle for business purposes. The taxpayer can keep records documenting the miles driven for business and the annual expenses for the vehicle and deduct the percentage of expenses based on the percentage of miles that were driven for business. Or, the taxpayer can use the standard deduction of 55 cents per mile. Jason Selvidge calculated his annual deductible expenses for his car to be $4,800. He drove his car a total of 16,000 miles during the year, and 10,000 of those miles were for business. How much of his actual expenses can he deduct? How much could he deduct if he takes the standard deduction?

Mixed Review

25. $12 \div \frac{1}{4}$

26. Rewrite 1.75 as a percent.

27. $1,236 + $0.78 + $12.37 + $672.99 + $5.82

28. Round to the nearest cent: $0.876, $15.027, $45.514

29. Marcus Ridley signed a promissory note for $10,200 at 9% ordinary interest for 180 days. Find the interest and amount due he will pay when the note is due.

30. Find the number of days from March 18 to July 5.

Delphine Salmer's credit card statement for August showed a previous balance of $658.18, new purchases and fees of $583.10 posted on August 10, and payments and credits of $218.40 posted on August 22. Her charge card company's monthly APR is 1.5%, and the company uses the previous balance method to figure the finance charge.

31. What is Delphine's finance charge for August and her new balance?

32. What would be the finance charge and new balance if the company used the adjusted balance method of computing finance charges?

33. Josie Lamas earns an annual wage of $63,470. Josie estimates her benefits at 32% of her wages. She also estimates that her job expenses are insurance, 6% of wages; commuting, $438; dues, $175; other, $225. Find her annual net job benefits.

34. What are the gross earnings of an employee who works 43 hours and is paid $11.23 an hour?

35. Xavier Morrero paid for a cable connection to StarNet, an ISP. The ISP charged a $15 installation fee, $65 for a network connection card for his computer, a monthly rental fee of $4 for a modem, and a monthly online access fee of $49.95 for an unlimited connection. Xavier also bought antivirus software for $29.99. What will be Xavier's total cost to connect to the Internet for the first year?

Last week, Greg Derkaz worked 4 hours a day Monday through Friday at his after-school job. He is paid $8.20 an hour.

36. How may hours did Greg work last week?

37. What was his gross pay for the week?

Chapter Review

Vocabulary Review

Find the term, from the list at the right, that completes each sentence. Use each term only once.

assessed value
bodily injury
closing costs
collision
comprehensive damage
depreciation
down payment
homeowners insurance
lease
manufacturer's suggested retail price (MSRP)
negotiation
mortgage loan
premium
principal
property damage
property taxes
renters policy
resale value
security deposit
trade-in value

1. The money paid to purchase an insurance policy is called the __?__.
2. A contract that allows you to use property, such as a car, for a certain period is known as a(n) __?__.
3. A type of car insurance that covers your liability for injury to other persons is called __?__.
4. The money paid in addition to a down payment to complete the purchase of a home is __?__.
5. The gradual reduction in the value of a home due to aging and use is referred to as __?__.
6. A home's estimated worth used for tax purposes is called __?__.
7. The insurance usually obtained by tenants to cover the things they own and to provide personal liability protection is known as a(n) __?__.
8. The protection you get from car insurance that covers possible damage to your car is known as __?__.
9. An amount of money kept by a landlord to cover any damage you may cause when renting property is called the __?__.
10. The money you pay to a government unit based on the value of the property you own is called __?__.

6-1 Borrowing to Buy a Home

11. Chester Thornton plans to buy a home for $105,700 with a 15% down payment. He estimates his closing costs as inspections, $360; property survey, $250; legal fees, $1,300; title insurance, $220; loan administration fee, $115; and recording fee, $180. What amount will Chester have to borrow to buy the home? What amount of cash will he need?
12. LuAnne Wiggins is buying a home for $167,000. She will make a 10% down payment and borrow the balance for 30 years at 8.23%. Her monthly mortgage payments will be $1,127.04. What total amount of interest will she pay over 30 years?
13. Robbie Whitaker's monthly mortgage payment is $823. His new monthly payment will be $694 if he refinances the mortgage loan. The refinancing costs are closing costs of $1,074 and a prepayment penalty of $421. How much will Robbie save in the first year by refinancing his mortgage?

6-2 Renting or Owning a Home

14. Muriel Voegel plans to buy a home. She estimates her annual home ownership expenses to be: mortgage interest, $4,873; property taxes, $2,156; home insurance, $418; depreciation, $1,800; utilities, $1,835; and maintenance and repairs, $825. She will lose $1,410 interest on her down payment and save $640 in income taxes. What will be Muriel's net cost of owning the home in the first year?

15. Rafael Gonzalez rents a home for $1,400 a month. The security deposit he paid is two months' rent. Renters insurance costs $210 a year. Rafael expects utilities to average $136 a month. What will be the cost of renting the home for the first year?

16. Heather Rayburn estimates that she would pay $4,150 a year for taxes, insurance, and maintenance on a home she bought. Other annual home costs would be $8,800 in mortgage loan interest and $1,300 in estimated depreciation. During the year, Heather would lose $820 interest on her down payment, but her income taxes would be $1,145 less. Heather could have rented the home for $1,300 a month and had annual expenses of $205 for insurance and $1,780 for utilities. Her security deposit would have been $1,200. For the first year, would it have been less expensive to buy or rent the house, and how much less?

6-3 Property Taxes

17. The Wabeek County Library System's total budgeted expense for a year is $8,240,000. Of that total, $1,970,000 will come from various sources. The rest must be raised by a property tax. The total assessed value of all property in the county is $8,500,000,000. What tax rate is needed to cover budgeted expenses, to the nearest hundred thousandth?

18. What property tax is due on a home assessed at $210,000 if the tax rate is $4.23 per $100?

19. The property tax rate is $34.67 per $1,000. Find the tax due on a business whose property is assessed at $540,000.

20. The property tax rate to maintain a park is 0.18 mills per $1 of assessed value. What tax will be paid on a home assessed at $42,500?

21. A vacant lot is assessed at $26,000. What property tax must be paid on the lot if the tax rate is 2.8 cents per $1?

6-4 Property Insurance

22. Virgil Tulley insured his home for $123,000 at an annual rate of $0.76 per $100. What premium did he pay?

23. Cecilia Emeg's apartment is located 6 miles from a fire station. She insured the apartment's contents for $15,000 and paid $49 extra for insuring a diamond ring. Use the renter's premium table to find the total insurance premium paid.

24. Jeremy O'Brien's homeowners policy has a face value of $56,400 and a $750 deductible. How much will the insurance company pay on a $1,612 loss?

25. A warehouse with a value of $240,000 is insured for $144,000 under an 80% coinsurance policy. A fire loss of $84,000 occurs. How much of the loss will the insurance company pay?

6-5 Buying a Car

26. Find the MSRP of a MR35 sports sedan, Model CTM with these options: power seats and windows, ski rack, and 36/45,000 warranty.

27. Libby Coulson agreed to a purchase price of $32,568 for a new SUV. She made a down payment of $7,500. Libby is charged 4% sales tax on the purchase price and paid $183 in registration costs. She also received a $1,500 manufacturer's rebate. Find the delivered price and the balance due on this purchase.

28. The purchase price of a used car Wade Hatcher bought was $6,140. Sales tax of 6.5% was charged on the purchase. Registration fees were $116. Wade paid cash for the car. What was the delivered price and balance due on this purchase?

6-6 Car Purchases and Leases

29. Archie Beane bought a used van for $8,127. He paid for the van with a $1,500 down payment and 24 monthly payments of $327.86. What total amount did he pay for the van and what was the finance charge on the loan?

30. Pattie Truitt made a $2,000 down payment on a 60-month SUV lease contract. Her monthly payments were $485.30. At the end of the lease she was charged $0.19 per mile for 1,800 excess miles. What was Pattie's total cost of leasing?

31. A car that costs $25,007 can be leased for $414 monthly over 4 years with a $2,100 down payment. The car's residual value is estimated to be $11,050. If the car is purchased with $2,100 down, the monthly payments will be $594 for 4 years. Which is less expensive, leasing or buying, and how much less?

6-7 Depreciating a Car

32. A new car bought for $28,240 is estimated to have a value of $6,100 after 6 years. What is the car's estimated average annual depreciation?

33. A used car bought for $16,230 was sold for $1,200 after 9 years of use. What was the car's rate of depreciation, to the nearest tenth percent?

34. A sports car bought for $24,995 was worth for $15,500 after 5 years of use. What was the car's rate of depreciation, to the nearest tenth percent?

35. A new car bought for $19,465 is estimated to have a value of $8,750 after 4 years. What is the car's estimated average annual depreciation?

6-8 Cost of Owning a Car

36. Kendra Busby drives her car to work. She wants to insure the car with $100 deductibles for collision and comprehensive coverage. Kendra also wants to carry $100/$300,000 bodily injury and $100,000 property damage coverage. Use the premiums table to find the premium she must pay.

37. Bernie Kuykendall bought a used car for $7,300. His expenses for the first year were: gas and oil, $1,184; insurance, $620; repairs, $217; lost interest, $296; license plates, $56; and depreciation, 11%. What was Bernie's annual cost of operating the car?

38. James Craig drives his car for business. The company's insurance carries $50/100,000 bodily injury and $100,000 property damage. The deductibles are $500 for collision and $100 for comprehensive. Use the premiums table to find the premium the company pays to insure James' car.

Technology Workshop

Task 1 Calculating Mortgage Payments

Enter data into a template that calculates the monthly mortgage loan payments and the total interest paid on the loan. You may use the template to compare the effects of changes in the interest rate and loan term on the total interest paid.

Open the spreadsheet for Chapter 6 (tech6-1.xls) and enter the data shown in blue (cells B3-5) into the spreadsheet. The spreadsheet will calculate the monthly loan payment, total amount paid on the loan, and the total interest paid on the loan.

Your computer screen should look like the one shown below when you are done.

	A	B
1	**MORTGAGE LOAN CALCULATOR**	
2	**Mortgage Loan Data**	
3	Amount	$110,000.00
4	Interest Rate (%)	8.160
5	Term (in years)	30
6	**Mortgage Payment Data**	
7	Mortgage Factor	0.9128142
8	Number of Payments	360
9	Monthly Payment	$819.44
10	Total Amount Paid	$294,998.40
11	Less Original Mortgage	$110,000.00
12	Total Interest Paid	$184,998.40

Task 2 Analyze the Spreadsheet Output

Answer these questions about the mortgage loan calculations.

1. For how many years was the loan made?
2. What amount was borrowed?
3. What was the monthly payment?
4. What total amount was paid on the mortgage loan?
5. What total amount of interest was paid on the loan?

Now move the cursor to cell B4, which holds the mortgage interest rate. Enter the rate 8.66%, which is $\frac{1}{2}$% higher than the rate you first entered. Enter the new rate of 8.66% without the percent symbol.

Answer these questions.

6. What total interest would be paid on the loan at the higher interest rate?
7. Approximately how much more would be paid in interest over 30 years at the $\frac{1}{2}$% higher interest rate?

8. Assume you changed the loan term to 25 years and kept the rate at 8.66%. Over which term, 25 years or 30 years, do you think you would pay the greatest total amount of interest? Now, change the term to 25 years and check your thinking.

Task 3 Design an Insurance Loss Payment Spreadsheet

You are to design a spreadsheet that will calculate the amount of loss paid by an insurance company under a coinsurance policy. Also calculate the required amount of coinsurance.

The spreadsheet should have two sections, one for input data, and another for calculated data. Design your spreadsheet so the amount of loss paid is never greater than the insurance carried on the property.

SITUATION: Dewayne Clayton owns a home worth $50,000. He insures it for $35,000 under an 80% coinsurance policy. His roof was damaged by high winds, and its repair will cost $2,000. Find the amount of this loss that will be paid by the insurance company.

Task 4 Analyze the Spreadsheet Output

Answer these questions about your completed spreadsheet:

9. What amount of loss did the insurance company pay?
10. What coinsurance amount should have been carried on the home?
11. How much of the loss will Dewayne have to pay?
12. If the policy had a deductible, what change would you have to make in your spreadsheet?

Photodisc/Getty Images

Chapter *Assessment*

Chapter Test

Answer each question.

1. $1,057 + $186.20 + $595.86
2. $248,112 − $162,905.70
3. 360 × $918.47
4. 2.45% × $137,000
5. $1\frac{3}{5}$ × $24,000
6. $6,840 ÷ 0.5%
7. 18 is what percent of 360?
8. $542 increased by $\frac{1}{4}$ of itself is?
9. $6\frac{1}{2} + 1\frac{3}{5} + 4\frac{3}{4}$

Solve.

10. The home that Hilda Vaughan wants to buy sells for $213,000. She plans to make a 5% down payment and borrow the balance at 7.67% for 25 years. Her monthly mortgage payments will be $1,517.80. What total interest will Hilda pay over 25 years?
11. Eldon Hudspeth estimates that the cost of taxes, insurance, and maintenance on a home he bought is $4,980 a year. His other yearly costs would be $12,760 in mortgage loan interest and $2,050 in estimated depreciation. For a year, he would lose $1,030 interest on his down payment and save $2,400 on his income taxes. Eldon could have rented a similar home for $1,400 a month and had annual expenses of $302 for insurance and $2,760 for utilities. The security deposit on the rented home would have been $500. For the first year, which would have been less expensive, buying or renting a home? How much less?
12. Find the property tax due on a building assessed at $360,000 if the tax rate is $37.40 per $1,000.
13. Wanda Frey insured her home for $156,400. Her annual insurance rate is $0.83 per $100. What annual premium did she pay?
14. Kerry Molloy's homeowners policy has a face value of $78,200 and a $250 deductible. How much of a $4,783 loss will the insurance company pay?
15. Janese Mosby bought a new car for $27,437. She made a down payment of $4,200. Janese must pay 5% sales tax on the purchase and $121 in registration costs. She received a $500 rebate from the manufacturer when she bought the car. Find the delivered price and the balance due on this car purchase.
16. A car bought for $18,216 was sold for $12,870 after 2 years of use. What was the car's rate of depreciation, to the nearest tenth percent?
17. Edward Laffin uses his car for pleasure driving. The insurance coverages he wants and their costs are: collision, $368.12; comprehensive, $146.89; bodily injury, $164.14; property damage, $186.34. Edward receives a 5% discount on the total premium because he drives the car less than 7,500 miles a year. What annual premium will Edward pay?
18. A car costing $34,000 can be leased for $473 monthly over 5 years with a $2,420 down payment. The car's residual value is estimated to be $16,400. If the car is purchased with a $5,000 down payment, the monthly loan payments will be $623 for 60 months. Is buying or leasing less expensive, and how much less?

Planning a Career in Architecture and Construction

Careers in architecture and construction can be as varied as an engineer building a bridge to a surveyor verifying boundary lines in a home sale. If you choose a career in architecture you can specialize in designs for homes, buildings, ships, planes, bridges, highways, or landscaping. A job in construction can mean you build skyscrapers, install plumbing, run wires for electricity, manage construction projects or crews, or inspect buildings or structures. If you have the ability to envision an idea that does not yet exist, or the skills to bring that idea into existence, a career in architecture and construction may be a good avenue for you.

Photodisc/Getty Images

Job Titles

- Civil engineer
- Brick layer
- Drafter
- Architect
- Aerospace engineer
- Landscape architect
- Electrician
- Boilermaker mechanic

Needed Skills

- strong organizational and leadership skills
- an exceptional eye for detail

- mathematical and scientific skills
- excellent problem-solving skills
- technical and computer skills
- the disciple to work independently, as well as in a team

What's it like to work in Architecture?

Architects specialize in designs of structures of either interiors or exteriors. Landscape architects design outdoor areas around houses, shopping centers, roadways, schools, and buildings. Landscape architects can work with engineers, scientists, and surveyors to plan the locations for buildings, roads, and walkways. These plans include sketches, reports, cost estimates, material lists, and the use of specialized software programs. Many landscape architects have their own businesses. Forty-nine states require landscape architects to be licensed. This is achieved by examination.

What About You?

What aspect of architecture and construction appeals to you? How might you best prepare for a career in this field?

How Times Have Changed

For Questions 1–2, refer to the timeline on page 207 as needed.

1. It is common for a homebuyer to pay a down payment of 10%, 15% or 20%. If a home in Baltimore, Maryland cost 315,000 in 2007, what is the range of money a buyer would likely pay as a down payment?
2. The number of households consists of the number of homeowners and the number of renters. The number of households in 2007 was about 111 million. About how many households owned homes in 2007? About how many 2007 households were renters?

Chapters 5–6 *Cumulative Review*

MULTIPLE CHOICE

Select the best choice for each question.

1. What is the finance charge per $100 on a loan of $5,300 with a finance charge of $954?

   A. $0.18 B. $1.80 C. $5.50
   D. $18 E. $555

2. What is the interest due at maturity for a $900 note borrowed for 1.5 years at a rate of 16.5%?

   A. $222.75 B. $148.50 C. $99
   D. $22.75 E. $14.85

3. You paid $55 interest on a 6-month promissory note of $1,000. What rate of interest, to the nearest percent, did you pay?

   A. 5.5% B. 5.8% C. 9.1%
   D. 11% E. 12.2%

4. How many days are between March 2 and May 9?

   A. 64 B. 65 C. 66
   D. 67 E. 68

5. You are thinking of buying a home for $104,300 and making an 18% down payment. You estimate closing costs will be 2.5% of the home's purchase price. How much cash will you need to buy the home?

   A. $16,166.50 B. $2,607.50 C. $18,774
   D. $21,381.50 E. $20,800

6. A home worth $105,000 is insured for $85,000 under a 90% coinsurance policy. How much of a $28,000 loss would an insurance company pay?

   A. $20,000 B. $28,000 C. $26,444.44
   D. $25,200 E. $25,185.19

7. Vicky Zielinski bought a truck on these terms: purchase price, $18,240; down payment, $2,500; rebate, $1,200; sales tax, 5%; and registration fees, $117. What was the balance due?

   A. $20,579 B. $15,569 C. $13,745
   D. $17,805 E. $17,979

8. Charles Codwell bought a used car for $7,195. He made an $800 down payment and paid the balance in 24 payments of $318.47 a month. What was the finance charge on the loan?

   A. $448.28 B. $1,258.32 C. $1,248.28
   D. $351.72 E. $1,428.28

9. A car bought new for $27,118 is sold for $950 after 14 years of use. What is the average annual depreciation?

   A. $1,869.14 B. $67.86 C. $1,937
   D. $2,854.52 E. $1,689.14

OPEN ENDED

10. Larry Pons has a 3-year, $10,000 installment loan at 8%. His monthly payment is $313.36. After making 22 payments, his balance is $4,175.32. If Larry decides to pay off the loan with his next payment, how much should he pay?

11. Yoko Soga borrowed $450 on a loan with a finance charge of $94.50. Find the finance charge per $100 of the amount financed.

Nathan Bogart borrows $1,540 on a simple interest installment loan at $11\frac{1}{2}$% agreeing to repay it in 12 equal monthly payments of $136.47.

12. What was the total finance charge on the loan?

13. What was the interest paid for the first month?

14. What was the amount applied to principal at the end of the first month?

15. The basic annual cost of an auto insurance policy on Kara Malgren's car is $726. She gets a 2% discount for having a theft alarm and side impact air bags. She also gets a 4% discount for her safe driving record. What annual premium will Kara pay?

16. A home may be bought for $72,000 with a 15% down payment. The monthly payments on a 20-year loan at 7.61% will be $497.15. What total interest will be paid on the loan?

17. A town needs $580,000 to maintain its parks. Park use fees raise $124,000 of that amount. The total assessed value of property in the town is $10,900,000. What tax rate is needed to provide enough money to maintain the parks? Round to four decimal places.

18. A home's assessed value is $185,410. If the property tax rate is $29.18 per $1,000 of assessed value, what tax is due on the home?

19. A tax rate of 24 mills per $1 of assessed value is equivalent to what rate in dollars?

20. Bess Ambrose owns a car with an average monthly cost of gas, oil, and repairs of $184. The annual costs include insurance, $842; depreciation, $1,080; license plates, $68; and lost interest, $48. What is the total annual cost of operating the car?

21. Yong's Flower Shop purchased a delivery van for $32,599. The salesperson guaranteed that the dealership would buy back the van after four years for $12,550. What is the expected rate of depreciation?

CONSTRUCTED RESPONSE

22. A car can be leased for $289 a month for 36 months with no money down. The same car could be bought for $15,680 with $1,700 down and 36 monthly payments of $453. Write a note to a friend explaining why even with the $164 monthly payment difference, leasing the car may be more expensive than buying.

Chapter 7

Insurance and Investments

Statistical Insights

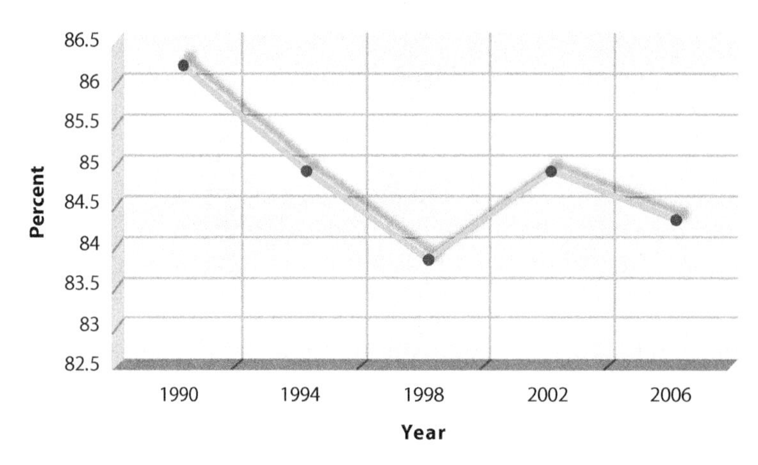

Source: Census Report

Workers who are not offered health insurance benefits through their place of employment sometimes purchase private health insurance, but most often do without health insurance. Use the line graph to answer Questions 1–4.

1. What does the point in the line graph over the year 2002 mean?
2. In what year shown in the graph did the greatest percentage of Americans have health insurance? About what percent?
3. In what year shown in the graph did the least percentage of Americans have health insurance?
4. **Explain** what the slope or trend of the line graph shows.

How Times Have Changed

The origin of the New York Stock Exchange can be traced to 1792. Since 1896, the Dow Jones Industrial Average (the Dow) has been used to indicate the trend of the NYSE. At its inception in 1896, the Dow was 40.74 points. About 100 years later on July 19, 2007, the Dow reached a record high of 14,000.41 points.

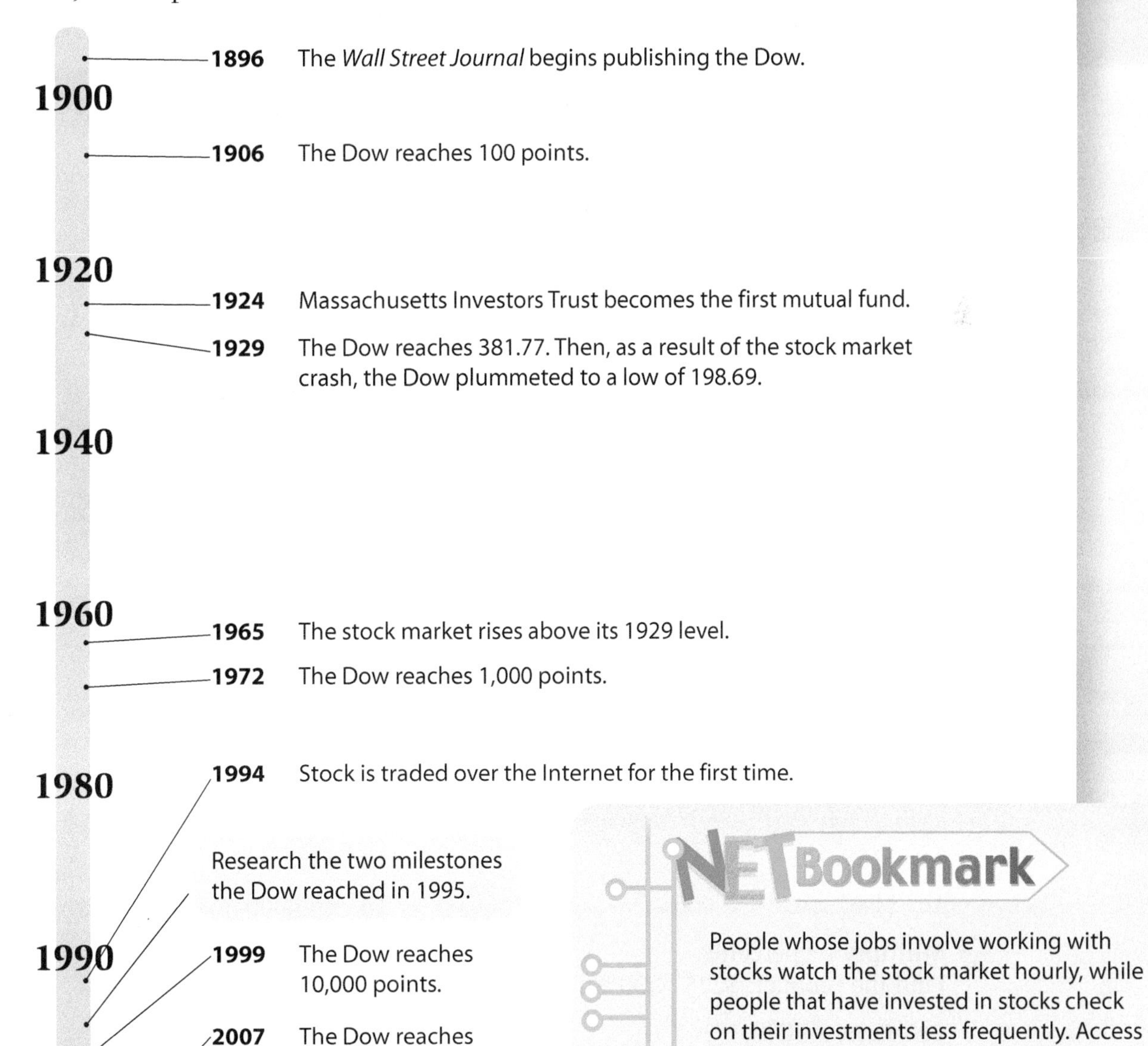

1896 The *Wall Street Journal* begins publishing the Dow.

1906 The Dow reaches 100 points.

1924 Massachusetts Investors Trust becomes the first mutual fund.

1929 The Dow reaches 381.77. Then, as a result of the stock market crash, the Dow plummeted to a low of 198.69.

1965 The stock market rises above its 1929 level.

1972 The Dow reaches 1,000 points.

1994 Stock is traded over the Internet for the first time.

Research the two milestones the Dow reached in 1995.

1999 The Dow reaches 10,000 points.

2007 The Dow reaches 14,000 points.

2008 The greatest points lost in one day, 777.68, in the history of the Dow.

NET Bookmark

People whose jobs involve working with stocks watch the stock market hourly, while people that have invested in stocks check on their investments less frequently. Access www.cengage.com/school/business/businessmath and click on the link for Chapter 7. How do individual investors keep track of the performance of their stocks? How easy is it to buy and sell stocks online?

7-1

Life Insurance

GOALS

- Calculate life insurance premiums
- Calculate the net cost of life insurance
- Calculate the cash and loan values of a life insurance policy

KEY TERMS

- life insurance
- cash value

Start Up

How important, on a scale of 1–10, is life insurance to the following people:

a. A single person with no dependents who attends school.

b. A single person who supports an aging parent.

c. A working couple with one child.

d. A working couple with no children.

e. A couple with two children, where one parent works and the other takes care of the children.

f. A widow with two self-supporting, adult children.

Photodisc/Getty Images

Math Skill Builder

Review these math skills and solve the exercises that follow.

1 **Divide** by 1,000.
Find the quotient. $250,000 ÷ 1,000 = $250

1a. $125,000 ÷ 1,000 **1b.** $335,000 ÷ 1,000

1c. $1,678,200 ÷ 1,000 **1d.** $89,000 ÷ 1,000

2 **Multiply** by percents.
Find the product. $225 × 30% = $225 × 0.30 = $67.50

2a. $189 × 10% **2b.** $210 × 15%

2c. $549 × 25% **2d.** $92 × 5%

3 **Divide** to find percents.
Find the percent. $350 is what percent more than $300?
$350 − $300 = $50; $50 ÷ $300 = $16\frac{2}{3}$%

3a. $300 is what percent more than $250?

3b. $280 is what percent more than $240?

Life Insurance Premiums

Life insurance is a way of protecting your family from financial hardship when you die. If your income supports your family, they will need to replace that income when you die. If you are a homemaker, your surviving spouse may need to pay someone to care for your children and home. In both cases, money is needed to pay funeral costs. Life insurance may also be bought to repay debts when you die, such as a home mortgage or a car loan.

A life insurance *policy* is a contract between the *insured,* the person whose life is covered, and the insurer, the insurance company. The contract states the amount of insurance to be paid upon the death of the insured, or the *death benefits* of the policy. The death benefits are usually equal to the face amount (or *face value)* of the policy.

The money paid to an insurance company for life insurance is the *premium.* When the insured dies, death benefits are paid to the beneficiary. The *beneficiary* is the person named in the policy to receive the death benefits.

There are two basic types of life insurance policies: term life insurance and permanent life insurance. Insurance companies have designed many variations of these two types for people with different needs and budgets.

Business Tip

If your reason for buying life insurance is to protect your family from the loss of your income, one way to estimate how much insurance you need is to estimate how much you would have received in take-home pay from the time of your death until a retirement age of 65 or 70.

Term life insurance offers protection for a fixed period of time, such as 1, 5, or 10 years. If you die within that time, your beneficiary receives the face value of the policy. Term insurance can usually be renewed after the fixed term expires, but usually the policy premiums will be higher because you are older and more likely to die. Term insurance is the least expensive kind of life insurance.

One variation of term life insurance is *decreasing term life insurance.* With decreasing term life insurance, the face amount of the policy decreases over time. Decreasing term life insurance is popular with homeowners who use the policies to cover their mortgage loans. Because the amount of insurance declines over time, the premiums are lower than with standard term life insurance.

Whole life insurance insures you for your whole life. Premiums usually are paid from the time you take out the policy until the time of your death.

A variation of whole life insurance is universal life insurance. *Universal life insurance* allows you limited ability to change the amount of the death benefit and how much you pay in premiums. A certain amount of each premium payment is invested and earns tax-free income. In years when you can, you may pay more than the premium. The overpayment is invested. In years when your money is tight, you can pay less than the premium or skip it entirely. The premium is then paid from the invested funds.

Annual Premiums per $1,000 of Life Insurance

Age of Insured	10-Year Term		Whole Life	
	Male	Female	Male	Female
20	1.12	1.08	9.84	8.92
25	1.14	1.10	11.61	10.56
30	1.17	1.13	14.08	12.81
35	1.30	1.26	17.44	15.86
40	1.54	1.49	22.60	20.55
45	1.97	1.91	27.75	25.24

The table at the bottom of the previous page shows the premiums an insurance company might charge for each $1,000 of life insurance. Different rates are given for men and women at different ages. The rates shown are for nonsmokers. To find the annual premium for a policy, divide the face amount by $1,000 and then multiply the result by the cost per $1,000 in the table.

Business Tip

Most insurance companies have web sites that let you get rate quotes for different types of life insurance policies.

$$\textbf{Premium} = \frac{\textbf{Face amount of Policy}}{\textbf{\$1,000}} \times \textbf{Cost per \$1,000}$$

EXAMPLE 1

Shelly Burnam buys a $25,000 whole life insurance policy at age 25. Shelly does not smoke. What is her annual premium?

SOLUTION

Find the cost per unit in the table.

25-year old, female, whole life: 10.56

Use the formula to find the premium.

$$\text{Premium} = \frac{\$25,000}{\$1,000} \times 10.56 = \$264$$

✔ **CHECK YOUR UNDERSTANDING**

A. Bob Walzcek bought a $135,000 term life insurance policy. He is 35 years old and does not smoke. Find his annual premium.

B. Risa Belvador bought a $150,000 whole life insurance policy at age 20. Risa does not smoke. Find her annual premium.

Net Cost of Insurance

Some insurance companies may return part of your premium to you as a *dividend*. You may deduct the dividend from the premium due or leave the dividend with the company to buy more insurance or to earn interest. The total premium for the year less the dividend is the net cost of the insurance for the year.

Net Cost of Insurance = Total Premiums − Dividends

EXAMPLE 2

Tom Fisher paid $57 quarterly for a life insurance policy. His policy also paid a dividend of $14.80 at the end of the year. What was the net cost of his insurance policy for the year?

SOLUTION

Multiply the quarterly premium by 4.

$\$57 \times 4 = \$228$ total premium for a year

Subtract the dividend from the total premiums to find the net cost of insurance for the year.

$\text{Net Cost of Insurance} = \$228 - \$14.80 = \$213.20$

✔ CHECK YOUR UNDERSTANDING

C. Ricardo Ballas received a premium notice from his insurance company for his life insurance policy. The policy listed an annual premium of $856 and also a dividend of $38.56. Ricardo decided to deduct the dividend from the premium and pay the difference. What was the net premium he paid?

D. Yolanda Pagan doesn't smoke. Five years ago, when she was 30, she bought a whole life insurance policy for $150,000. This year her policy paid a dividend of $23.16. Using the insurance premium table, find her annual premium and the net cost of her policy for this year.

Photodisc/Getty Images

Life Insurance Cash Values

If you cancel a term policy, you get nothing. Whole life policies build cash value after premiums have been paid for a few years. **Cash value** is the money that you get if you cancel the policy. The policy may give you a choice of taking the cash, or using it to buy a small amount of whole life insurance that is totally paid up, or to buy term insurance.

The policy may also allow you to borrow up to the total amount of the cash value, often at a lower interest rate than that offered by other lenders. If you don't pay back the loan, it will be subtracted from the amount paid to your beneficiaries when you die.

A whole life policy that builds cash value would have a table much like the one shown at the right. The cash values of universal life insurance policies will vary with the current value of the investments that have been made.

Cash Value Table

Year	Cash/Loan Values per $1,000
1	0
5	10
10	42
15	80
20	124
25	174

EXAMPLE 3

Using the cash value table, find the maximum amount you can borrow against your $100,000 policy if you had the policy for 20 years.

SOLUTION

Divide the face value of the policy by $1,000.

$100,000 ÷ $1,000 = 100   number of $1,000 units in the policy

Multiply the number of units by the appropriate table amount.

100 × $124 = $12,400   maximum loan amount of policy

✔ CHECK YOUR UNDERSTANDING

E. Using the cash value table, how much can you borrow against a 10-year policy with a face value of $300,000?

F. Using the cash value table, how much would you receive if you cancelled a 15-year policy with a face value of $150,000?

Wrap Up ▶▶▶

When you rated the importance of insurance, did you consider the (a) number of dependents, (b) need for insurance to cover a mortgage, (c) need for insurance to protect against the loss of income for both working partners, and (d) need for money to cover the cost of burial?

Communication

Investigate the interest rates charged for a car loan from one credit union and one bank. Then find the interest rate charged for borrowing from a whole life insurance policy's cash surrender value. Prepare a chart of your results and present your findings to the class.

Some sources for gathering information include:

- Internet
- a credit union
- a bank
- an insurance agent

Exercises

Find the quotient or product.

1. \$104,580 ÷ 1,000
2. \$59,320 ÷ 1,000
3. \$28 × 25%
4. 412 × \$0.76
5. \$339 × 20%
6. 78 × \$1.54

Divide to find the percent.

7. \$500 is what percent more than \$400?
8. \$750 is what percent more than \$500?

Use the insurance premium table to solve Exercises 9–14.

	Policy Type	Age and Sex	Policy Face	Annual Premium
9.	Whole Life	20, female	\$200,000	
10.	10-Year Term	30, male	\$150,000	
11.	10-Year Term	40, female	\$400,000	
12.	Whole Life	25, male	\$50,000	

13. How much more is the annual premium on a \$50,000, whole life policy for a male at age 45 than at age 25?
14. How much more is the annual premium on a \$100,000, whole life policy than a 10-year term policy for a 30-year old female?

Solve.

15. Paul Poncelli, age 30, is comparing the premium for a $100,000, whole life policy he may take now and the premium for the same policy taken out at age 35. Find the difference in total premium costs over 20 years for this policy at the two age levels.

16. Melva Davis paid an annual premium of $12.80 per $1,000 for an $8,000 life insurance policy. Her policy also paid a dividend of $18.90, which she used to reduce her premium. What net premium did Melva pay?

17. Trent Coleman pays a premium of $18.80 per $1,000 for a $120,000 life insurance policy. During the year, his policy paid a dividend of $11.90. What is the net cost of the policy for the year?

18. Alicia Ronzetti bought a 30-year decreasing term life insurance policy to repay her 30-year, $150,000 mortgage in the event of her death. The insurance company's annual rate per $1,000 for the policy was $0.76. What was her annual premium?

19. Silvia Olivares takes out a universal life policy for $50,000 and pays $490 in annual premiums. What total amount will she pay in premiums in 20 years?

20. **STRETCHING YOUR SKILLS** Because she smokes, Valerie Corini pays 20% more for life insurance. How much more will Valerie pay for $150,000 of 10-year term insurance at age 45 than a nonsmoker would pay at the same age?

21. **STRETCHING YOUR SKILLS** To the nearest percent, what percent greater is the cost of a whole life policy taken out by a male at age 45 than at age 35?

STRETCHING YOUR SKILLS Rollie Collins, age 25, wants to pay no more than $600 a year in life insurance. In even thousands of dollars, what is the largest policy he can buy without spending more than $600 annually on a

22. whole-life policy?
23. 10-year term insurance policy?

Use the table of cash values to solve Exercises 24–28.

24. How much cash would you get if you canceled a $50,000 whole life policy after paying premiums for ten years?

25. You have paid annual premiums of $242 on a $75,000 policy for ten years. What amount could you borrow on your policy?

26. What amount could you borrow on a $200,000 policy that was 25 years old?

27. How much could you borrow on a 15-year-old policy with a face value of $50,000?

28. What amount would you receive if you canceled a $150,000 policy that you had held for 20 years?

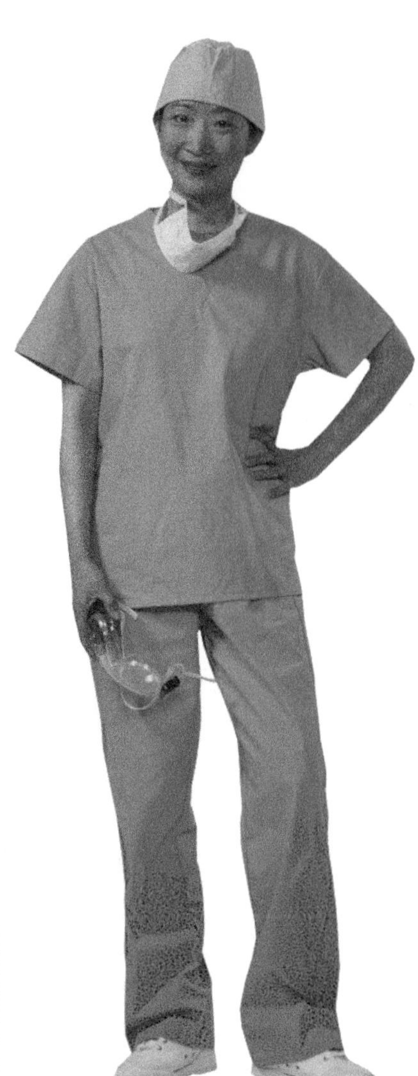

Photodisc/Getty Images

29. **STRETCHING YOUR SKILLS** Bill Woolsley paid annual premiums on a $50,000 whole life policy at a rate of $17.20 per $1,000. After ten years, he canceled the policy and found that its cash value was $49 per $1,000. Over the ten years, he received dividends of $318.55. For the time Bill had the policy, find the net cost of the insurance he held.

30. **CRITICAL THINKING** Four different types of life insurance were discussed in this lesson. Return to the list of insurance situations presented at the start of this lesson. Which type of life insurance, if any, would you recommend in each situation and why?

31. **FINANCIAL DECISION MAKING** What concerns should a person have for borrowing money from the cash surrender value of their insurance policies?

Mixed Review

32. Write $128\frac{3}{5}\%$ as a decimal.

33. Find $12\frac{1}{2}\%$ of $184.

34. Multiply $217.80 by 22%, to the nearest hundredth.

35. $30.20 increased by 30% of itself is?

36. The average of 14, 28, 19, 22, and 18 is?

37. $\frac{2}{5} \div \frac{1}{8}$

38. Charlie Evers is paid $12.56 an hour and time-and-a-half for overtime. Last week he worked 40 regular and 4.5 overtime hours. What was his gross pay for the week?

39. Sonia Ortiz earned $1,500 last year at her part-time job. Her parents claimed her as a dependent on their federal income tax return. What taxable income did Sonia have last year?

40. Rosie McFarland's check register balance on October 31 was $374.60. In making a reconciliation statement, she found that a check for $17 was incorrectly recorded in the register as $71; she had no record in her register of a service charge of $2.80, earned interest of $0.71, and a deposit of $68.74. What was her correct check register balance?

41. Umeki Akita repaid a loan of $3,200 in 15 monthly installments of $232.80 each. Use the annual percentage rate table in Chapter 5 to find the APR on his loan.

42. The regional library system tax rate in Odell County is 2.5 mills per dollar of assessed value. Find the tax to be paid on property assessed at $87,000.

Vi Foe is 25 and doesn't smoke. She owns a $200,000 whole life insurance policy.

43. Using the insurance premium table, what is her annual premium?

44. During this year, her policy paid a dividend of $62.10. What was the net cost of her policy for the year?

45. Roger canceled his $250,000 whole life policy. The cash/loan value for his policy at the time was $95 per $1,000 of insurance. How much did Roger receive when he cancelled his policy?

7-2

Health Insurance

GOALS

- Calculate health insurance premiums
- Calculate health insurance benefits and coinsurance

KEY TERMS

- health insurance
- major medical insurance
- coinsurance

Start Up ▶▶▶

Wilbur Bradley has just entered college as a freshman. As part of the enrollment procedures, he is offered a health plan that covers him while he is enrolled as a student. Wilbur doesn't think he should spend the money on the premiums because he is young and healthy. He asks you for advice. What would you tell him?

Photodisc/Getty Images

Math Skill Builder

Review these math skills and solve the exercises that follow.

1. **Add** dollar amounts.
   Find the sum. $378 + $108 + $2,823 = $3,309

   1a. $4,298 + $218 + $48 **1b.** $376 + $294 + $1,397

2. **Subtract** dollar amounts from dollar amounts.
   Find the difference. $3,298.28 − $1,089.27 = $2,209.01

   2a. $1,703.93 − $727.19 **2b.** $7,319.29 − $3,519.07

3. **Multiply** dollar amounts by whole numbers and percents.
   Find the products. $45 × 12 = $540 and $798 × 35% = $798 × 0.35 = $279.30

   3a. $59.20 × 12 **3b.** $197 × 6

   3c. $488 × 65% **3d.** $1,497 × 72%

Health Insurance Premiums

Health insurance, like other insurance, protects against financial loss. In this case, the financial loss is from medical bills. Employers often provide *group health insurance* as a job benefit for employees and their families. The employee usually pays part of the cost of the group policy. If you are not covered by a group policy, you may buy individual health insurance for yourself and your family, but it is usually more expensive.

Group health policies usually provide *basic health coverage*, including:

- *Hospitalization insurance*, which helps pay expenses of a hospital stay, such as hospital room, medicine, lab tests, X-rays, operating room.
- *Surgical insurance*, which covers the fees of doctors who do surgery or who help with surgery in or out of a hospital.
- *Medical insurance*, which pays the fees of other doctors who see you in or out of the hospital, as well as some other medical expenses, such as physical therapy.

You may supplement your basic health coverage with **major medical insurance**. It helps pay for hospital, surgical, medical, or other health care expenses due to a major illness or an injury. Often basic health insurance and major medical insurance policies are combined into one comprehensive health package.

Your employer may also have group plans for other health areas, such as dental insurance and vision insurance.

Problem Solving *Tip*

To find the employee share of the premium, deduct the employer's percentage share from 100%. The difference is the employee's percentage share. Then multiply the total premium by the employee's percentage share.

EXAMPLE 1

Lela Wendt's employer offers a health insurance plan that covers Lela, her husband, and their child. The total monthly premium is $285, of which the employer pays 26%. How much does Lela pay for the health insurance for one year?

SOLUTION

Multiply the total monthly premium by 12.

$285 × 12 = $3,420 total annual premium

Multiply the total annual premium by 26%.

$3,420 × 0.26 = $889.20 part of annual premium paid by employer

Subtract the employer's share of the premium from the total premium.

$3,420 − $889.20 = $2,530.80 Lela's share of health insurance for year

✔CHECK YOUR UNDERSTANDING

A. Ted Larkin's employer pays 50% of his annual health insurance premium. If the total monthly premium for the insurance is $57, what is Ted's share of the annual premium?

B. An employer provides dental health insurance to employees. The monthly premium cost per employee is $36. If the employees pay 65% of the premium, what is the total annual premium paid by an employee for the dental insurance?

Health Insurance Benefits and Coinsurance

Basic health insurance plans usually include an *annual deductible amount* for each insured person. When the health bills for a person covered by the plan exceed the deductible amount for that person, the insurance company begins to pay benefits.

Major medical insurance plans usually have a deductible amount for *each* treated illness or injury. For example you may be required to pay the first $500 of a health bill for an injury before the insurance company begins to pay benefits.

Indemnity plans allow you to choose your own doctors and hospitals but the coinsurance amounts are usually higher than HMOs. HMOs usually require you to use specified doctors and hospitals.

Once the deductible amount has been met, you usually must pay part of the remaining health bills out of your own pocket. These partial payments are called **coinsurance**, or *co-payments*. For example, you may be required to pay as coinsurance 15% of a surgery bill. Or, you may be required to pay a $25 co-payment for each visit to a doctor's office.

Usually coinsurance is stated as a percent and co-payments as a dollar amount. Whether you pay coinsurance or a co-payment, they are both your share of the bill for services rendered.

Finally, the total health bill may not be covered by your policy. For example, a psychiatrist may charge $85 for each office visit but your insurance policy may set a maximum benefit of $70 for such visits.

To find how much of a health bill you will have to pay, you must first determine how much of the bill is covered by your policy. Next, you must determine if you have any deductible amount left to pay for the year. Then you must know what coinsurance percent you are responsible for.

Uncovered Amount = Total Bill − Covered Amount

Coinsurance Amount = (Covered Amount − Deductible) × Coinsurance Rate

Amount Insured Must Pay = Uncovered Amount + Deductible + Coinsurance

EXAMPLE 2

Jolene Ridgeway underwent surgery for an injury. The hospital portion of the bill was $5,298, of which only $4,875 was covered by Jolene's group medical insurance policy. In addition, the coinsurance amount of the bill was 18%, and the remaining deductible she had for the year was $300. How much of the hospital bill must Jolene pay?

Photodisc/Getty Images

SOLUTION

Subtract the covered portion from the total bill.

$5,298 − $4,875 = $423 uncovered amount

Subtract the deductible from the covered amount.

$4,875 − $300 = $4,575

Then multiply by the coinsurance rate.

$4,575 × 0.18 = $823.50 coinsurance amount

Add the uncovered amount, the deductible amount, and the coinsurance amount.

$423 + $300 + $823.50 = $1,546.50 amount Jolene must pay

Business *Tip*

Group policies are issued on a group of people, such as all the employees in a firm. Group policies usually do not require the employee to take a physical examination.

✔ CHECK YOUR UNDERSTANDING

C. The surgery portion of Ruiz Alicea's total medical bill was $2,964. Only $2,583 was covered by his group medical insurance policy. Ruiz's coinsurance for the surgery was 21%, and his remaining annual deductible was $500. What amount of the surgery bill must Ruiz pay?

D. Elisa Renteria's dental insurance plan pays a maximum of $450 for a crown. It also requires her to pay 10% coinsurance. Her policy does not have a deductible. The bill she receives for a crown from her dentist is $525. How much of that bill will Elisa pay?

Wrap Up ▶▶▶

You might tell Wilbur that accidents can happen to anyone at anytime. Also, while he is less likely than older people to become seriously ill, the possibility still exists. By taking out the policy, he will be insuring against the risk of large health bills with a relatively small amount of insurance premium.

TEAM Meeting

With two other students, investigate the advantages and disadvantages of health maintenance organization health plans, or HMOs, and indemnity health plans. You should use Internet search tools and talk to at least one health insurance agent to obtain your information.

You need to define a health maintenance organization plan. Name the requirements and the general premise under which they operate. Also explain an indemnity health plan, its requirements, and how they function.

List the advantages and disadvantages of each. It is a good idea to question adults who participate in each type of program and get their opinions.

Exercises

1. $298 + $12,216 + $4,228 **2.** $9,039 − $457.89 **3.** $3,158 × 12%

4. Bella Melino elects to be covered by her employer's vision health insurance program. The program covers part of the expense of eye examinations, eye glasses or contact lenses, and office visits. The total monthly premium is $105.36, of which the employer pays 45%. Bella's share of the monthly premium is deducted from her monthly paycheck. What is the amount of the deduction?

Photodisc/Getty Images

The annual premium for Chi Kuo's health insurance plan is made up of $2,844 for hospitalization and medical; $1,649 for surgery; and $428 for major medical. Chi's employer pays 42% of the premium.

5. What is the total annual premium for Chi's health plan?

6. What is Chi's share of the total annual premium?

Rosa Suarez pays for a general health plan, a dental health plan, and a vision health plan through her employer. The monthly premiums are: general health, $209; dental health, $265; vision health, $59. Her employer's share of these plans is: general health, 35%; dental health, 45%; vision health, 75%.

7. What is the total monthly premium for all of Rosa's health plans?
8. What is her employer's share of that total monthly premium?

Stanislov Pulkin works for a county agency as an accountant. His employer provides group health policies for basic health care, major medical health care, and dental health care. The annual premium for Stanislov consists of $3,190 for basic health, $518 for major medical, and $2,875 for dental. The county picks up 57% of his basic, 60% of his major medical, and 32% of his dental health premiums.

9. What is the total annual premium for all of Stanislov's health plans?
10. What is his share of the total annual premium?
11. If the county deducts his share from his weekly paycheck, what is the amount of the deduction?

Zena Tubicek can buy a group major medical insurance plan from her employer at a monthly premium of $56 or buy an individual policy with similar coverage from another insurance company for $938 annually.

12. What is the difference in annual premiums between the employer's group policy and the individual policy?
13. By what percent, to the nearest tenth of a percent, does the individual policy premium exceed the group policy premium?

Davey's major medical policy has a $1,000 deductible feature and he must pay 10% coinsurance. He is injured in an accident and his health care bills amount to $21,700.

14. What amount will be paid by his insurance company?
15. What amount will he pay?

Photodisc/Getty Images

Molly Nairah had three x-rays taken at a total cost of $230. Under her major medical coverage, the insurance company paid 80% of the cost of x-rays after a $25 deductible fee for each x-ray.

16. What was the company's share of the cost of the x-rays?
17. What was Molly's share of the cost?

Eve and Ollie Dunbar's major medical policy pays 90% of covered expenses for each of them in any year. A $500 deductible feature applies to each person's claim. Last year the Dunbars made two medical claims. Eve's claim was for $1,230; Ollie's claim was for $1,870. The insurance company did not cover $170 of Eve's claim and $225 of Ollie's claim. What amount did the Dunbars receive from their insurance company for

18. Eve's claim?
19. Ollie's claim?

Solve.

20. Frank Duval was hospitalized for an illness for 12 days. The cost of his hospital room was $458 a day. The cost of medical services was $2,492. His insurance company covered the full amount of the medical services but allowed only 8 days of hospitalization for his illness. The policy also requires $250 in deductible and 15% coinsurance for the remaining hospital room and medical services costs. What amount did Frank have to pay for his illness?

21. Julie Crane required lengthy hospital and medical care. The fees of her doctors were $128,700 and covered at 90% by her major medical policy. Her hospital expenses were $44,460, and the policy covered 85% of the hospital bills beyond a $500 deductible. After Julie left the hospital, a physical therapist made 15 visits to her home at $95 a visit. Julie's policy paid 70% of the therapy bills. Of the total expenses, what amount did Julie have to pay?

22. Mehta Goldberg was hospitalized for 12 days. Her total bill for medical care was $21,570. Mehta's major medical coverage pays for 85% of medical expenses above a $750 deductible. How much of the bill does Mehta owe after the insurance company pays its share?

23. **STRETCHING YOUR SKILLS** An employee's share of a vision insurance policy premium was $130. This was 65% of the policy premium. What was the policy premium?

24. **FINANCIAL DECISION MAKING** What factors should you consider as you evaluate a health insurance policy?

Mixed Review

25. Add $34.12 + $19.98 + $108.29 + $72.07 + $2,781.

26. Multiply 208.108 × 0.28, to the nearest hundredth.

27. Divide 642 by 46 to the nearest thousandth.

28. Subtract: $1\frac{5}{8} - 1\frac{1}{4}$

29. Subtract: $4\frac{1}{4} - 1\frac{7}{8}$

30. $13.30 is what percent greater than $10.64?

31. A truck which originally cost $16,450 is traded in eight years later for $3,290. What was the average annual depreciation on the truck?

32. After 10 years, Lisa Myers canceled her $35,000 life insurance policy and took the cash value of $57 per $1,000. The annual premiums on the policy were $370. While the policy was in effect, she received a total of $217.90 in dividends. What was the net cost of the policy?

Ben Arnold worked 48 hours last week. He earned $11.72 per hour for the first 37.5 hours. For time worked over 37.5 hours, he earned time-and-a-half. Deductions of $172.29 were made from his paycheck.

33. Find Ben's gross pay for the week.

34. Find his net pay for the week.

7-3

Disability Insurance

GOAL

- Calculate disability insurance benefits

KEY TERM

- disability insurance

Start Up ▸▸▸

Ghayda Meguid is 35 and has 2 children. She has purchased as much term life insurance as she thinks she needs. However, she has no disability insurance. She works as a systems programmer for a software company and doesn't believe she is at risk for injuries. What would you advise her to do?

Maxim Bolotniktov/ Shutterstock.com

Math Skill Builder

Review these math skills and solve the exercises that follow.

1. **Divide** dollar amounts by whole numbers.
Find the quotient. $280,000 ÷ 4 = $70,000

1a. $105,000 ÷ 3 **1b.** $325,000 ÷ 4 **1c.** $1,286,700 ÷ 20

2. **Multiply** dollar amounts by percents.
Find the product. $25,000 × 60% = $25,000 × 0.6 = $15,000

2a. $18,290 × 30% **2b.** $21,800 × 45% **2c.** $54,790 × 65%

3. **Subtract** dollar amounts.
Find the difference. $1,560 − $253 = $1,307

3a. $2,250 − $314 **3b.** $1,230 − $428 **3c.** $1,548 − $217

Disability Insurance Benefits

Disability insurance pays you a portion of the income you lose if you cannot work due to a health condition or an injury.

One form of disability insurance is *short-term disability insurance.* This policy pays you a portion of your income for a short period of time, such as 13–26 weeks. Usually there is a maximum amount that can be paid out per week or month.

Business Tip

The portion of your income paid by insurance is usually stated as a percentage of your income. It is called the *benefits percentage.*

Another form of disability insurance is *long-term disability insurance.* This type of policy may cover you for several years or until you reach retirement age. The longer the term of coverage, the higher the premium.

Disability insurance is usually bought through a group plan offered through your employer. You can also buy an individual plan rather than group disability insurance, but it is usually more expensive.

If your job is covered by social security, you also may be eligible for disability insurance through the federal government. Your annual social security statement shows how much monthly disability benefits you are eligible to receive.

If you are injured on the job, you may be covered by *Worker's Compensation insurance.* This insurance covers lost wages and medical expenses from on-the-job injuries. It is usually required by state governments and for work on federal contracts and paid for by the employer.

The benefits you receive from disability insurance depend on whether you are totally or partially disabled, how long you have worked, your wages, percent of your wages that are paid as benefits, and other factors. Also, the benefits you receive may be reduced by the benefits you receive from worker's compensation and social security disability insurance.

EXAMPLE 1

Phil Kustin injures himself in an accident and cannot work. His disability policy pays 60% of his average annual wages for the last 3 years. His wages were $28,500, $29,070, and $29,940. What is Phil's monthly disability benefit?

SOLUTION

Add the annual wages. Divide the total by 3.

$28,500 + $29,070 + $29,940 = $87,510

$87,510 ÷ 3 = $29,170 average wages for last 3 years

Multiply the average wages by the benefit percentage.

$29,170 × 0.6 = $17,502 annual disability benefit

Divide the annual disability benefit by 12.

$17,502 ÷ 12 = $1,458.50 monthly disability benefit

Business Tip

The Social Security web site address is www.ssa.gov.

✔ CHECK YOUR UNDERSTANDING

A. Yan Kaponovich becomes totally disabled. His group disability policy's benefit percentage is 65% of his average annual salary for the last 3 years. His annual salaries were $45,200; $48,300; and $49,900. What monthly benefit amount will he receive?

B. Kelly O'Hara is injured and is covered by a group disability policy and by social security disability insurance. Her group policy pays 45% of her average wages for the last 4 years, less any other disability benefits she receives from other policies. Social security disability will pay her $225 a month. If her wages for the last 4 years were $34,600; $34,900; $40,800; and $42,000, what is the total amount Kelly will receive a month from her group policy?

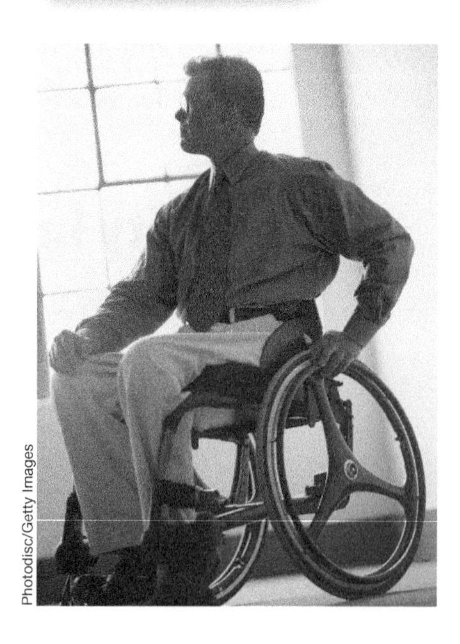

Wrap Up ▶ ▶ ▶

You might advise Ghayda that she can be injured in her home or in her car. About one third of all Americans will suffer a serious disability between the ages of 35 and 65 and 1 in 5 of all Americans will become disabled to some extent during their lives. If she becomes totally or partially disabled, her family will lose her income. She should consider buying disability insurance and she should check to see if she has coverage through worker's compensation and social security.

Communication

Life insurance policies provide income to your survivors when you die. Disability insurance policies provide income to you and your dependents if you become unable to work due to injury or illness. Use the Internet to find answers to these questions:

1. What percent of employees of U.S. companies are offered group life insurance policies and what percent are offered group disability insurance?
2. What is the chance that a person who is 21 years of age will die versus the chance that that person will become disabled?

Prepare an information flyer that might be posted in a lunch or break room at a place of employment. The flyer should include the web page addresses of your sources of information.

Exercises

Perform the indicated operation.

1. $569,400 ÷ 12
2. $45,800 ÷ 5
3. $5,884 − $885
4. $22,509 × 10%
5. $31,770 × 60%
6. $12,929 − $4,228
7. Fayad Mehkta is insured under a disability policy that calculates his benefits percentage at 2.25% for each year he has worked for his company. If Fayad has worked 12 years at the company, what is his benefits percentage?

Louisa Tibaldi's group disability policy has a benefit percentage of 65% of her average annual salary for the last 2 years. Her annual salary for the last 2 years was $65,200 and $68,300.

8. What is Louisa's average annual salary for the last 2 years?
9. What monthly benefit amount will she receive?

Trisha Dabney's disability insurance policy will pay her 45% of her average annual wages for the last 4 years. It will reduce the benefits paid by any amounts Trisha receives from worker's compensation insurance. Trisha's wages for the last 4 years were $24,560; $25,100; $25,820; and $26,200.

10. What was Trisha's average annual wage for the last 4 years?
11. If she receives $467 monthly from worker's compensation, what monthly income will her disability insurance company pay her?

Renaldo Rodriguez's group disability policy pays a benefit percentage of 2.5% for each year that he has worked for his company. The benefit percentage is applied to his average monthly compensation for the last 36 months. Renaldo has worked for 17 years for the company at an average monthly wage of $3,500 for the last 36 months.

12. What is his disability benefit percentage?

13. What monthly benefit amount would he receive?

A college provides short-term disability insurance for its employees. The benefits depend on years of service at the college and the benefit percentage the employee chooses. The chart below shows the number of weeks employees would receive benefits for depending on whether they chose to receive 100%, 80%, or 60% of their weekly income.

The Number of Weeks an Employee Will Receive Benefits			
	Benefit Percentage Chosen		
Years of Service	**100% of Wages**	**80% of Wages**	**60% of Wages**
Less than 5	1	4	5
5 but less than 10	5	5	10
10 but less than 15	5	10	15
15 but less than 26	5	15	20

14. How many weeks of benefits will an employee who has worked for 7 years receive if the employee chooses a benefit percentage of 80% of his or her salary?

15. Kim Lucas has worked at the college for 12 years. How many weeks of coverage would she receive if she chose a benefit percentage of 60%?

16. Jose Fuentes worked for 21 years at the college before being injured. If Jose needs to receive benefits for 20 weeks, what benefits percentage should Jose choose?

Photodisc/Getty Images

STRETCHING YOUR SKILLS A worker receives $27,830 in monthly disability benefits based on the average of her last 3 years of salary. If her salary for the last 3 years was $49,300; $50,300; and $52,200

17. What was her average salary for the last 3 years?

18. What benefits percentage did she receive?

19. **CRITICAL THINKING** What reasons might be used to explain why disability payments are usually less than the last salary or wages of an employee?

Mixed Review

20. Write $1\frac{1}{3}$ as a percent.
21. What percent is 77.25 of 618?
22. Divide 592 by 1.06, to the nearest tenth.
23. $60.40 is what percent less than $78.52, to the nearest tenth percent?
24. Find the average of $1,250, $1,280, $1,297, and $1,320.
25. Divide $\frac{3}{8}$ by $\frac{2}{5}$.
26. Multiply: $5\frac{2}{3}$ by $2\frac{1}{2}$
27. Alberto Viña is paid $13.20 an hour for the first eight hours of work each day and time-and-a-half for overtime past 8 hours. Last week Alberto worked these hours: Mon., 8; Tues., 10; Wed., 6; Thurs., 7; Fri., 10. What was Alberto's gross pay for the week?
28. Ivy Washington had a balance of $259.12 in her checking account at the start of the day. During the day she wrote a check for $35, withdrew $100 using her ATM card, and deposited a rebate check for $25 she received from a recent purchase. What is her new balance after her bank processes the items?
29. Tom Wilson received a $2,300 trade-in for a car that he originally paid $12,500 for six years ago. What was the average annual depreciation on the car?
30. Sean O'Leary bought a $200,000 term insurance policy. The annual premium was $1.07 per $1,000 of insurance. What was Sean's annual premium for the policy?
31. Rosa Carlotto's employer pays 65% of her health insurance premium. If the total monthly premium for the insurance is $127, what is Rosa's share?
32. Roger Tulane invested $10,000 in a 4-year CD that paid 6% annual interest. When he cashed out the CD at the end of 3 years, he was charged an early withdrawal penalty of 6 months' interest. What was the amount of the penalty?
33. June Riebold's taxable income last year was $45,380. She paid state income taxes of 5.5% and city income taxes of 1.3%. What was the total of her state and city income taxes for the year?
34. On July 1, Laura Knolls deposited $540 in a bank account that paid 4% interest per year, compounded semiannually. Interest was added on January 1 and July 1. Find her balance on July 1 of the next year if she made no other deposits or withdrawals.
35. Joelle had a balance of $274.35 in her checking account. She went to the bank with these deposits: $125.99, $98.45, and $303.30. She asked for $75 cash back. What was her total deposit? What is the balance in her account after the deposit?

Photodisc/Getty Images

7-4

Bonds

GOALS
- Calculate the market price of bonds
- Calculate the total investment in bonds

KEY TERMS
- bond
- premium
- discount

Start Up ▶▶▶

Vanessa Olemkov earns $44,500 a year. She is considering investing all of her savings in the stock market because she has been told that the market provides the most return for your dollar over the long run. What advice might you give Vanessa?

Photodisc/Getty Images

Math Skill Builder

Review these math skills and solve the exercises that follow.

1. **Multiply** dollar amounts by whole numbers.
   Find the product. $1,075 × 4 = $4,300

   1a. $504 × 6 **1b.** $1,085 × 9 **1c.** $509 × 8

2. **Multiply** dollar amounts by percents.
   Find the product. $1,000 × 106% = $1,000 × 1.06 = $1,060

   2a. $500 × 97% **2b.** $1,000 × 95%

   2c. $500 × 101% **2d.** $1,000 × 103.2%

   2e. $500 × 6.5% **2f.** $1,000 × 7.8%

3. **Add** dollar amounts.
   Find the sum. $1,050 + $55 = $1,105

   3a. $512 + $28 **3b.** $1,040 + $65 **3c.** $519 + $632

4. **Write** percents as decimals.
   Write 103% as a decimal. 103% = 1.03

   4a. 106% **4b.** 108.9%

   4c. 98% **4d.** 96.4%

Market Price of Bonds

A bond is a form of long-term promissory note. **Bonds** are a written promise to repay the money loaned on the due date. *Bondholders,* or the people who own the bonds, may keep them until the due date or sell them to other investors.

Bonds are usually issued with a face, or par value of $1,000. Bonds may also be issued with other par values, such as $500, $5,000, or $10,000. Par value is the amount of money that the *issuer,* or the organization that sells the bonds, agrees to pay the bondholder on the due date.

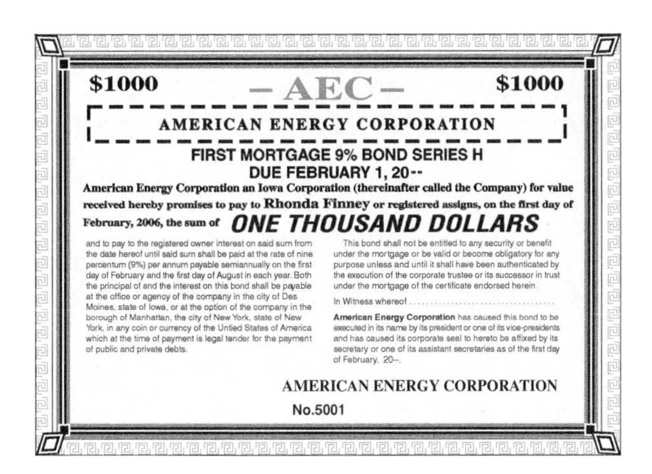

$1000 — AEC — $1000

AMERICAN ENERGY CORPORATION

FIRST MORTGAGE 9% BOND SERIES H
DUE FEBRUARY 1, 20--

American Energy Corporation an Iowa Corporation (thereinafter called the Company) for value received hereby promises to pay to Rhonda Finney or registered assigns, on the first day of February, 2006, the sum of ONE THOUSAND DOLLARS

and to pay to the registered owner interest on said sum from the date hereof until said sum shall be paid at the rate of nine percentum (9%) per annum payable semiannually on the first day of February and the first day of August in each year. Both the principal of and the interest on this bond shall be payable at the office or agency of the company in the city of Des Moines, state of Iowa, or at the option of the company in the borough of Manhattan, the city of New York, state of New York, in any coin or currency of the Untied States of America which at the time of payment is legal tender for the payment of public and private debts.

This bond shall not be entitled to any security or benefit under the mortgage or be valid or become obligatory for any purpose unless and until it shall have been authenticated by the execution of the corporate trustee or its successor in trust under the mortgage of the certificate endorsed herein.

In Witness whereof ....

American Energy Corporation has caused this bond to be executed in its name by its president or one of its vice-presidents and has caused its corporate seal to hereto be affixed by its secretary or one of its assistant secretaries as of the first day of February, 20--.

AMERICAN ENERGY CORPORATION

No.5001

The *market value* of a bond is its selling price and may be different from par value. If the market value is more than par value, the bond is selling at a **premium**. If the market value is less than the par value, the bond is selling at a **discount**. The amount of the premium or discount is the difference between the market value and the par value.

The market price, or market value, of a bond is quoted as a percent of the par value. For example, a price quotation of 97.056 means 97.056% of the par value. This bond is selling at a discount. To find a bond's market price, multiply its par value by the percent. If the price of a $500 par value bond is 102.182, the market price is 102.182% of $500, or $510.91. This bond is selling at a premium.

EXAMPLE 1

Edgewood School Bonds are selling at 97.223. What is the price of one of the school's $1,000 bonds? Are these bonds selling at a discount or a premium?

SOLUTION

Convert the market price to a decimal and multiply by the bond's par value.

$97.223 = 0.97223$; $0.97223 \times \$1{,}000 = \$972.23$ market price

The bond is selling at a discount because the price is less than the par value.

✔ CHECK YOUR UNDERSTANDING

A. AGL Industries $1,000 bonds are quoted at 103.883. What is the market price of the bonds? Are they selling for a discount or premium?

B. Millville Water District bonds are selling at 96.225. What is the market price of one of their $500 bonds? What is the amount of discount on each bond?

Total Investment in Bonds

Bonds are usually bought and sold through a broker, who is a dealer in stocks and bonds. *Full service* brokers provide advice on what and when to buy and sell. They charge a *broker's commission* or brokerage fee but the commission is usually included as part of the price the buyer pays for a bond and not shown separately. *Discount* and *online brokers* offer less financial help but also charge less commission. They usually show their commission rates on their web sites.

To find the total investment in bonds, you must find the market price of one bond, add the commission if it is known, and multiply by the number of bonds bought.

Total Bond Investment = (Market Price + Commission) × Number of Bonds

EXAMPLE 2

Leroy Walker bought 10, $1,000 Regis, Inc. bonds at 104.113 from a broker. The broker charges $4 per bond commission. What was Leroy's total investment?

SOLUTION

Convert the market price to a decimal and multiply by the par value.

104.113 = 1.04113; $1,000 × 1.04113 = $1,041.13   market price of 1 bond

Add the commission charge to the market price.

$1,041.13 + $4 = $1,045.13   total price of bond plus commission

Multiply the total price of 1 bond by the number of bonds bought.

$1,045.13 × 10 = $10,451.30   total investment

✔ CHECK YOUR UNDERSTANDING

C. Clara Maliszewski bought 10, $1,000 Xnet Corporation bonds at 97.297. No commission was shown. What is the total investment Clara has in the bonds?

D. Taylor Wilson bought 15, $1,000 Maryville Sewer District bonds at 103.228. The broker charged $3 per bond commission. What total investment did Taylor make in the bonds?

Wrap Up ▶▶▶

While it is true that the stock market has been a good investment over the long run, Vanessa may need money in the short run for emergencies or job loss. Placing all her savings in the stock market may mean that she must make withdrawals for emergencies when the market is low. Many financial experts advise keeping enough money in savings accounts, money market accounts, and short-term CDs to cover at least 3–6 months of your income. They may also recommend putting some savings into income producing investments like bonds.

Exercises

Find the product or sum.

1. $629 × 10
2. $2,198 × 9
3. $1,000 × 102.7%
4. $500 × 93.6%
5. $1,108 + $77
6. $521 + $48

State the market price, in dollars and cents, of each $1,000 bond below. Also state whether the bond is selling at a discount or premium.

7. 92.877
8. 89.231
9. 103.088
10. 102.662
11. 109.836
12. 88.114

Find the amount of money invested in each bond purchase below.
Madison County, $1,000 bonds:

13. 5 @ 91.445
14. 12 @ 88.331
15. 20 @ 106.292

Eggleston Power Company, $1,000 bonds:

16. 6 @ 99.323
17. 15 @ 94.494
18. 18 @ 102.313

Teasdale Transportation Authority, $500 bonds:

19. 24 @ 114.673
20. 8 @ 92.555
21. 12 @ 79.447

Solve.

22. Rob Adams invested in 25 bonds with a par value of $1,000 each. The quoted price for each bond was 110.345. His broker charged $3.50 per bond commission. How much did Rob invest?
23. Simone Tremont bought 8, $1,000 bonds at 88.563. No commission was shown. What was her total investment in the bonds?
24. Olaf Hansen bought 4, $1,000 bonds at 88.559. Two days ago the price of the bonds was 87.443. What was the amount of Olaf's investment in the bonds?

Leslie Ikwelugo buys a $1,000 bond at 105.874. The broker charges $3 commission per bond with a minimum of $30 commission per order.

25. What commission was Leslie charged?
26. What was the total amount she invested in the bond?
27. What is the amount of premium she paid on the bond?

Henry Schmidt bought 6, $500 bonds at 98.580.

28. What is his total investment in the bonds?
29. What discount did he receive on each bond?
30. **CRITICAL THINKING** If a bond promises a bondholder a guaranteed price at the maturity date, why may the market value be more or less than the par value?
31. **FINANCIAL DECISION MAKING** Corporations often use their land, buildings, or equipment as collateral for the money they borrow. If the loan is not repaid, the bondholders may take over the corporation's property. Since the collateral of a company backs these bonds, does that mean there is no risk involved in purchasing a bond?

Mixed Review

32. Find the sums and grand total.

$$\begin{array}{l} 24.0 + 16.0 + 12.5 + 9.2 \\ \underline{12.6} + \underline{22.7} + \underline{18.6} + \underline{6.3} \end{array}$$

33. Divide 1,560 by 27 to the nearest hundredth.

34. Divide 23.67 by 100.

35. Add: $\frac{3}{8} + \frac{5}{6}$.

36. What fractional part of 56 is 8?

37. Multiply $2\frac{3}{4}$ by $2\frac{2}{7}$.

38. Janice Ludlow bought the following number of bonds in a week: Mon., 4; Tues., 6; Wed., 6; Thurs., 5. How many bonds did she buy on Friday if the average number of bonds she bought each day was 5?

39. Tomas Reynoso spent 35.6% of his total income last year on housing. If his income last year was $67,800, how much did he spend on housing?

40. Vic Davis paid an annual premium of $11.80 per $1,000 for a $30,000 life insurance policy. His policy also paid a dividend of $29.50, which he used to reduce his premium. What net premium did Vic pay?

41. Risa Levine was injured and hospitalized for 8 days. Her total medical bill was $16,360. Risa's major medical coverage pays for 80% of medical expenses above a $500 deductible. How much does Risa have to pay?

42. Juan Imalgo was billed $125 for an office visit and eye examination and $339 for new eyeglasses. His group insurance policy had a deductible of $75 and a coinsurance rate of 25% on the visit and exam, but only allowed a maximum amount of $275 for glasses. What amount of the bill did Juan pay?

Photodisc/Getty Images

43. Suba Jaidka is a waitress and earned these tips last week: Tuesday, $55; Wednesday, $75; Thursday, $80; Friday, $120; Saturday, $135. What is the average amount of tips she earned per day?

44. Velma Reese's March 31 bank statement balance was $281.73. Her outstanding checks were for these amounts: $25.17, $19.75, $27.89. Velma deposited $125 late on March 31 in the bank's night deposit box. It was not opened by bank employees until the next day and did not appear on the bank statement. Reconcile the check register.

Reiko Wakui wants to buy a home for $175,000. She expects to make a 20% down payment and estimates her closing costs as: legal fees, $1,350; title insurance, $331; property survey, $275; inspection, $175; loan processing fee, $96; recording fee, $540.

45. What amount of mortgage loan will she need?

46. What amount of cash will she need when she buys the house?

7-5

Bond Interest

GOALS

- Calculate bond income
- Calculate bond yield
- Calculate total cost of bonds

KEY TERM

- current yield

Start Up

You can buy a $1,000, 9.5% bond for 97.000 or a $1,000, 10.5% bond for 108.000. Which offers the highest true rate of interest?

Photodisc/Getty Images

Math Skill Builder

Review these math skills and solve the exercises.

1 **Rewrite** percents as decimals.
Rewrite 95.141% as a decimal. 0.95141

1a. 102.597% **1b.** 97.289%

2 **Add** dollar amounts.
Find the sum. $\$2,890.80 + \$395 = \$3,285.80$

2a. $\$5,190 + \$56.16$ **2b.** $\$11,390.56 + \$298.88$

3 **Multiply** dollar amounts by percents, whole numbers, and fractions.
Find the product. $\$500 \times 99.567\% = \$500 \times 0.99567 = \$497.835$, or \$497.84
Find the product. $\$1,000 \times 8\% \times \frac{1}{2} = \$1,000 \times 0.08 \times \frac{1}{2} = \$40$
Find the product. $\$26.59 \times 2 = \$53.18$

3a. $\$1,000 \times 105.295\%$ **3b.** $\$49.50 \times 2$

3c. $\$500 \times 97.114\%$ **3d.** $\$1,000 \times 9.5\% \times \frac{1}{2}$

4 **Divide** dollar amounts by dollar amounts to find percents.
Find the percent, to the nearest tenth. $\$83 \div \$946.80 = 0.0876$, or 8.8%

4a. $\$94.60 \div \$1,200$ **4b.** $\$37.94 \div \$10,000$

Bond Income

Investors in bonds receive interest payments as income. Bond interest is often paid semiannually. The interest rate of a bond is based on the bond's par value. Since the par value is the principal, the interest formula is:

Interest = Par Value × Rate × Time

To find your bond income, find the interest you receive for one bond. Then multiply that result by the number of bonds you own.

Algebra Tip

To find the interest on a bond, you can use the simple interest formula:.

$I = P \times R \times T$

where *P* is the principal or the par value of the bond, *r* is the rate of interest, and *t* is the time in years.

EXAMPLE 1

Find the interest for one year on 5, $1,000 par value, 9% bonds. Find the semiannual interest paid on 1 bond.

SOLUTION

Rewrite the interest rate as a decimal and multiply by the par value.

$I = \$1,000 \times 0.09 \times 1 = \$90$ interest for 1 year on 1 bond

Multipy the interest for 1 bond by the number of bonds owned.

$5 \times \$90 = \$450$ interest for 1 year on 5 bonds

If the interest is paid semiannually, the amount of each interest payment for this bond would be $45.

$I = \$1,000 \times 0.09 \times \frac{1}{2} = \$45$ semiannual interest on 1 bond

✔CHECK YOUR UNDERSTANDING

A. Alif Guilak owns 10, $1,000, 9% bonds. What is his semiannual interest on the bonds?

B. Beatrice Grezlak bought 20, $500, 8.5% bonds. What is her annual income from the bonds?

Bond Yields

One way to compare bond investments is to find the current yield of bonds. The **current yield** of a bond is found by dividing the bond's annual interest income by the bond's price.

Current Yield = Annual Income ÷ Bond Price

EXAMPLE 2

What is the current yield on a $1,000, 7% Elgin Transit Company bond priced at 96.462? Round your answer to the nearest tenth of a percent.

SOLUTION

Multiply the face value of the bond price times the bond's interest rate.

$I = \$1,000 \times 0.07 = \$70$ annual income

Multiply the bond's face value times the bond price.

$0.96462 \times \$1{,}000 = \$964.62$ bond price

Divide the annual income by the bond price.

$\$70 \div \$964.62 = 00.0725 = 7.3\%$ current yield

✔ **CHECK YOUR UNDERSTANDING**

C. Crescent Company $1,000, 9% bonds are offered at 101.585. What is the current yield, to the nearest tenth percent?

D. The semiannual interest on $1,000 Lancaster Housing bonds is $47.50. If you buy the bonds at 94.598, what is the current yield?

Total Cost of Bonds

When a bond is sold, whoever owns the bond on the next interest date receives the full amount of interest for the entire past interest period. When you buy a bond, you may have to pay the market price of the bond plus any interest that the bond has earned from the last interest date.

For example, if on April 1 you buy a bond that pays interest semiannually, you have to pay the seller for the accrued interest that he or she has already earned on the bond from January 1 through March 31. On July 1, the bond will pay you interest for the full 6 months. The interest for that first three months is called *accrued* interest.

Business *Tip*

Accrued interest is interest that has been earned but not yet paid.

EXAMPLE 3

Ed Martin buys 5, $1,000, 8% bonds through a dealer at 102.797 plus accrued interest of $20 per bond. The dealer charged $3 commission per bond. What is the total cost of the bonds to Ed?

SOLUTION

Change the market price to a decimal and multiply by the par value.

$102.797\% = 1.02797$; $1.02797 \times \$1{,}000 = \$1{,}027.97$ price of one bond

Add the accrued interest, commission, and the price of one bond.

$\$1{,}027.97 + \$20 + \$3 = \$1{,}050.97$ cost of each bond

Multiply the total cost of each bond by the number of bonds purchased.

$\$1{,}050.97 \times 5 = \$5{,}254.85$ total cost of bond purchase

✔ **CHECK YOUR UNDERSTANDING**

E. Reba Neel buys 10, $1,000, 9.7% bonds through a dealer at 97.272 plus accrued interest of $24.25 per bond and commission of $4 per bond. What is the total cost of the bonds to Reba?

F. Julio Pujols buys 20, $1,000, 8% bonds on April 1 at 105.288 plus accrued interest from January 1. Commission was not shown. What is Julio's total cost for each bond?

Wrap Up

The interest rate of both bonds is based on the par value of the bonds. You will receive $\$1,000 \times 9.5\%$, or \$95 as annual interest on the first bond. You will receive $\$1,000 \times 10.5\%$, or \$105 on the second bond. However, the market price of the first bond is only \$970. If you buy that bond, the true rate of interest is \$95 divided by the market price of the bond, or 9.8%, rounded to the nearest tenth percent. The true rate of interest on the second bond is \$105 divided by \$1,080, or 9.7%, rounded to the nearest tenth percent.

Communication

Municipal bonds are bonds that are issued by state and local governments and government agencies. Write complete sentences to answer each of the following questions.

1. Why is the interest rate paid on municipal bonds often lower than the interest rates paid on bonds of comparable quality issued by corporations?
2. What is the advantage of buying municipal bonds?

Write a brief memo to another person that includes the answers to these questions.

Exercises

Find the annual income in each problem.

1. 14, \$1,000, 6% bonds
2. 6, \$1,000, 10% bonds
3. 8, \$1,000, 12% bonds
4. 2, \$500, 7.25% bonds
5. 20, \$500, 9% bonds
6. 15, \$1,000, $8\frac{1}{2}\%$ bonds
7. 5, \$1,000, 9.25% bonds
8. 12, \$1,000, 11.5% bonds
9. What is the semiannual income from six \$1,000, 5.5% bonds?
10. You own 30 bonds with a par value of \$1,000 each and paying 9.75% interest. Find your semiannual income from these bonds.

Find the total investment and the total annual income from the investment.

	Bonds Owned	Par Value per Bond	Price Paid	Total Investment	Interest Rate	Annual Income
11.	5	\$1,000	99.246		9%	
12.	10	\$1,000	104.932		12%	
13.	30	\$1,000	107.253		$12\frac{1}{2}\%$	
14.	45	\$1,000	94.342		8.7%	

Find the current yield on each bond to the nearest tenth of a percent.

	Par Value	Interest Rate	Price Paid	Current Yield
15.	$1,000	10%	91.899	
16.	$1,000	8.5%	104.363	
17.	$1,000	$9\frac{1}{2}$%	112.008	
18.	$500	7.8%	92.826	

19. How much is each interest payment on a $500, 8% bond if the interest is paid semiannually on June 1 and December 1?

20. What estimated and actual annual interest would you get from 10, $1,000 par value bonds that pay 6.8% interest?

21. A $1,000 bond, paying 8% interest, was bought at 78.569. What is the current yield to the nearest tenth percent?

22. What is the current yield, to the nearest tenth percent, on a $1,000, 11.2% bond bought at 108.289?

23. What is the current yield, to the nearest tenth percent, on a $500, 9% bond bought at 95.976?

24. **STRETCHING YOUR SKILLS** E.Print, Inc. $1,000, $9\frac{1}{4}$% bonds can be purchased at 103.976. How much money must be invested in the bonds to produce an annual income of $1,850?

25. **STRETCHING YOUR SKILLS** What amount must be invested in Poe County $1,000, 9% bonds at 80.360 in order to earn an annual income of $2,250?

26. Denny Lensing bought 10 Brittle Company $1,000 par value, 13% bonds at 150.883. Interest on these bonds is paid semiannually on January 1 and July 1. What semiannual interest payment will Denny receive from this investment?

27. On April 1, Alan Durston buys 20, $1,000, 8% bonds at 95.089, plus accrued interest from January 1. No commission is shown. What total amount does Alan spend for the bonds?

28. **FINANCIAL DECISION MAKING** You can invest in a 15-year bond with a current yield of 8.5% or a 6-month CD with an APR of 6.5%. Why might you invest in the CD? Why might you invest in the bond?

Mixed Review

29. Subtract $\frac{2}{5}$ from $\frac{4}{9}$.

30. Divide $5\frac{1}{5}$ by $2\frac{1}{2}$.

31. What number is 10% greater than 55?

32. June Silva borrows $450 and agrees to repay the loan in 36 payments of $15.50 each. How much is the finance charge on the loan?

7-6

Stocks

GOALS

- Calculate the cost of stock purchases
- Calculate annual stock dividends
- Calculate the yield on stock investments
- Calculate the proceeds from the sale of stock

KEY TERM

- market price

Start Up ▸▸▸

You and a friend are reading a newspaper article describing an Internet music provider. The article states that the company's stock started last year at $21 and rose to $53 at the end of the year. Your friend said, "Boy, if I had bought 100 shares of that stock, I would have made $3,200." Is your friend right?

Photodisc/Getty Images

Math Skill Builder

Review these math skills and solve the exercises that follow.

1 **Multiply** dollar amounts by percents and whole numbers.
Find the product. $14.50 × 200 = $2,900
Find the product. $100 × 2.4% = $100 × 0.024 = $2.40

1a. $72.58 × 100 **1b.** $117 × 500

1c. $1,000 × 4.5% **1d.** $100 × 2.14%

2 **Divide** dollar amounts by dollar amounts to find percents, to the nearest tenth.
Find the quotient. $14 ÷ $350 = 0.04, or 4%

2a. $8.25 ÷ 200 **2b.** $15.23 ÷ $428

Purchasing Stock

Companies issue shares of stock to raise money, which might be used to expand or to offer new products. Investors who buy the shares are called *stockholders.* Each stockholder gets a stock certificate that shows on its face the number of shares it represents. A stock certificate is shown at the top of the next page.

Business *Tip*

Stockholders are also called shareholders.

COMMON STOCK

1079 NUMBER

CAPITAL STOCK
$10,000.00
100 SHARES PAR VALUE $100 EACH

100 SHARES

Vertig Printing, Inc.

This Certifies that Ivan Probieski *is the owner of* One hundred - - - - - - - - - - - - - - - - *Shares of the Capital Stock of* **Vertig Printing, Inc.,** *transferable only on the Books of the Corporation by the said owner, in person or by duly authorized Attorney, upon the surrender of this Certificate properly endorsed.*

In Witness Whereof *We have hereunto set our hands and affixed the seal of the Company at* 10:30 a.m *this* 5th *day of* October *20* - - -

Carl Mohler — SECRETARY

Vanna Charles — PRESIDENT

SHARES $100 EACH

Stockholders usually buy and sell their shares through a broker. The price at which a stock sells is called the **market price** or *market value* and is shown in stock tables on the Internet and in some daily newspapers.

52 Weeks High	52 Weeks Low	Stocks	Div.	Sales in 100s	High	Low	Last	Net Change
45.28	29.33	Amicio Corp.	2	387	34.40	32.82	33.50	−.50
15.25	6.98	Bellin-Carr, Inc.		483	12.34	10.50	11.23	−.75
105.10	88.98	Crest Industries	4.75	825	103.23	99.46	101.54	+.27
22.34	8.25	Technet Co.	1.25	346	10.55	10.35	10.35	−.16
33.37	12.45	Unser, Inc.	.45	872	33.37	31.68	33.37	+1.05
87.28	68.41	Gen. Prod. Co.	2.50	226	82.45	82.13	80.42	+.54

Highest and lowest prices for past 52 weeks

Net difference between last price for this day and last price of yesterday

Corporation name usually abbreviated

Current dividend rate in dollars per year

Shares sold (in hundreds) for the day

Highest and lowest prices at which a sale was made for the day

Last or closing price at which a sale was made

When you buy stock through a broker, the total cost of the stock is the market price of the stock plus the broker's commission.

Total Cost = Market Price + Commission

The amount of a broker's commission depends on the services the broker provides, the price of the stock, and the number of shares bought. As in bonds, discount and online brokers usually charge lower commissions but give less service to customers than full service brokers.

Business *Tip*

Share prices can be found at many web sites. For starters, visit the sites for the *Wall Street Journal* (www.wsj.com) or the New York Stock Exchange (www.nyse.com).

EXAMPLE 1

Velma D'Anglico bought 500 shares of Vesta stock at $15. The broker charged her $106 commission. Find the total cost of the stock.

SOLUTION

Multiply number of shares times price.

$500 \times \$15 = \$7,500$ total price of shares

Add the broker's commission to the price of the shares.

Total Cost $= \$7,500 + \$106 = \$7,606$ total cost of shares

✔ CHECK YOUR UNDERSTANDING

A. Trent Vallow purchased 200 shares of stock from his broker at $24.50. The broker charged $123.51 commission. What is the total cost of the stock?

B. A broker sold Lisa Colon 400 shares of Roly Plastics, Inc. stock at $31.89. Lisa's broker charged $246.89 commission. Find the total cost of the stock to Lisa.

Stock Dividends

Unlike bondholders' investments, the money invested in stock does not have to be repaid. Stockholders are owners of the company, not lenders. However, stockholders have a right to share in company profits. These profits are distributed to shareholders as *dividends* and are usually paid quarterly.

Many corporations issue two classes of stock—common stock and preferred stock. A corporation sets a *preferred stock*'s dividends when it is first issued. *Common stock* is the ordinary stock of a corporation and does not have a set dividend. There is no guarantee that dividends will be paid to either class of stockholder. When dividends are paid, they go first to shareholders of preferred stock. Dividends may be shown either as a percent of a stock's par value or as an amount of money per share. For stock with no par value, the dividend is always an amount per share.

Photodisc/Getty Images

EXAMPLE 2

Cecile Ware owns 100 shares of Teleos Communications common stock, par value $100. If a dividend of 2.5% is declared, how much should Cecile get in dividends?

SOLUTION

Change the dividend percent to a decimal and multiply by the par value.

$2.5\% \times \$100 = 0.025 \times \$100 = \$2.50$ dividend on one share

Multiply the dividend on one share by the number of shares.

$100 \times \$2.50 = \$250$ total dividend

✔CHECK YOUR UNDERSTANDING

C. Emile Van Tassel owns 600 shares of Brunis Imaging Co. common stock with a par value of $100. Brunis Imaging declares a dividend of 3.7% on the par value to shareholders. How much in dividends will Emile receive?

D. Dome Petroleum preferred stock pays a quarterly dividend of $1.50 per share. If Sara Karadic owns 800 shares of the stock, what amount will she receive in dividends for the year?

Stock Yields

The yield, or rate of income, received from an investment is found by dividing the annual income from the investment by the amount invested. For stocks, the investment is the total cost of the stock, including any expenses or commission paid in obtaining the stock. The income is the amount of annual dividends.

Yield = Annual Dividends ÷ Total Cost of Stock

EXAMPLE 3

Sandra bought 10 shares of Calcon, Inc. stock at $25. Her broker charged her $28 commission. If the stock pays an annual dividend of $1.20, what is its yield?

SOLUTION

Multiply the number of shares by share price and add commission.

$(10 \times \$25) + \$28 = \$278$ total cost of the stock

Multiply the annual dividend by the number of shares.

$10 \times \$1.20 = \$12$ total annual dividend

Divide the total annual dividend by the total cost of the stock. State the yield as a percent to the nearest tenth percent.

Yield $= \$12 \div \$278 = 0.0431$, or 0.043, or 4.3% annual yield to nearest tenth percent

✔CHECK YOUR UNDERSTANDING

E. Vladan Kortic bought 600 shares of stock at $11.35. His broker charged $182.60 commission. The stock pays a quarterly dividend of $0.12. What is the annual yield, to the nearest tenth percent?

F. Penina Kabamba purchased 1,000 shares of Peltor, Inc. stock at $20.80. Her broker charged her $408.23 commission. If the stock pays an annual dividend of $0.62, what is its annual yield, to the nearest tenth percent?

Stock Sales

When you sell stock through a broker, you pay a commission. You may also pay charges such as a service fee and a *Securities and Exchange Commission (SEC) fee.* Your state may charge a transfer tax. When you buy stock, you do not pay a transfer tax or an SEC fee. When you sell stock, the net proceeds is the market price less the commission and all other charges (service fee, SEC fee, transfer tax).

Net Proceeds = Market Price − (Commission + Other Charges)

EXAMPLE 4

Find the net proceeds from the sale of 100 shares of Danbury Corporation stock at \$30.25 with commission and other charges of \$86.

SOLUTION

Multiply number of shares by market price. $100 \times \$30.25 = \$3{,}025$ total sale

Subtract the commission and other charges from the total sale.

Net Proceeds $= \$3{,}025 - \$86 = \$2{,}939$

The profit or loss on a sale of stock is the difference between the total cost of purchasing the stock and the net proceeds. If the amount of the net proceeds is greater than the total cost, there is a profit. If it is less than the total cost, the result is a loss.

Profit = Net Proceeds − Total Cost or Loss = Total Cost − Net Proceeds

EXAMPLE 5

Find the profit or loss from the sale of the Danbury Corporation stock in Example 4. You bought the 100 shares of stock originally at \$21.50 a share and paid a commission of \$68.55.

SOLUTION

Multiply the price per share by the number of shares and add the commission.

$100 \times \$21.50 = \$2{,}150$ total price of stock

$\$2{,}150 + \$68.55 = \$2{,}218.55$ total cost of stock

The net proceeds, \$2,939, is greater than the total cost, \$2,218.55, so find the profit.

Profit $= \$2{,}939 - \$2{,}218.55 = \$720.45$ profit from sale of stock

✔ CHECK YOUR UNDERSTANDING

G. You bought 100 shares of Pendel preferred stock at \$14.70. The commission charge was \$54. You sold the same shares later at \$21.45. The commission and other fees on the sale were \$71. What was the profit or loss on the stock?

H. Jorge Venteria bought 300 shares of Silver Forge common stock at \$41.80 and was charged a commission of \$243. He later sold the stock for \$39.12. The commission and fees on the sale were \$235. What was Jorge's profit or loss?

Wrap Up ▶▶▶

Your friend overlooks the commission on the purchase of the shares and commission and fees on the sale of the shares. The profit would be reduced by these amounts. Also, you don't make a profit or loss until you actually sell the shares. If your friend checked the current price of the stock, he or she might find that the price had fallen to \$15 a share. Unless the stock was sold at year-end, your friend would have made no profit. Unless the stock is sold now, your friend would suffer no loss on the stock.

Financial Responsibility

Managed Funds

When you have money to invest and would like an expert to make decisions to help you make your money grow, a managed fund is a good choice. A *managed fund* is an investment fund that is managed for you by a professional investments manager. This manager constantly researches the market to know when to buy and when to sell, who to buy and who to sell.

The greatest benefit of a managed fund is that your money is pooled with other investors into a single account that creates a large sum of money to invest into a diversified group of funds. If you had $1,000 to invest, you would likely only be able to purchase the stock of one company. If that company performs poorly, you will lose your money. If you invest that $1,000 into a managed fund, you will likely invest in as many as 20–50 companies, both domestic and international companies. And, more importantly, a professional will monitor your investments and make changes to minimize your losses and maximize your gains.

Exercises

Perform the indicated operation.

1. $5,478 + $56.88
2. $45,298 − $1,497
3. $2,500 × 1.5%
4. Find the percent, to the nearest tenth: $5.60 ÷ $500

Find the total cost of each stock purchase.

	Number of Shares	Name of Stock	Market Price	Commission	Total Cost
5.	300	Reinhold	$41.55	$241.55	
6.	100	Seibold	$22.44	$69.80	
7.	100	Danville	$17.52	$39.20	
8.	50	Net Managers	$61.28	$76.30	
9.	200	BPM	$110.58	$143.90	
10.	800	Monterey	$14.75	$268.85	
11.	400	Nextsand	$5.88	$42.20	
12.	150	Streiser	$32.94	$71.20	
13.	35	Tolker	$44.26	$40.80	
14.	200	Newdays	$58.44	$116.40	

15. **STRETCHING YOUR SKILLS** A discount broker offers 40% off the $117 commission charged by a full service broker to handle the purchase of 200 shares of Epsonique at $22.50. How much could a buyer save by purchasing from the discount broker?

Find the total annual dividend received by each shareholder.

	Shares Owned	Par Value per Share	Annual Dividend Rate	Dividend
16.	100	$100	5%, annually	
17.	800	$100	8%, annually	
18.	120	$50	4.5%, annually	
19.	500	—	$0.75 per share, quarterly	
20.	100	—	$1.50 per share, quarterly	

Solve.

21. Sun Fabrications declares a dividend of $3.50 per share on its common stock. If Jules Kortel owns 400 shares of the stock, what amount will he receive in dividends?

22. Karen Hedrick owns 300 shares of Tilden Electronics preferred stock. If the stock pays a quarterly dividend of $1.20 a share, what is the total amount that Karen will receive in a year?

Find the yield on the investment in each of the following stocks. Round all answers to the nearest tenth of a percent.

	Total Cost per Share	Par Value per Share	Dividend Rate	Dividend Payable	Yield
23.	$120	$100	6%	Annually	
24.	$80	$100	3%	Annually	
25.	$50	—	0.62 per share	Quarterly	
26.	$26	—	0.40 per share	Quarterly	
27.	$32	—	0.55 per share	Quarterly	

Tien Niu bought 500 shares of stock at $47.75 a share. She paid a discount broker a commission of $26 for the purchase. The stock pays an annual dividend of 5% on a par value of $100.

28. What total dividend does Tien get from the 500 shares?

29. What yield, to the nearest tenth percent, does Tien earn on this investment?

Photodisc/Getty Images

Sobal Chemical Company preferred stock sells for $38 and pays an annual dividend of 2.7% on a par value of $100. Elgin Equipment Corporation preferred stock has a market price of $18.25 and pays a quarterly dividend of $0.19.

30. Which stock earns a higher yield?

31. How much higher, to the nearest tenth percent?

32. **STRETCHING YOUR SKILLS** What is the rate of income on a share of preferred stock that costs $180 and pays a semiannual dividend of $7.20 a share?

INTEGRATING YOUR KNOWLEDGE Aegis Security Company common stock pays a regular annual dividend of $2.50 a share.

33. How many shares must you buy to get an annual income of $1,000 from the investment?

34. What total investment will you make in the stock if you buy the amount of shares in Exercise 35 at $52.50 and pay $78 per 100 shares for commission?

Find the profit (+) or loss (−) in each of these sales.

	Name of Stock	Shares Traded	Selling Price	Commission & Other Fees	Total Cost of Purchase	Profit/Loss
35.	RMB Plastics	200	42.44	$172.88	$7,628.12	
36.	eBuy, Inc.	300	11.50	115.73	$1,809.25	
37.	Solaris Corp.	400	17.10	165.97	$7,118.90	
38.	DigiNet	50	28.25	58.60	$1,178.34	
39.	Roces Mfg.	100	37.40	29.95	$3,165.22	

40. **CRITICAL THINKING** Explain the difference between a bond holder and a stockholder of a corporation.

STRETCHING YOUR SKILLS You bought 400 shares of stock for a total cost of $8,120 and kept the stock for 3 years during which time you received quarterly dividends of $0.26 per share. You sold the stock and received net proceeds of $9,215.

41. What total dividends did you receive while you owned the stock?

42. What is your profit on the sale of the stock?

43. What is your total gain from owning and selling the stock?

Mixed Review

44. Divide 3,600 by 21.4, rounded to the nearest tenth.

45. $482 is what percent of $24,800, to the nearest tenth percent?

46. 42 increased by 11.25% of itself is what number?

47. Louise Alvarez bought 200 shares of 8% preferred stock with a par value of $100. What is her annual dividend income?

48. The annual premium for Lee Nomi's health insurance plan is $3,870 for hospitalization and basic medical; $1,845 for surgery and major medical; and $1,048 for pharmacy benefits. Lee pays 39% of the premium. What is the total amount that Lee pays each month for health insurance coverage?

7-7

Mutual Funds

GOALS

- Calculate the total investment in a mutual fund
- Calculate the amount and rate of commission
- Calculate profit or loss from mutual fund investments

KEY TERMS

- mutual fund
- net asset value

Start Up

Assume you collect baseball cards. Identify strategies you might use to have a good chance of ending up with the trading card of a future superstar.

Jason Figert/Shutterstock.com

Math Skill Builder

Review these math skills and solve the exercises that follow.

1. **Subtract** dollar amounts.
   Find the difference. $1,542 − $1,312 = $230

   1a. $2,078 − $1,993 **1b.** $12,307.87 − $4,110.57

2. **Multiply** dollar amounts by decimals.
   Find the product. $35.17 × 45.135 = $1,587.397, or $1,587.40

   2a. $14.29 × 325.112 **2b.** $8.31 × 28.922

3. **Divide** dollar amounts to find a percent.
   Find the quotient as a percent, to the nearest tenth.
   $32.28 ÷ $460 = 0.07, or 7.0%

   3a. $34.80 ÷ $625 **3b.** $19.77 ÷ $822

Total Investment in a Mutual Fund

Mutual fund investment companies use the money from investors to buy stock in many companies. By investing in many companies, the mutual fund increases its chances of buying stocks that will be profitable.

There are many different kinds of mutual fund companies. Some have aggressive growth goals. Others choose to maximize the income from shares held. Some mutual funds specialize in certain sectors of the economy, such as the health sector. Others buy stock from many different types of organizations and from many different sectors of the economy.

Mutual fund shares are traded based on their net asset values. The net assets are the total value of the fund's investments less any debts it has. The **net asset value**, or NAV, is found by dividing the net assets by the number of shares held by stockholders.

For example, a fund with net assets of $10,000,000 and 500,000 shares issued will have a net asset value of $20 ($10,000,000 ÷ 500,000 = $20).

Mutual fund NAVs are published on the financial pages on the Internet and in some daily newspapers.

Fund Name	NAV	Offer Price
Ameris Growth	10.12	10.96
Banner Income	29.70	30.88
Jantzen High Yield	11.85	12.38
Mercer International	16.58	N.L.
Overland Technology	7.27	N.L.
Parkson Equity	9.27	N.L.
Strand Balanced	10.15	10.47
Tubman Health	32.23	N.L.

Two types of mutual funds are shown in the table: *no-load funds* and *load funds.* The term *load* means commission. No-load funds are sold without a commission and have the abbreviation "N.L." in the Offer Price column. When you buy load funds, you pay the amount shown in the Offer Price column, which includes a commission charge.

To find the total investment made in no-load funds, multiply the number of shares by the NAV. For load funds, multiply the number of shares by the offer price.

Total Investment in No-Load Funds = No. Shares × NAV

Total Investment in Load Funds = No. Shares × Offer Price

EXAMPLE 1

Alberto Allende bought 300 shares of Parkson Equity Fund (see table above) and 500 shares of Ameris Growth Fund. What is Alberto's total investment in these mutual funds?

SOLUTION

Multiply the number of shares of Parkson Equity by its NAV.

300 × $9.27 = $2,781   total investment in no-load fund

Multiply the number of shares of Ameris Growth by its offer price.

500 × $10.96 = $5,480   total investment in load fund

Add the investments in the two mutual funds.

$2,781 + $5,480 = $8,261   total investment in mutual funds

✔CHECK YOUR UNDERSTANDING

A. Shizu Wakichi purchased 400 shares of no-load Mellon Technology Fund at its NAV of $17.35. What was her total investment in the fund?

B. Nicolas Davila bought 1,000 shares of a load fund, Allied Industries Fund, at its offering price of $26.74. What was his total investment in the fund?

Amount and Rate of Commissions

When you buy no-load funds, you are not charged a commission. For load funds, the commission is the difference between the net asset value and the offer price. To find the rate of commission on a load fund's purchase, divide the commission by the offer price.

Commission = Offer Price − Net Asset Value

Rate of Commission = Commission ÷ Offer Price

EXAMPLE 2

What is the rate of commission, to the nearest tenth percent, on Banner Income Fund with a net asset value of $29.70 and an offer price of $30.88?

SOLUTION

Subtract the net asset value from the offer price.

$30.88 − $29.70 = $1.18   amount of commission

Divide the commission amount by the offer price.

$1.18 ÷ $30.88 = 0.0382, or 3.8%   rate of commission

✔CHECK YOUR UNDERSTANDING

C. Rosser Midcap Fund has a NAV of $9.45 and an offer price of $9.87. What is the rate of commission, to the nearest tenth percent?

D. Tanella Yaou bought Security Real Estate Fund for $45.18. The fund's NAV was $43.86 at the time. What rate of commission, to the nearest tenth percent, did Tanella pay?

Profit or Loss from Mutual Fund Investments

When shares are *redeemed,* or sold back to the mutual fund company, the investor is paid the net asset value. The proceeds from the sale are found by multiplying the net asset value by the number of shares redeemed.

Proceeds = Number of Shares × Net Asset Value

The profit or loss from owning mutual fund shares is calculated by finding the difference between the proceeds and the total amount invested. If the proceeds exceed the investment, there is a profit. If the amount invested is larger than the proceeds, there is a loss.

Profit = Proceeds − Amount of Investment

Loss = Amount of Investment − Proceeds

EXAMPLE 3

Find the proceeds from the sale of 100 shares of a mutual fund with a net asset value of $8.50. Find the amount of the profit or loss if the total investment in the 100 shares is $675.

Photodisc/Getty Images

SOLUTION

Multiply the number of shares by the net asset value.

$100 \times \$8.50 = \$850$ proceeds

The proceeds, $850, is greater than the amount of investment, $675, so find the profit.

Profit $= \$850 - \$675 = \$175$ Profit

✔ CHECK YOUR UNDERSTANDING

E. Upon graduation, Todd redeemed 300 shares of a mutual fund at $17.22. His total investment in the shares was $4,391.10. What was his profit or loss?

F. Ping-lin Sheng redeemed his 500 shares of Madison Capital Fund for $39.45 a share. His total investment in the shares was $19,922.25. What was his profit or loss on the redemption?

Wrap Up ▶▶▶

One strategy might be to purchase every trading card for a season, which is similar to the diversification strategy practiced by many mutual funds. Another strategy might be to buy only the cards of players on those teams that were contenders for the playoffs, which is similar to mutual funds investing in businesses that are in sectors that are likely to do well in a year. Another strategy might be to purchase the trading cards of players who show early success, which is similar to a mutual fund's practice of investing in companies with demonstrated potential.

TEAM Meeting

Some problems are simplified if you can find a pattern and use it to answer questions. Suppose you own stock and are thinking of selling it. You could look for patterns that might help you make a decision. You have tracked its selling price and dividends.

	Jan. 1	April 1	July 1	Oct. 1	Dec. 31
Selling Price	25.25	22.50	20.87	18.38	19.25
Dividends	2.1%	1.4%	0.5%	3.2%	3%

Form a team with another student to discuss the following questions: (a) What patterns do you see? (b) Would you have sold the stock in July? (c) If today were January 2, would you sell the stock? Explain your reasoning.

Exercises

Find the difference.

1. $4,089 − $3,415
2. $34,551 − $29,796

Find the product.

3. $16.32 × 108.55
4. $7.24 × 44.784

Find the percent, to the nearest tenth.

5. $45.12 ÷ $193
6. $62.19 ÷ $525

Using the information given in the Mutual Fund NAV table, find the total investment in these mutual fund purchases.

	Mutual Fund	Number Shares	Total Investment
7.	Ameris Growth	400	
8.	Banner Income	132	
9.	Jantzen High Yield	800	
10.	Mercer International	260	
11.	Overland Technology	100	
12.	Parkson Equity	200	
13.	Strand Balanced	500	

14. Allie Jenkins bought these technology sector shares of New Economy Fund, a no-load mutual fund, on different days: 70 shares, NAV $39.78; 200 shares, NAV $41.52; and 80 shares, NAV $42.15. What was her total investment in the fund?

15. Wilson Growth, a no-load fund, has a net asset value of $42.52. Wilson Income, a load fund, is quoted with a net asset value of $30 and an offer price of $32.12. What total investment would be made if 100 shares each of Wilson Growth and Wilson Income were purchased?

16. Ridge Health Sector Fund is quoted at 12.16 NAV, 12.48 offer price. To the nearest tenth of a percent, what commission rate is charged?

17. Your broker quotes Stamper International Fund at these prices: NAV, 7.03; offer price, 7.40. Find the rate of commission.

Find the amount of commission and the rate of commission, to the nearest tenth percent, for each of these mutual fund shares.

	Mutual Fund	NAV	Offer Price	Amount of Commission	Rate of Commission
18.	Cyber Net Fund	18.87	19.26		
19.	Draper Energy Fund	51.12	53.76		
20.	Ethan Pacific Fund	88.75	97.00		
21.	Wu Equity Fund	68.64	71.84		

22. A commission of $3.52 is charged for a mutual fund with an offer price of $45.25. What is the commission rate, to the nearest tenth percent?

Photodisc/Getty Images

The Bradford Energy Fund has a current NAV of $47.48. Edward Carter redeemed the 1,280 shares that he had bought for a total cost of $42,792.

23. What were Edward's proceeds?

24. Find his profit or loss.

The Columbia Fund had a NAV of $47.64 and an offer price of $48.60 on the day that Sally Worth bought 500 shares of the fund. Six months later, the fund was quoted at these prices when Sally redeemed her shares: NAV, $48.66; offer price, $49.68.

25. What proceeds did Sally receive?

26. What amount of profit or loss did Sally make?

27. **FINANCIAL DECISION MAKING** Many financial advisers suggest that you evaluate your personality and your tolerance for risk to help you decide what kind of investments you should make. Describe the risk differences between purchasing stocks and shares in a mutual fund. Evaluate which investment your personality is best suited for.

Mixed Review

28. 250 is what part greater than 200?

29. 420 is what percent of 4,800?

30. Write $\frac{44}{30}$ as a decimal, rounded to the nearest tenth.

The Roswell Income Fund trades at a net asset value of $34.88 and an offer price of $36.46.

31. What commission amount is paid per share?

32. To the nearest tenth percent, find the commission rate.

33. Manfred Tulley bought 300 shares of Tempe Growth Fund for $21,577.50. Two years later he sold his shares at a $65.25 NAV. What was his profit or loss?

Gazi Ecevit purchased 200 shares of common stock in the Rivalo Company at $43.68 through an online broker. The commission was $29.95. He later sold the same stock at $41.25, paying commission and fees of $32.25.

34. What was Gazi's total cost for the stock?

35. What amount did Gazi receive as net proceeds from the sale?

36. What was Gazi's profit or loss on the sale?

Killean paid one and a half month's rent as a security deposit to rent an apartment for $695 a month. His utility bills are estimated to be an average of $129 a month. His annual renter's insurance premium is $239.

37. What is the annual cost Killean will pay to live in this apartment?

38. What is the monthly cost for Killean? Round to the nearest dollar?

7-8

Real Estate

GOALS

- Calculate net income from real estate investments
- Calculate the rate of return on real estate investments
- Calculate the monthly rent to be charged

KEY TERM

- capital investment

Janice and Jim Canter would like to move to a larger house. They are considering keeping and renting their present house as an investment. They wonder whether renting the house would be a good investment for them financially. What factors might they consider in making this decision?

Photodisc/Getty Images

Math Skill Builder

Review these math skills and solve the exercises that follow.

1 **Add** dollar amounts.
Find the sum. $450 + $2,398 + $941 = $3,789

1a. $1,290 + $4,208 + $638 + $2,108

1b. $289 + $6,792 + $749 + $3,184

2 **Subtract** dollar amounts from dollar amounts.
Find the difference. $9,600 − $7,895 = $1,705

2a. $11,893 − $9,215

2b. $6,182 − $4,297

3 **Multiply** dollar amounts by whole numbers and percents.
Find the product. $249.20 × 12 = $2,990.40
Find the product. $246,500 × 3% = $246,500 × 0.03 = $7,395

3a. $487 × 12

3b. $134,800 × 2.5%

3c. $629 × 12

3d. $89,345 × 2.7%

4 **Divide** dollar amounts by dollar amounts to find a percent.
Find the quotient, to the nearest tenth percent.
$2,400 ÷ $45,000 = 0.0533, or 5.3%

4a. $3,680 ÷ $56,700

4b. $2,734 ÷ $38,400

Net Income from Real Estate

When you invest in real estate, the rent you receive is your gross income from the investment. Your net income is the amount left after you pay all the expenses of owning the property.

The following table shows that only part of the money you collect as rent is profit. The rest of the money is usually used to pay for taxes, repairs, insurance, and interest on a mortgage. Since buildings wear out because of age and use, you will also have to calculate depreciation.

You usually calculate your income and expenses on an annual basis. The annual net income is the amount left after deducting the annual expenses from the annual rental income.

Annual Net Income = Annual Rental Income − Annual Expenses

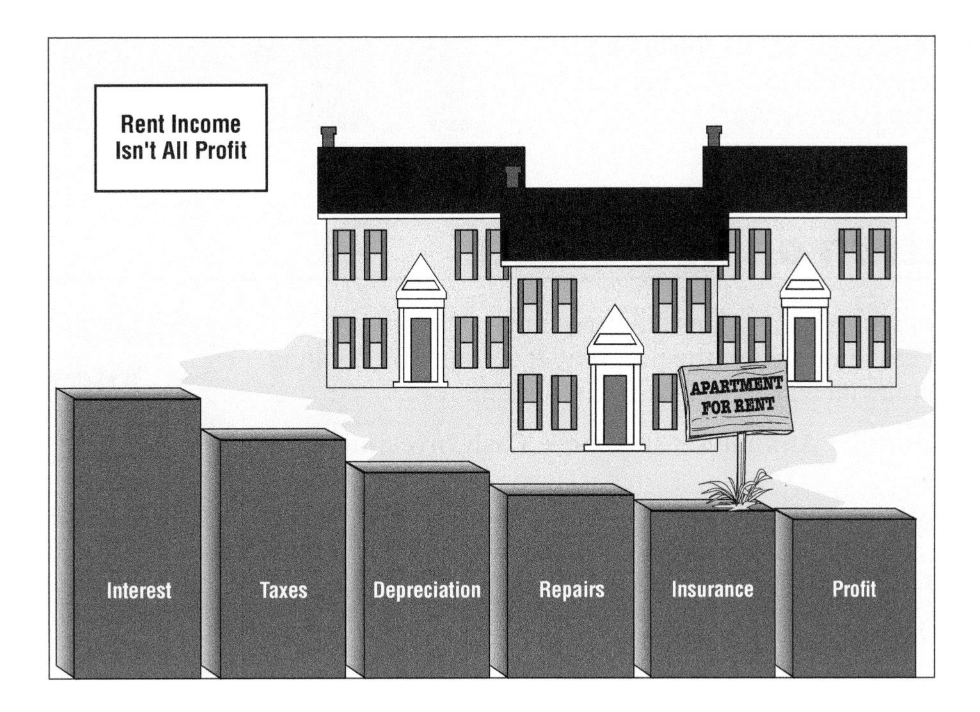

EXAMPLE 1

Chieko Beppu bought a house and lot for $150,000. She made a $30,000 cash down payment and got a $120,000 mortgage for the balance. She rented the house to a tenant for $1,500 a month. Her annual payments for taxes, repairs, insurance, interest, depreciation, and other expenses totaled $14,100. What annual net income did she earn?

SOLUTION

Multiply rent by 12.

$12 \times \$1{,}500 = \$18{,}000$ annual rental income

Subtract the annual expenses from annual rental income.

Annual Net Income $= \$18{,}000 - \$14{,}100 = \$3{,}900$ annual net income

✔CHECK YOUR UNDERSTANDING

A. Ursula Pavlok bought a building with four stores in it and the lot on which it stands for $850,000. During the first year, she received $2,300 a month in rent for each store unit. Her expenses for the year were: mortgage interest, $102,000; 2% depreciation on the building valued at $650,000; taxes, repairs, insurance, and other expenses, $74,500. Find her net income or loss for the year.

B. Melvin Weisbaum bought a house and lot in a resort area for $288,000. He paid $72,000 in cash and got a mortgage for the rest. He rented the house to a tenant for $3,800 a month. For the first year, Melvin's expenses were: mortgage interest, $19,200; 3% depreciation on the house, valued at $198,000; taxes, repairs, insurance, and other expenses, $15,750. What was his net income for the year?

Photodisc/Getty Images

Rate of Return on Real Estate

The rate of return on a real estate investment is based on the money or cash invested in the property. It is found by dividing the annual net income by the cash investment.

Rate of Return = Annual Net Income ÷ Cash Investment

EXAMPLE 2

Find the rate of return for Chieko Beppu's house in Example 1.

SOLUTION

Divide the annual net income by the cash invested.

Rate of Return = $3,900 ÷ $30,000 = 0.13 or 13% rate of return

> **Business Tip**
>
> The rate of return on an investment is also called the *yield.* It is also called the *rate of income* on an investment.

As a property owner, you spend money on both capital investments and expenses. **Capital investment** is the amount of cash you originally invested plus money you spend for improvements that increase the property's value. Adding a room or a garage to a rental house are examples of capital investments.

Money spent for repairing or replacing broken items does not increase the value of property. It is an expense of owning the property. The money paid returns the property to its original condition. Expenses include repainting the house, replacing a broken sidewalk, and repairing leaking faucets.

To find the return on investment after you have made capital investment improvements in a property, you must add the cash spent on the improvements to your original cash investment.

> **Business Tip**
>
> Capital investment is an investment in anything that cannot easily be turned into cash and which is usually held by the investor for longer than one year.

EXAMPLE 3

Suppose that on buying the rental house, Chieko (see Example 1) spent an additional $2,700 in cash to add a deck to her rental house. What would the rate of return on her real estate investment be?

SOLUTION

Add the two cash amounts invested in the house.

$30,000 + $2,700 = $32,700   capital investment in house

Divide the annual net income by the capital investment.

Rate of Return = $3,900 ÷ $32,700 = 0.119, or 11.9%   rate of return

✔ CHECK YOUR UNDERSTANDING

C. Talibu Mbasa bought a house with a $25,000 down payment. He added a basement room and bath for $14,600. He spent $2,600 to paint the exterior and $2,300 to repair the plumbing. What is Talibu's capital investment in the house?

D. Ida Silvers purchased a cottage and lot for $85,000, paying $17,000 down and using a mortgage for the rest. The lot is estimated to be worth $5,000. She added a room by remodeling the attic for $12,600. She estimates that she will be able to rent the cottage for $1,200 a month. Her annual expenses will be $5,400, mortgage interest; 2.5% depreciation on the cottage and its capital improvements, and $4,800 other expenses. What is Ida's estimated return on her investment, to the nearest tenth percent?

Finding What Monthly Rent to Charge

To find the amount to charge for rent, first find the total annual expenses. Then find the amount you want to earn as net income on your investment. Add those two amounts and divide the sum by 12 to find the monthly rent to charge.

Annual Rental Income = Annual Net Income + Annual Expenses

Monthly Rent = Annual Rental Income ÷ 12

EXAMPLE 4

Ezra wants to earn 15% annual net income on his $10,000 cash investment in property. His annual expenses of owning the property are $2,700. What monthly rent must Ezra charge?

SOLUTION

Change the percent to a decimal and multiply by the cash investment.

15% × $10,000 = 0.15 × $10,000 = $1,500   desired annual net income

Add the desired annual net income and the annual expenses.

$1,500 + $2,700 = $4,200   annual rental income

Divide the annual rent income by 12. $4,200 ÷ 12 = $350   monthly rent

✔ CHECK YOUR UNDERSTANDING

E. Sid Weisbaum bought a vacant warehouse with a $24,000 down payment. He estimates that expenses will be $8,640 the first year. Sid wants to earn a rate of income of 8% on his investment. To do that, what monthly rent must he charge?

F. Ellie Burns wants to earn 15% annual net income on her $30,000 cash investment in a property. Her annual expenses of owning the property are $8,100. What monthly rent must Ellie charge?

Wrap Up ▶▶▶

One factor to consider is how the return on the rental property compares to the return on other investments they might make with the cash from the sale of the old house. The Canters might estimate their gross income using rents charged for similar properties in their neighborhood. They can subtract from that gross income the expenses they know they will have with the house and find the likely return on their investment. They should then compare their return on the house to returns on other investments, such as bonds, stocks, or mutual funds.

Communication

Visit or call a real estate agent to determine the typical return on investment percentages for rental housing in your area. Also, call a local bank to find the APR for a 3-year CD, and the APR for a savings account. Finally, call a local stockbroker to find the annual yield on a high quality corporate bond. Prepare a chart showing the various percentages you found.

Exercises

Perform indicated operation. Round to the nearest tenth, if needed.

1. $19 + $927 + $638
2. $842 + $72 + $604
3. $18,598 − $16,223
4. $8,450 ÷ $97,922
5. $362 − $53
6. $361 × 14
7. $187,526 × 1.8%
8. $422 ÷ $7,121

Find the annual net income for each real estate owner.

	Monthly Rent Income	Annual Expenses						Annual Net Income
		Taxes	Repairs	Insurance	Interest	Depreciation	Other	
9.	$420	$1,204	$329	$350	$1,470	$910	$35	
10.	$620	$1,900	$720	$760	$2,200	$1,120	$80	
11.	$420	$1,400	$105	$295	$1,500	$1,025	$45	
12.	$1,080	$3,260	$1,680	$720	$3,740	$2,400	$150	

Solve.

13. Marvin bought a house and lot for $72,000. He paid $18,000 cash and got a mortgage for the balance. He rented the house to a tenant for $950 a month. For the first year, Marvin's expenses were: mortgage interest, $4,800; 3% depreciation on the house valued at $58,000; and taxes, repairs, insurance, and other expenses, $3,750. What was his net income for the year?

14. Fatou Keita bought an eight-unit apartment building for $165,000. During the first year of ownership, she received $480 a month for the rent of each apartment unit. Her expenses for the year were: mortgage interest, $20,700; 2.5% depreciation on the building valued at $125,000; and taxes, repairs, insurance, and other expenses, $14,200. Find her net income for the year.

Find the rate of return, to the nearest tenth percent, on each cash investment. Show losses, or negative rates of return with a minus sign.

Photodisc/Getty Images

			Annual Expenses		
	Cash Investment	Monthly Rental Income	Interest on Mortgage	Other	Rate of Income
15.	$23,000	$600	$2,800	$2,100	
16.	$35,000	$800	$1,800	$3,100	
17.	$41,000	$1,600	$10,200	$9,800	
18.	$13,000	$400	$2,100	$1,550	
19.	$33,000	$700	$2,500	$2,050	

Solve.

20. Mario Valente took out a $58,400 mortgage on a two-family house after making a down payment of $14,600. During the first year, he rented one unit at $550 a month and the other unit at $580 a month. For the year, he paid $5,800 in mortgage interest and $6,300 in other expenses. To the nearest tenth of a percent, what rate of income did Mario earn on his cash investment?

21. For $390,000, Bella Leipsen can buy an office building and lot that rents for $2,600 a month. Taxes, insurance, and repair expenses average $19,500 annually. Depreciation is estimated at $6,000 a year. Bella plans to pay $78,000 cash as a down payment. To the nearest tenth of a percent, find the rate of income she will make on her cash investment.

22. Bill Walsh made a $7,000 down payment on a condominium apartment that cost $35,000. He rented the condo at $510 monthly for the first year. During the year he had these expenses: taxes, $1,090; insurance, $370; interest, $2,400; repairs, $1,200; and depreciation, $700. What was Bill's rate of income on his cash investment, to the nearest tenth percent?

23. Marla Rios bought a six-unit apartment house for $140,000 and made a cash down payment of $40,000. The first year, she rented each of the 6 apartments at $500 a month. Her expenses for the year were: mortgage interest, $9,600; depreciation at 3% of the house's value of $110,000; and taxes, insurance, and other expenses, $12,600. Find Marla's rate of income on her cash investment, to the nearest tenth percent.

Marvin makes a down payment of $45,600 to buy a house. He also spends $14,000 to improve the property by adding a room and paving the driveway. Marvin estimates the total annual expenses of owning the house to be $16,200.

24. What would be Marvin's capital investment in the house?

25. What monthly rent will he have to charge in order to make a net income of 7% on his total capital investment?

Find the monthly rent the owner must charge.

	Cash Investment	Desired Annual Net Income	Annual Expenses	Monthly Rent
26.	$25,000	12% of investment	$3,900	
27.	18,000	8% of investment	2,990	
28.	35,000	11% of investment	9,430	
29.	46,000	7% of investment	8,600	

Bea Tompkins bought two lots for $16,000 at a tax sale. She estimates that her yearly expenses of owning these lots will be $740. A nearby factory wants to use the lots for parking trucks overnight. To earn a 12% rate of income on her investment:

30. What annual rent should Bea charge?

31. What monthly rent should Bea charge?

32. **CRITICAL THINKING** Assume you want to buy a small apartment building. What problems do you expect to have by being a landlord? Can you think of ways to prevent or solve those problems? Will you have to work full time at being a landlord? Based on your answers to these questions, what do you think are some of the advantages and disadvantages of investing in real estate?

33. **FINANCIAL DECISION MAKING** Which rate of return would you rather earn: 7% from a rental house or 7% from a bond? Explain your answer.

Mixed Review

34. Divide 96 by $\frac{3}{4}$.

35. Find 105% of 120.

36. Rewrite 4.5 as a percent.

37. 24 is what percent of 75?

38. Nellie O'Brien earns a salary of $1,500 a month plus 5% commission on all sales. Last month her sales were $40,000. What was her gross income for the month?

39. Zed Hargrove cashed in a $1,200 time-certificate account that paid 7% annual interest. Because he withdrew his money before the end of the term, the bank charged a penalty of 2 months' interest. What was the amount of the penalty?

7-9

Retirement Investments

GOALS

- Calculate your retirement income
- Calculate your pension income
- Calculate the required minimum payout from a pension fund
- Calculate the penalty for early withdrawal from an individual retirement account

KEY TERM

- individual retirement account (IRA)

Start Up ►►►

Some working people do not save money for retirement, counting on their Social Security benefits to provide their retirement income. Is this a wise plan?

Photodisc/Getty Images

Math Skill Builder

Review these math skills and solve the exercises that follow.

1. **Subtract** dollar amounts.
   Find the difference. $4,290 − $2,978 = $1,312

   1a. $5,398 − $3,148

   1b. $11,408.23 − $8,971.93

2. **Multiply** percents by whole numbers and dollar amounts.
   Find the product. 2.4% × 40 = 0.024 × 40 = 0.96
   Find the product. 96% × $75,000 = 0.96 × $75,000 = $72,000

   2a. 1.7% × 30

   2b. 2.3% × 40

   2c. $84,500 × 82.4%

   2d. $49,300 × 75.9%

3. **Divide** dollar amounts by decimals and dollar amounts.
   Find the quotient. $360,000 ÷ 21.8 = $16,513.76
   Find the quotient to the nearest tenth percent. $5,500 ÷ $420,000 = 1.3%

   3a. $459,600 ÷ 18.4

   3b. $277,500 ÷ 12.7

   3c. Divide $3,670 by $248,900 to the nearest tenth percent.

Retirement Income

You may receive retirement income from several sources: Social Security benefits, your own pension plan, a company, union, or organization pension plan, and income from other investments.

The income taxes on the amount you invest and the income your investment earns may be deferred until you retire, depending on the type of investment. This allows your retirement funds to grow much faster.

For example, one type of retirement investment is the **individual retirement account**, or IRA. There are many types of IRAs. The traditional IRA allows individuals earning less than a certain annual amount to invest up to $5,000 plus a cost-of-living adjustment each year. The tax on the money invested and any earnings may be deferred until you retire. When you retire, the money you withdraw from your IRA will be taxed at the current tax rate.

Advantage of Tax Deferred Investments

Years	Nontaxed Investment	Investment Taxed at 20%
5	$28,187	$22,016
10	$65,906	$49,846
15	$116,382	$85,028
20	$183,931	$129,505
25	$274,327	$185,731
30	$395,297	$256,810
35	$557,182	$346,667
40	$773,820	$460,261

Business Tip

The maximum amount a person under 50 can contribute to an IRA is $5,000 plus a cost-of-living adjustment, which adjusts for inflation.

The table shows the results of investing $5,000 each year from 5 to 40 years. The chart assumes that a $5,000 investment is made at the beginning of each year and that the investment grows at 6% a year. Notice that if you invest in a tax deferred IRA, your total investment would grow to $773,820 in 40 years. If you place your money in a taxable investment, your balance would only be $460,261. Your money would have grown 68% more by being in a tax deferred IRA instead of an investment taxed yearly!

Business Tip

Taxpayers earning less than a certain amount a year may deduct from taxable income up to $5,000 plus a cost-of-living adjustment by investing in a traditional IRA. The amount of investment that can be deducted is phased out as annual income rises.

EXAMPLE 1

Sam Weisbrunner is retiring at age 65. At age 65, his company pension will pay him $1,560 a month and social security will pay him $800 a month. He has $120,000 in an IRA fund. What percent, to the nearest tenth, of the fund must Sam withdraw each month to raise his monthly retirement income to $3,600?

SOLUTION

Add the monthly company pension and social security.

$1,560 + $800 = $2,360   retirement income

Subtract the non-IRA income from the desired monthly income.

$3,600 − $2,360 = $1,240   amount needed from IRA

Divide the amount needed from the IRA by the amount of the IRA fund.

$1,240 ÷ $120,000 = 0.0103, or 1.0%   percent to withdraw each month

✔ CHECK YOUR UNDERSTANDING

A. Jane Eiler will receive $1,620 in pension and $900 from social security each month when she retires at age 65. She wants her monthly retirement income to be $4,000. What percent, to the nearest tenth percent, of her $350,000 IRA must she withdraw monthly to reach the monthly income she wants?

B. Abner Duncan started investing $5,000 in an IRA when he was 30. The account has earned a steady 6% growth each year. Find how much money Abner has in his IRA at age 60.

Photodisc/Getty Images

Pension Income

There are two basic types of pension funds: *defined contribution plans* and *defined benefit plans.* Under defined contribution plans, you, your employer, or both contribute to your pension fund, often a percent of your annual wages. The amount you receive depends on how well the investments in your plan do over the years.

A defined benefit plan pays you a specific amount on retirement. The amount you receive from your pension fund usually depends on a number of factors, including how old you are when you retire, how many years you contribute to that pension plan, and how much money you put in the pension fund over the years.

You may be penalized by retiring early. For example, you may receive a reduced benefit for retiring before age 65 or 67. You may also have to contribute for a minimum number of years to a pension plan to collect any money. You may also be given a choice of how much money you can put into the fund each year.

EXAMPLE 2

John Baker's pension fund will pay him a pension rate of 2.2% of his average salary for the last four years for each year of service with his organization. If John plans to retire after 30 years of service, what percent of his final salary will he receive? If his final average salary is $56,000, what annual amount will he receive?

SOLUTION

Multiply the yearly pension rate by the number of years of service.

2.2% × 30 = 66%   total pension rate

Multiply the total pension rate by the final average salary.

66% × $56,000 = 0.66 × $56,000 = $36,960   amount of pension

✔ CHECK YOUR UNDERSTANDING

C. Keisha Turner is retiring after 30 years with her firm. Her pension fund pays 1.9% of her average salary for the last four years for each year of service. Her annual salary for the last four years was $71,000, $74,000, $75,000, and $77,000. What is her average salary for the last four years? What is the total pension rate? What monthly pension amount will she receive?

D. Juan Vellano's union pension fund provides for $5 a month at age 65 for each $130 he has contributed. When Juan retired at age 65, he had contributed $96,000 to the fund. What is his monthly pension amount?

Withdrawals from a Retirement Investment

With few exceptions, you can't withdraw funds from a traditional IRA before you reach age $59\frac{1}{2}$. If you do, you will pay a 10% penalty on the amount you withdraw. You will also have to pay federal and state income taxes on the withdrawal.

A person who has a private pension fund, such as an IRA, must withdraw a minimum amount each year from the fund when that person reaches $70\frac{1}{2}$. The minimum amount of the withdrawal required by law depends on the age of the person withdrawing funds and is found using the chart shown below. To use the chart, find your age at retirement. Then divide the divisor for that age into the total value of your IRA. The result is the required minimum you must withdraw each year.

Age	Divisor	Age	Divisor
70	27.4	78	20.3
71	26.5	79	19.5
72	25.6	80	18.7
73	24.7	81	17.9
74	23.8	82	17.1
75	22.9	83	16.3
76	22.0	84	15.5
77	21.2	85	14.8

Business *Tip*

Some exceptions to the early IRA withdrawal penalty are if the money is used for a first-time home purchase, higher education, health insurance premium, certain medical expenses, or if you become disabled. The money withdrawn will still be subject to federal and state income taxes.

EXAMPLE 3

Lu-yin Huang's IRA balance is $428,000 at age 75. What amount must she withdraw from her IRA during the year?

SOLUTION

Divide the IRA balance by the divisor in the table for age 75.

$428,000 ÷ 22.9 = $18,689.96   minimum amount to be withdrawn

EXAMPLE 4

Tito Carlocci withdrew $3,000 from his IRA at age 41. What penalty did he pay?

SOLUTION

Multiply the amount withdrawn before the age of $59\frac{1}{2}$ by 10%.

$\$3,000 \times 10\% = \$3,000 \times 0.10 = \$300$ amount of penalty

✔ CHECK YOUR UNDERSTANDING

E. Prasam Shinawatra is 72 and his IRA investment totals $396,100. What minimum amount must he withdraw this year?

F. Karl Schmidt withdrew $12,500 from his IRA at age 46. What penalty did Karl pay?

Wrap Up ▶▶▶

The Social Security Act provides only *supplemental* retirement benefits and should not be counted on to pay enough for a person to live on in retirement. It is important that you save money in a retirement account or pension plan other than social security if you wish to have enough money to retire.

Exercises

Solve.

1. Eileen Rustio started investing $5,000 in an IRA when she was 40. The account has earned a steady 6% growth each year. Use the table in the lesson to find how much money Eileen has in her IRA at age 65.

2. Use the table in the lesson to find, to the nearest whole percent, by how much the nontaxed investment amount is greater than the investment taxed at 20% after 30 years.

3. Rod Schweiger receives $1,150 in monthly pension and $650 monthly from social security at retirement. He wants his monthly retirement income to be $3,000 a month. What percent of his $275,000 IRA must he withdraw monthly to reach the monthly income he wants?

4. Salvador Nuncio's monthly retirement income is made up of $1,900 from his employer-based pension fund and $900 from social security. He also has an IRA worth $150,000. What percent of his IRA must he withdraw if he wants his total retirement income to be $42,600 a year?

5. Tony Conte is retiring after 40 years of work at age 65. He will receive the following monthly amounts: $1,400 from his union pension and $840 from social security. What is his annual retirement income?

Photodisc/Getty Images

Kelly O'Malley's pension fund pays 2.1% of her average salary for the last three years for each year of service. Her annual salary for the last three years was $87,000, $90,000, and $95,000. Kelly has contributed to her pension fund for 25 years.

6. What is her average salary for the last three years?
7. What is Kelly's total pension rate?
8. What monthly pension amount will she receive if she retires this year?

Pablo Gonzales contributed $48,000 to his pension fund over 20 years of service to his company. The fund pays $4 a month for each $100 of pension funds contributed if he retires at age 65.

9. What is his monthly pension amount at age 65?
10. **STRETCHING YOUR SKILLS** How many months will it take Pablo to recover the amount of his contributions?

Photodisc/Getty Images

Solve.

11. Peter True's pension fund reduces the total pension rate he is to receive by $\frac{1}{2}$% for each year that he retires before the age of 67. If Peter retires at age 62, by what percent will he be penalized?
12. Marla Pezweski is age 82 and has a $1,289,350 IRA. What amount must she withdraw, as a minimum, from her IRA this year?
13. Jake Reilly has a total of $529,200 in his IRA. If Jake is 76, what minimum amount must he withdraw from his IRA this year?
14. Amoni Rahum is age 78. Her IRA investment is $738,300. What minimum amount must she withdraw this year?
15. Rich Vanegan withdraws $4,350 from his IRA when he is only 35. What penalty must he pay?
16. **FINANCIAL DECISION MAKING** Bea Roche is 48 years old. She wants to withdraw $5,125 from her IRA for a period of one year. Her friend suggests that she borrow the money from the cash value of her life insurance company. The loan would be at a rate of 7.5%. Which plan will cost Bea less?

Mixed Review

17. Divide 569 by 16 to the nearest tenth.
18. Divide 104.68 by 4.5 to the nearest hundredth.
19. Tomas Mendosa borrowed $250 for 30 days on his credit card using a cash advance. His card company charged a cash advance fee of $25 and a daily periodic interest rate of 0.0514%. What was the total finance charge on the cash advance?

Chapter *Review*

Vocabulary Review

Find the term, from the list at the right, that completes each sentence. Use each term only once.

bonds
capital investment
cash value
coinsurance
current yield on bonds
disability insurance
discount
health insurance
individual retirement account (IRA)
life insurance
major medical insurance
market price
mutual fund
net asset value
premium

1. If you cannot work due to a health condition or an injury, __?__ pays a portion of the income you lose.
2. __?__ protects your family against financial loss due to your death.
3. __?__ helps pay for hospital, surgical, medical, or other health care expenses due to a major illness or an injury.
4. __?__ provides basic protection against financial loss from medical bills.
5. The money you get if you cancel a life insurance policy is called __?__.
6. __?__ are a form of long-term promissory note.
7. Annual interest income divided by a bond's price is called __?__.
8. The price at which a stock sells is called the __?__.
9. __?__ companies use money from investors to buy stock in many companies.
10. The price at which mutual fund shares are traded is called __?__.
11. The amount of cash you originally invest plus what you spend for improvements in real estate is called a(n) __?__.
12. A popular form of retirement investment is called __?__.
13. When the market value of a bond is more than its par value, the bond is selling at a(n) __?__.
14. When the market value of a bond is less than its par value, the bond is selling at a(n) __?__.
15. After the deductible is met, the __?__ is the amount of money that the insured must pay out of his or her own pocket for health care services.

7-1 Life Insurance

16. Bob Walzcek bought a $135,000 term life insurance policy. He paid an annual premium of $1.65 per $1,000 of insurance. What annual premium did he pay?
17. Ava Leland pays $145.80 quarterly for a whole life insurance policy. This year, her policy paid a dividend of $43.71. Find her annual premium and the net cost of her policy for the year.
18. Using the cash value table, how much can you borrow against a 15-year policy with a face value of $100,000?
19. Using the cash value table, how much would you receive if you cancelled a 10-year policy with a face value of $50,000?

7-2 Health Insurance

20. Alan Lester's employer pays 60% of his annual health insurance premium. If the total monthly premium for the insurance is $86, what is Alan's share of the annual premium?

21. Louisa Corita's total medical bill was $3,989. Only $3,583 was covered by her group medical insurance policy. Louisa's coinsurance for the bill was 21%, and her remaining annual deductible was $300. What amount of the medical bill must Louisa pay?

22. Trisha Longworth has to pay 35% of her family's monthly health insurance premium. Her employer pays the rest of the $419 monthly premium. What is the annual amount that Trisha's employer pays for health insurance for Trisha and her family?

7-3 Disability Insurance

23. Boris Raskonov is totally disabled. His group disability policy's benefit percentage is 60% of his average annual salary for the last 3 years. His annual salary for the last 3 years was $55,300; $58,600, and $59,700. What monthly disability benefit amount will he receive?

24. Emma Abbot has a debilitating disease that has left her able to work. Her disability policy's benefit is 55% of her average annual salary for the last 5 years. Her annual salaries were $48,450; $51,300; $52,875; $52,875, and $55,090. What benefit amount will she receive each month?

7-4 Bonds

25. Tenco Foundry $1,000 bonds are quoted at 101.379. What is their market price?

26. Tessalee Lightner bought 10, $1,000 Montgomery County bonds at 102.682. The broker charged $3 per bond commission. What total investment did Tessalee make in the bonds?

27. Alltown School Bonds are selling at 98.54. What is the price of one of the school's $1,000 bonds? Are the bonds selling at a discount or a premium?

7-5 Bond Interest

28. Victor Armound bought 50, $1,000, 7.5% bonds. What is his annual income from the bonds?

29. Tara Sebring owns 100, $1,000, 8.25% bonds. How much does she earn semiannually from the bonds?

30. The semiannual interest on Laclede Airport $1,000 bonds is $62.50. If you buy the bonds at 98.487, what is the current yield, to the nearest tenth percent?

31. Vaughnie Kinder buys 20, $1,000, 9.4% bonds through a dealer at 102.775 plus accrued interest of $23.50 per bond and commission of $4 per bond. What is the total cost of the bonds to Vaughnie?

32. Timothy O'Hane buys 10, $1,000, 7% bonds on April 1 at 101.528 plus accrued interest from January 1. Commission was not shown. What is Timothy's total cost for each bond?

7-6 Stocks

33. A broker sold Octavia Relenza 200 shares of stock at $23.19. Octavia's broker charged $199.13 commission. Find the total cost of the stock to Octavia.

34. Reality Films common stock pays a quarterly dividend of $3.50 per share. If Vincente Guillermo owns 500 shares of the stock, what amount will he receive in dividends for the year?

35. Diane Limbaugh bought 100 shares of stock at $10.78. Her broker charged her $47.30 commission. If the stock pays a quarterly dividend of $0.14, what is its annual yield, to the nearest tenth percent?

36. Pedro Lamas bought 200 shares of common stock at $31.50 and was charged a commission of $134. He later sold the same stock for $38.20. The commission and fees on the sale were $143. What was Pedro's profit or loss?

7-7 Mutual Funds

37. Hisako Matsunaga purchased 400 shares of a no-load fund at its NAV of $11.73. She also purchased 300 shares of a load fund at its offer price of $23.77. What was her total investment in the funds?

38. A mutual fund has a NAV of $6.94 and an offer price of $7.22. What is the rate of commission, to the nearest tenth percent?

39. You redeemed 300 shares of a mutual fund at its NAV of $14.72. Your total investment in the shares was $3,910.10. What was your profit or loss on the sale?

7-8 Real Estate

40. Zebulon Poke bought a house and the lot it is on for $250,000. During the first year, he received $2,150 a month in rent. His expenses for the year were: mortgage interest, $12,800; 2% depreciation on the building valued at $200,000; and taxes, repairs, insurance, and other expenses, $5,500. Find his net income or loss for the year.

41. Bineka Zaheer bought a house with a $20,000 down payment. She also added a carport for $4,600. She estimates that she will be able to rent the house for $1,000 a month and that her first-year expenses will be mortgage interest, $5,600; taxes, $896; depreciation, $1,450; and other expenses, $1,288. What is Bineka's estimated return on her investment, to the nearest tenth percent?

42. Ollie Breem wants to earn 10% annual net income on his $25,000 cash investment in a property. His annual expenses of owning the property are $8,100. What monthly rent must Ollie charge?

7-9 Retirement Investments

43. Jeanne Wilder is retired and receives $1,870 in monthly pension and $1,100 monthly from social security. She wants her monthly retirement income to be $4,500. What percent, to the nearest tenth, of her $250,000 IRA must she withdraw monthly to reach the monthly income she wants?

44. Keiko Yoshino is retiring after 30 years with her firm. Her pension fund pays 2% of her average salary for the last three years for each year of service. Her annual salary for the last three years was $57,400, $59,200, and $61,600. What is her average salary for the last three years? What is the total pension rate? What monthly pension amount will she receive?

45. Karl Lamour withdrew $22,000 from his IRA at age 49. What penalty did he pay?

Technology Workshop

Task 1 Enter Data in an IRA Calculator Template

Complete a template that estimates the future value of investments you make in an IRA.

Open the spreadsheet for Chapter 7 (tech7-1.xls) and enter the data shown in the blue cells (cells B3 through B5). Your computer screen should look like the one shown below when you are done.

	A	B
1	**IRA Calculator**	
2		
3	Estimated Annual Growth Rate	8.0%
4	Annual Investment	$5,000
5	Number of Years	40
6		
7	Future Value	$1,295,304.32

The spreadsheet calculates the value of your investments in the future. For example, suppose that you invest $5,000 each year for 40 years starting at age 25. Suppose also that you estimate that the annual earnings of your investment will be 8%. The spreadsheet calculates that the *future value* of your IRA at the end of 40 years will be $1,295,304.32.

Calculating the future value of investments is no different than calculating compound interest when deposits are made at the end of each year. Spreadsheets make these laborious calculations quick and easy to do.

Task 2 Analyze the Spreadsheet Output

It is important to recognize that you must contribute as early as possible to have a large IRA balance when you retire. Small amounts invested over long periods of time are equal to much larger amounts invested over short periods of time. Unfortunately, some people wait until they are in their forties to save for retirement. Enter 20 years in cell B5 to show what the future value of the IRA would be if a person started saving at age 45.

Answer the following questions.

1. What is the future value of the 20-year IRA?
2. Estimate the difference in the balances between the 20-year and 40-year contributions.
3. What is the exact difference in the balances?
4. Keep raising the annual investment amount until you get close to the future value, $1,295,304 shown in the original spreadsheet. What approximate annual amount must be contributed to some type of retirement investment starting at age 45 to reach the same value (about $1,295,000) as a $5,000 annual investment started at age 20?

5. What spreadsheet function is used in cell B7?
6. Re-enter 40 years in cell B5 and $5,000 in cell B4. What would be the future value of the contributions if the estimated growth rate was 10% instead of 8%?

Task 3 Design a Major Medical Coverage Spreadsheet

You are to design a spreadsheet that will calculate the amount of medical expenses paid by the insurer and the amount paid by the insured for a major medical insurance policy.

SITUATION: You work in the personnel office of a small company that provides major medical insurance for its employees. The major medical policy pays 85% of all bills after a $1,000 deductible is paid. An employee has $50,000 in major medical expenses after a serious accident.

Task 4 Analyze the Spreadsheet Output

Answer these questions about your completed spreadsheet.

7. How much of the employee's major medical expenses does the insurance company pay?
8. How much does the employee pay?
9. What formula did you use to find the amount that the insurance company paid?
10. What formula did you use to calculate the employee share?
11. Suppose an employee had major medical expenses of $25,500. How much of the employee's major medical expenses does the insurance company pay?
12. How much does the employee pay?

Trutta55/Shutterstock.com

Chapter *Assessment*

Chapter Test

Answer each question.

1. $235,000 ÷ $1,000
2. 26 × $38.89
3. 125% × $78
4. 80% × $1,600
5. $1,354 − $600
6. 1,050 × $0.49
7. $\frac{1}{4}$ × $42,100
8. $37 + $28 + $376 + $73

Solve.

9. Pamela Hopkins has a life insurance policy for $15,000. She pays a premium rate of $23.48 per $1,000 annually. This year the insurance company paid a dividend of $33.60. What is Pamela's premium for the year after deducting the dividend?

10. Tia's major medical policy has a $1,000 deductible feature and a 15% coinsurance feature. Last month she spent $10,300 for medical expenses, $700 of which were not covered under her policy. How much did Tia pay?

11. Tito Tortelli becomes totally disabled. His group disability policy's benefit percentage is 60% of his average annual salary for the last 3 years. His annual salaries were $35,800; $38,500; and $39,800. What monthly benefit amount will he receive?

12. Find the yield, to the nearest tenth percent, on a $1,000, 11.5% bond bought at 106.

13. Park Lee bought 300 shares of stock at 26.45, plus commission of $182.50. What was the total cost of the purchase?

14. Jolene Williams owns 800 shares of TRI, Inc. common stock paying a quarterly dividend of $0.57 per share and 400 shares of TRI preferred stock, $100 par value, paying an annual dividend of 7.2%. Find the total annual dividend she gets from the common and preferred stock.

15. Bev Jorald bought 500 shares of a stock at a total cost of $12,762. She sold the shares at 40.76 and was charged a commission and other costs of $374. What was her profit or loss from the sale of the stock?

Ted Ling made a $37,000 down payment on a resort condominium that sold for $185,000. His average monthly rental income will be $1,850. His total annual expenses will be $18,800.

16. Find his annual net income.

17. Find his rate of income earned, to the nearest tenth of a percent.

Laura Barn's pension fund pays her 1.9% of her average salary for the last three years for each year of service. In the last three years, she earned $57,000, $59,800, and $62,500.

18. What is her average salary for the last three years?

19. If she retired after 30 years of service, what is the total pension rate?

20. What monthly pension amount will she receive?

Planning a Career in Health Sciences

Many jobs in health sciences are directly in patient care, like nurses, radiology technicians, physicians, and phlebotomists. Other opportunities include areas of medical records, pharmaceuticals, laboratory work, biochemical engineering, and emergency medicine. Workers in the health sciences can be found in hospitals, universities, private offices, laboratories, industrial settings, schools, and retail markets. If you are detail oriented, have good communication skills, and work well under pressure, health sciences may right for you.

Jaimie Duplass 2008/Shutterstock.com

Job Titles

- Physician
- Sonographer
- Optometrist
- Nurse
- Lab technician
- Medical records technician
- Pharmacist

Needed Skills

- good communication skills
- excellent problem solving and deduction skills
- strong mathematical and scientific background

- ability to use technology
- outstanding leadership skills
- ability to work independently, as well as with others

What's it Like to Work in Medical Records

As a medical records technician, for every doctor's appointment, medical test, or treatment that a person receives, a record is generated. Personal information, medical history, current symptoms, and treatment plans must all be documented accurately to ensure that complete information is available when medical personnel or insurance companies access a record. One critical component of the job of health information technicians is the coding of medical information that is required for insurance purposes. Accuracy of all records and coding is of utmost importance.

What About You?

Does dealing with patients or medical tests and record appeal to you? What field in health sciences might you be most interested?

How Times Have Changed

For Questions 1–2, refer to the timeline on page 271 as needed.

1. In 1925, the Massachusetts Investors Trust mutual fund had $392,000 in assets and about 200 individual investors. By 1969, there were about 270 mutual funds with $48 billion in assets. Today there are over 10,000 mutual funds with over $7 trillion assets and about 83 million individual investors. By what percentage have the assets of all mutual funds increased since 1969? By what percentage has the number of mutual funds increased since 1969?
2. What was the approximate percent of increase in the Dow Jones Industrial Average from 1906 to 1972? What was the approximate percent of increase in the Dow Jones Industrial Average from 1972 to 2007?

Chapter 8

Budgets

Statistical Insights

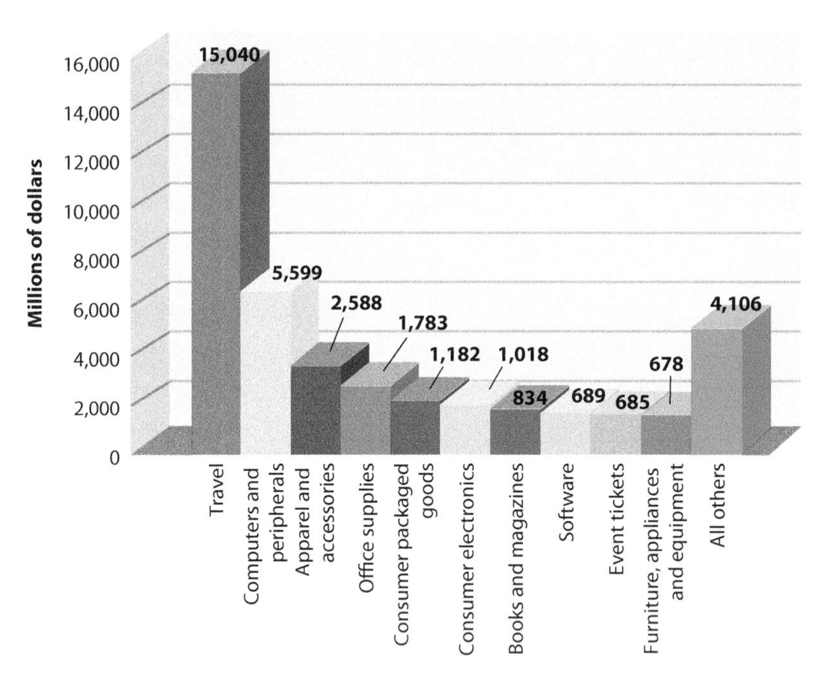

Source: comScore Media Metrix

Online purchasing has changed the retail industry. Use the bar graph to answer Questions 1–4.

1. How much did consumers spend on apparel and accessories?
2. How much more did consumers spend on computers and peripherals than software?
3. In how many categories did consumers spend more than 1,000,000,000?
4. **Explain** why you think more money is spent on travel online than any other category.

How Times Have Changed

Individuals and businesses alike increasingly depend on wireless communication. One hundred and fifty years ago, there were no telephones. Even 40 years ago, there were no cell phones. Wireless communication has come a long way from the days of smoke signals, the Pony Express, and Morse code.

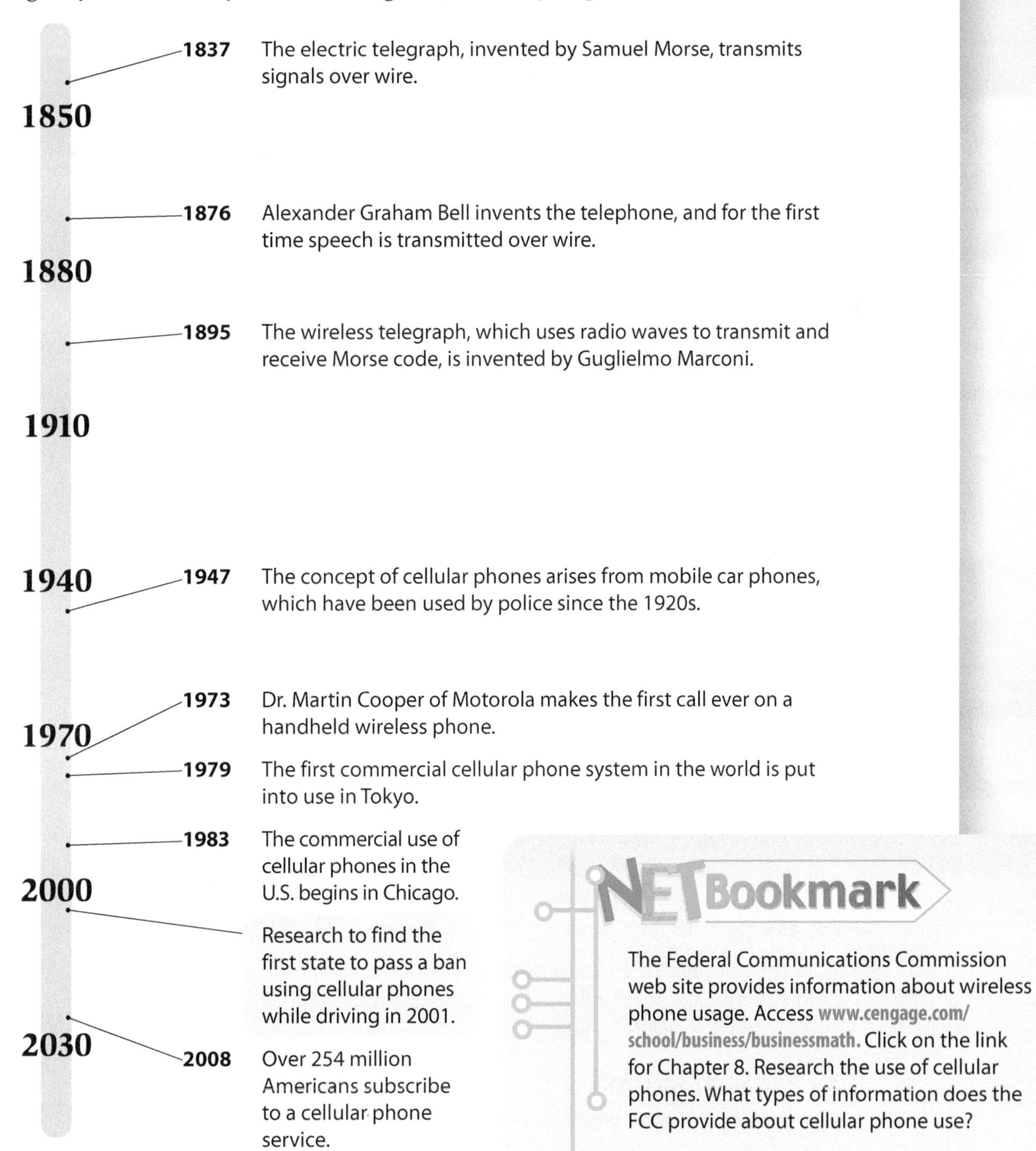

1850
1880
1910
1940
1970
2000
2030
1837 The electric telegraph, invented by Samuel Morse, transmits signals over wire.
1876 Alexander Graham Bell invents the telephone, and for the first time speech is transmitted over wire.
1895 The wireless telegraph, which uses radio waves to transmit and receive Morse code, is invented by Guglielmo Marconi.
1947 The concept of cellular phones arises from mobile car phones, which have been used by police since the 1920s.
1973 Dr. Martin Cooper of Motorola makes the first call ever on a handheld wireless phone.
1979 The first commercial cellular phone system in the world is put into use in Tokyo.
1983 The commercial use of cellular phones in the U.S. begins in Chicago.
Research to find the first state to pass a ban using cellular phones while driving in 2001.
2008 Over 254 million Americans subscribe to a cellular phone service.
NET Bookmark
The Federal Communications Commission web site provides information about wireless phone usage. Access www.cengage.com/school/business/businessmath. Click on the link for Chapter 8. Research the use of cellular phones. What types of information does the FCC provide about cellular phone use?

8-1

Average Monthly Expenses

GOALS
- Calculate total expenses
- Calculate average monthly expenses

KEY TERM
- average monthly expenses

Start Up

On what kind of purchases do you spend the money you earn or receive? Make a list of the things you do with your money in a typical week. In your list include how often you spend money on each item, where you buy it from, and an estimate of how much you spend.

Math Skill Builder

Review these math skills and then answer the questions that follow.

1 **Add** money amounts.
Find the sum. \$450 + \$82.96 + \$22.29 + \$123.67 = \$678.92

1a. \$225 + \$63.86 + \$145.86 + \$45.21

1b. \$153.07 + \$93.76 + \$124.97 + \$282.71

2 **Multiply** money amounts by a whole number.
Find the product. \$45.63 × 7 = \$319.41

2a. \$55.82 × 7

2b. \$223.82 × 52

2c. \$12.82 × 12

2d. \$4.78 × 26

3 Find the **average** of a group of money amounts.
Find the average of \$45.86, \$56.72, and \$49.22.

$$\frac{\$45.86 + \$56.72 + \$49.22}{3} = \frac{\$151.80}{3} = \$50.60$$

3a. \$62.54, \$75.89, \$55.22

3b. \$78.49, \$0, \$125.83

3c. \$156.83, \$125.92, \$143.66

Calculator *Tip*

When finding the average, use the [=] to find the total of the numbers before dividing.

Tracking Expenses

Many people have trouble managing their money because they do not keep records of how the money is spent. Tracking expenses is the first step in managing your money. To get a snapshot of where you spend your money, keep a log of every penny you spend for a month.

EXAMPLE 1

Amy Tracer is a student who is tracking her expenses for a month. In the first week of the month, she has spent the following amounts.

Date	Explanation	Amount
6/1	Lunch with friends	$7.86
6/1	Drink machine	$2
6/2	Drink machine	$2
6/3	New music CD	$15.36
6/3	Gasoline for car	$35
6/3	Drink machine	$2
6/4	Drink machine	$2
6/5	Drink machine	$2
6/6	Drink machine	$2
6/7	Drink machine	$2
6/7	Movie and snacks	$15

How much money has Amy spent in one week at the drink machine? If she spends this much every week, how much will she have spent in a year at the drink machine?

SOLUTION

Amy spent $2.00 per day for 7 days at the drink machine.

$\$2 \times 7 = \$14$ amount spent in one week

$\$14 \times 52 = \$728$ amount spent annually

✔ CHECK YOUR UNDERSTANDING

A. If Amy buys gasoline once a week and spends an average of $35, how much will she spend in a year on gasoline?

B. How much money did Amy spend in the first week of June?

Average Monthly Expenses

With several months' records of tracked expenses, you can determine categories in which your expenses can be grouped. Amy's categories might be food, drink, gas, entertainment, and miscellaneous. A person with housing expenses might have categories of housing, utilities, food, entertainment or other.

You can find **average monthly expenses** for each category or for each month by finding the sum of the monthly expenses and dividing by the number of months.

$$\textbf{Average Monthly Expense} = \frac{\textbf{Total Expenses}}{\textbf{Number of Months}}$$

EXAMPLE 2

Consuelo Valdez has tracked her expenses for 3 months. She grouped her expenses into categories and made a table to show how much she spent each month. What is the average amount that Consuelo spends each month?

Category	January	February	March
Housing	$550	$550	$550
Utilities	$175.18	$180.23	$145.16
Insurance	$540	$280	$280
Transportation	$120.67	$340.02	$105.93
Food	$240.26	$225.14	$238.29
Clothing	$55.36	$0	$122.18
Entertainment	$36.29	$122.18	$75.82
Savings	$120	$120	$120
Health Care	$0	$105	$22.18
Miscellaneous	$75.46	$22.38	$56.82

Math Tip

A monthly expense of $0 is still included in the average monthly expense formula.

SOLUTION

Find the total expenses for each month.

Jan: $550 + $175.18 + $540 + $120.67 + $240.26 + $55.36 + $36.29 + $120 + $0 + $75.46 = $1,913.22

Feb: $550 + $180.23 + $280 + $340.02 + $225.14 + $0 + $122.18 + $120 + $105 + $22.38 = $1,944.95

March: $550 + $145.16 + $280 + $105.93 + $238.29 + $122.18 + $75.82 + $120 + $22.18 + $56.82 = $1,716.38

$$\text{Average Monthly Expenses} = \frac{\$1{,}913.22 + \$1{,}944.95 + \$1{,}716.38}{3} = \frac{\$5{,}574.55}{3} = \$1{,}858.18$$

✔CHECK YOUR UNDERSTANDING

C. What is Consuelo's average monthly expense for food?

D. What is Consuelo's average monthly expense for health care?

Problem Solving Tip

An annual premium is paid once a year. A semi-annual premium is paid twice a year, or every 6 months. A quarterly premium is paid every 3 months, or 4 times a year.

Many expenses are paid on a monthly basis, such as rent, utilities, and loan payments. However, there are other payments that are paid on different schedules.

Insurance premiums, property taxes, and car registration fees are examples of payments that may be made on an annual basis. Tracking expenses for just a few months does not give an accurate picture of these expenses. If you save money each month to make an annual payment when it is due, the annual expense is averaged over the 12 months of the year.

EXAMPLE 3

Vince Bartolemo's homeowners insurance is due each year in December. His annual premium is $1,050. How much should he save each month to pay the annual premium in December?

SOLUTION

Find the average monthly expense.

$$\text{Average Monthly Expense} = \frac{\$1{,}050}{12} = \$87.50$$

✔ CHECK YOUR UNDERSTANDING

E. Vince's auto insurance premium is $850/year. What amount should he save each month for his insurance?

F. Vince's life insurance policy has a semi-annual premium of $420. What is the average monthly expense for his life insurance?

Wrap Up ▸▸▸

Look back at the list of expenses you made at the beginning of the lesson. Create a list of categories for your typical expenses. On what categories do you feel you may be spending too much money? Does your list of categories include savings?

Exercises

Find the sum.

1. $24 + $12.30 + $67.12
2. $57 + $132.21 + $228.14 + $23.49
3. $26.03 + $132.41 + $24.85
4. $55.42 + $312.71 + $28.14 + $112.79

Find the product.

5. $523.19 × 7
6. $456.12 × 12
7. $21.36 × 52
8. $2,107.88 × 12

Find the average.

9. $234.56, $862.22, $586.31
10. $15.76, $18.99, $14.87
11. $663.78, $568.23, $0
12. $56.73, $125.49, $226.92

The table shows the first week of Andrea Turnman's expense tracking for the month of July.

Date	Explanation	Amount
7/1	Lunch	$7.86
7/1	Gas for auto	$42.38
7/2	Lunch	$5.62
7/3	Lunch	$15.36
7/3	Groceries	$68.22
7/4	Lunch	$8.32
7/5	Lunch	$6.92
7/6	Lunch	$10.23
7/6	Auto Service	$156.83
7/7	Lunch	$7.82
7/7	Groceries	$35.91
7/7	Movie and snacks	$20.00

13. How much did Andrea spend on lunches in one week?

14. If she spends this much every week for lunch, how much will she have spent on lunches in one year?

15. How much did Andrea spend on groceries in one week?

16. If she spends this much for groceries every week, how much will she spend on groceries in a year?

17. How much did Andrea spend on transportation for the week?

The McGregor's have tracked their expenses for three months and categorized the expenses. Find the average monthly expenses for each category and for each month.

	Category	January	February	March	Average
18.	Housing	$625	$625	$625	
19.	Utilities	$220.83	$183.56	$256.78	
20.	Insurance	$380	$380	$640	
21.	Transportation	$176.83	$224.36	$186.22	
22.	Food	$453.62	$402.36	$410.86	
23.	Clothing	$22.18	$126.83	$0	
24.	Entertainment	$86.74	$126.83	$74.18	
25.	Savings	$250	$250	$250	
26.	Health Care	$125	$140	$62.85	
27.	Miscellaneous	$259.80	$141.06	$94.11	
28.	TOTAL				

Copy and complete the table.

	Type of Insurance	Premium	Frequency	Monthly expense
29.	Auto	$365	Semi-Annual	
30.	Life	$750	Quarterly	
31.	Homeowners	$1,200	Annual	
32.	Health	$5,220	Annual	
33.	Renters	$450	Semi-Annual	

34. **CRITICAL THINKING** Why is it important to track your spending?

35. **FINANCIAL DECISION MAKING** Your auto insurance has an annual premium of $500, or you can pay 12 monthly payments of $45. Is it cheaper to make one annual payment or the 12 monthly payments? Which option would you choose and why?

Mixed Review

36. $\frac{7}{10} \times \frac{25}{14}$

37. $\frac{4}{5} \times \frac{15}{16}$

38. $11 - 6\frac{2}{3}$

39. What is $\frac{2}{3}$ of $456.21?

40. Rewrite 0.675 as a fraction.

41. Lavina McLean insured her home for $116,000 at an annual rate of $0.42 per $100. Find her annual premium.

42. Mickey Fantini plans to buy a home for $135,000 with a 10% down payment. He estimates his closing costs as: loan origination fee, $250; legal fees, $650; property survey, $225; inspection fee, $300; title insurance, $175; recording fee $275. Find these amounts: down payment, mortgage loan, and cash needed to buy the home.

43. Find the interest paid for 6 months on $1,500 at 3.2% annual interest, compounded quarterly.

44. Curtis Vanover has disability insurance that pays him 55% of the average of his last three year's annual earnings. His earnings for the past three years were: $28,500, $35,100, and $34,800. How much will Curtis receive from his disability policy in one year?

45. Brandon earns $13.50 an hour. He is paid time-and-a-half for any hours beyond 40 hours per week. What is Brandon's gross wages for a week that he works 57 hours?

46. Bea Rosenthal is single with a taxable income of $23,616. Her employer withheld $3,796 from her wages for income tax during the year. Using tax tables in Lesson 2-2, find how much her refund should be.

8-2

Creating a Budget

GOALS
- Calculate the percent of income spent on expenses
- Prepare a budget

KEY TERM
- budget

Start Up ▶▶▶

Choose a club or organization in which you participate. What types of income and expenses are typical of that club? What are the actual or estimated annual amounts for each income and payment type? Make a list of the types and amounts of the income and expenses.

Photodisc/Getty Images

Math Skill Builder

Review these math skills and solve the exercises that follow.

1. **Divide** to the nearest hundredth.
   Find the quotient. $12,600 ÷ $56,230 = 0.224, or 0.22

   1a. $10,500 ÷ $35,400 **1b.** $5,200 ÷ $24,700 **1c.** $3,740 ÷ $67,390

2. **Rewrite** decimals as percents.
   Rewrite as a percent. 0.5234 = 52.34%

   2a. 0.67 **2b.** 0.245 **2c.** 0.5698 **2d.** 0.2791

3. **Multiply** and **round** to the nearest dollar.
   $28,000 × 15% = $28,000 × 0.15 = $4,200

   3a. $56,000 × 14% **3b.** $73,400 × 25% **3c.** $29,700 × 12.5%

Evaluate Spending

After you have tracked expenses and calculated average monthly expenses, you can take the next step in the budget process, evaluate spending habits. It is helpful to answer the following questions about your spending:

- Am I spending more than I earn?
- Am I spending too much in a certain area?
- Am I saving at a rate that will meet my financial goals?

When you track your expenses, you may find categories in which you spend more than you realized. After examining spending habits, if you are spending more than you earn or more on a category than you want, you need to adjust your spending habits. If you want to save money for a large purchase, college or other future expenses, you must be saving money at a rate that will enable you to meet your goals.

Another way to evaluate your spending habits is to evaluate the percent of income spent in each category. Many financial advisors suggest the following spending guidelines:

- Save 10–20% of your income in a savings account or other investments.
- Spend no more than 25–30% of your net pay on housing.
- Spend no more than 20% of your net pay on transportation, including car payment, insurance, service and gasoline, tolls, and parking.

EXAMPLE 1

The Tracer family has tracked their expenses and accounted for annual payments. The table represents their average monthly income and spending. Find the percent of income that the Tracers spent on housing. Using the guidelines above, do the Tracers spend a reasonable amount on housing expenses?

Tracer Monthly Spending and Income	
Category	**Monthly Average**
Housing	\$650
Utilities	\$175
Life Insurance	\$280
Transportation	\$550
Food	\$450
Clothing	\$55
Entertainment	\$85
Savings	\$20
Health Care	\$150
Miscellaneous	\$185
Income	\$2,600

SOLUTION

Divide the average monthly housing expense by their monthly income. Round to the nearest whole percent.

$\$650 \div \$2,600 = 0.25 = 25\%$

Using the guidelines above, the Tracers spend a reasonable amount on their monthly housing expenses.

✔ CHECK YOUR UNDERSTANDING

A. Find the percent of income that the Tracers spend on transportation. Based on the guidelines above, is this amount reasonable?

B. Find the percent of income that the Tracers save. Draw a conclusion about the amount the Tracers save.

Budgeting

Budgets are your future spending plans. They help you allocate your future income to meet your future needs and save for the things you want.

A budget should reflect your average monthly spending analysis as well as incorporate any changes that you want to make in your spending based on your evaluation of your spending habits.

A budget may include the categories of your choosing. If you are focusing on reducing spending in a certain area, it may help to make several, very specific categories. However, if your budget includes too many specific categories it will be time-consuming to work with your budget.

EXAMPLE 2

Last year the Tracers did not save much of their income. In the coming year, they want to save 15% of their income. How much should they budget for savings if their expected income is $3,000 per month?

SOLUTION

Rewrite the expense percent as a decimal: $15\% = 0.15$

Multiply by the estimated income: $0.15 \times 3{,}000 = \$450$

The Tracers should budget $450 for savings each month.

✔ CHECK YOUR UNDERSTANDING

C. Tri-county Supply budgets 5% of the previous year's total revenue for utilities. The receipts from last year showed revenue of $4,500,000. How much will be budgeted for utilities for the year?

D. The news reports that next year, gasoline prices will increase. Mr. Jones knows that he needs to increase the auto budget to 15% of his monthly income to cover the increase. Mr. Jones' gross pay each month is $5,489. What will his monthly auto budget be to the nearest dollar?

To develop a budget you must identify not only your past income and expenses but estimate your expected future income and expenses.

You should also try to plan for income and expenses that cannot be scheduled or accurately predicted, such as income tax refunds and car repairs.

Budgets include *fixed income* and *expenses* and *variable income* and *expenses.* With fixed income and expenses, a set amount of money is allocated each month for that income or expense.

Fixed income may include weekly wages or monthly salary. Fixed expenses may include money set aside to pay the insurance, mortgage, rent, car payment, or a savings plan.

Other expenses, such as telephone bills, transportation expenses, and personal expenses, while they occur regularly, vary in amount. Those expenses are called variable, or *flexible,* expenses.

Calculator *Tip*

On many calculators, you can multiply by a percent directly without rewriting it. Enter the number to be multiplied, $3,000, press the multiplication sign [×], enter the percent 15 and press the percent sign [%]. The answer, $450 will appear in the calculator display.

Budgets can be written to cover a month or a year. Typically a budget for a business would be expressed as a yearly budget, while a family will often find a monthly, or quarterly budget more helpful.

EXAMPLE 3

For the next year, the Tracer's expect to make $3,000 per month. Based on their past spending and their future goals, the Tracers have identified the percentage of their monthly income that they would like to budget for each category of expenses. How much money have they budgeted monthly for housing?

Tracer Family Budget	
Category	**Percent of Income**
Housing	25%
Utilities	5%
Life Insurance	10%
Transportation	15%
Food	15%
Clothing	2%
Entertainment	3%
Savings	15%
Health Care	5%
Miscellaneous	10%
Income	$3,000

SOLUTION

Rewrite the expense percent as a decimal: 25% = 0.25

Multiply the decimal by the income:
0.25 × $3,000 = $750

The Tracers have budgeted $750 per month for housing.

✔ CHECK YOUR UNDERSTANDING

E. Find the amount of money budgeted per month for clothing.

F. Find the amount of money budgeted per month for food.

Wrap Up ▶▶▶

Look at the list you prepared for the school club or organization at the start of this lesson. Assume that the club will have a 5% growth in total income. Prepare a budget for the next school year for the club. Allocate amounts to each budget item as you wish but make sure that the budgeted total does not exceed 105% of last year's total.

Exercises

Divide to the nearest hundredth.

1. $6,750 ÷ $36,250
2. $15,730 ÷ $44,900

Rewrite decimals as percents.

3. 0.39
4. 0.641
5. 0.705
6. 1.089

Multiply and round to the nearest dollar.

7. $36,500 × 12%
8. $87,400 × 14.7%

For 9–12, use the data from the Tracer family's monthly spending and income.

9. To the nearest whole percent, what percent of the Tracer's monthly spending was for utilities?
10. To the nearest whole percent, what percent of the Tracer's monthly spending was for life insurance?
11. To the nearest whole percent, what percent of the Tracer's monthly spending was spent on clothing?
12. To the nearest whole percent, what percent of the Tracer's monthly spending was for health care?

13. **STRETCHING YOUR SKILLS** Last year Tylon Company's expenses totaled $951,600. If $677,446 was spent for salaries, what percent of the total expenses was salary expense, to the nearest whole percent?

14. **STRETCHING YOUR SKILLS** Relco Internet Services, Inc., had total sales for one year of $1,891,720. Their advertising expenses were $114,740. Find the percent of total sales that advertising expenses were to the nearest tenth of a percent.

Use the data from the Tracer family budget to answer Exercises 15–18.
Based on their new monthly income included in their budget, find the amount of money budgeted for each category of spending.

15. Transportation
16. Utilities
17. Entertainment
18. Miscellaneous

19. **FINANCIAL DECISION MAKING** Suppose the Tracers decide they want to adjust their budget so they can save for a special summer vacation that will cost $2,000. How would you suggest they go about revising their budget to meet their goal?

20. **CRITICAL THINKING** Why should items not usually paid monthly, such as insurance and taxes, be included in a monthly budget?

21. **CRITICAL THINKING** Tyrone's net annual income is $28,000 a year. On his 25th birthday, he spent $1,500 to celebrate with friends and family. Rosita makes $200,000 a year. When she celebrated her 50th birthday, she spent $5,000 for her party. Which one of them do you think overspent for the party? Explain your answer.

Mixed Review

22. Multiply $3\frac{1}{3}$ by $2\frac{1}{6}$.

23. Subtract $3\frac{1}{2}$ from $7\frac{1}{8}$.

24. $2\frac{3}{4} + 6\frac{1}{2} + 3\frac{2}{3}$

25. $3\frac{3}{5} \div \frac{2}{3}$

26. 344 increased by 20% of itself is what number?

27. $324 increased by $\frac{1}{2}$ of itself equals what amount?

28. $350 is what percent of $500?

29. Regis Computer Supply, Inc., had total sales of $2,975,000 last year and spent $1,950,000 on salaries and wages. What percent, to the nearest percent, of total sales was the amount they spent on salaries and wages?

30. Regina Wilson spent the following amounts for transportation during the month of July, August, and September: July: $250 car payment, $55 gas, $55 insurance, and $100 repair; August: $250 car payment, $58 gas, and $55 insurance; September: $250 car payment, $75 gas, $55 insurance, and $40 repair. What is Regina's average monthly expense for transportation?

Photodisc/Getty Images

8-3

Best Buys

GOALS

- Find and compare unit costs
- Calculate savings
- Compare the cost of rental options
- Compare the cost of renting vs. buying

KEY TERM

- unit price

Start Up

You can buy a set of irons (golf clubs) for $89.99 at a discount store in a nearby city or for $98.88 at a local sporting goods store. Why might you choose to buy the irons locally, even at a higher cost?

SNEHIT 2008/Shutterstock.com

Math Skill Builder

Review these math skills and solve the exercises that follow.

1. **Multiply** dollar amounts by whole numbers and decimals.
   Find the product. $12 \times \$0.95 = \$11.40$

   1a. $\$1.23 \times 144$ **1b.** $\$45.99 \times 3$ **1c.** $\$1.89 \times 5.2$

2. **Subtract** dollar amounts.
   Find the difference. $\$45.99 - \$39.89 = \$6.10$

   2a. $\$1.49 - \$1.28$ **2b.** $\$25.89 - \$19.88$

3. **Divide** dollar amounts by whole numbers and decimals, and round the quotients to the nearest cent.
   Find the quotient. $\$2.98 \div 5.2 = \$0.573$, or $0.57

   3a. $\$3.89 \div 5$ **3b.** $\$12.79 \div 6.2$

Unit Price Comparisons

Many items are packaged in ways that make it difficult to compare them easily with competing brands. For example, one product may be packaged in an 8-oz jar while another brand uses a 12-oz jar.

To help shoppers compare the costs of products, many stores post unit prices on their shelves. The **unit price** is the price of one item or one measure of the item. It may be an ounce, a pound, a quart, a dozen, a hundred feet, or some other measure. If unit prices are not posted, you have to calculate them to compare the costs.

EXAMPLE 1

Gelo toothpaste costs \$1.28 for a 6-oz tube. GloWite toothpaste costs \$1.99 for a 4.6-oz tube. Which brand costs more per ounce? How much more?

SOLUTION

Divide Gelo price by 6 to find price per ounce:
$\$1.28 \div 6 = \$0.213$

Divide GloWite price by 4.6 to find price per ounce:
$\$1.99 \div 4.6 = \$0.433$

Subtract the per-ounce cost of Gelo from the per-ounce cost of GloWite.

$\$0.433 - \$0.213 = \$0.22$ difference in cost per ounce

Dividing prices (\$1.99) by units (4.6) may not result in an even number of cents. To compare the unit price of products, it is usually sufficient to round off the unit price to the nearest tenth of a cent (\$0.433).

Math *Tip*

1 lb = 16 oz
1 T = 2,000 lb
1 ft = 12 in.
1 yd = 3 ft
1 qt = 2 pt
1 gal = 4 qt

✔ CHECK YOUR UNDERSTANDING

A. A store sells an 18-oz. box of National Mills corn flakes for \$3.29 and a 24-oz. box of Keller's corn flakes for \$3.99. Which box sells for less per ounce? How much less per ounce?

B. You can buy a package of four D batteries for \$5.99 or a package of two D batteries for \$3.49. Which package costs less per battery? How much less per battery?

Calculating Savings

To save money, you should know how to calculate the amount you can save by buying in large amounts or at discount stores. Many of the items you buy come in different sizes and at different prices.

EXAMPLE 2

How much will you save by buying 12 rolls of film now at \$3.98 rather than buying 1 roll at \$4.39 at a time?

SOLUTION

Multiply the quantity by the unit price to find the total price of buying the specified number of products.

$12 \times \$3.98 = \$47.76$ cost at \$3.98 each

$12 \times \$4.39 = \$52.68$ cost at \$4.39 each

Subtract the lower total price from the higher total price.

$\$52.68 - \$47.76 = \$4.92$

Amount saved by buying the larger quantity is \$4.92.

Photodisc/Getty Images

✔CHECK YOUR UNDERSTANDING

C. A 53.7 square foot roll of Country paper towels costs $1.49. An 80.6 square foot roll of the same paper towels costs $1.89 on sale. Which is less expensive per square foot, the small or large roll?

D. After the holidays you can buy boxes of greeting cards @ 2 for $32. The same boxes sold during the season for $23.99 each. If you bought 6 boxes after the holidays, how much would you save?

Rental Options

You can rent items you do not use often or only need for a short amount of time. For example, since a homeowner is not likely to refinish the floors in a home very often, renting a floor sander instead of buying it makes sense.

Rental companies offer different rental rates for different time periods, such as hourly, daily, weekly, and monthly. To determine the best rental option, you should know how long you will need the item and then find the most economical rate.

EXAMPLE 3

A floor sander can be rented for $7 an hour or $40 a full day. If you estimate that it will take you 6 hours to sand the floors in two rooms, which rate would be the least expensive?

SOLUTION

Calculate the cost of renting the item by the hourly rate.

$6 \times \$7 = \$42$ hourly rate total

Find the amount saved by renting at the least expensive rate.

$\$42 - \$40 = \$2$ amount saved by renting at the full-day rate

In this case, it is more economical to choose the day rate.

✔CHECK YOUR UNDERSTANDING

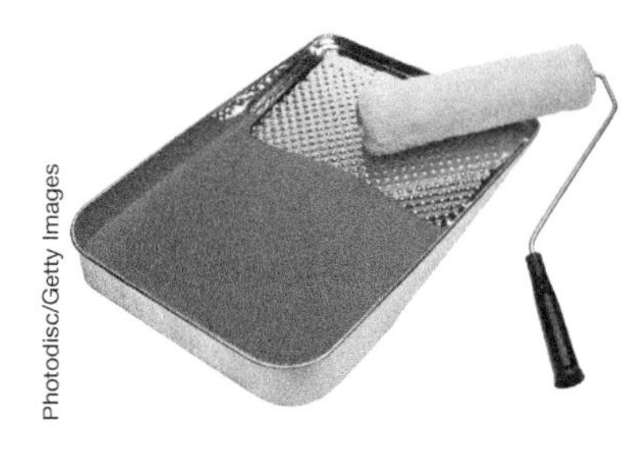
Photodisc/Getty Images

E. Cary wants to rent a large screen video projection system to see games 3, 4, and 5 of the World Series. The projection system rental price by the day is $130 and by the week, it is $600. If the three games are played over 4 days, which rental rate will be the least expensive for him?

F. Jill Sun plans to paint her garage and estimates that it will take about 5 hours to do it with a paint sprayer. She can rent a sprayer at an hourly rate of $18, at a half-day (8 hours) rate of $56, or a full-day (24 hours) rate of $80. Which is the least expensive rental rate for the task?

Photodisc/Getty Images

Rent or Buy

If you need something on a regular basis, it may be less expensive to buy than rent. To make this decision, you need to:

- Calculate the annual cost of renting the equipment.
- Calculate how many years it would take for the rental price to equal or exceed the purchase price.

Business *Tip*

There are other factors to consider in any decision to rent or buy. In this case, you may want to consider purchasing the rug cleaner because in about two years you would "break even" on the purchase. But, if you don't have room to store the cleaner or if the machine is likely to break down after 1 or 2 years, you may decide to continue renting.

EXAMPLE 4

You rent a rug-cleaning machine four days a year at a cost of \$25 a day. You see a rug cleaner on sale for \$225. Do you think you should buy the rug cleaner or continue to rent?

SOLUTION

$4 \times \$25 = \$100$ cost of renting the cleaner each year

Divide the cleaner's purchase price by the annual rental cost to find the number of years it would take for the cost of the rental to equal or exceed the cost of the purchase.

$\$225 \div 100 = 2.25$ years to cover purchase price

✔ CHECK YOUR UNDERSTANDING

G. A backpack leaf blower rents for \$25 a day. The same leaf blower sells new for \$169.99. Meka Jackson thinks she will use the leaf blower for 5 days a year to clean her garage of dust and dirt and her yard of leaves. How many days of renting, to the nearest tenth of a day, will it take for the rental cost to equal or exceed the purchase price?

H. Joe Fiorelli rents a 40-foot extension ladder for a total of 4 days each year to clean the windows in his home in the spring and fall. The daily rental for the ladder is \$23.99. The same ladder costs \$349.99 new. How many years, to the nearest tenth of a year, would it take for the rental charges to equal or exceed the cost of buying the ladder?

Wrap Up ▸▸▸

You might buy the set of irons locally because it is convenient to do so and it will save you time and gas money. The quality of the clubs in the local store might be better than in the discount store. You may also need advice about the product and you think that you will get better service locally than at a discount store. You may also want to support local businesses.

TEAM Meeting

With a team of one or two other students, develop a comparison of rental prices in your area for a home or garden product. For example, you might compare rental prices on floor waxers or garden tillers. Your comparison should describe the stores visited, the differing features of the product found at each store, the different rental terms offered, and the prices charged. Terms might include time of day, length of rental, and whether supplies are included in the rental price. Prepare a presentation for the class.

Exercises

Find the product.

1. $4.69 × 8
2. $5.39 × 5.5
3. $12.77 × 12

Find the difference.

4. $45.29 − $38.39
5. $139.26 − $108.66

Find the quotient, rounded to the nearest cent.

6. $2.59 ÷ 3.8
7. $83.66 ÷ 8

Find the unit price. Round up to the next higher cent.

8. 4 batteries for $5.99
9. 3 rolls of film for $12.99
10. 3 large cans of dog food for $2.00
11. 5 cans of tuna for $2.19
12. 16-oz box of dog treats for $1.79
13. 5 suit hangers for $2.29

Solve.

14. One radial tire sells for $89.99. A set of 4 is priced at $305.96. How much would you save if you bought a set of 4 now instead of one tire at a time over the next few months?
15. A digital camera is priced at $369.89 at a department store. The same camera can be bought for $310.88 plus $12.50 shipping and handling over the Internet. How much would you save by buying the camera over the Internet?
16. **BEST BUY** A 13-oz. dispenser of Coral moisturizing lotion sells for $7.29. A 12-oz. dispenser of Dino moisturizing lotion sells for $6.99. Which is less expensive? How much less expensive per ounce?
17. **BEST BUY** You can buy a bottle of 250 Vitamin C pills for $11.99 or a bottle of 100 Vitamin C pills of a competing brand for $5.99. Which cost less, the larger or the smaller bottle? How much less per pill?
18. Tricia Lanley can buy DVD movies at the regular price of $14.95 each, or at a sale price of 5 for $66.95. How much would she save by buying five movies now instead of one at a time?
19. At an end of season sale, Vicky Charles bought a room air conditioner that was reduced from $398.99 to $349.99 and a fan that was reduced from $39.95 to $32.99. What total amount did she save?

Randy Wilson tills his garden once each year. He rents a garden tiller for one day at $57 per day to do the job. He pays 5% tax on the rental price and uses 4 gallons of gas at $1.65 per gallon.

Photodisc/Getty Images

20. What is his total cost for using the tiller?
21. If a similar tiller costs $346.99 including tax, for how many whole days could Randy rent it before renting would cost more than buying?

A $1\frac{1}{4}$ lb box of Brand A cereal sells for $4.39.

A $1\frac{1}{2}$ lb box of Brand B cereal sells for $4.69.

22. What is the difference in the price per pound?
23. Which brand costs less per pound?

BEST BUY A notebook computer can be rented for $105.95 a week or $29.89 a day. You need the computer for 4 days.

24. At which rate, daily or weekly, would it be cheaper to rent?
25. How much would you save by renting it at that rate?
26. If a similar computer costs $1,997 to buy, for how many full weeks could you rent the computer at the weekly rate before renting would cost more than buying?
27. **FINANCIAL DECISION MAKING** A friend sees an ad that would allow him to rent a large screen TV for $20 a week, for 152 weeks. At the end of the rental period, he would own the TV. He could buy the same TV at the store for $2,000. He thinks the rent-to-own price is a good buy. You don't. What are two comparisons between renting and buying the TV that you can make to support your position?

Mixed Review

28. Find the average of 13, 15, 14, 19, and 20.
29. Subtract $\frac{1}{4}$ from $\frac{5}{8}$.
30. How many days are there between July 12 and September 23?
31. Billy Pinkus worked a total of 50 hours in one week. Of that time, 40 hours was at the regular rate of $11.25 an hour and 10 hours was at time-and-a-half for overtime. What was Billy's gross pay for the week?
32. Josie Lamas earns an annual wage of $63,470. Josie estimates her benefits at 32% of her wages. She also estimates that her job expenses are insurance, 6% of wages; commuting, $438; dues, $175; and other, $225. Find her annual net job benefits.
33. On October 1, Maria Moya's bank statement balance was $371.03, and her check register balance was $307.65. While comparing the statement and her check register, she found a service charge of $4.23; earned interest of $0.25; and outstanding checks for $17.87, $2.97, $45.33, and $1.19. Prepare a reconciliation statement for Maria.

8-4

Optional Personal Expenses

GOALS

- Calculate and compare the costs of connecting to the Internet
- Calculate and compare the cost of wireless phone service
- Calculate and compare the cost of expanded television service

KEY TERMS

- home coverage area
- roaming charges
- airtime
- Internet Service Provider
- access fees
- cable television
- satellite television

Start Up ▸▸▸

Enrico Williams has just moved from his parents' home into his first apartment. When he lived at home, his parents provided a cell phone, Internet access for his computer and cable TV. Now that he is on his own, he must make decisions about which of these expenses he can afford. What should he consider to make a decision?

© RTimages 2008/Shutterstock.com

Math Skill Builder

Review these math skills and then answer the questions that follow.

1. **Add** money amounts.
   Find the sum. $135.23 + $29.99 + $18.50 = $183.72

   1a. $277.13 + $15.32 + $20.86 + $19.11

   1b. $43.07 + $13.26 + $81.68 + $42.19

2. **Subtract** money amounts.
   Find the difference. $289.39 − $176.34 = $113.05

   2a. $77.12 − $52.50

   2b. $153.07 − $136.22

3. **Multiply** money amounts by whole numbers.
   Find the product. $24.99 × 12 = $299.88

   3a. $19.99 × 12

   3b. $49.89 × 12

4. **Multiply** money amounts by percents.
   Find the product. $285.36 × 12% = $285.36 × 0.12 = $34.24

   4a. $127.83 × 15%

   4b. $723.82 × 11%

Math *Tip*

To multiply a number by a percent, change the percent to a decimal by moving the decimal point two places to the left.

Wireless Phone Service

Wireless phone service, or *cell phone* service, is provided by many different carriers, offering different service plans, or fee schedules. Some carriers require you to sign a contract for a year or more. If you cancel your contract, you may pay a cancellation fee. The table below shows the features of several wireless service plans.

Airtime is counted from the time you press the SEND button until you press the END button.

Most wireless carriers in the U.S. have a **home coverage area** that includes the entire country, and selected regions of Mexico and Canada. These carriers offer nationwide calling without long distance charges. If you travel outside the service region, you will pay **roaming charges** to place and receive calls. Roaming charges are higher than charges for calls made within your home coverage area.

Most carriers offer service plans with a minimum number of calling minutes included in a flat monthly rate. If you use minutes beyond the included minutes, you are charged a rate per minute, which is higher than the average per minute rate of the service plan.

The minutes you spend on a cell phone are called **airtime**. Many service plans split airtime into *peak* hours and *off-peak* hours. Peak hours are typically during the business day, Monday through Friday. Off-peak hours are evenings and weekends.

Some carriers charge a one-time activation fee to start your service plan. Many carriers offer discounted equipment and other services, such as text messaging, email, Internet access, and digital cameras. You should shop all carriers in your area to find the deal that fits your needs best. All carriers add federal, state, county, or city taxes to your bill.

Business *Tip*

Many carriers do not charge you airtime for calling another cell phone that uses the same carrier as you. These are called mobile-to-mobile calls.

Individual Cell Phone Plans				
	Telco	**Wyrless**	**Loadstar**	**Vega**
Activation Fee	$35	$0	$45	$15
Basic Monthly Rate	$39.99	$40	$49.99	$69.99
Minimum Phone Price	$0	$109	$0	$0
Peak Minutes	450	Unlimited	750	450
Off-Peak Minutes	5,000	Unlimited	Unlimited	Unlimited
Each Extra Minute	$0.45	$0	$0.40	$0.25
Roaming Charge per minute	$0	$0.79	$0.59	$0
Long Distance per minute	$0	$0	$0.40 when roaming	$0
Contract Period	2 years	None	2 years	2 years
Cancellation Fee	$200	$0	$20 for each remaining month	$150

EXAMPLE 1

Juanita Callara got her first cell phone from Loadstar. Juanita used 850 minutes of peak airtime during February. Of those minutes, 128 minutes were made outside her home coverage area, and 80 minutes of those roaming minutes were for long distance calls. Taxes and other charges were 15% of the total airtime charges. How much was Juanita's phone bill for February?

SOLUTION

$850 - 750 = 100$ number of extra minutes used

$100 \times \$0.40 = \$40$ charge for extra minutes

$128 \times \$0.59 = \$75.52$ roaming charge

$80 \times \$0.40 = \$32$ long distance charges

$\$49.99 + \$40 + \$75.52 + \$32 = \$197.51$ total airtime charges

$\$197.51 \times 0.15 = \$29.63$ taxes and other charges

$\$197.51 + \$29.63 = \$227.14$ February phone bill

Math *Tip*

Notice that cell phone charges are additive. Extra minutes that are used out of the home coverage area for long distance are charged at a rate of $0.40 + $0.59 + $0.40 = $1.39 per minute from Loadstar.

✔ CHECK YOUR UNDERSTANDING

Use the individual cell phone plan table to solve problems A and B.

A. Farshid Meguid's company provides him with a cell phone and service plan from Telco. During May, he used the phone for 500 peak minutes. Taxes and other charges were $12.50. What was Farshid's Telco phone bill for May?

B. Telron, Inc. buys a service plan for 3 cell phones from Wyrless. There is no charge for activation, but each of the phones cost $109. The 3 phones each used 250 minutes of roaming minutes. Taxes and other charges were 12%. What was Telron's total phone bill from Wyrless for the first month?

Internet Connection

To connect your home computer system to the Internet, you must open an account with an **Internet Service Provider**, or ISP. This is a company that provides access to the Internet in your area.

ISPs charge a variety of fees, including installation or set-up fees, fees for special equipment, and monthly **access fees**.

ISPs may offer different ways to connect to the Internet. One is a *dial-up* connection where a modem is used and the connection is through your telephone system. Faster access speeds are available using a *cable* connection, a *digital subscriber line (DSL) connection*, a *satellite connection* or a *wireless connection*.

You should protect your computer from viruses and from *hackers*. To do so, many people use antivirus programs and *firewall* hardware and/or software.

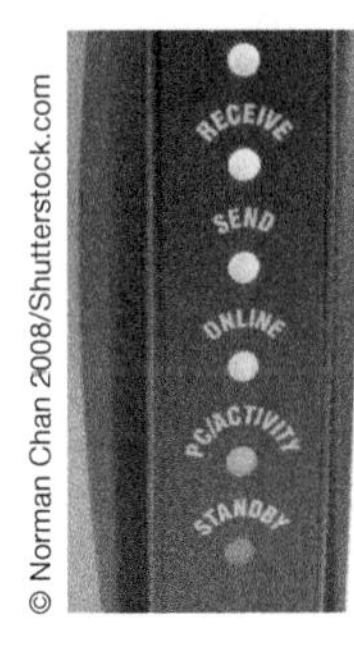

Firewalls protect your system from unauthorized access by outside sources. The cost of the antivirus and firewall software usually includes free updates for a subscription period. You need to update the software regularly because new viruses are created every day.

EXAMPLE 2

TeleMet is an ISP that offers Clara Figueroa a DSL connection to the Internet. TeleMet charges an installation fee of $15 and an access fee of $12 a month. Clara also pays $29.95 a year to another firm for antivirus software and updates. What will Clara's total cost be for Internet connection for the first year?

SOLUTION

$12 \times \$12 = \$144$ Multiply the monthly fee by 12 to find yearly access fee.

Add the yearly access fees to the installation charge and other costs to find the total annual cost of the connection.

$\$144 + \$15 + \$29.95 = \$188.95$ total cost of the connection

✔CHECK YOUR UNDERSTANDING

C. Sal Bonacci paid for a cable connection to AreaNET, an ISP. The ISP charged a $25 installation fee, $75 for a network connection card for his computer, a monthly rental fee of $5 for a cable modem, and a monthly online access fee of $39.95 for an unlimited connection. Sal also bought antivirus software for $36.99. What will Sal's total cost be to connect to the Internet for the first year?

D. Townetwork, an ISP, offers you free installation of a DSL connection and unlimited connection to the Internet. Access fees are $40 per month, $450 for a whole year, or $875 for 2 years. How much will you save by paying a 2-year access fee instead of the monthly access fee?

Business Tip

The speed at which you access the Internet is usually stated in *thousands* or *millions* of bits per second, or kbps and mbps. Two download speeds commonly available for each type of home Internet connection include:
Dial-up 3.3 kbps, 56 kbps
Cable 128 kbps, 256 kbps
DSL 144 kbps, 768 kbps
Satellite 150 kbps, 500 kbps

Expanded Television Service

Local television programming can be received free with a television and an antenna. Expanded television programming requires a paid service such as cable or satellite.

Cable television provides television service by radio frequency signals transmitted through cables. **Satellite television** delivers television service via communication satellites.

Cable and satellite television providers usually charge a monthly fee depending on the plan that you choose. Typically, the more stations you receive, the higher the monthly costs. In addition, there may be set-up or installation fees, as well as equipment fees.

EXAMPLE 3

Melissa Flint wants expanded television service. The cable company that serves her area charges $52.50 per month with free installation. A satellite company charges $37.99 per month and a $49.99 set-up fee. For the first year, which option is less expensive? How much less?

SOLUTION

$\$52.50 \times 12 = \$630$ yearly cost for cable

$\$37.99 \times 12 = \$455.88$ yearly subscription rate for satellite

$\$455.88 + \$49.99 = \$505.87$ total first year cost for satellite

$\$630 - \$505.87 = \$124.13$

Satellite television is $124.13 less for the first year.

✔CHECK YOUR UNDERSTANDING

E. After the first year, which option, cable or satellite, will be less expensive for Melissa? How much less?

F. A satellite television provider charges $29.99 per month for service. If you sign a 2-year contract, they will eliminate the $59.99 set-up fee and charge only $24.99 per month. How much money will you save by committing to a 2-year contract?

Wrap Up ▶▶▶

Now that Enrico is living on his own, he needs to consider what he can afford. If Enrico has a roommate, some of these expenses, such as Internet or expanded television service, could be shared. He must create a budget and evaluate whether he earns enough money to pay for these optional personal expenses.

Exercises

Find the sum.

1. $\$89.23 + \$1.50 + \$15.99$
2. $\$34.99 + \$56 + \$12.45$

Find the difference.

3. $\$458 - \$150$
4. $\$217 - \$175$
5. $\$239.28 - \$69.52$
6. $\$482.12 - \$178.43$

Find the product.

7. $\$0.25 \times 26$
8. $\$12.99 \times 12$
9. $\$24.89 \times 12\%$
10. $\$18.55 \times 8\%$

Vincent Trucano can get an ISP account with WorldVu. The standard plan costs $9.99 per month and the premiere plan costs $14.99 a month. Taxes are 10%.

11. What is the annual cost of the standard plan for the first year?
12. What is the annual cost of the premiere plan for the first year?

Use the Individual Cell Phone Plans table given to solve Exercises 13–16.

13. Yale MaGoo's company lets him use a cell phone and service plan from Loadstar. During October he used 825 peak minutes. Of those minutes, 112 minutes were out of his home coverage area and he used 55 minutes of long distance while he was roaming. Taxes and other charges were 13% of his basic monthly rate and airtime charges. What was Yale's Loadstar phone bill for October?

14. What would Yale's (see Exercise 13) October phone bill be if his carrier was Wryless instead of Loadstar and his taxes and other charges were still 13%?

15. What would Yale's (see Exercise 13) October phone bill be if his carrier was Vega instead of Loadstar and his taxes and other charges were still 13%?

16. Biutta Corporation cancels its service plan for 12 cell phones from Loadstar. There were 4 months to go on the plans. What cancellation fee will Loadstar charge Biutta?

An ISP sells Sara James an unlimited Internet connection for a $39.99 monthly access fee, rental of a cable modem for $3.50 a month, and an installation fee of $125. Sara also spends $19.95 for antivirus software and $29.99 for firewall software.

17. What is the monthly cost of the connection?

18. What is the total of the other connection costs?

19. What is the total cost for the first year?

20. **BEST BUY** Ben Morganstein could pay an ISP $27 for a set-up fee for a dial-up Internet connection. He also would need to pay his telephone company a $29 activation fee for a second phone line for the dial-up connection. His monthly fees would include $15.89 for the phone line and $19.95 for an unlimited Internet connection. His cable company offers him free installation, and a $45 monthly access fee for an unlimited Internet connection that includes rental of the cable modem. Which plan would provide the least total cost?

The cable company that serves the area in which Barry Convers lives charges $29.99 per month with free installation. A satellite company charges $24.99 per month and a $49.99 set-up fee.

21. For the first year, which option is less expensive? How much less?

22. Which option is less expensive after the first year?

Mixed Review

23. Round $0.9607 to the nearest cent

24. Round $10.3049 to the nearest cent

25. Suroyo Wahyudi deposited $3,500 in a money market account that pays interest compounded quarterly. For the first 3 months, the account paid 4.3% annual interest. For the next 3 months, an annual interest rate of 4% was paid. What total interest did Suroyo earn for the six months?

8-5

Adjusting a Budget

GOALS
- Transfer budgeted money between categories
- Adjust a budget after cutting expenses
- Account for changes in income

KEY TERMS
- over-budget
- under-budget

Start Up

After tracking expenses for several months, Dai created a budget. Several months have passed and Dai has spent more than she has budgeted each month. What should she do?

Math Skill Builder

Review these math skills and answer the questions that follow.

1. **Add** money amounts.
   Find the sum. $50 + $250 + $85 = $385

   1a. $75 + $185 + $25 **1b.** $155 + $125 + $750

2. **Subtract** money amounts from money amounts.
   Find the difference. $585 − $25 = $560

   2a. $354 − $125 **2b.** $875 − $423 **2c.** $623 − $235

3. **Calculate** percents, and round to the nearest percent.
   What percent of $5,000 is $325? $325 ÷ $5,000 = 0.065 = 7%

   3a. What percent of $2,200 is $250?

   3b. What percent of $3,550 is $300?

   3c. What percent of $4,000 is $800?

4. **Multiply** money amounts by percents, and round the product to the nearest whole cent.
   Find the product. $4,255 × 7% = $4,255 × 0.07 = $297.85

   4a. $1,249 × 27%

   4b. $2,876 × 18%

   4c. $3,682 × 5%

Math *Tip*

To rewrite a percent as a decimal, move the decimal point two places to the left and drop the percent sign. To rewrite a decimal as a percent, move the decimal two places to the right and add the percent sign.

Transferring Money Between Budget Categories

After creating and working with a budget for several months, you may find that your spending does not match your budget. You are **over-budget** if you are spending more in a category or in total than was budgeted. You are **under-budget** if you are spending less in a category or in total than was budgeted.

A *balanced budget* is a budget where expenses match revenue. When your spending matches your budget or is under-budget, but one or more categories are over-budget, you need to re-allocate budgeted money. Re-allocating money means that you transfer projected spending from one category to another category. The result of transfers is revised percentages for categories.

Business Tip

A budget is *in the red* when expenses exceed revenue. A budget is *in the black* when revenue exceeds expenses.

Financial Responsibility

Saving for the future

A common budgeting error is to continue to re-allocate money from the savings category to pay for other budget items. In doing so, you may meet your total budget and not spend more than you make, but no money will be saved. It is estimated that almost half of Americans live like this, paycheck-to-paycheck. What are some strategies that you can use so that you are not living paycheck-to-paycheck?

EXAMPLE 1

After three months, the Tracer family calculated the average amount they actually spent for each category. Identify how they can adjust their budget to reflect their spending, while keeping a balanced budget.

Tracer Family Budget			
Category	**Percent Budgeted**	**Dollars Budgeted**	**3 Month Average**
Housing	25%	$750	$750
Utilities	5%	$150	$200
Life Insurance	10%	$300	$300
Transportation	15%	$450	$500
Food	15%	$450	$500
Clothing	2%	$60	$60
Entertainment	3%	$90	$90
Savings	15%	$450	$450
Health Care	5%	$150	$100
Miscellaneous	5%	$150	$50
Monthly Income	100%	$3,000	$3,000

© Supri Suharjoto 2008/Shutterstock.com

SOLUTION

Identify the categories that were under- or over-budget. Find the total amount under or over budget.

Utilities: $50 over-budget

Transportation: $50 over-budget

Food: $50 over-budget

$50 + $50 +$50 = $150 over-budget

Health care: $50 under-budget

Miscellaneous: $100 under-budget

$50 + $100 = $150 under-budget

Re-allocate the money to each category. Increase budgets for Utilities, Transportation, and Food by $50. Decrease budgets for Miscellaneous by $100 and Health Care by $50. Modify the family's budget to reflect the new dollar amounts for each category.

Tracer Adjusted Family Budget		
Category	Revised Percent Budgeted	Revised Dollars Budgeted
Housing	25%	$750
Utilities*	7%	$200
Life Insurance	10%	$300
Transportation*	17%	$500
Food*	17%	$500
Clothing	2%	$60
Entertainment	3%	$90
Savings	15%	$450
Health Care*	3%	$100
Miscellaneous*	2%	$50
Monthly Income	101%	$3,000

Calculate new budget percentages to the nearest percent.

Utilities: $200 ÷ $3,000 = 0.067 = 7%

Transportation: $500 ÷ $3,000 = 0.167 = 17%

Food: $500 ÷ $3,000 = 0.167 = 17%

Health Care: $100 ÷ $3,000 = 0.033 = 3%

Miscellaneous: $50 ÷ $3,000 = 0.0167 = 2%

*__This category rounded to the nearest percent.__

✔ CHECK YOUR UNDERSTANDING

A. Jim Rhodes evaluates his spending and finds that he is spending about $100 more per month on entertainment than he has budgeted. He can transfer money from other categories to increase his entertainment budget to $200 per month. If his total monthly income is $2,800, to the nearest percent, what percent of his monthly income is budgeted for entertainment?

B. Ravi Aves budgeted $5,000 to donate to charity in a year. If his annual income is $45,000, to the nearest percent, what percent of his income is budgeted for charity?

Cutting Expenses

Many times adjusting a budget requires cutting expenses. If you are spending more money than you make each month, then you need to cut expenses in order to stay within your budget, and avoid creating debt.

Cutting expenses may involve eliminating or spending less on optional personal expenses or other items, such as clothing and entertainment. Another way to cut expenses is to find ways to economize on fuel or grocery bills.

EXAMPLE 2

The Tracer family used their adjusted budget for three months and then evaluated their budget again. They found that due to increasing prices, they were still over-budget in food and transportation costs. They want to maintain the same level of saving, so they plan to cut expenses. After having an energy audit done, they estimate they can cut their utility expenses by $50/mo. They changed their homeowner's insurance policy to cut $25/mo on housing costs. They plan to drive less and make wiser choices in their food shopping to cut transportation and food costs by $25 each. Find the adjusted budget percentages.

Tracer Family Budget		
Category	**Dollars Budgeted**	**3 Month Average**
Housing	$750	$750
Utilities	$200	$200
Life Insurance	$300	$300
Transportation	$500	$575
Food	$500	$550
Clothing	$60	$60
Entertainment	$90	$90
Savings	$450	$450
Health Care	$100	$100
Miscellaneous	$50	$50
Monthly Income	$3,000	$3,125

SOLUTION

Adjust the amounts in each category that are affected, and find the revised budget percentages. Verify that the revised amounts equal the monthly income.

Housing: $750 – $25 = $725 $725 ÷ $3,000 = 0.242 = 24%

Utilities: $200 – $50 = $150 $150 ÷ $3,000 = 0.05 = 5%

Transportation: $575 – $25 = $550 $550 ÷ $3,000 = 0.183 = 18%

Food: $550 – $25 = $525 $525 ÷ $3,000 = 0.175 = 18%

✔CHECK YOUR UNDERSTANDING

C. The Tracers are considering cutting their entertainment budget by $25 and budget $550 per month for food to adjust their budget. Find the adjusted budget percentages for these changes.

D. Giselle budgeted 10% of her $4,000 monthly income for clothing. She must cut expenses and decides to decrease her monthly clothing budget to $250. To the nearest percent, what is her adjusted budget percentage for clothing?

Tracer Family Budget	
Category	**Percent Budgeted**
Housing	25%
Utilities*	7%
Life Insurance	10%
Transportation*	17%
Food*	17%
Clothing	2%
Entertainment	3%
Savings	15%
Health Care*	3%
Miscellaneous*	2%
Monthly Income	101%

Income Changes

A budget will also have to be adjusted as income changes. Revise the dollar amounts for each category to reflect monthly income change. After a few months of spending under the revised budget, you should evaluate spending and make adjustments as needed.

EXAMPLE 3

The Tracer family income increased to $3,600 per month. Using the current budget percentages, how much should they budget per month for food?

SOLUTION

Multiply the new monthly income amount by the percent budgeted for food.

$3,600 × 17% = $3,600 × 0.17 = $612

The Tracers should budget $612 per month for food.

✔CHECK YOUR UNDERSTANDING

E. How much should the Tracers budget for utilities per month?

F. How much should the Tracers budget for food for the year?

Wrap Up ▶▶▶

Dai should look for ways to cut expenses so that she is not spending more than she earns. She should evaluate each category to see if money should be re-allocated to different budget categories.

Exercises

Calculate the percent to the nearest percent.

1. What percent of $2,000 is $120?
2. $450 is what percent of $3,000?
3. What percent of $65,000 is $22,000?
4. $220 is what percent of $1,800?

Find the product.

5. $1,400 × 22%
6. $2,725 × 35%

Use the budget at the right.

7. Identify the categories that the Jacksons are over-budget.
8. Identify the categories that the Jacksons are under-budget.

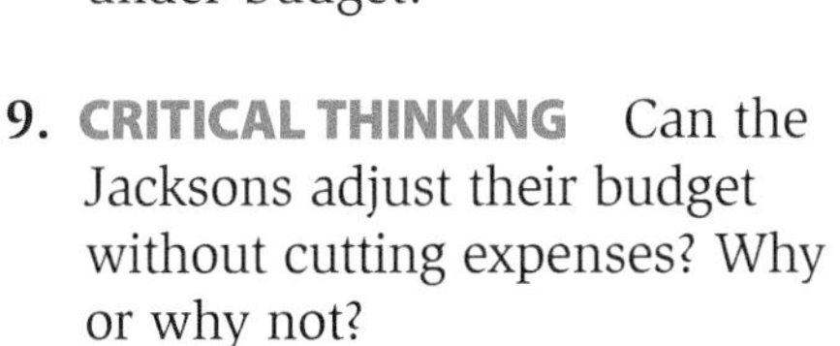

9. **CRITICAL THINKING** Can the Jacksons adjust their budget without cutting expenses? Why or why not?
10. If the Jacksons cut their Transportation budget to $400 per month, to the nearest percent, what percentage of their budget will be for Transportation?
11. If the Jacksons cut expenses and adjust their Miscellaneous budget to $300 per month, to the nearest percent, what percentage of their budget will be for Miscellaneous?

Jackson Family Budget			
Category	Percent Budgeted	Dollars Budgeted	3 Month Average
Housing	30%	$1,350	$1,350
Utilities	15%	$675	$675
Transportation	10%	$450	$400
Food	15%	$675	$650
Clothing	5%	$225	$175
Entertainment	10%	$450	$500
Savings	5%	$225	$150
Health Care	5%	$225	$225
Miscellaneous	5%	$225	$485
Monthly Income	100%	$4,500	$4,610

The Macias have been working with a budget for the last year when their annual net income was $48,000. They are expecting a $300 per month increase in their net earnings. Complete their budget for the coming year.

	Category	Percent Budgeted	Monthly Budget	Yearly Budget
12.	Housing	24%		
13.	Utilities	12%		
14.	Transportation	10%		
15.	Food	18%		
16.	Clothing	5%		
17.	Entertainment	10%		
18.	Savings	10%		
19.	Health Care	6%		
20.	Miscellaneous	5%		
21.	Total	100%		

22. **CRITICAL THINKING** How can you adjust a budget if you experience a decrease in income?

23. **FINANCIAL DECISION MAKING** After working with your budget for several months, you find that you are spending more money than you earn. You spend several months cutting expenses in housing, food, and transportation categories, but you are still not meeting your budget. You have identified the following areas that you can cut expenses: eliminate cell phone, Internet service or cable TV, reduce clothing or entertainment categories or reduce savings. What would you do to cut expenses so you do not spend more than you earn?

Mixed Review

24. Write 87.6% as a decimal

25. 2.9×41.25

26. $4\frac{1}{6} + 3\frac{1}{4}$

27. $\frac{3}{4} \times \frac{4}{7}$

28. $308{,}000 - 109{,}266$

29. $6.4 \div 1{,}000$

30. Find the quotient, to the nearest cent: $612.19 ÷ 11

31. 96 increased by $\frac{1}{4}$ of itself is what number?

32. The county tax rate is 18 mills per $1 of assessed value. Find the tax due on property assessed at $62,500.

33. Marietta and Albert Bilicki, a married couple, have an adjusted gross income of $64,100. They are filing a joint return. Their itemized deductions are $11,450 and they claim six exemptions, one for each spouse and four for their children. What is their taxable income?

34. A satellite television provider charges $39.99 per month for service. If you commit to a 2-year contract, they will waive the $40 set-up fee and charge $30.00 per month. How much money will you save by committing to a 2-year contract?

8-6

Economic Statistics

GOALS

- Interpret consumer price index data
- Calculate rates of inflation
- Calculate the purchasing power of the dollar
- Analyze unemployment data

KEY TERMS

- Consumer Price Index (CPI)
- base period
- inflation
- purchasing power of the dollar
- unemployment rate
- labor force

Start Up

There is one Law of Statistics which states: If the statistics do not support your viewpoint, you obviously need more statistics. What does this statement mean?

Photodisc/Getty Images

Math Skill Builder

Review these math skills and solve the exercises that follow.

1. **Subtract** decimal amounts.
Find the difference. 148.7 − 100.0 = 48.7

1a. 237.8 − 100.0 **1b.** 107.9 − 100.0

1c. 315.4 − 250.0 **1d.** 118.8 − 10.8

2. **Divide** decimals and round to the nearest tenth percent.
Find the percent. 4.8 ÷ 140.5 = 0.0341 = 3.4%

2a. 3.7 ÷ 168.2 **2b.** 8.4 ÷ 191.3

2c. 2.5 ÷ 35 **2d.** 10.2 ÷ 100

3. **Divide** decimals and round to the nearest thousandth.
Find the quotient. 100 ÷ 140.2 = 0.7132 = 0.713

3a. 100 ÷ 118.6 **3b.** 100 ÷ 183.5

3c. 100 ÷ 268.4 **3d.** 100 ÷ 215.5

Consumer Price Index

The **Consumer Price Index (CPI)** is a widely reported measure of how much the prices of goods and services typically bought by consumers have changed when compared to a base period. A **base period** is a period of time with which comparisons are made. The base period for most of the items in the CPI is the 1982–84 period.

The CPI uses a single number, called an index number, to compare price changes to the base period. The index number for the base period is always equal to 100.

The Historical Report of the CPI shows index numbers for various categories of consumer goods and services. The All Items column gives an average number considering all categories and is the number commonly used when referring to the CPI. Note that two categories, Recreation and Education & Communication, were added to the CPI in 1997, so 1997 is their base period.

Historical Report—Consumer Price Index, 1994–2007
Categories of Goods and Services

Years	All Items	Food & Beverages	Housing	Apparel	Transportation	Medical Care	Recreation	Education & Communication	Other
1982–84	100.0	100.0	100.0	100.0	100.0	100.0	—	—	100.0
1994	148.2	147.2	145.4	130.5	137.1	215.3	93.0	90.3	202.4
1995	152.4	150.3	149.7	130.6	139.1	223.8	95.6	93.9	211.1
1996	156.9	156.6	154.0	130.3	145.2	230.6	98.5	97.1	218.7
1997	160.5	159.1	157.7	131.6	143.2	237.1	100.0	100.0	230.1
1998	163.0	162.7	161.3	130.7	140.7	245.2	101.2	100.7	250.3
1999	166.6	165.9	164.8	130.1	148.3	254.2	102.0	102.3	263.0
2000	172.2	170.5	171.9	127.8	154.4	264.8	103.7	103.6	274.0
2001	177.1	175.3	177.6	124.8	149.0	278.3	105.3	106.9	287.0
2002	179.9	177.8	181.1	121.5	154.2	291.3	106.5	109.2	295.8
2003	184.0	184.1	185.1	119.0	154.7	302.1	107.7	110.9	300.2
2004	188.9	188.9	190.7	118.8	164.8	314.9	108.5	112.6	307.8
2005	195.3	193.2	198.3	117.5	172.7	328.4	109.7	115.3	317.3
2006	201.6	197.4	204.8	118.6	175.4	340.1	110.8	118.9	326.7
2007	207.3	206.9	210.9	118.3	190.0	357.7	111.7	121.5	337.6

The CPI may be expressed in several ways. For example, the CPI for "All Items" in 2007 is 207.3. This means that the cost of goods in 2007 was 207.3% of their cost in the base period. The percent increase in prices from the base period to 2007 is 107.3% (207.3 − 100.0). Looking at the relationship in another way, it cost $207.30 in 2007 to buy the same goods for which you would have paid $100 in the base period.

EXAMPLE 1

Use the table above to find the CPI for Housing for 2007 and the percent the 2007 CPI increased from the base period.

SOLUTION

Copy the index number from the box in the table where the year, 2007, and the "Housing" column meet.

210.9   CPI index number for 2007

Subtract the base period index number from the 2007 index number and add a percent sign to the number.

$210.9 - 100.0 = 110.9$ percent CPI increased from base period to 2007

✔ CHECK YOUR UNDERSTANDING

A. Use the CPI table. By what percent did the CPI for Apparel increase from the base period to 1994? to 2007?

B. Refer to the CPI table. Of all the categories whose base period is 1982–84, which one showed the greatest percent increase to 2007? What is the percent increase?

Rate of Inflation

For consumers, business firms, organizations, and the government, **inflation** means that the prices of goods and services they buy are rising. The U.S. Department of Labor publishes the Consumer Price Index report that tells how much inflation has occurred within the past year. A calculation in the report, called the *rate of inflation,* shows the percent increase in prices from the previous year.

EXAMPLE 2

The CPI table shows an index number of 201.6 for 2006 and an index number of 207.3 for 2007 for food and beverages. What was the rate of inflation for 2007, to the nearest tenth percent?

SOLUTION

Find the difference between the two index numbers.

$207.3 - 201.6 = 5.7$

Divide the difference by the index number for the earlier year.

$5.7 \div 201.6 = 0.0282 = 2.8\%$

The rate of inflation of food and beverages for 2007 was 2.8%.

✔ CHECK YOUR UNDERSTANDING

C. Find the rate of inflation for 1999, to the nearest tenth percent.

D. Find the rate of inflation for the Medical Care category for 2002.

Scholarship Scams

Tuition rates for colleges and universities have been increasing at almost double the general inflation rate. That increase is leading more and more students to seek scholarships to help pay for college. Many unscrupulous companies offer promises of scholarships, grants, or fantastic financial aid packages with high pressure sales pitches and upfront fees that must be paid immediately. Legitimate companies do not guarantee scholarships or grants. You can find more information on scholarship scams from the Federal Trade Commission at www.ftc.gov.

Purchasing Power of the Dollar

When inflation occurs, each dollar buys less than it did in the past. The **purchasing power of the dollar** is a measure of how much a dollar now buys compared to what it could buy during some base period. The base period is a time period with which all purchasing power of the dollar comparisons are made.

Suppose a Department of Labor report shows that the purchasing power of a dollar in 2007 was $0.48. In the base period, 1982–1984, the dollar was worth its full value of $1.00. In 2007, the dollar was worth $0.48 compared to the base value. This means that a 2007 dollar could buy only about $0.48 worth of the same goods that could have been bought in 1982–84. The 2007 dollar is worth less because of inflation.

Using the Internet or the library, find what types of goods and services are included in the Education and Communication category of the Consumer Price Index. You can begin your search with any of following keywords. For a more refined search use any of these words joined with the word "and".

- consumer price index
- CPI
- education
- communication
- goods
- services

Identify three types of goods and services within the Education and Communication category whose prices have risen most since the base year.

EXAMPLE 3

Use the CPI table to calculate the purchasing power of the dollar for 2005, to the nearest tenth of a cent.

SOLUTION

Divide the CPI index number for the base period by the CPI index number for the year with which a comparison is made, to the nearest thousandth.

$195.3 = \text{CPI index for 2005}$

$100 = \text{CPI index for base period}$

$100 \div 195.3 = 0.5120 = 0.512$ decimal rate

Multiply the decimal rate by $1.

$\$1 \times 0.512 = \$0.512$ purchasing power of the dollar in 2005

✔ **CHECK YOUR UNDERSTANDING**

E. Use the CPI table to find the purchasing power of the dollar in 2007, to the nearest tenth of a cent.

F. Use the CPI table to find how much a 1997 dollar was worth compared to 1982–84, to the nearest tenth of a cent.

Unemployment Rate

The **unemployment rate** tells the percentage of the total labor force that is not working. The **labor force** consists of all people who are of working age and who either have a job or are looking for a job.

The table at the top of the next page shows the unemployment rate for different persons for one month as estimated by the U.S. Department of Labor.

Math *Tip*

One meaning of the term "rate" is percent.

EXAMPLE 4

Which workers shown in the table above had the highest rate of unemployment? What was the rate?

SOLUTION

Locate the highest rate in the Unemployment Rate column: 16.8%.

Teen workers had the highest unemployment rate.

April 2008 Unemployment Rates by Age, Sex, and Race

Worker Classification	Unemployment Rate
All	4.9
Teen	16.8
Men	4.4
Women	4.3
White	4.4
Black or African American	8.8
Hispanic or Latin ethnicity	6.5

✔ **CHECK YOUR UNDERSTANDING**

G. Refer to the table above to find the workers with the lowest unemployment rate.

H. What was the difference in the unemployment rate of male and female workers in data shown?

Wrap Up ▶▶▶

The Law of Statistics tells consumers to be wary of the statistics they hear or read. Almost any viewpoint can be supported by statistics. Individuals, business firms, labor unions, non-profit organizations, and all levels of government use statistics that support their position and ignore those that do not. To believe the statistics used by others, you need to know how the data were collected and analyzed to determine their truthfulness.

Exercises

Find the difference.

1. 132.9 − 100.0
2. 287.7 − 100.0

Find the percent. Round to the nearest tenth percent.

3. 5.8 ÷ 142.7
4. 2.6 ÷ 115.4

Find the quotient. Round to the nearest thousandth.

5. 100 ÷ 236.3
6. 100 ÷ 107.1

Use the CPI table to solve Exercises 7–15.

7. What was the CPI for Transportation in 1996?
8. By what percent did the CPI for Food and Beverages increase from the base period to 2007?
9. Which category of goods and services showed the smallest price increase between the base period of 1982–84 and 1994?
10. Which 3 categories of consumer goods and services had price increases greater than those of the CPI from the base period to 2007?
11. By what percent did prices for Housing increase from the base period to 2006?
12. What was the rate of inflation in 1995 and 2005, rounded to the nearest tenth percent?
13. Of these categories, Food and Beverages, Housing, and Apparel, which had the highest rate of inflation in 2007? What was the rate, to the nearest tenth percent?
14. During the years 1999–2003, in which year was the highest rate of inflation reported, to the nearest tenth percent? What was the percent?
15. What was the purchasing power of the dollar in 2001 and what amount did it drop from the previous year, to the nearest tenth of a cent?

Math *Tip*

The purchasing power of any year's dollar is calculated from the base period.

By what amount, to the nearest tenth of a cent, did the purchasing power of the dollar drop?

16. From the base period through 1998?
17. From 1994 through 2007?
18. In the Unemployment Rate Table, which gender, male or female, had the highest unemployment rate?
19. **CRITICAL THINKING** The CPI index number for Apparel was 130.5 in 1994 and 118.3 in 2007. What might be some reasons for the decrease in Apparel compared to other categories?

STRETCHING YOUR SKILLS Marjorie earned $12,000 a year and Timothy earned $5 an hour in the base period of 1982–84. Since then they both have received wage increases equal to the increase in the CPI through 2007.

20. Marjorie should have been earning what annual wage?
21. Timothy should have been earning what hourly wage, to the nearest cent?

Photodisc/Getty Images

22. **INTEGRATING YOUR KNOWLEDGE** Calculate the rates of inflation from 1994 through 2007. Make a vertical bar graph to show the inflation rates.

23. **FINANCIAL DECISION MAKING** When you are planning a new budget for your family, how can the CPI influence your decisions?

Mixed Review

24. Rewrite as a fraction: 0.005

25. 7.8% of $265

26. $\frac{3}{4} \div \frac{3}{8}$

27. $1\frac{1}{3}$ of $183.42

28. 10 × $45.18

29. What percent of 7,500 is 150?

30. $\frac{3}{4} + \frac{1}{2}$

31. Round to the nearest cent: $19.9951

32. The Groat family has a fixed-rate mortgage with monthly payments of $682.40. They can refinance their current mortgage with a new loan at a lower interest rate and longer term. The monthly payments on the new loan will be $519.18. To get the new mortgage loan, they had to pay a prepayment penalty of $680 and closing costs of $1,050. How much will they save in the first year by refinancing?

33. Angie Grosbeck's gross pay is $617.50 a week. How much will Angie earn in one quarter of a year at her current pay?

34. A 13.5 oz. size of body lotion sells for $8.67. A 5.4 oz. size of the same lotion sells for $4.39. Which size of lotion costs more per ounce? How much more per ounce, to the nearest tenth of a cent?

35. Cassie Siebert borrowed $5,200 for 24 months to landscape her home. The promissory note carried 14% interest. Find the amount of interest Cassie must pay. Also find the amount she must repay on the due date of the note.

36. The tip income of the seven food servers at a buffet restaurant averages 7% of the food bill. The total of the food bills for one evening was $8,200. How much did each food server earn for the evening?

37. After surgery, Gwen Schoeffler spent 4 days in the hospital. The hospital room cost of $375 a day was covered for only the two-day stay that her insurance company allowed for the type of surgery Gwen had. Other hospital costs related to this surgery were $6,278, of which the insurer approved $5,690. Gwen still had $500 of deductible to use and her policy required 12% coinsurance. What amount did Gwen have to pay for her surgery?

38. Diane was charged a commission of $119.60 to buy 200 shares of Kirby Products stock at $63.50. What was her total investment in the stock?

Chapter Review

Vocabulary Review

Find the term, from the list at the right, that completes each sentence. Use each term only once

- access fees
- average monthly expenses
- airtime
- base period
- audget
- cable television
- Consumer Price Index
- home coverage area
- inflation
- internet service provider
- labor force
- over-budget
- purchasing power of the dollar
- roaming charges
- satellite television
- under-budget
- unemployment rate
- unit price

1. A rise in the price of goods and services is known as __?__.
2. A company that provides access to the Internet is called a(n) __?__.
3. A single number used to measure changes in the prices of goods and services is called the __?__.
4. Future spending plans are called __?__.
5. A statistical measure of how much a dollar will buy is known as the __?__.
6. The minutes you spend calling on a cell phone are called __?__.
7. The __?__ is the price of one item or one measure of an item.
8. The period of time with which comparisons are made is a(n) __?__.

8-1 Average Monthly Expenses

9. If Daja spends an average of $140 per month on gas for her car, how much will she spend in a year on gas?
10. Jeannine spent the following amounts on transportation for 3 months: $330, $285.96, and $553.22. What is her average monthly transportation expense?
11. Driscoll's semi-annual auto insurance premium is $483. How much should he budget each month for auto insurance?

8-2 Creating a Budget

12. Jon Marlow's total income last year was $31,300. Of that amount, he spent $7,900 for food. What percent of Jon's total income did he spend for food?
13. Cesar Guerr's total income last year was $35,000. He expects to earn $4,000 more this year and wants to budget 24% of that income for food. What yearly amount should Cesar budget for food? What monthly amount should he budget for food?
14. What yearly amount should Cesar budget for transportation if it is 17% of his income? What monthly amount, rounded to the nearest dollar?
15. Joanna earns $750 per week. She plans to save 15% of her income. How much will she save per week? How much will she save per year?

8-3 Best Buys

16. Shampoo A costs $3.89 for a 12-ounce bottle. Shampoo B costs $7.29 for a 25-ounce bottle. Which shampoo costs less per ounce? How much less? Round to the nearest tenth of a cent.

17. You can buy 2, 16-oz. jars of pasta sauce on sale for $1.99. Each 16-oz. jar regularly sells for $1.29. How much will you save if you buy 6 jars on sale?

18. A trencher can be rented for $27 an hour or $145 a day. If you need the trencher for 6 hours, how much will you save by renting it for the day instead of for 6 hours?

19. Leslie Wickam can rent an air compressor for $240 a week or buy it for $1,499.99. How many weeks of renting, to the nearest tenth of a week, will it take for the rental cost to equal or exceed the purchase price?

8-4 Optional Personal Expenses

20. Use the Individual Cell Phone Plan table from Lesson 8-4. Mia Ropa has a cell phone with Telco. In the month of March, she used 500 peak minutes and 2,000 off-peak minutes. If taxes were 12% of the total airtime charges, what was her total cell phone bill for March?

21. Val Myer's ISP charged a $20 installation fee, $95 for a network connection card, a monthly rental fee of $3 for a modem, and a monthly access fee of $49.95 for an unlimited connection. In addition, Val bought antivirus and firewall software for $39.98 total. What will be Val's total cost to connect to the Internet for the first year?

22. A satellite television provider charges $19.99 per month for service. If you commit to a 2-year contract, they will waive the $39.99 set-up fee and charge $15.99 per month. How much money will you save by committing to a 2-year contract?

8-5 Adjusting a Budget

23. The Rutladge family finds that they are over-budget $100 in food spending. They increase their $750 food budget by $100 per month, and decrease their $300 entertainment budget by $100. If their monthly income is $5,000 per month, what percent of income is budgeted for food? for entertainment?

24. The Settle family earns $45,000 per year. In the coming year, they will have an 8% increase in their income. If they budget 15% for savings, how much of their new income should they be saving per year? per month?

8-6 Economic Statistics

25. The Medical Care category for 1994 showed a CPI index number of 215.3. By what percent did medical care prices increase from the base period?

26. The CPI index number in 1996 for the Housing category was 154.0. In 1997 the index number for Housing was 157.7. What was the rate of inflation in Housing for 1997, to the nearest tenth percent?

27. The CPI index for 2000 was 172.2. What was the purchasing power of the dollar in 2000, compared to the base period 1982–1984? Round to the nearest tenth of a cent.

28. If the unemployment rate for all workers shown in the Unemployment Rate Table doubled because of an economic slowdown, what would the new unemployment rate be? At the new rate, how many workers would be unemployed out of every one million workers?

Technology Workshop

Task 1 Calculating Inflation Indicators

Enter data into a template that calculates the purchasing power of the dollar and the annual inflation rate. You may use the template to study how the CPI is converted to two common inflation indicators, the purchasing power of the dollar and the rate of inflation.

Open the spreadsheet for Chapter 8 (tech8-1.xls) and enter the data shown in blue (cells B5-20) into the spreadsheet. The spreadsheet will calculate the purchasing power of the dollar for each year, the annual change in the purchasing power of the dollar, and the annual inflation rate. Your computer screen should look like the one shown below when you are done.

	A	B	C	D	E
1			INFLATION INDICATORS		
2				Annual Change	Annual
3			Purchasing	In Purchasing	Inflation Rate
4	Year	CPI Index	Power of Dollar	Power of Dollar	Based on CPI
5	1982-84	100.0	$1.000	$0.000	0.0%
6	1985	107.6	0.929	−0.071	7.6%
7	1986	109.6	0.912	−0.017	1.9%
8	1987	113.6	0.880	−0.032	3.6%
9	1988	118.3	0.845	−0.035	4.1%
10	1989	124.0	0.806	−0.039	4.8%
11	1990	130.7	0.765	−0.041	5.4%
12	1991	136.2	0.734	−0.031	4.2%
13	1992	140.3	0.713	−0.021	3.0%
14	1993	144.5	0.692	−0.021	3.0%
15	1994	148.2	0.675	−0.017	2.6%
16	1995	152.4	0.656	−0.019	2.8%
17	1996	156.9	0.637	−0.019	3.0%
18	1997	160.5	0.623	−0.014	2.3%
19	1998	163.0	0.613	−0.010	1.6%
20	1999	166.6	0.600	−0.013	2.2%
21	2000	172.2	0.581	−0.020	3.4%
22	2001	177.1	0.565	−0.016	2.8%
23	2002	179.9	0.556	−0.009	1.6%
24	2003	184.0	0.543	−0.012	2.3%

Task 2 Analyze the Spreadsheet Output

Answer these questions about the inflation indicator calculations.

1. What do the minus signs in front of the output in Column D mean?
2. What was the purchasing power of the dollar in 1992?
3. What was the year with the highest annual inflation rate. What was the rate?
4. In which year did the purchasing power of the dollar drop the most? How much was the drop, in cents?

5. In which year since 1985 was the annual inflation rate the lowest? What was the rate?
6. From 1985 to 1995, in which years was the inflation rate greater than 4.5%?

Now insert rows for the years 2004 to 2007. Insert the CPI index from the table in Lesson 8-6.

Answer these questions.

7. What was the purchasing power of the dollar for 2006?
8. Did the purchasing power of the dollar increase or decrease from 2004 to 2005? By how much did the purchasing power of the dollar change?
9. What was the annual rate of inflation in 2007?
10. Since 2001, in what year did the purchasing power of the dollar change the most? How do you know?

Task 3 Design a Spreadsheet to Graph Sales Data

You are to design a spreadsheet that will use the charting features of your software to create a bar graph. The graph will show monthly sales by product line for a two-year period. The monthly sales for each product line are to be printed side-by-side. Round the sales data to the nearest $500 or a multiple of $500 before entering the data into a worksheet. The sales data and the graph are to appear on separate worksheets.

SITUATION: The Clayton Door & Window Company wants a bar graph that shows current year and previous year sales data for its product line. Sales data follows:

Clayton Door & Window Company
Comparative Sales Data
January 2008 and 2009

	January, 2008	January, 2009
Entry Doors	$21,500	$21,050
Garage Doors	$18,900	$15,150
Security Systems	$24,600	$34,025
Windows	$31,112	$25,236

Task 4 Analyze the Spreadsheet Output

Answer these questions about your completed graph.

11. Which product line's 2009 sales were greater than 2008 sales?
12. Which product line's sales were almost equal in both years?
13. Which product line had the greatest amount of sales in 2008?
14. Which product line had monthly sales less than $20,000, and in which year?

Chapter *Assessment*

Chapter Test

Answer each question.

1. Rewrite 15% as a decimal.
2. Rewrite 0.85 as a percent.
3. Find what percent $160 is of $2,000
4. What is 12% of $120?
5. Find the average: $120.56, $362, $250.54
6. Divide: $400 ÷ 12

Solve.

7. Jacquelyn has the following housing expenses for 3 months: $820, $760, and $785. What is the average monthly housing expense?
8. Janelle's auto insurance has an annual premium of $540. If she pays monthly, she is charged $48 per month. How much cheaper is it to pay the annual premium?
9. Your budget says that savings should be 11% of income. If you earn $1,500 a month, how much should you save in a year?
10. Re-writable CDs, regularly priced at $3.99 a box, are on sale, 3 boxes for $1.99. How much would you save by buying 12 boxes at the sale price?

A camcorder can be rented for $35 per day, or $125 a week. You need it for 6 days.

11. How much cheaper is it to rent it for a week instead of by the day?
12. If you could buy the camcorder for $1,089, how long would it take, to the nearest tenth of a week, for the weekly rental charges to equal or exceed the purchase price?
13. Regatta, Inc. cancels its service plans for 8 cell phones from a company that charges a cancellation fee of $10 a month for each remaining month on the service plan. There were 3 months to go on the plans. What is the cost of the cancellation fee?
14. An ISP's access fees are $14.99 a month or $159.99 a year. How much would you save over three years by buying the 1-year deal than a monthly deal?
15. Will Brandt's ISP charges $16.99 a month for unlimited access. He also pays $21 for a start-up fee, 19.89 for antivirus software, and $79.99 for a modem. What is the total cost of the Internet connection for the first year?
16. The Simpson family increases their $325 transportation budget by $75 per month. If they earn $4,500 per month, what percent, to the nearest percent, of their income have they budgeted for transportation?
17. The Recreation CPI index number was 108.5 in 2004 and 111.7 in 2007. What was the percent increase in the prices of recreation goods and service for that three year period, to the nearest tenth percent?
18. During a recession, the unemployment rate was 34.7% for teen workers and 8.7% for male workers. Out of every 1,000 workers in each category, how many more teens were unemployed than males?

Planning a Career in Government and Public Administration

Rafa Irusta 2008/Shutterstock.com

Career choices in government and public administration present unique opportunities that are may not be found in the private sector. If you choose a career in government and public administration, you might find yourself working as an urban planner, a city manager, a legislator, a magistrate, or as support staff for government positions. If you have the ability to communicate well with other, and want to work in public service, then a career in government and public administration may be a career path for you.

Job Titles

- Judge
- State Senator
- Mayor
- Public works administrator
- Law clerk
- Tax collector
- Deputy clerk
- City planner
- Corrections officer
- City councilperson

Needed Skills

- outstanding organizational, leadership, and communication skills
- honesty, integrity, and commitment to public welfare

- legal expertise and background
- excellent problem solving and decision making skills
- business or accounting background
- technology and computer skills

What's it like to work in the Judicial Branch?

The Judicial branch of the government interprets and applies the laws written by the legislative branch. This occurs on the local, state, and federal levels. Judges, magistrates, and other judicial personnel run the legal process in courts with cases that range from traffic tickets, family law, criminal offences, civil cases, to constitution violations. Judicial employees are responsible for making certain that the legal rights of the parties involved are protected by ensuring hearings and trials are fair. Appellate courts provide a place for litigants to appeal, or ask for review of court decisions.

What About You?

What aspect of government public service appeals to you? How might you best prepare for a career in this field?

How Times Have Changed

For Questions 1–2, refer to the timeline on page 335 as needed.

1. When cellular phones were introduced in the U.S. in 1983, each phone weighed 28 ounces and cost $3,995. By 1984, the number of cell phone subscribers was approximately 300,000. If each subscriber bought a $3,995 phone, about how much would have been spent on the phones alone?
2. If each cell phone subscriber pays an average monthly rate of $50 for the service and an additional $25 per month in roaming charges, extra minutes, text messaging, and other fees, how much was spent in the year 2008 for cell phone service plans?

Chapters 7–8 *Cumulative Review*

MULTIPLE CHOICE

Select the best choice for each question.

1. The monthly health insurance premium provided by Lawanda's employer is $448, and Lawanda pays 20% of the cost. What annual amount does she pay for health insurance?

   A. $1,075.20 **B.** $4,300.80 **C.** $89.60
   D. $985.60 **E.** $1,057.20

2. Crest County 8%, $1,000 par value bonds sell at 95.314. What is the price of 5 bonds?

   A. $7,625.12 **B.** $476.57 **C.** $953.14
   D. $5,000 **E.** $4,765.70

3. Dean Grigsby was injured on his job. He is insured by a disability policy that pays 35% of his average annual salary of $36,800. What amount does Dean collect monthly?

   A. $23,920 **B.** $12,880 **C.** $1,993.33
   D. $1,073.33 **E.** $2,142.83

4. The semiannual interest on a $500 par value bond is $20. If the bond now sells at $98, what is its current yield, to the nearest tenth percent?

   A. 8.0% **B.** 8.2% **C.** 20.0%
   D. 4.1% **E.** 20.4%

5. Kristy Chancellor owns 540 shares of FiberQueue $100 par value common stock. The stock pays a dividend of 4%. What total amount will Kristy receive in dividends?

   A. $2,016 **B.** $21.60 **C.** $5,184
   D. $2,160 **E.** $1,350

6. The Grogan family spent the following amounts for food for three months: $945.86, $876.39, $702.43. What is their average monthly spending for food?

   A. $841.56 **B.** $922.31 **C.** $985.28
   D. $1,012.22 **E.** $2,524.68

7. Jillian Tia's total income last year was $25,000. Jillian spent $3,000 on entertainment. What percent of Jillian's total income did she spend on entertainment?

   A. 5% **B.** 8% **C.** 10%
   D. 12% **E.** 13%

8. The Medical Care CPI index for 2006 was 340.1 and 357.7 in 2007. What was the rate of inflation in Medical Care for 2007, to the nearest tenth percent?

   A. 4.9% **B.** 5.2% **C.** 18%
   D. 5.1% **E.** 17.6%

9. The CPI for the Food category in 1997 was 159.1. What was the purchasing power of the 1997 Food dollar compared to the 1982–84 Food dollar, to the nearest tenth of a cent?

   A. $1.591 **B.** $0.371 **C.** $0.629
   D. $0.591 **E.** $0.831

OPEN ENDED

10. Walter Griggs bought a store for $22,000 cash. He paid cash of $36,000 for renovations. The annual expenses of operating the store are estimated at $13,000. What monthly rent must Walter charge to earn a 14% rate of income on his cash investment?

11. Ann Lee can buy term life insurance for $1.20 per $1,000 less 5% for being a non-smoker. What annual premium would she pay for $350,000 of term life insurance?

12. Virgil Simmons' oral surgery cost $850. His insurance company approved $775 of the cost. Virgil's policy has 25% co-insurance. He has $80 remaining on his deductible for the year. What amount must Virgil pay for this surgery?

13. Regina Upshaw bought 12, $500 Olan School District bonds at $102.864 through a broker. The broker's charge was $2.50 per bond. The bonds had accrued interest of $14 per bond. What was Regina's total investment in the bonds?

14. Rodney Branch bought 500 shares of Exastent stock at $41.24 plus $186 commission. He later sold the stock for $54.18 and was charged $216 commission. What net profit did Rodney make on this stock investment?

15. Juanita Denson bought a home for $95,000 with a $12,000 down payment and a mortgage loan for the rest. The lot on which the home stands is valued at $15,000. Juanita spent $11,000 in cash to remodel the home. Annual expenses of owning the home are 1.5% for depreciation, $6,200 in mortgage interest, and $3,900 in other expenses. Juanita estimates she can rent the home for $1,400 a month. What is Juanita's expected return on her investment, to the nearest percent?

16. Find the rate of commission, to the nearest tenth percent, charged by a fund whose shares are quoted as NAV, 29.80, and Offer Price, 30.85.

17. Jason Braddock will receive a $1,050 monthly pension from Social Security and $1,700 monthly from a company pension when he retires at age 65. He wants an income of $4,800 a month when he retires. His IRA's value is $380,000. What percent of the IRA must he withdraw each year to get the income he wants, to the nearest percent?

18. A backhoe can be rented for $220 per day, for 24 hours, or $48 per hour. If you estimate that you will use the backhoe for 6 hours, should you rent by the hour or the day? How much will you save?

19. Hisako Akita budgeted $675 per month to pay his monthly rent. His landlord increased his rent to $900 per month. If Hisako earns $2,700 per month, what percent of his income will his new rent payment be? Round to the nearest percent.

20. Helen got a new cell phone. She signed a two-year contract. She paid $54.99 for her phone, $18 activation fee, and $39.99 per month for 800 minutes per month. Extra minutes are $0.35 per minute. In the first year, Helen used a total of 120 extra minutes. What was her total first year cost for her cell phone, before taxes?

CONSTRUCTED RESPONSE

21. Karl has a job where his hourly pay will be adjusted each year for changes in the CPI. Willa works at a job similar to Karl's. Her employer does not guarantee regular pay increases, but gives merit pay raises instead. These raises often exceed the change in the CPI. Explain in writing whether you would prefer to work for Karl's or Willa's employer.

Chapter 9

Business Costs

Statistical Insights

Carriers' Restrictions for Weight and Dimension		
Carrier/Service	**Maximum Weight**	**Maximum Dimensions**
FedEx Ground	Up to 150 lb	Length: 108 in. Length + Girth: 130 in.
FedEx NextDay	Up to 150 lb	No side can exceed 48 in.
FedEx Priority Overnight	Up to 150 lb	Length: 119 in. Length + Girth: 165 in.
United Parcel Service (UPS)	Up to 150 lb	Length: 108 in. Length + Girth: 165 in.
United States Postal Service Express Mail	Up to 70 lb	Length + Girth: 108 in.
United States Postal Service Parcel Post	Greater than 1 lb Up to 70 lb	Length + Girth: 130 in.
United States Postal Service Priority Mail	Up to 70 lb	Length + Girth: 108 in.

Sources: ups.com, fedex.com, and usps.gov

Length is the measurement of the longest side. **Girth** is the distance all the way around the package, or twice the height plus twice the width of the package.

Envelopes and packages are delivered door to door by several U.S. carriers. The more expeditious the delivery is, the higher the cost. Restrictions on items that are shipped allow carriers to fulfill their delivery guarantees. Use the data to answer Questions 1–3.

1. What carriers can be used if a package weighs 72 lb and has dimensions of 5.5 ft long, 10 in. wide, and 8 in. deep?
2. FedEx and UPS have recently added fuel surcharges to their rates to contend with rapidly rising fuel costs. FedEx adds an 8.5% fuel charge to domestic packages when diesel fuel prices are between $4.06 and $4.14 a gallon. If the basic cost of sending a package is $16.75, what will be the total cost including the fuel charge?
3. **Explain** Some carriers have added fuel surcharges to their rates to contend with rising fuel costs. Why would a carrier choose to add on a fuel charge instead of increasing their rates?

How Times Have Changed

Manufacturing is the act of transforming raw materials into finished products, usually on a large scale. Manufacturing processes have changed dramatically over the years due to the use and improvement of machines, and the availability of inexpensive resources and methods of shipment. More recently, the use of computers has impacted manufacturing.

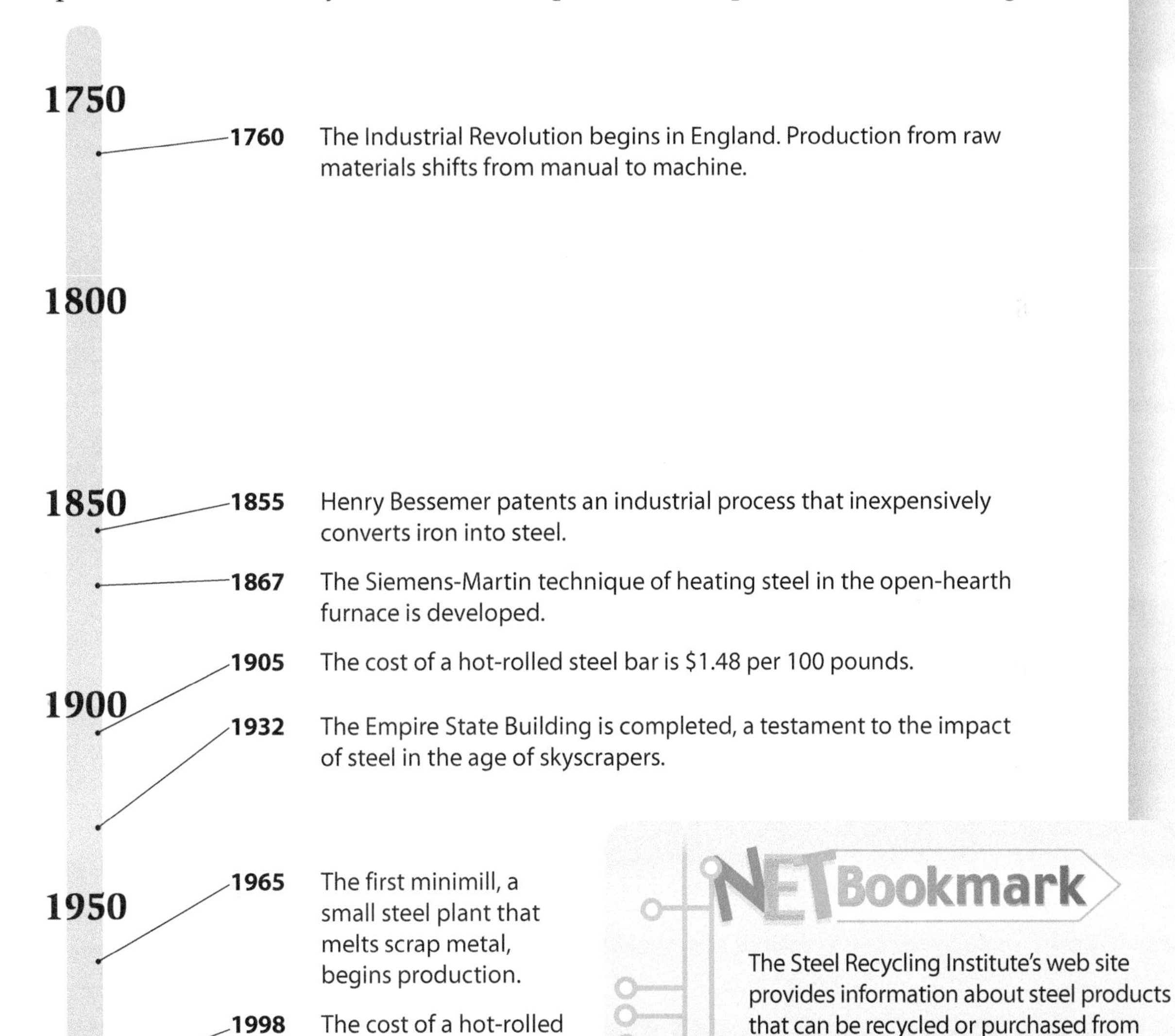

1750
1760 The Industrial Revolution begins in England. Production from raw materials shifts from manual to machine.
1800
1850
1855 Henry Bessemer patents an industrial process that inexpensively converts iron into steel.
1867 The Siemens-Martin technique of heating steel in the open-hearth furnace is developed.
1905 The cost of a hot-rolled steel bar is $1.48 per 100 pounds.
1900
1932 The Empire State Building is completed, a testament to the impact of steel in the age of skyscrapers.
1950
1965 The first minimill, a small steel plant that melts scrap metal, begins production.
1998 The cost of a hot-rolled steel bar is $18.75 per 100 pounds.
2000

Research to find the approximate tons of steel recycled In 2007.

NET Bookmark

The Steel Recycling Institute's web site provides information about steel products that can be recycled or purchased from recycled materials. Access www.cengage.com/school/business/businessmath and click on Chapter 9. Research material efficiency. What is the byproduct of iron making? How much byproduct is produced for each ton of steel? What are some uses for this byproduct?

9-1

Manufacturing Costs

GOALS

- Calculate prime cost and total manufacturing costs
- Distribute factory expenses to units

KEY TERMS

- prime cost
- factory overhead
- total manufacturing cost

Start Up ▶▶▶

Martin Balcerzak/Shutterstock.com

Mr. Dabney, your friend's dad, walked by you on his way home from work one day. Trying to be friendly, you asked him what he did at work that day. He answered, "Oh, I just tried to keep out of the way today." Puzzled, you asked him what he meant by that and he answered: "You see, I am part of the overhead down at the plant." Now you were really puzzled, but before you could ask him another question, he had walked on. What did he mean when he said he was part of the "overhead" at the plant?

Math Skill Builder

Review these math skills and solve the exercises that follow.

1. **Divide** money amounts by whole numbers.
   Find the quotient, to the nearest cent. $238,674 ÷ 4,200 = $56.827, or $56.83

   1a. $560 ÷ 30 **1b.** $372,000 ÷ 600

2. **Multiply** money amounts by fractions.
   Find the product. $\frac{10,000}{50,000} \times \$358,000 = \$71,600$

   2a. $\frac{5,000}{15,000} \times \$138,400$ **2b.** $\frac{12,000}{72,000} \times \$598,300$

Prime Cost and Total Cost of Manufacturing

Manufacturers make the products they sell. They must keep records of their factory costs so that they can control those costs and set selling prices that will produce a net income instead of a net loss.

There are three kinds of factory costs:

1. *Raw materials* costs are the costs of materials which are used in manufacturing and which become part of the finished product.
2. *Direct labor costs* are the wages of all the workers who work directly on the products as they move through the factory.

3. *Factory overhead* includes the expenses that cannot be directly tied to producing a product. For example, it includes salaries and wages of the factory managers, supervisors, inspectors, and other workers who do not work directly on the manufactured products. It also includes building rent, depreciation of equipment, heat, power, insurance, and factory supplies.

The costs of raw materials and direct labor are called the **prime cost** of manufacturing a product. The prime cost plus **factory overhead** are the **total manufacturing cost** of a product.

Prime Cost = Raw Materials + Direct Labor

Total Manufacturing Cost = Prime Cost + Factory Overhead

EXAMPLE 1

The costs of manufacturing 1,000 computer monitors are $15,000 for raw materials, $25,000 for direct labor, and $5,000 for factory overhead. What is the prime cost of the computer monitors? What is the total manufacturing cost of the monitors?

SOLUTION

Add the costs for raw materials and direct labor.

Prime Cost = $15,000 + $25,000 = $40,000

Add prime cost and factory overhead.

Total Manufacturing Cost = $40,000 + $5,000 = $45,000

✔ CHECK YOUR UNDERSTANDING

A. The records of a bicycle factory show these costs for the goods produced in the first quarter of a year: raw materials, $861,980; direct labor, $1,976,200; factory overhead, $387,950. What was the prime cost of the goods produced during the quarter? What was the total manufacturing cost of the goods?

Photodisc/Getty Images

B. To make 150 telephones, ComTech, Inc. had these manufacturing costs: materials, $1,282.29; labor, $1,975.26; factory overhead, $1,234.31. What was the prime cost of making the telephones? What was the total manufacturing cost of each phone, on average, to the nearest whole cent?

Distribute Factory Overhead to Units

A manufacturer needs to know the costs of running each of its divisions, departments, or other units. So, factory expenses, or overhead, are often distributed or charged to each unit.

The way they are distributed varies with the company and the kind of expense. For example, rent may be distributed in proportion to the floor space used by the units. Taxes and insurance on equipment may be distributed based on the value of the equipment in each unit. Cleaning expenses may be distributed on the basis of floor space. Management salaries may be distributed based on the number of factory workers in a unit.

EXAMPLE 2

Mayforge, Inc. pays $30,000 a month rent for its factory and distributes the rent on the basis of floor space. What amount should be charged to each of its three departments? Department A has 2,000 sq. ft of floor space; Department B, 5,000 sq. ft; and Department C, 3,000 sq. ft.

SOLUTION

Add the floor space for each department.

$2{,}000 + 5{,}000 + 3{,}000 = 10{,}000$ sq. ft. total floor space

Divide the floor space for each department by the total floor space. Then multiply the result by the monthly rent

$\frac{2{,}000}{10{,}000} \times \$30{,}000 = \$6{,}000$ monthly rent charged to Dept. A

$\frac{5{,}000}{10{,}000} \times \$30{,}000 = \$15{,}000$ monthly rent charged to Dept. B

$\frac{3{,}000}{10{,}000} \times \$30{,}000 = \$9{,}000$ monthly rent charged to Dept. C

Math Tip

Square feet may be abbreviated to sq. ft or ft^2.

✔ CHECK YOUR UNDERSTANDING

C. The four divisions of Panasol Electronics, Inc., use this floor space: Printers, 6,000 ft^2; Scanners, 2,800 ft^2; Copiers, 5,000 ft^2; and Faxes, 2,200 ft^2. The yearly maintenance cost of the building, $28,000, is distributed based on the floor space of each division. How much is each division charged annually?

D. Biosfeer, Inc. pays its managers a total of $500,000 a year. This expense is charged to the four sections of the company on the basis of the number of workers in each section. The number of workers is: Fabrication, 120; Testing, 20; Painting, 46; and Assembly, 64. What amount is charged to each section?

Wrap Up ▸▸▸

Your friend's dad probably meant that in his job he did not directly produce products. He might be an accountant, manager, supervisor, or any of a number of people who provide support for those who actually build and assemble products. The wages and salaries of such workers go to make up the overhead costs of a manufactured product.

Communication

A term that is likely to be heard in discussions of manufacturing costs is *cost accounting*.

- What is cost accounting?
- What are common job titles of people who perform cost accounting?
- How does cost accounting relate to the topics included in this lesson?

Answer these questions in a brief report. Attach to the report a list of the sources you used to complete the report.

Exercises

Find the quotient to the nearest cent.

1. $40,908 ÷ 112
2. $586,883 ÷ 105
3. $382,400 ÷ 420
4. $1,483,800 ÷ 720

Find the product.

5. $\frac{45{,}000}{90{,}000} \times \$890{,}300$
6. $\frac{25{,}000}{125{,}000} \times \$308{,}800$

In March of last year, manufacturing costs of Eder Stamping Company were: raw materials, $529,926; direct labor, $756,416; and factory overhead, $157,344.

7. What was Eder's prime cost?
8. What was Eder's total manufacturing cost?

For August, Santoni, Inc. had these manufacturing costs: raw materials, $139,648; direct labor, $324,814; and overhead, $106,584. In that month, Santoni manufactured 4,000 units of their product.

9. What was Santoni's prime cost for manufacturing the units?
10. What are Santoni's total manufacturing costs?
11. Estimate the average cost per unit.
12. What was the actual average cost per unit, to the nearest cent?

Photodisc/Getty Images

Solve.

13. An auto parts factory had this overhead for September: supervisory wages, $108,342; rent, $7,278; depreciation, $23,207; power, $7,725; maintenance and repairs, $12,465; and other, $3,674. What was the total factory overhead?
14. A firm's $4,500 electric power bill is distributed by the number of horsepower-hours used by the equipment in each division. This is found by multiplying the horsepower of each motor by the number of hours it is used. The horsepower-hours of each division are: Division X, 3,000; Division Z, 7,500; Division Y, 4,500. What is the amount to be charged to each division?
15. The annual sales of Eberle Press, Inc. are $4,300,000. The cost of Eberle's insurance on its manufacturing equipment is $2,890, which is 27% of the total insurance costs of the company. The equipment insurance cost is distributed in proportion to the value of the equipment in each department. Those equipment values are: Department G, $36,125; Department R, $86,700; and Department K, $21,675. How much insurance should be charged to each department?

The factory records of Glaser Sheetrock show these costs for the last quarter: raw materials, $517,912.80; direct labor, $635,724.80; supervisory salaries and wages, $59,538.47; rent, $27,235; depreciation and repairs, $32,105.67; power, $18,181.30; factory supplies, $15,945.74; and other factory expense, $7,209.82.

16. What was the total factory overhead for the quarter?

17. What was the prime cost for the quarter?

18. What was the total manufacturing cost?

INTEGRATING YOUR KNOWLEDGE During June, Radion Products had these manufacturing costs: raw materials, $419,754; direct labor, $1,329,784; and overhead, $384,598. In that month, Radion manufactured 24,600 units of their product, 1% of which were found to be defective by their quality control department.

19. How many defective units were there?

20. What was the total manufacturing cost for June?

21. What was the average manufacturing cost for each non-defective item?

22. If Radion produces 28,900 units in July, how many units are likely to be defective?

23. **CRITICAL THINKING** One of the ways to lower manufacturing costs is to improve worker productivity. How might you improve worker productivity in the factory?

Mixed Review

24. Add 275 + 0.04 + 7.202 + 28.1.

25. Multiply $\frac{3}{5} \times \frac{8}{9}$.

26. Rewrite 60% as a fraction and simplify.

27. Divide $81.20 by 140%.

28. Divide 10 by $\frac{2}{5}$.

29. 45 is what percent of 150?

30. The Knabe family's income on which they must pay state tax is $68,540. The tax rate is 5.8%. What is their state tax?

Photodisc/Getty Images

31. Tom Wu bought a light truck for $12,900 and drove it 15,000 miles in the first year. His expenses in that year were: depreciation, 20% of the cost of the truck; interest at 12% of the cost of the truck; and gas, oil, insurance, and other expenses, $1,860. Find the operating cost of the truck per mile, to the nearest tenth of a cent.

32. After Ursala retired she starting receiving a Social Security check for $835 each month. She needed $20,100 to meet her yearly expenses. How much did she need to withdraw from her IRA account each month to cover these expenses?

9-2

Breakeven Point

GOALS

- Calculate the breakeven point for a product in units
- Calculate the breakeven point for a product in sales dollars

KEY TERMS

- breakeven point
- fixed costs
- variable costs

Start Up ▶▶▶

Imagine that your company has just invented a product that is so new it has never been offered on the market before. What factors might you use to help you set the price of the product?

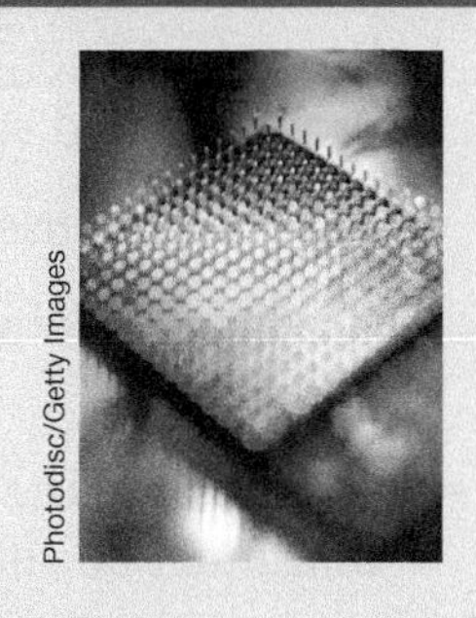
Photodisc/Getty Images

Math Skill Builder

Review these math skills and solve the exercises that follow.

1 **Subtract** dollar amounts from dollar amounts.
Find the difference. $45.49 − $23.78 = $21.71

1a. $108.39 − $74.19 **1b.** $358.12 − $286.87

2 **Multiply** dollar amounts by whole numbers.
Find the product. $16.88 × 3,500 = $59,080

2a. $3.89 × 38,970 **2b.** $98.30 × 42,670

3 **Divide** dollar amounts by dollar amounts.
Find the quotient. $34,500 ÷ $12 = 2,875

3a. $174,000 ÷ $15 **3b.** $83,300 ÷ $35

Breakeven Point in Units and Sales Dollars

To plan their operations, manufacturing firms must decide:

- How many units they expect to sell
- How many units to produce
- How much to spend to produce and sell these units
- At what price they must sell the units to make the profit they want

To make these decisions, firms may calculate the breakeven point. The **breakeven point** is the point at which income from sales equals the total cost of producing and selling goods. It is the point at which the business will make no profit or suffer a loss.

When sales exceed the breakeven point, there is a profit. When sales are less than the breakeven point, there is a loss.

To find the breakeven point, you need to know the:

- Fixed costs for manufacturing the product
- Variable costs for manufacturing each unit of the product
- Expected selling price of each unit of the product

Fixed costs are costs such as rent, salaries, heat, insurance, advertising, and other overhead costs that remain the same no matter how much of the product is manufactured or sold. **Variable costs** are costs such as raw materials, direct labor, and energy that vary or change directly with the amount of product produced and sold.

The formula used to calculate the number of units that must be sold to breakeven is:

Breakeven Point in Units =
Fixed Costs ÷ (Sales Price per Unit − Variable Cost per Unit)

To calculate the amount of sales needed to breakeven, multiply the number of units that must be sold to breakeven by the sales price of each unit.

Breakeven Point in Dollars = Breakeven Point in Units × Sales Price per Unit

EXAMPLE 1

BF Corporation plans to produce porcelain bowls that will be sold at $10 per unit. Manufacturing any quantity of bowls will cost an estimated $12,000 in fixed costs. The variable costs of producing each bowl are estimated to be $5. How many bowls must they sell to breakeven? What sales must BF earn on the bowls to reach the breakeven point?

SOLUTION

Divide the fixed costs by the difference between the selling price of each bowl and the variable costs for each bowl.

$12,000 ÷ ($10 − $5) = $12,000 ÷ $5 = 2,400 bowls that must be sold to breakeven

Multiply the number of bowls needed to breakeven by the selling price of each bowl.

2,400 × $10 = $24,000 sales needed to breakeven

✔CHECK YOUR UNDERSTANDING

A. Rally Co.'s fixed costs to produce toy trucks are $200,000. The variable costs to produce each truck are $4. They will price the trucks at $20. How many trucks must they sell to breakeven? What sales must they reach to breakeven?

B. Grossi Corporation estimated their fixed cost of producing aluminum baseball bats at $105,900 and their variable costs at $39 per bat. If they plan to sell the bats at $89, how many bats do they need to sell to breakeven? How much more than the breakeven point will their income be if they sell 3,000 bats for $89?

Wrap Up ▶▶▶

Setting a price on a new product is not easy. Setting the price too high might drive customers away. Setting the price too low might result in a loss. Factors to use in setting the price are the cost of producing the product, how many units you expect to sell, and how much people can be expected to pay for it. These factors are also interdependent. The more you produce, the lower the cost per unit. The lower the price, the more of it you will sell.

Consumer Alert

Low-Cost Guarantee

Information technology is changing manufacturing. One change is the ability to provide for mass customization. Manufacturers are combining mass production with the flexibility of computer-aided manufacturing systems to produce custom output at a relatively low price. The custom products may actually have features custom designed for a particular retailer, or the product may only be "custom" because of a name or number printed on the item.

Both manufacturers and retailers can benefit from mass customization. Often, when consumers are comparison shopping, they find look alike products in several different stores and assume the products are the same. The consumer is often unaware that a product is customized.

Some retailers use this type of mass customization in their marketing and sales approaches. Have you ever heard of a "low price guarantee"? These guarantees are common among electronic and appliance retailers. An example of such a guarantee is, "If you find the same model elsewhere for a lower price, we will beat that price by 10%."

The catch for the consumer is that the model number sold by that particular retailer is only sold by that particular retailer. A product may have the same appearance and same features of products you see in other stores, but the model number is different. The different model number means that the low-price guarantee does not apply because it is not the "same model."

Exercises

Find the difference.

1. $74.56 − $42.88
2. $126.90 − $89.56

Find the product.

3. $28.70 × 10,200
4. $5.87 × 25,800

Find the quotient.

5. $108,800 ÷ $8.50
6. $20,160 ÷ $4.20

Boldfinch, Inc. plans to make patio chairs and sell them at $75 each. They estimate their fixed costs to produce the chairs at $300,000 and variable costs at $45 per chair.

7. How many chairs must Boldfinch sell to breakeven?
8. What sales amount must they reach to breakeven?
9. If they sold 25,600 chairs, how much over the breakeven point will their sales income be?

A manufacturer produced 6,000 video games and sold them at $20 each. Fixed costs were $18,000 and the variable cost of each game was $5.

10. What were the production costs to produce the games?

11. How many games did they need to sell to breakeven?

12. How much over the breakeven point was their sales income?

13. **STRETCHING YOUR SKILLS** To produce and sell 40,000 copies of a novel, Beardsley-Martin Publishing Company will have fixed and variable costs totaling $500,000. The company wants to make a profit of $250,000 on the books. At what price must the books be sold to make the profit they want?

14. **STRETCHING YOUR SKILLS** Newlet Tracker Corporation plans to manufacture and sell 25,000 units of a computer printer. They estimate their fixed costs will be $3,750,000 and their variable costs $4,900,000. At what price each must they sell the printers to breakeven?

Photodisc/Getty Images

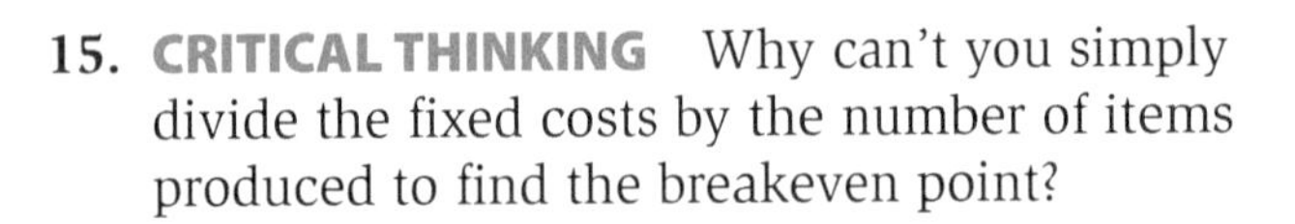

15. **CRITICAL THINKING** Why can't you simply divide the fixed costs by the number of items produced to find the breakeven point?

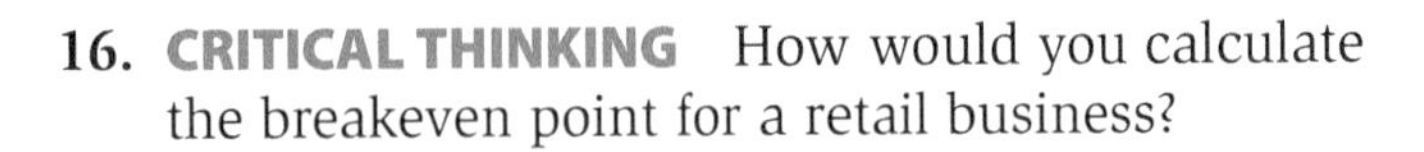

16. **CRITICAL THINKING** How would you calculate the breakeven point for a retail business?

Mixed Review

17. $82 \times 21\frac{1}{2}$

18. $6.95 × 1,000

19. $2\frac{1}{4} \div 18$

20. $160\frac{1}{2}\% \times \$280$

21. $25\frac{3}{4}\%$ of 800

22. $3\frac{1}{4}$ of 80

23. Muhammed Rahum owns 50 bonds with a par value of $1,000 each that pay 8.75% interest. Find his semiannual income from these bonds.

24. Juanita's lot and house are assessed at $76,800. The school tax rate in her district is 2.13 cents per $1. What is Juanita's school tax?

25. A factory had this overhead for June: supervisory wages, $218,834; rent, $16,726; depreciation, $45,187; power, $15,125; maintenance and repairs, $25,165; other, $7,367. What was the total factory overhead?

26. Three departments of a store use this floor space: home furnishings, 2,400 ft^2; patio furniture, 1,200 ft^2; lighting, 400 ft^2. Cooling costs are charged to departments on the basis of floor space. What is each department's share of the $1,200 summer cooling costs?

Blend Images/Jupiter Images

27. A building valued at $200,000 was insured for $120,000 under an 80% coinsurance clause policy. A fire caused a loss of $25,600. How much did the insurance company pay?

9-3

Depreciation Costs

GOALS

- Calculate depreciation using declining-balance method
- Calculate depreciation using sum-of-the-years-digits method
- Calculate depreciation using modified accelerated cost recovery system method

KEY TERMS

- book value
- declining-balance method
- sum-of-the-years-digits method
- modified accelerated cost recovery system (MACRS)

Start Up ▶▶▶

Business property loses value over time from wear and tear. Can you think of another reason why business property might lose value over time?

Harald Holland Tjostheim/ Shutterstock.com

Math Skill Builder

Review these math skills and solve the exercises that follow.

1. **Subtract** money amounts.
   Find the difference. $13,500 − $2,560 = $10,940

   1a. $24,850 − $6,520 **1b.** $14,208 − $11,078

2. **Multiply** money amounts by percents.
   Find the product. $24,800 × 20% = $24,800 × 0.20 = $4,960

   2a. $14,780 × 8.5% **2b.** $86,480 × 19.5%

Business *Tip*

Taking large amounts of depreciation early on more closely reflects the real loss in resale value of many business properties. It also provides large, early tax deductions.

Declining-Balance Method

The depreciation of property that has a life of more than one year is a major expense for many businesses. The Internal Revenue Service regulates the calculation of depreciation. They allow depreciation to be calculated in several ways, including the straight-line method that you have already learned.

Businesses may use depreciation methods that deduct greater amounts in the early years than in later years. Two such methods are the declining-balance method and the sum-of-the-years-digits method. When calculating depreciation, the term book value is used. **Book value** is the original cost of the property less the total depreciation to date.

The **declining-balance method** uses a fixed rate of depreciation for each year. Because the rate is applied to a declining or decreasing balance, the amount of depreciation decreases each year.

EXAMPLE 1

A van costing $16,000 is estimated to depreciate 20% each year. What is the estimated book value of the truck at the end of the first and second years?

SOLUTION

$0.20 \times \$16,000 = \$3,200$ first year depreciation

$\$16,000 - \$3,200 = \$12,800$ book value, end of first year

$0.20 \times \$12,800 = \$2,560$ second year depreciation

$\$12,800 - \$2,560 = \$10,240$ book value, end of second year

✔CHECK YOUR UNDERSTANDING

A. A forming machine that cost $45,000 is depreciated each year at the rate of 8%. What is the book value of the machine at the end of the second year?

B. LB Forging buys a drill press for $7,500. The estimated life is 12 years, and the annual depreciation rate used is 8%. What will be the book value of the press at the end of 3 years?

Sum-of-the-Years-Digits Method

Another way to calculate depreciation is the **sum-of-the-years-digits method**. This is a variable-rate method. Like the declining-balance method, the sum-of-the-years-digits method provides the greatest amount of depreciation in the first year and smaller amounts of depreciation after that.

For example, if you estimate you will use a machine for five years, the amount of depreciation is calculated by adding the years together: $1 + 2 + 3 + 4 + 5 = 15$ years. Then you depreciate the machine $\frac{5}{15}$ of the total depreciation for the first year, $\frac{4}{15}$ of the total depreciation for the second year, $\frac{3}{15}$ for the third year, $\frac{2}{15}$ for the fourth year, and $\frac{1}{15}$ for the fifth year.

Algebra Tip

You can find the sum of the digits using the following formula:

$$S = \frac{n \times (n+1)}{2}$$

where S is the sum of the digits from 1 to n.

For example, to find the sum of the digits 1 to 5,

$$S = \frac{5 \times (5+1)}{2} = \frac{5 \times 6}{2}$$
$$= 15$$

EXAMPLE 2

A lathe costing $16,000 will be used for 5 years, and then traded in for an estimated $8,500. Find the book value of the lathe at the end of the second year.

SOLUTION

Subtract the estimated trade-in value from the original cost.

$\$16,000 - \$8,500 = \$7,500$ total depreciation

Find the sum-of-the-years-digits: $1 + 2 + 3 + 4 + 5 = 15$

Find the depreciation for the first year: $\frac{5}{15} \times \$7,500 = \$2,500$

Find the book value at the end of the first year: $\$16,000 - \$2,500 = \$13,500$

Find the depreciation for the second year: $\frac{4}{15} \times \$7,500 = \$2,000$

Find the book value at the end of the second year: $\$13,500 - \$2,000 = \$11,500$

✔ CHECK YOUR UNDERSTANDING

C. Yankee Products bought a hydraulic winch for $23,600. They plan to use it for 6 years and then trade it for $4,500. Find the book value of the winch at the end of the second year.

D. Roberto Bros. bought a loader for $80,000. The firm estimates that the loader will be used for 7 years and then traded in for $6,000. What will the book value of the loader be at the end of the first and third years?

Modified Accelerated Cost Recovery System Method

For federal income tax purposes, the **modified accelerated cost recovery system (MACRS)** must be used to calculate depreciation for most business property placed in service after 1986. The MACRS method allows you to claim depreciation over a fixed number of years depending on the *class life* of the property. The class life means how long the Internal Revenue Service (IRS) will let you depreciate the property.

The Internal Revenue Service (IRS) categorized the lives of different types of property into a number of classes, including 3, 5, 7, 10, 15, and 20 years. For example, the IRS puts cars, trucks, and most office equipment into a 5-year class life and office furniture into a 7-year class life. The rate of depreciation to be used for each year of a property's life is set by the IRS and varies with each class life.

You calculate the depreciation deduction for any one year by multiplying the original cost by the rate of depreciation for that year. Trade-in value, or salvage value, is not used in the MACRS method.

Business *Tip*

The MACRS method allows businesses to depreciate their property fully and at a faster rate than they might otherwise. Congress passed the MACRS to encourage business to invest in new equipment.

The table of depreciation rates for properties with class lives of 5 and 7 years is shown. Notice that 5-year properties and 7-year properties are depreciated over six and eight years, respectively.

	Cars, Trucks, Office Equipment	Office Furniture and Fixtures
Class Life	*5-Years*	*7-Years*
First Year	20.0%	14.29%
Second Year	32.0%	24.49%
Third Year	19.2%	17.49%
Fourth Year	11.52%	12.49%
Fifth Year	11.52%	8.93%
Sixth Year	5.76%	8.92%
Seventh Year		8.93%
Eighth Year		4.46%

EXAMPLE 3

Elena Suarez bought a business truck for $19,000. Find the depreciation for each year. What is its book value at the end of its class life?

SOLUTION

Multiply the original cost by the rate of depreciation for each year.

First Year	0.20	× $19,000 =	$ 3,800.00
Second Year	0.32	× 19,000 =	6,080.00
Third Year	0.192	× 19,000 =	3,648.00
Fourth Year	0.1152	× 19,000 =	2,188.80
Fifth Year	0.1152	× 19,000 =	2,188.80
Sixth Year	0.0576	× 19,000 =	1,094.40
Total Depreciation			$ 19,000.00

Subtract the total depreciation from the original cost.

$19,000 − $19,000 = $0 book value at end of class life

> **Math Tip**
>
> To check your calculations for MACRS, add the depreciation for each year. If you did your work correctly, the total depreciation should equal the original cost.

✔ CHECK YOUR UNDERSTANDING

E. LaBeck, Inc. paid $14,200 for a copier that had a class life of 5 years. What amount of depreciation was allowed on the copier for each of the first 5 years?

F. A suite of office furniture cost $15,900. What total depreciation is allowable for the first year's use? Using the MACRS table, what is the book value of the furniture after the third year?

Stephen Coburn/Shutterstock.com

Wrap Up ▸▸▸

One reason that business property might lose value over time is obsolescence. New equipment might do the work faster and cheaper, for example. New equipment might also perform functions that were not performed by the older equipment.

Exercises

1. 1 + 2 + 3 + 4 + 5 + 6

2. $3,108 + $1,397

3. $180,380 − $83,700

4. $16,338 − $11,089

5. $245,700 × 19.2%

6. $375,990 × 20%

Copy and complete the table using the declining-balance method of depreciation.

				Book Value	
	Property	Original Cost	Rate of Depreciation	At End of First Year	At End of Third Year
7.	Sprayer	$1,400	8%		
8.	Fork Lift Truck	$11,200	15%		
9.	Press	$8,600	12%		
10.	Die Cutter	$45,000	14%		
11.	Hydraulic Punch Press	$22,800	13%		

12. L. Diaz, Inc., bought a boring machine for $18,900. The company plans to use the borer for 7 years and then sell it. The company estimates the resale value will be $3,800. Find the depreciation for each of the first 3 years using the sum-of-the-years-digits method.

13. Bellevue Farms bought 5 items of office equipment, weighing 14 metric tons, for $84,200. The company plans to use the equipment for 16 hours daily for 12 years and then scrap it for no value. Using the sum-of-the-years-digits depreciation method, what will be the book value of the equipment at the end of 3 years?

Use the MACRS table given in the lesson to solve Exercises 14–17.
A property cost $250,000. It is classified as a 5-yr class life property under MACRS.

14. What is the total amount of depreciation allowed for its first year of use?

15. What is the total amount of depreciation allowed for the life of the property?

16. What is the book value of the property after the first year's use?

17. What is the book value of the property after two years' use?

18. **CRITICAL THINKING** Another method of depreciation is called the *unit of performance* (or production) *method.* What do you think this means and how do you think you calculate depreciation using this method?

19. **FINANCIAL DECISION MAKING** Why is there an advantage to a company to have a larger tax deduction for depreciation in the early years of a property than having depreciation spread evenly over each year of a property's life?

Mixed Review

20. Red Frankel deposited $20,000 in a three-year certificate of deposit that pays simple interest at a fixed annual rate of 6.7%. What total interest will Red have earned at the end of three years?

21. To make 600 book covers, the costs are: materials, $7,289; labor, $12,828; factory overhead, $1,723. What is the average cost of each cover, to the nearest cent?

9-4

Shipping Costs

GOALS
- Calculate shipping charges
- Calculate freight charges

KEY TERMS
- freight
- f.o.b. (free on board)

Start Up ▶▶▶

Julia Martino wants to ship a present to her dad. When she calls a shipping service for help they ask her for the weight and girth of the package. She doesn't know how to measure for girth and she doesn't have a scale for weighing packages. What help can you give her?

Craig Hill/Shutterstock.com

Math Skill Builder

Review these math skills and solve the exercises that follow.

1. **Add** dollar amounts.
   Find the sum. $148 + $12.50 + $34.89 = $195.39

   1a. $24.50 + $89.10 + $1.56   **1b.** $108 + $38.22 + $16.75

2. **Multiply** dollar amounts by dollar amounts and whole numbers.
   Find the product. $0.76 × 456 = $346.56 $4.87 × 34 = $165.58

   2a. $0.53 × $580   **2b.** $1.56 × $750

   2c. $6.78 × 125   **2d.** $2.16 × 1,560

3. **Divide** dollar amounts and whole numbers by 100.
   Find the quotient. $1,250 ÷ 100 = $12.50

   3a. $25,900 ÷ 100   **3b.** 145,800 ÷ 100

4. **Round** whole numbers up to the next 100.
   Round up to the next 100. 14,356 or 14,400.

   4a. 289   **4b.** 1,978   **4c.** 28,084   **4d.** 87,219

Shipping Charges

Businesses use the U.S. Postal Service, private carriers, and other shippers to deliver products to their customers. Businesses may also pay shipping costs when they purchase items. Before deciding how to ship, a firm must consider the speed and distance the package will travel, and the size, weight, and shape of the package. The firm may also want insurance, special handling, C.O.D. (collect on delivery), weekend delivery, and door-to-door pickup and delivery. These services raise the cost of shipping the goods.

A package may be sent by the U.S. Postal Service by *parcel post* service if the package weighs not more than 70 pounds and is not more than 130 inches in combined length and *girth*. Other shipping services use different limitations. When measuring weight, most services consider a fraction of a pound as a full pound.

A table of rates for a shipping company is shown below. These rates are for delivery within two business days and are charged for each package delivered. Distance is shown in *zones*, or circular bands of miles around the shipping point. The number of miles wide each zone is may vary with each shipper.

Math *Tip*

Girth is the measurement around a package at its thickest part. If a package is 30 inches long, 10 inches wide, and 4 inches deep, the girth is 10 + 4 + 10 + 4, or 28 inches. The combined length and girth is 30 + 28, or 58 inches.

EXAMPLE 1

Clarion Corporation ships a 14-lb package to zone 6. Insuring the package costs $7.50 more. What is the total cost to ship the package using the shipping rate table below?

SOLUTION

Find the shipping charge by moving down to the row for the package weight and over to the right for the column of the destination zone.

$33.25 shipping charge

Add the shipping charge and the insurance charge.

$33.25 + $7.50 = $40.75 total cost of shipping

National Shipping Company Rates for 2-Day Delivery

Package weight (lb)	Destination Zones						
	1 & 2	3	4	5	6	7	8
1	$7.50	$8.25	$9.00	$9.75	$10.75	$11.25	$11.50
2	$8.00	$8.75	$10.00	$11.00	$12.00	$12.75	$13.00
3	$8.50	$9.25	$11.00	$12.75	$13.75	$14.25	$14.75
4	$9.25	$10.00	$12.00	$13.75	$15.25	$16.00	$16.75
5	$10.00	$11.00	$13.25	$15.25	$17.25	$18.25	$18.75
6	$10.75	$11.75	$14.25	$16.75	$19.25	$20.50	$21.00
7	$11.25	$12.75	$15.50	$18.00	$21.25	$22.25	$23.00
8	$12.00	$13.50	$16.50	$19.50	$23.00	$24.00	$25.00
9	$12.50	$14.25	$17.75	$21.00	$25.00	$26.00	$26.75
10	$13.00	$15.00	$18.50	$22.25	$26.50	$27.75	$29.00
11	$13.75	$15.75	$19.50	$23.50	$28.00	$29.75	$30.50
12	$14.50	$17.00	$20.50	$25.00	$30.00	$30.75	$32.00
13	$15.25	$17.75	$21.50	$26.50	$31.25	$32.25	$33.75
14	$15.75	$18.25	$22.75	$27.75	$33.25	$34.25	$35.25
15	$16.50	$19.00	$23.75	$28.75	$34.25	$35.50	$36.50
16	$16.75	$19.75	$24.75	$29.50	$35.25	$36.50	$38.25

✔ **CHECK YOUR UNDERSTANDING**

A. Tosco Imaging, Inc. ships 3 packages weighing $5\frac{1}{2}$, 12, and 15 pounds to zone 3 using National Shipping's 2-Day service (see shipping rate table). Insurance costs were $31.89. Since the second day falls on Saturday, $15 extra per package is charged. What was the total cost of the shipment?

B. Ridgeway, Inc. ships and insures 5 packages worth $1,800 each by SpeediAir for same day delivery. SpeediAir charges $159 for each package. They also charge $0.50 per $100 of value for insurance. What is the total cost of the shipment?

Freight Charges

Freight is often used for shipping heavy, bulky goods. Freight shipments may be sent by airplane, truck, train, barge, or ship. Many shipping companies that deliver letters and small parcels handle freight, too. Sample rates for a freight company are shown.

Photodisc/Getty Images

Curry Freight and Express Company							
3-Day Delivery—Rates Per Pound							
Weight in Pounds	Destination Zones						
	1 & 2	3	4	5	6	7	8
151–499	0.50	0.90	1.10	1.19	1.66	2.08	2.34
500–999	0.49	0.88	1.07	1.19	1.61	1.97	2.34
1000–1999	0.47	0.85	1.04	1.14	1.56	1.97	2.28
2000 +	0.45	0.83	1.03	1.09	1.51	1.92	2.23

The term **f.o.b.**, or **free on board**, may be used by sellers to show who pays the shipping costs and the point at which the responsibility for the goods transfers from buyer to seller. For example, a seller in New York quotes a buyer in St. Louis a price, f.o.b. St. Louis. This means that the seller will pay the transportation charges to St. Louis, or that the goods will travel "free on board" to the buyer in St. Louis and the seller assumes responsibility for the goods until they arrive.
If the seller quotes a price, f.o.b. New York, the buyer pays the transportation costs from New York to St. Louis and the buyer is responsible for the goods during transportation.

Business *Tip*

Knowing when the responsibility for a shipment changes is important for determining the liability if there is a loss during shipment.

Shipping charges may be calculated on the basis of weight per 100 pounds, or *cwt (hundredweight)*. To find the charge, divide the weight of the shipment by 100 and multiply by the rate per cwt. Any amount less than 100 pounds is usually charged at the full 100-pound rate. So, a 532-pound parcel would be charged as if it were 600 pounds.

EXAMPLE 2

A 1,500-pound package is to be delivered in three days by Curry Freight and Express to a firm in zone 5. What is the freight charge?

Photodisc/Getty Images

SOLUTION

Using the freight table, find the price per pound by moving down to the row for the package weight and then over to the column for destination zone.

$1.14   price per pound

Multiply the weight of the package by the price per pound.

1,500 × $1.14 = $1,710   freight charge

✔CHECK YOUR UNDERSTANDING

C. Gomez Bros. ships two 850-pound packages to zone 4 by Curry Freight in three days. What is the freight charge?

D. Louisa Montez, Inc. of Mayfield ships 42,350 pounds of steel rods by freight to Carlson Forging Corporation of Toledo, f.o.b. Toledo. The freight company charges $19.45 a cwt. What is the freight charge? Who pays it?

Wrap Up ▸▸▸

Girth is measured by the distance around a package. For cartons or boxes, a ruler can be used to measure all four sides around the package. For round packages or oddly shaped packages, a tape measure will usually do. To weigh a package without a regular parcel scale, Julia can use a standard home scale. Julia should step on the scale and weigh herself. Then she should hold the package and weigh herself again. The difference in weights is the weight of the package.

Exercises

1. $2.59 × 359

2. $0.24 × 5,480

3. $4,810 ÷ 100

4. 89,340 ÷ 100

5. Round up to the next 100: 32,876

6. $369.28 + $12.89 + $83.19

Which of the packages below may be sent by the U.S. Postal Service's parcel post service?

	Length	Width	Depth	Weight
7.	23 inches	8 inches	14 inches	16 lb
8.	50 inches	12 inches	7 inches	75 lb
9.	30 inches	6 inches	8 inches	$42\frac{1}{2}$ lb
10.	60 inches	15 inches	12 inches	50 lb
11.	70 inches	55 inches	14 inches	70 lb

Use the shipping table given in the lesson to find the shipping charges for each package in Exercises 12–21.

	Weight in Pounds	Destination in Zone	Shipping Charge		Weight	Destination in Zone	Shipping Charge
12.	6	8		**13.**	$13\frac{1}{2}$ lb	6	
14.	10	3		**15.**	3.25 lb	4	
16.	16	7		**17.**	9 lb 4 oz.	7	
18.	5	2		**19.**	4 lb 8 oz.	5	
20.	8	5		**21.**	7 lb 3 oz.	8	

22. The U.S. Postal Service charges $6.40 each to ship 12 parcels of reel film by parcel post. The company that is shipping the film insures the contents with a private insurer for $48.90. What is the total cost of the shipment?

23. A shipping service charges $2.25 per pound to deliver a 16-pound package on a weekend. Insurance costs an additional $0.35 per $100, or fraction of $100 of value. The company values its parcel at $590. Weekend delivery costs $25 additional. What is the cost to ship the package?

BEST BUY A company wants to ship 3 parcels weighing 12 lbs. each to zone 7. They can have the packages delivered in three days using National Shipping (see shipping rate table) or a week to ten days using the U.S. Postal Service's parcel post. The parcel post rates are $15.15 per package.

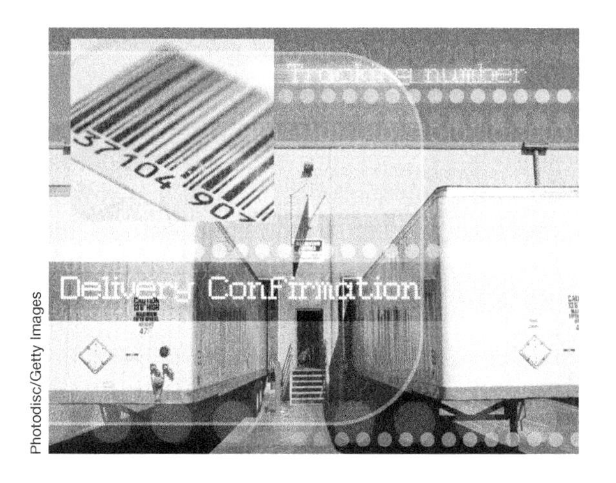

Photodisc/Getty Images

24. What will it cost to ship the packages in 3 days using National?

25. What will it cost to ship the packages in 7–10 days using parcel post?

26. How much will the company save by shipping the packages by parcel post?

Use the freight table to calculate the freight charges for each shipment.

	Weight in Pounds	Destination in Zone	Freight Charge		Weight in Pounds	Destination in Zone	Freight Charge
27.	386	2		**28.**	513	8	
29.	1,370	6		**30.**	725	4	
31.	160	3		**32.**	919	7	
33.	584	8		**34.**	499	5	
35.	1,809	5		**36.**	2,598	3	

The freight rate to Hammond for a certain type of item is $38.87 per cwt or any remaining fraction of the total shipment. A customer wants to ship 9 crates weighing 95 pounds each and 11 cartons weighing 22 pounds each.

37. Estimate the total weight of the shipment.

38. Estimate the freight charge.

39. What is the exact freight charge?

40. **INTEGRATING YOUR KNOWLEDGE** Johnson Electric Company buys 380 electrical switches @ $1.88 and 20 per-cut cables @ $2. The switches and cables are shipped in one box weighing 14 pounds to zone 6 using National Shipping. What is the total cost of the switches to Johnson?

41. **FINANCIAL DECISION MAKING** You are sending a package valued at $150 through the mail. The cost to ship the package is $8.55. You can add insurance for $1.20. Would you choose to insure the package? Why or why not?

Mixed Review

42. Multiply 28 by 5.

43 What amount is 140% greater than $34?

44. Find the number of days from May 1 to October 5.

45. Show 0.0975 as a percent.

The Bernsteins' gross income for a year was $79,104. From that amount they subtracted a standard deduction of $10,900, and 3 exemptions at $3,500 each to find their taxable income.

46. What was their taxable income?

47. What was the amount of their state income tax in a state with an income tax rate of 3.5% on taxable income?

48. Yang Su's online checking account balance was $437. In the past seven days, she had these transactions in her account: deposit, $362.18; debit card purchase of $43.68; one check written for $162. Today Yang is making online payments for car insurance, $278; charge card, $48.92; power bill, $164.55. What will be the balance of her online checking account after all the transactions and online payments are entered?

49. Ahmed Tahil bought a car for $15,600. He estimates its trade-in value will be $5,200 at the end of 4 years. What is the estimated average annual depreciation of the car?

50. A computer that cost $2,400 is estimated to depreciate 20% each year. Using the declining-balance method, find the book value of the computer at the end of the third year.

51. Tolson, Ltd. ships and insures 3 packages worth a total of $840 for same day delivery. The shipper charges $125 to deliver each package and $0.35 per $100, or fraction of $100 of the value of the total shipment for insurance. What is the total cost of the shipment?

9-5

Office Costs

GOALS
- Calculate the costs of office space
- Calculate the cost of a unit of office work

Start Up ▶▶▶

Ruby Lake recently found that the firm for which she works can monitor the e-mail messages that she receives and sends. She felt this was a violation of her privacy and was illegal. What do you think?

Photodisc/Getty Images

Math Skill Builder

Review these math skills and solve the exercises that follow.

1. **Subtract** money amounts from money amounts.
   Find the difference. $89.12 − $77.64 = $11.48

   1a. $159.35 − $140.98 **1b.** $207.44 − $56.29

   1c. $542.36 − $258.14 **1d.** $187.00 − $68.96

2. **Divide** money amounts by whole numbers.
   Find the quotient. $108,245 ÷ 36,000 = $3.007, or $3.01

   2a. $29,800 ÷ 2,500 **2b.** $310,780 ÷ 5,500

   2c. $20,100 ÷ 1,300 **2d.** $724,550 ÷ 44,000

Office Space

Office space is a major office expense. The costs for providing office space may include rent, light, heat, air conditioning, insurance, cleaning, and maintenance. If the office is owned instead of rented, the costs may include depreciation of the building, mortgage interest, and real estate taxes.

EXAMPLE 1

A 4,500 sq ft office area rents for $62,250 a year. Utilities cost $4,800 a year. What is the total yearly cost of this office space per square foot? If a clerk's workstation is 64 sq ft, how much does this workstation space cost per year?

SOLUTION

Add the costs for rent and utilities.

$62,250 + $4,800 = $67,050   total annual cost of office space

Divide the total annual cost of the office space by the square feet.

$67,050 ÷ 4,500 = $14.90   cost of office space per square foot

Multiply the cost per square foot by the workstation square feet.

$14.90 × 64 = $953.60   annual cost of workstation space

✔ CHECK YOUR UNDERSTANDING

A. An office with 2,600 ft^2 rents for $49,400 a year. Power, natural gas, and maintenance are estimated to cost $4,940 a year. What is the total monthly cost of this office space per square foot? If one workstation's area is 32 ft^2, how much does this workstation cost per year?

B. A 40,000 ft^2 building contains a warehouse and offices. The offices occupy 10% of the building. The rent, power, maintenance, insurance, and other costs for the entire building are $1,289,600 per year. How many square feet does the office occupy? What is the annual cost of the office space per square foot?

Unit Costs of Office Work

Owners and managers often need to know the cost of one job, of each workstation, or of a unit of work, such as a letter or report. These costs can then be compared to the costs of other firms and industry averages.

EXAMPLE 2

Last year an office spent these amounts on three workstations: salaries, $42,824; rent and utilities, $8,200; depreciation of equipment, $7,140; supplies, repairs, postage, and telephone, $9,420. Estimate the total amount spent on the three workstations. What was the average annual amount spent on each workstation?

SOLUTION

Round off and add the amounts spent on the workstations.

$40,000 + $8,000 + $7,000 + $9,000 = $64,000   estimated total spent

Add the exact amounts spent on the workstations.

$42,824 + $8,200 + $7,140 + $9,420 = $67,584   total spent on 3 workstations

Divide the total spent on workstations by the number of workstations.

$67,584 ÷ 3 = $22,528   average spent on each workstation

✔ CHECK YOUR UNDERSTANDING

C. Three workers in a document center produced this number of lines of text in a week: Emma LeCono, 15,600; Edwin Rosen, 13,600; Jose Cruziero, 12,800. The total cost of running the center for the week was $2,257. How much did each document line cost, to the nearest cent?

D. Copy paper costs $4.15 per ream (500 sheets). When bought in a case of 10 reams, the same paper costs $38 per case. What is the cost per ream when buying a case? What is the savings per ream when buying a case?

Wrap Up ▶▶▶

Unless the employer promises employees that their e-mail messages are off limits to the firm, it is legal for the firm to view the messages. The e-mail messages were created using the firm's computer system, network, and the other resources, including paid time. If the employee has private messages on the computer, then the employee used the firm's workstation for personal use.

TEAM Meeting

An office manager wants to increase the number of workstations that can be put in a space 40 feet wide by 30 feet long. That space now contains 4 rows of 5 workstations each. Each workstation requires a minimum of 30 square feet of space. Aisles must be 3 feet wide. Movable partitions can be used to separate workstations, but you must be able to get to each workstation without going through another.

With another student, use graph paper to draw a diagram that shows how you would rearrange the workstations. Then answer these questions.

1. If the office space rents for $22,400 per year, what is the annual rent per workstation with the old arrangement?
2. What is the annual rent per workstation with your new arrangement?
3. By what percent has your new arrangement increased the number of workstations in the same space?

Exercises

1. $56.78 − $45.49
2. $1,597 − $997.45
3. $288,560 ÷ 10,800
4. $83,480 ÷ 3,280

One office in an office building has an area of 5,200 ft². The office costs are $87,360 a year including rent, heating, cooling, lighting, and maintenance.

5. What is the annual cost of the office per square foot?
6. What is the annual cost of a supervisor's workstation that is 100 ft² in area?

Tomas Figuero works in an advertising copy center and keys in 585 lines of text per day, on average.

7. In 7.5 hours of work per day, how many lines per minute does he key in?
8. If he earns $10.80 an hour, what is the labor cost per line, to the nearest tenth of a cent?

A policy clerk can sort and file 170 original, signed insurance policies per hour.

9. How many policies would the clerk file in $7\frac{1}{2}$ hours of work?
10. At a wage rate of $9.20 per hour for the clerk, what does it cost to file each policy, to the nearest tenth of a cent?

A manager wants to increase the number of workstations that can be put in a space 43 feet wide by 58 feet long. That space now contains 6 rows of 10 workstations each. The manager wants to increase floor space by placing every two rows of desks together so that the aisle between them is eliminated.

11. How many aisles will be eliminated?

12. If the aisles are 3 feet wide, how much floor space, in square feet, is gained?

13. If two more rows of 10 workstations are added, by what percent has the manager increased the number of workstations in the same space?

14. If the office space rents for $49,680 per year, what is the annual rent per workstation with the old arrangement?

15. What is the annual rent per workstation with the new arrangement?

16. What is the total cost of equipping a workstation with this equipment?

1 desk @ $359.89	1 letter tray @ $19.50
1 chair @ $119.50	1 computer system @ $1,659
1 calculator @ $15.99	1 telephone @ $86.99
1 lamp @ $69.95	1 floor mat @ $29.95

17. **FINANCIAL DECISION MAKING** What are some ways that adding more workstations to this space will impact the workers? How would you weigh these factors into your decision?

18. An office manager supervises five data-entry workstations. The cost of each workstation is as follows: wages, 1,950 hours per year @ $10.80; fringe benefits, 25% of wages; space, 60 square feet @ $18.50 per square foot per year; data terminal rental, $50 per month; supplies, $275 per year; and other costs, $1,300 per year. What is the total yearly cost of the five stations?

Photodisc/Getty Images

STRETCHING YOUR SKILLS When mailing 5,000 letters to customers, a mail clerk put 58 cents postage on each letter instead of the right amount, 55 cents.

19. What amount of postage was wasted on this mailing?

20. If the clerk was paid $10 per hour, how many hours of wages were wasted?

Mixed Review

21. Multiply $16\frac{1}{8}$ by 12.

22. What is 12% of $4.30?

23. Divide 0.042 by 0.006.

24. 16 is what percent of 20?

25. Find the number of days from March 10 to May 6.

26. Find exact interest on $800 at 12% for 3 months.

9-6

Travel Expenses

GOALS
- Calculate mileage reimbursement
- Calculate business travel expenses

KEY TERMS
- travel expense
- per diem

Start Up

Beverly O'Hara used her car to deliver a package for her employer. However, she parked in a no parking zone and also dented her car when she accidentally hit the no parking sign. Since she was using her car for her employer, she feels the employer should pay for these expenses. Do you think she is right?

Photodisc/Getty Images

Math Skill Builder

Review these math skills and solve the exercises that follow.

1. **Add** money amounts.
   Find the sum. $398 + $12.58 + $98.31 + $43 + $3.75 = $555.64

   1a. $598 + $18.90 + $125 + $315.78 **1b.** $31.64 + $397 + $22.65

2. **Multiply** money amounts by whole numbers.
   Find the product. $112 × 4 = $448 2,498 × $$0.38\frac{1}{2}$ = $961.73

   2a. $135.78 × 8 **2b.** 5,078 × $$0.40\frac{1}{2}$

3. **Multiply** money amounts and whole numbers by percents.
   Find the product. $389 × 85% = $330.65 24,875 × 38% = 9,452.50

   3a. $1,970 × 76.5% **3b.** 32,089 × 42.5%

Mileage Reimbursement

Companies may pay back, or *reimburse* employees on a per mile basis when personal cars are used for business. The IRS also allows people to deduct car mileage when they use their car for business if the employer doesn't reimburse them fully.

Often, the rate per mile paid by businesses matches the rate allowed by the IRS. If the firm pays employees less per mile than the IRS, employees can claim the difference on their tax returns.

NET Bookmark

Use the IRS web site to find the current IRS mileage rate.

EXAMPLE 1

Rosa Lazare is a salesperson for Trumble Publishing Company and uses her own car to make sales calls. Trumble reimburses her for the use of her car at $\$0.54\frac{1}{2}$ per mile. If Rosa drove 1,268 miles on company business last month, how much did Trumble reimburse her for the use of her car?

SOLUTION

Multiply the business miles by the rate per mile.

$1,268 \times \$0.545 = \$686.70$ reimbursement amount

> **Business Tip**
>
> The IRS allows you to keep records of your actual car expenses and use these for the tax deduction instead of using the per mile rate.

✔ CHECK YOUR UNDERSTANDING

A. Erica Rosenberg drove her car 986 miles in May, 35% of which were for her job. The company reimburses Erica $0.55 a mile for business use. What did Erica receive as reimbursement for the use of her car in May?

B. Sateesh Alwash drove his car 19,308 miles last year. Of those miles, 28% were reimbursed by his company at $0.52 a mile. The IRS mileage rate for that year was $0.55 a mile. How much did Sateesh receive from his company for mileage? What amount might Sateesh use as a tax deduction?

Business Travel Expenses

In addition to mileage, often employees are reimbursed for other **travel expenses**. These include hotel fees, meal charges, airfares, taxi fares, rental car charges, expenses for entertaining customers, and related expenses.

Some businesses reimburse employees for their actual expenses, and some reimburse employees on a **per diem**, or per day basis regardless of how much was spent.

Many organizations set limits on certain types of expenses, such as maximums for hotel charges and meals. Many organizations also do not reimburse employees for personal expenses that they consider unrelated to the employee's travel, such as the cost of personal entertainment.

> **Business Tip**
>
> The IRS also allows you to deduct mileage when you use your car in the service of a charity. The rate per mile that they allow for charitable use is lower than for business use.

EXAMPLE 2

Marta Guzman is a technical support specialist for 100-Com Communications. 100-Com sent Marta to a desktop operating system seminar in another city. 100-Com paid the seminar fees directly to the seminar company. They also paid Marta $459 to reimburse her for airfare and $120 per diem for the three days she attended the seminar. How much did Marta receive as reimbursement?

SOLUTION

Multiply the per diem rate by the number of days.

$3 \times \$120 = \$360$ per diem reimbursement

Add the per diem and airfare reimbursements.

$\$360 + \$459 = \$819$ total amount of reimbursement

✔ CHECK YOUR UNDERSTANDING

C. Po-ling Shen attended a business conference in another state. His approved expenses were: airfare, \$582; meals, \$224; taxis, \$38; airport parking, \$42; porterage, \$12; conference registration fee, \$275; mileage to and from airport, 32 miles; hotel charges, \$216; other expenses, \$54. If his company pays \$0.50 a mile for use of personal cars, how much was Po-ling reimbursed?

D. Susan O'Bannion spent $3\frac{1}{2}$ days at a conference. She was reimbursed for traveling the 315 miles to and from the conference by personal car at \$0.52 a mile. She was also paid \$125 per diem. What was her total reimbursement for the conference?

Business *Tip*

Porterage is the tip amount you give to have your bags carried by airport limousine drivers, hotel bellhops, and others.

Wrap Up ▸▸▸

Most organizations will not pay for parking fines or car accidents that are the direct fault of the employee even though the car was being used for business purposes.

Financial Responsibility

When making travel plans, whether it is for business travel or personal travel, you should consider buying *trip insurance*. Without insurance, if you become ill, have a family emergency, or some other reason that you cannot make your trip as planned, any nonrefundable payments made are forfeited. You cannot get your money back.

With insurance, the money that you have prepaid can be refunded. There are many different types of trip insurance ranging in price. You should buy a policy that meets the level of coverage you desire.

The cost of insurance is minimal compared to the cost of the trip.

Exercises

1. Add: \$189 + \$42.16 + \$15 + \$68.26
2. Multiply 8,497 by \$0.42.
3. Multiply \$109 by $6\frac{1}{2}$.
4. Multiply 13,597 by 75%.

Solve.

5. Clarice Johnson put 1,489 miles on her car in June. Of those miles, 27% were for her job. The company reimburses Clarice \$0.48 a mile for business use. What did Clarice receive as reimbursement for the use of her car in June?
6. Julio Valliente drove his car 23,188 miles last year, 32% of which were reimbursed by his company at $\$0.53\frac{1}{2}$ a mile. The IRS mileage rate for that year was \$0.55 a mile. How much did Julio receive from his company for mileage? What amount might Julio use as a tax deduction?

7. LaDonna White works in the human resource department of Valatia Corporation and attended a conference on health insurance. Her approved expenses were: airfare, $768; meals, $267; taxis and limousines, $43; airport parking, $54; porterage, $16; registration fee, $175; mileage to and from airport, 26 miles; hotel charges, $368; and other expenses, $41. If her company pays 47¢ a mile for use of personal cars, how much was LaDonna reimbursed?

8. Felicia Ryan spent $2\frac{1}{2}$ days at a business show. She was reimbursed for traveling the 173 miles to and from the show by personal car at $\$0.48\frac{1}{2}$ a mile. She was also paid $146 per diem. What was her total reimbursement for the show?

9. Don Dykstra attended a business conference for his company. His approved expenses were: $734 for airfare; $257 for meals; $29 for limousines and taxis, $389 for his hotel room, $400 in conference fees, and $215 in other charges. What was the amount of his reimbursement?

10. Calvin Jones spent the following on a 3-day business trip: airfare, $586; airport limousines, $24; porterage and other tips, $28; taxis, $45; movie, $8.95; hotel, $145 a night for three nights; meals, $58 a day. His employer did not reimburse entertainment expenses and has a maximum hotel charge of $130 a night. What was Calvin's total reimbursement amount?

Last year Yolanda paid $25,800 for a light truck and drove it 21,000 miles in the first year. She spent these amounts: $1,120, gas; $578, repairs and maintenance; $1,277, insurance; $64, truck washes; $1,200, interest on truck loan; and other, $156.

11. **INTEGRATING YOUR KNOWLEDGE** Using a 20% rate and the declining-balance method, what is the depreciation for the first year?

12. What was the total cost of owning and operating the truck last year, including depreciation?

13. What was the cost per mile of owning and operating the truck last year, to the nearest tenth of a cent?

14. Yolanda used her truck 23% of the time to deliver products for her employer, who reimbursed her $0.56 a mile. How much did Yolanda receive for the use of her car from her employer?

15. **FINANCIAL DECISION MAKING** Elvis Clarke is self-employed and uses a new van that cost $22,500 exclusively in his business. The van was driven 10,000 miles. When preparing his income tax return, Elvis had a choice of declaring his actual expenses for the car or using that year's IRS per mile rate of $0.55. Which method is likely to offer Elvis the largest tax deduction for the first and second years of use? Why?

Mixed Review

16. Divide 56.7 by 8.9 rounded to three decimal places.

17. What percent is 136 of 1,700?

18. Yallow Appliances, Inc. owns 680 desktop and notebook computer systems. Last year it spent $255,390 on replacement parts and equipment, $143,788 on software upgrades, and $493,714 on repairs to their systems, user training, and help for their computer users. What amount was spent on technical support per computer system?

Chapter Review

Vocabulary Review

Find the term, from the list at the right, that completes each sentence. Use each term only once.

book value
breakeven point
declining-balance method
factory overhead
fixed costs
free on board (f.o.b.)
freight
modified accelerated cost recovery system method (MACRS)
per diem
prime cost
total manufacturing cost
travel expenses
sum-of-the-years-digits method
variable costs

1. Prime cost plus factory overhead is called __?__.
2. Expenses not directly tied to producing goods are called __?__.
3. The point at which income from sales equals the total cost of producing and selling goods is called __?__.
4. The costs that remain no matter how much of a product is produced are called __?__.
5. Costs that change directly with the number of units produced are called __?__.
6. The __?__ of depreciation uses a fixed rate of depreciation for each year.
7. The depreciation method required by the IRS for most business properties is called __?__.
8. A method of shipping heavy, bulky goods, often by rail, truck, barge, or ship is called __?__.
9. __?__ is a term used to determine who pays shipping costs.
10. A daily reimbursement rate used by some employers is called __?__.

9-1 Manufacturing Costs

11. An electric appliance factory had these costs for the goods produced in the first quarter of a year: raw materials, $1,671,680; direct labor, $2,168,320; and factory overhead, $684,850. What was the prime cost of the goods produced during the quarter? What was the total manufacturing cost of the goods?

12. The 3 departments of Aranson's Auto Repair use this floor space: Body, 3,000 ft^2; Repair/Service, 2,000 ft^2; and Painting, 1,000 ft^2. The annual insurance cost of the building is $2,800. It is distributed on the basis of the floor space of each department. How much should each department be charged annually for insurance, to the nearest whole dollar?

9-2 Breakeven Point

13. Balen Operating Corporation plans to produce cowlings that will be sold at $20 per unit. Manufacturing any cowlings will cost an estimated $24,000 in fixed costs. The variable costs of producing each cowling are estimated to be $10. How many cowlings must they sell to breakeven? What sales must Balen earn on the cowlings to reach the breakeven point?

9-3 Depreciation Costs

14. A factory machine that cost $164,000 is depreciated each year at the rate of 12% using the declining-balance method. What is the book value of the machine at the end of the first year?

15. The original cost of a stamping press was $152,600. The factory plans to use it for 5 years and then trade it in for $44,600. Find the book value of the press at the end of the first year using the sum-of-the-years-digits method.

16. A business van costs $21,000. Using the MACRS table, find the depreciation for the first year. What is its book value at the end of the first year?

17. A truck that cost $15,000 is depreciated in value each year at the rate of 8%. What is the book value of the machine at the end of the second year using the declining-balance method?

9-4 Shipping Costs

18. A company wants to ship a package that weighs 10 pounds to zone 5, using National Shipping Company's two-day service (see the shipping table). Insuring the package will cost $6.75 more. What is the total cost to ship the package?

19. Belcro, Inc. in Madison ships 42,350 pounds of lumber by freight to True-Time, Inc. in Pine Valley, f.o.b. Pine Valley. The freight company charges $21.19 a cwt. What is the freight charge? Who pays it?

20. Crayco, Inc. of Madison ships 12,635 pounds of metal bars by freight to Sisco Manufacturing Company of Roswell, f.o.b. Madison. The freight company charges $12.49 a cwt. What is the freight charge? Who pays it?

9-5 Office Costs

21. An office with 2,500 ft^2 rents for $48,000 a year. Light, heat, insurance, and maintenance are estimated to cost $4,500 a year. What is the total annual cost of this office space per square foot? If one workstation's area is 30 ft^2, how much does this workstation cost per year?

22. The total annual cost of an office is $158,000, including wages, fringe benefits, rent, insurance, and power. If the office has 5 workstations, what is the average annual cost of each workstation?

23. A clerk can sort and file 1,125 invoices per day. If the clerk earns $10.20 per hour and works for $7\frac{1}{2}$ hours a day, how much does it cost to file each invoice, to the nearest tenth of a cent?

9-6 Travel Expenses

24. Sandra Duarte attended a business conference in another state. Her approved expenses were: airfare, $485; meals, $267; taxis, $28; airport parking, $34; porterage, $15; conference registration fee, $150; mileage to and from airport, 62 miles; hotel charges, $258; other expenses, $76. If her company pays $0.52 a mile for use of personal cars, how much was Sandra reimbursed?

25. Rick Belle spent $2\frac{1}{2}$ days at a business seminar. He was reimbursed for traveling the 250 miles to and from the seminar by personal car at $0.515 a mile. He was also paid $135 per diem. What was his total reimbursement for the seminar?

26. Seela Alwaif drove her personal auto 176 miles last year while helping a charity in a clothes drive. The IRS in that year allowed a deduction of $0.14 per mile for charitable use of a personal car. What was the amount of Seela's tax deduction?

Technology Workshop

Task 1 Enter Data in a MACRS Depreciation Template

Complete a template that calculates the amount of MACRS depreciation and book value for properties with 5-year and 7-year class lives.

Open the spreadsheet for Chapter 9 (Tech 9-1.xls) and enter 10,000 in the blue cell, B5. Your computer screen should look like the one shown below when you are done.

	A	B	C	D	E
1		MACRS Depreciation Calculator			
2					
3		Class Life			
4		5-Year	7-Year		
5	Cost	$10,000.00	$0.00		
6					
7		Depreciation and Book Value Calculation			
8		5-Year		7-Year	
9	Year	Depreciation	Book Value	Depreciation	Book Value
10	1	$2,000.00	$8,000.00	$0.00	$0.00
11	2	$3,200.00	$4,800.00	$0.00	$0.00
12	3	$1,920.00	$2,880.00	$0.00	$0.00
13	4	$1,152.00	$1,728.00	$0.00	$0.00
14	5	$1,152.00	$576.00	$0.00	$0.00
15	6	$576.00	$0.00	$0.00	$0.00
16	7			$0.00	$0.00
17	8			$0.00	$0.00
18	Total	$10,000.00		$0.00	

Notice that when you enter the cost of a 5-year class property into cell B5, the spreadsheet automatically calculates the annual depreciation and book values of the property.

Task 2 Analyze the Spreadsheet Output

Move your cursor to cell C5 and enter 20,000. Notice that the annual depreciation and book values of a 7-year class property are automatically calculated.

Answer these questions about your updated spreadsheet.

1. What function is used in cell D18?
2. What arithmetic is done in cell D11?
3. What arithmetic is done in cell E11?
4. Now enter 10,000 in both cells B5 and C5. Look at the row for year 3 for both the 5-year and 7-year properties. Which property depreciated the most that year? How much more? Why?

5. What is the book value for both properties, 5-year and 7-year, by the end of their class life?

Task 3 Design a Breakeven Point Spreadsheet

Design a spreadsheet that will calculate the breakeven point in units and in dollars.

SITUATION: Brainard Furniture, Inc. plans to make computer desks and sell them at $175 each. They estimate their fixed costs to produce the desks at $600,000 and variable costs of $90 per chair. You need to find the number of units Brainard must produce and sell to breakeven. You also need to find the total sales needed to breakeven.

Task 4 Analyze the Spreadsheet Output

Answer these questions about your completed spreadsheet.

6. How did you calculate the breakeven point in number of units?
7. How did you make the spreadsheet round up the number of units to the nearest whole unit?
8. How many desks must Brainard produce and sell to breakeven?
9. How did you calculate the amount of sales needed to breakeven?
10. What is the minimum amount of desk sales Brainard needs to breakeven?

Chapter *Assessment*

Chapter Test

Answer each question.

1. $1,574 − $560
2. $1,237 + $28.79 + $746 + $175.37
3. $0.26 × 1,589
4. 12,350 ÷ 100
5. 37% × 42,718
6. $1,599 ÷ $82
7. 1,000 × $34.49
8. $\frac{3}{4}$ × $64,800
9. $22,050 ÷ 2,100
10. 0.36 × $3,800

Solve.

11. A factory had these costs for the products produced in one year: raw materials, $1,468,190; direct labor, $2,365,100; factory overhead, $539,850. What was the prime cost of the goods produced? What was the total manufacturing cost of the goods?

12. Four departments of a company use this floor space: Dept. A, 12,000 sq. ft; Dept. B, 5,600 sq. ft; Dept. C, 10,000 sq. ft; and Dept. D, 4,400 sq. ft. The yearly maintenance cost of the building, $36,000, is distributed based on the floor space. How much of the annual maintenance cost is each department charged?

13. A company's fixed costs to produce a product are $150,000. The variable costs to produce each product are $6. They will sell the product at $18. How many products must they produce and sell to breakeven?

14. Trical Manufacturing buys a machine with an estimated life of 6 years for $28,500. They use an annual depreciation rate of 15%. What will be the book value of the machine at the end of one year using the declining-balance method?

15. A van costing $26,000 will be used for 5 years, and then traded in for an estimated $8,000. Find the book value of the van at the end of the first year using the sum-of-the-years-digits method.

16. Benton Corporation ships and insures 3 packages worth $600 each using a shipping company for next-day delivery. The shipper charges $56 for each package. They also charge $0.45 per $100 of value for insurance. What is the total cost of the shipment?

17. Travers, Inc. ships 3 packages weighing 400 lb each by freight, f.o.b., customer. The shipping company charges $36 per cwt plus $18.50 for insurance. What is the total cost to ship the packages?

18. An office has an area of 2,500 ft^2. The office costs $76,500 a year including rent, power, insurance, and maintenance. What is the annual cost of the office per square foot?

19. Grace Larke spent 5 days at a business conference. She was reimbursed for traveling 434 miles to and from the conference by personal car at 0.47\frac{1}{2}$ a mile. She was also paid $130 per diem. What was her total reimbursement for the conference?

Planning a Career in Manufacturing

Mark Aplet 2008/Shutterstock.com

Manufacturing jobs include the many aspects involved in turning materials into goods. These manufactured items include textiles, chemicals and petroleum, food items, plastics, electronics, and appliances. Employment in manufacturing consists of workers who assemble items or operate machinery and managers that ensure efficient manufacturing of goods. The needs in manufacturing include people who can competently follow directions and trouble shoot problems, or those that work in research and development. If you are mechanically inclined, work well under pressure, and are physically fit, a career in manufacturing might be for you.

Job Titles

- Material handler
- Machinist
- Cost estimator
- Production worker
- Mechanical engineer
- Inspector
- Chemist
- Purchasing agent
- Computer programmer

Needed Skills

- strong background in mathematics and science

- good eye sight, physical stamina and able to work in a fast-paced environment
- excellent observation and problem solving skills
- ability to receive and follow instructions
- able to adapt to new technologies
- a talent for multi-tasking

What's it like to work in Manufacturing?

A production manager oversees the manufacturing process and coordinates the resources to meet production goals. Production managers decide how to use personnel and machinery so that the desired quality and quantity of product is manufactured. Depending on the size of the plant, a production manager might supervise an area of production or an entire factory. Production managers meet with other company managers to determine goals, trouble shoot problems, manage employees, and coordinate the purchase, delivery, and use of materials.

What About You?

Can you see yourself working in manufacturing? What aspect of the field is most appealing to you?

How Times Have Changed

For Questions 1–2, refer to the timeline on page 381 as needed.

1. Before the mass-production of steel, railroad tracks had to be replaced or repaired often because the wrought iron would crack or deform. In 1905, if a company that manufactured new steel railroad tracks ordered 40 tons of hot-rolled steel bars, how much would the steel cost the company?
2. The steel skeleton of the Empire State Building is made up of about 57,000 tons of steel. Using $1.58 per 100 pounds, the 1932 price of a hot-rolled steel bar, how much did 57,000 tons of steel cost?

Chapter 10

Sales and Marketing

Statistical Insights

Restaurant Advertising Expenditures, U.S.

Category	Expenditures (Millions of Dollars)
Magazines	$37
Sunday Magazines	$3
Newspapers	$89
Network Television	$1,166
Spot Television	$1,296
Syndicated Television	$183
Cable Networks	$364
Radio	$24
Total	**$3,162**

Use the data shown above to answer each question.

1. On what type of ad is the least and most money spent? Name amount spent and category.
2. Which category has about 10% of the total amount spent of restaurant advertising?
3. To the nearest percent, what percent of all restaurant advertising appears in printed publications?

How Times Have Changed

Retailers are continually looking for new ways to improve the checkout procedures, both for themselves and for customers. With the help of technology, and the Universal Product Code (UPC) and the European Article Numbering (EAN) symbols, the process has become more efficient than before.

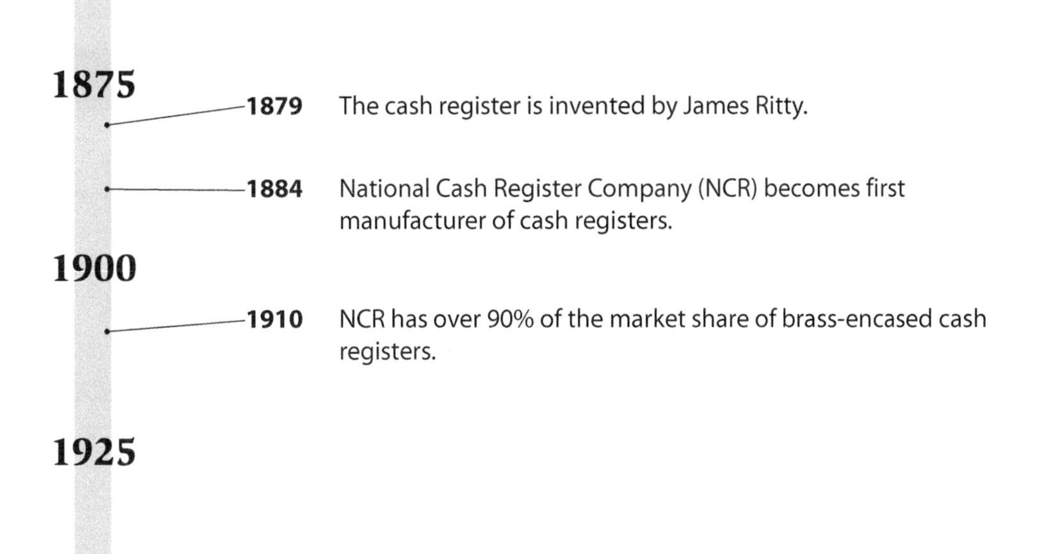

1875
1879 The cash register is invented by James Ritty.
1884 National Cash Register Company (NCR) becomes first manufacturer of cash registers.
1900
1910 NCR has over 90% of the market share of brass-encased cash registers.
1925

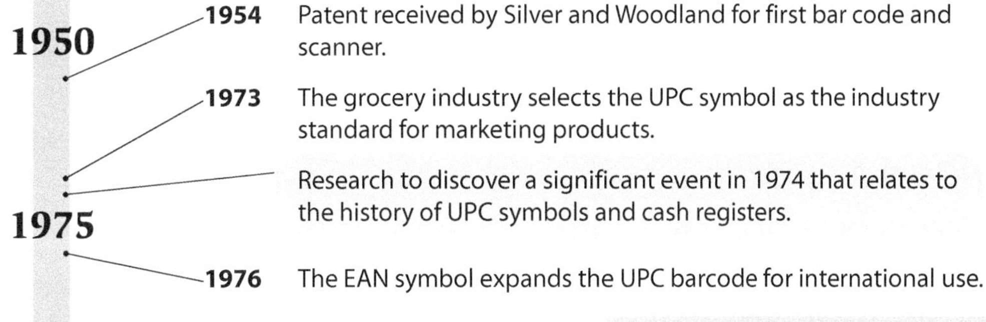

1950
1954 Patent received by Silver and Woodland for first bar code and scanner.
1973 The grocery industry selects the UPC symbol as the industry standard for marketing products.
Research to discover a significant event in 1974 that relates to the history of UPC symbols and cash registers.
1975
1976 The EAN symbol expands the UPC barcode for international use.

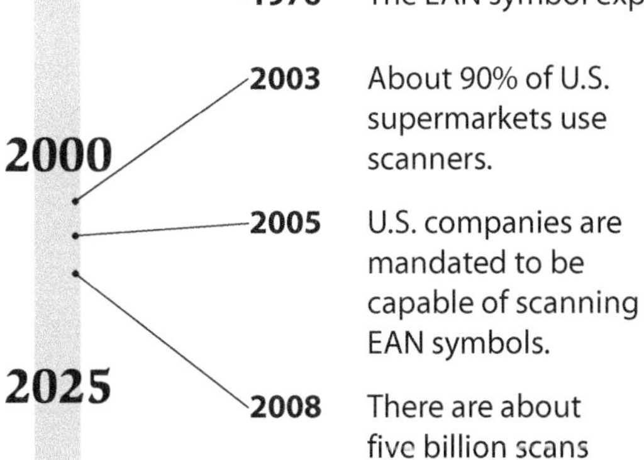

2000
2003 About 90% of U.S. supermarkets use scanners.
2005 U.S. companies are mandated to be capable of scanning EAN symbols.
2025
2008 There are about five billion scans worldwide each day.

Almost everything you buy has a UPC code on it. Access **www.cengage.com/school/business/businessmath** and click on the link for Chapter 10. Find out who issues UPC codes and what the digits mean. What type of information is stored as data for the UPC of an item that is sold in a retail store? How has the data changed over the years as UPC codes have become more universal?

10-1

Cash Sales and Sales on Account

GOALS

- Complete a cash proof form
- Calculate sales invoice and credit memo totals
- Calculate a customer account balance

KEY TERMS

- proving cash
- sales invoice
- credit memo
- on account

Start Up

You buy a new sweater and charge the purchase. A day later you decide you don't like the sweater and return it to the store. Will the store give you a cash refund if you request one, or will you get a credit against your charge account?

Photodisc/Getty Images

Math Skill Builder

Review these math skills and solve the exercises that follow.

1 **Add** money amounts.
Find the sum. $1,284.74 + $827.47 = $2,112.21

1a. $386.12 + $182.34 + $78.28

1b. $2,980 + $34.78

2 **Subtract** money amounts.
Find the difference. $7,492.92 − $234.87 = $7,258.05

2a. $18,472.87 − $12,852.48

2b. $1,273.48 − $28.47

3 **Multiply** money amounts by whole numbers.
Find the product. 18 × $12.87 = $231.66

3a. 125 × $18.40

3b. 65 × $118.72

Proving Cash

Cash registers provide a place to keep cash and a means to record cash sales and payments. Employees who use cash registers are called cash register clerks or cashiers.

Most cash registers are really computer terminals with a display screen and a scanner that is connected to a computer. The scanner reads bar codes printed on the items being purchased. The *bar codes* tell the computer the department, brand, size, and price of each item bought. This information is shown on the display screen and printed on a cash register receipt.

The computer also finds the sales tax, totals the sale, and updates inventory records. When the clerk keys in the amount received from the customer, the correct change is displayed on the screen. The computer also keeps a running total of sales.

Cashiers put money in the cash register drawer when they start work so they can make change. This money is called a *change fund.* While they work, they take in and pay out cash. At the end of their work period, cashiers take a reading of total sales for their register. Then they have to prove cash.

Proving cash means counting the money in the drawer and checking this amount against the cash register readings to see if the right amount is on hand. A *cash proof form* is used for this purpose. If you have less cash than you should, you are *cash short.* If you have more cash than you should, you are *cash over.*

EXAMPLE 1

Gretchen Roth works as a cashier for Tri-City Markets. When she started work on June 11, she put $100 in her cash register drawer as a change fund. At the end of her work period, her register readings showed total cash received, $4,672.98, and total cash paid out, $68.42. When she counted the cash in her drawer, she found $4,702.56. Prove cash using a cash proof form.

SOLUTION

Add the change fund to the register's record of cash received. Record the total cash received and then subtract the total cash paid out.

Compare the cash that should be in the drawer with the actual cash in the drawer. Enter the difference on the appropriate cash short or cash over line.

Gretchen's cash proof showed cash short of $2.

Tri-City Markets Cash Proof Form		
Change fund	*100*	*00*
+ *Register total of cash received*	*4,672*	*98*
Total	*4,772*	*98*
− *Register total of cash paid out*	*68*	*42*
Cash that should be in drawer	*4,704*	*56*
Cash actually in drawer	*4,702*	*56*
Cash short	*2*	*00*
Cash over		

Date : *June 11, 20--* Cash Register No.: *7*
Cash Register Operator: *Gretchen Roth*

✔ CHECK YOUR UNDERSTANDING

A. At the start of his morning work period on January 8, Avery Nugent put $120 in change into his cash register drawer. When his work period finished, Avery's cash register totals showed $5,289.76 received and $76.30 paid out. Avery counted the money in the register and found he had $5,334.81 cash. Was the cash short or over? How much?

B. On August 3, Coletta Butzel put an $80 change fund into her cash register. When she read the cash register totals at the end of her working day, they showed cash received of $3,102.45 and cash paid out of $34.87. The actual cash in the drawer was $3,144.83. Prepare a cash proof form. What does the form show?

Sales Invoice and Credit Memo

When a seller sells goods to a buyer on credit, the seller gives the customer a sales invoice. A **sales invoice** lists the goods sold and delivered to the buyer. The buyer calls this form a *purchase invoice.*

On the sales invoice the unit price is multiplied by the quantity to find the price extension. The price extensions are added together to find the total amount of the invoice.

A sales invoice that Garden Products, a wholesale yard equipment firm, sent to Yard-N-Stuff, a retailer, is shown.

Garden Products
7208 Central Avenue
Baltimore, MD 21207-3071

Date: *May 1, 20--*
Account No.: *45-3874-5*
Invoice No. *1311*

To *Yard-N-Stuff*
3078 Jefferson Street
Rockville, MD 20852-8671

Shipped: *Via Truck*
Terms: *n/30*

We have charged your account as follows:

Quantity	Description	Unit Price	Total
23	Mowers, Model R-237	286.50	6,589.50
6	Tillers, Model 23V	387.45	2,324.70
31	Trimmers, Model 24A	45.10	1,398.10
	TOTAL		10,312.30

When merchandise bought on credit is returned, the seller does not return the buyer's money. Instead, the seller reduces the buyer's account balance by the amount of the return. The seller notifies the buyer about the reduction by sending the buyer a credit memorandum, or **credit memo**.

For example, when Yard-N-Stuff returned two defective mowers from Invoice #1311, Garden Products sent a credit memo.

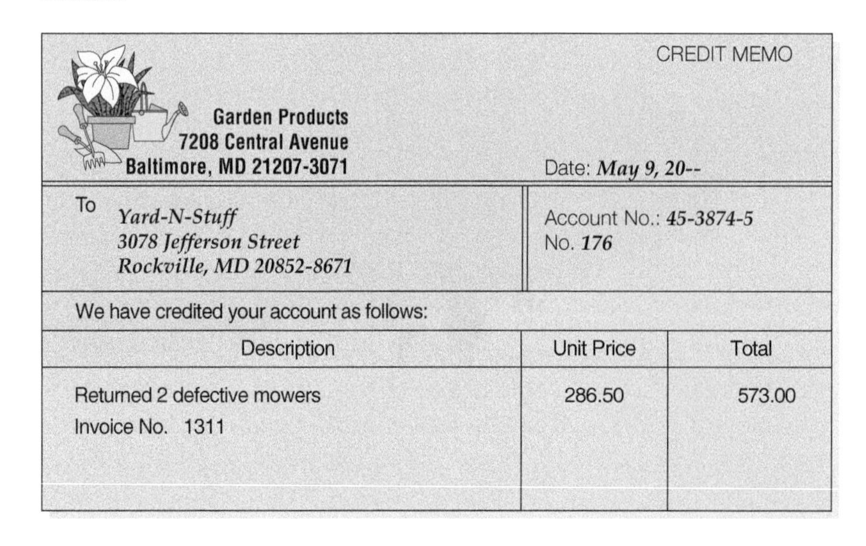

CREDIT MEMO

Garden Products
7208 Central Avenue
Baltimore, MD 21207-3071

Date: *May 9, 20--*

To *Yard-N-Stuff*
3078 Jefferson Street
Rockville, MD 20852-8671

Account No.: *45-3874-5*
No. *176*

We have credited your account as follows:

Description	Unit Price	Total
Returned 2 defective mowers Invoice No. 1311	286.50	573.00

Business *Tip*

A credit memo may be issued when the buyer returns defective goods to the seller or when the seller gives a price reduction for damaged goods reported by the buyer. Sometimes a buyer returns stock for other reasons. If so, the seller may charge a *restocking fee* that is deducted from the credit the buyer would normally receive.

EXAMPLE 2

Garden Products sold Yard-N-Stuff 45 watering cans @ $9.31 and 60 lawn sprinklers @ $7.89 on Invoice #1604. What was the total amount of the sales invoice?

SOLUTION

Multiply the unit cost of each item by its quantity. Then add the price extensions for each item.

$45 \times \$9.31 = \$418.95$ cost of watering cans

$60 \times \$7.89 = \$473.40$ cost of lawn sprinklers

$\$418.95 + \$473.40 = \$892.35$ sales invoice amount

EXAMPLE 3

Two watering cans and one lawn sprinkler bought on Invoice #1604 as described in Example 2 were defective. They were returned by Yard-N-Stuff to Garden Products. What was the total of the credit memo issued by Garden Products to cover these returns?

Photodisc/Getty Images

SOLUTION

Multiply the unit cost of each returned item by its quantity. Then add the price extensions.

$2 \times \$9.31 = \$18.62$ watering can credit

$1 \times \$7.89 = \$7.89$ sprinkler credit

$\$18.62 + \$7.89 = \$26.51$ credit memo total

✔ CHECK YOUR UNDERSTANDING

C. Cellamation, an electronics wholesaler sold these items to CellNow, a mall retailer: 150 Model NP-4 cell phones @ $85.15 and 75 Model CR-12 charging kits @ $14.80. What was the total of the sales invoice?

D. Of the items purchased in Problem C, CellNow returned 12 phones due to a wrong model being sent. CellNow also returned 2 charging kits because of damage. What is the amount of the credit memo that Cellamation issues for these returns?

Communication

Interview a store manager or head cashier to discuss the store's policies about proving cash.

Ask questions such as

1. How frequently is it done?
2. Who verifies a cashier's cash proof?
3. Does the cashier have to make up any cash shortages?
4. What happens to cash overages?

Prepare a summary to share with the entire class.

Customer Account Balance

Most people use credit cards when they do not pay cash for a purchase. Businesses whose customers are other businesses often let them buy on credit, or **on account**. The seller then keeps records for each customer to show how much each customer owes. A common form of *customer account* is shown on the next page. This account shows the transactions between Garden Products (the seller) and Yard-N-Stuff (the customer).

EXAMPLE 4

Prepare a customer account for Yard-N-Stuff to show: balance owed to Garden Products on May 1 was $12,763.87; Invoice #1311 for $10,312.30 charged on May 1; Credit Memo #176 for $573 credited on May 9; payment of $12,763.87 received on May 20; payment for the May 20 account balance received on May 31.

SOLUTION

Enter the May 1 balance in the balance column. Then enter the invoice amount in the charges column and the credit memo amount in the credits column, taking a balance after each entry.

Record the May 10 payment in the credits column. Find the account balance. Enter as the May 31 payment the May 20 balance.

Account:
Yard-N-Stuff
3078 Jefferson Street
Rockville, MD 20852-8671

Account No.: ***45-3874-5***

Date	Description	Charges	Credits	Balance
5/1	*Balance Forward*			*12,763.87*
5/1	*Invoice No. 1311*	*10,312.30*		*23,076.17*
5/9	*Credit Memo No. 176*		*573.00*	*22,503.17*
5/20	*Payment*		*12,763.87*	*9,739.30*
5/31	*Payment*		*9,739.30*	*0.00*

✔CHECK YOUR UNDERSTANDING

E. The customer account for Retro Hardware showed a balance of $2,384.76 on August 1. Sales to Retro were Invoice #1823 for $500.15 on August 3 and Invoice #1956 for $841.56 on August 9. A payment of $2,884.91 was received on August 13. Credit memo #112 was issued to Retro for $56.73 on August 16. There were no other transactions in August. What was Retro's account balance on August 31?

F. Mint Wholesale's customer account for Laurel Place Markets showed a balance of $29,379 on November 1. On November 13 a payment of $21,893 was received from Laurel. Laurel was charged $8,112 for purchases made on Invoice #1876 on November 16. On November 25 Mint issued Laurel a credit memo for $2,570 for stock that was recalled by the manufacturer. What was Laurel's account balance on November 25?

Wrap Up ▶▶▶

If you charge a purchase, stores will issue a credit against your charge. Since you did not pay cash, they will not give you a cash refund. If they gave you cash, you would have use of the cash until your charge bill had to be paid. This would be the same as getting an interest-free loan.

Exercises

Find the sum.

1. $34,281.82 + $23,183.09
2. $1,283 + $412.83

Find the difference.

3. $4,284.84 − $231.07
4. $12,899 − $8,243.18

Find the product.

5. 120 × $23.87
6. 512 × $3.76

Solve.

7. At the start of the day, March 6, Marsha Vaught put a $100 change fund in her cash register. The cash register readings at the end of the day showed total cash received, $3,283.86, and total cash paid out, $45.31. The cash in the register at the end of the day was $3,338.37. Prepare a cash proof form.

8. When Kyle Timmons opened his shop on December 3, he put $80 in change in his cash register drawer. At the end of the day, the cash register readings showed that $4,924.67 had been taken in, that $328.17 had been paid out in cash refunds, and that Kyle had taken out $40 for his personal use. The cash on hand in the drawer at the end of the day was $4,637.90. Was the cash short or over? How much?

9. Annette had $140 in change in her cash register at the start of work on March 19. At the end of her work period, there was $2,961.19 in the cash register. The register totals showed that she received $2,983.27 and paid out $159.76 during the period. How much was the cash over or short?

10. Risco Products sold these items to Sponson's Flooring:
    60 cases of floor tile @ $24.54 a case
    12 gallons of adhesive @ $31.89 a gallon
    12 tile cutters @ $48.23 each.
    What was the total of the sales invoice?

11. HairGlo shampoo was recalled for safety reasons. The manufacturer will issue a credit memo to each retailer for 120% of the original cost of the shampoo to cover the expenses of removing the item from store shelves. A retail chain of 8 stores returned 1,230 bottles of HairGlo whose original cost was $2.95 each. What amount of credit should the chain receive?

All-Marine Products sold 28 Model XV-450 outboard engines at $1,682.60 to Key Boats, Inc. on Invoice #4523. Key Boats had financial problems and returned 10 engines to All-Marine for credit. All-Marine issued Credit Memo #624 after deducting a 25% restocking charge for each returned engine.

Photodisc/Getty Images

12. What was the total of Invoice #4523?

13. For what amount was Credit Memo #624 issued?

14. Make an account form like the one shown in the solution to Example 4. Record these facts about Alexea Co.'s account with Derstin Supply Company. Alexea's address is 708 Gower St., Greenville, SC 29611-3381.

Nov 1	The balance in Alexea's account was $5,372.38
3	Alexea paid the November 1 balance
5	Sold goods to Alexea, $4,182.89; Inv. #7021
17	Alexea returned $619 of the goods from Inv. #7021; Credit Memo #671
18	Sold goods to Alexea, $2,278.36; Inv. #7287.
25	Alexea paid Inv. #7021, less Credit Memo #671
26	Alexea returned $89.01 of the goods on Inv. #7287; Credit Memo #709.

15. Make an account form like the one shown in the Example 4 solution. Record these transactions between Bay Sales and a customer, Earp Products, Inc. Earp is located at 5302 Post Road, Warwick, RI 02886-9898. Calculate all balances.

Mar 1	The balance of Earp's account was $3,145.50
8	Sold goods to Earp, $1,892.32; Inv. #3369; terms, 30 days.
16	Earp returned $71.45 of the goods sold on March 8; Credit Memo #277.
20	Earp paid the March 1 balance of $3,145.50.
26	Sold goods to Earp, $2,574.36; Inv. #3502; terms, 30 days.
31	Earp paid the invoice of March 8, less Credit Memo #277
Apr 12	Sold goods to Earp, $916.76; Inv. #3689; terms, 30 days.
15	Earp returned $12.56 of the goods sold on April 12; Credit Memo, #388.
25	Earp paid Inv. #3502

16. **CRITICAL THINKING** Up to 50% of businesses fail in their first year of operation. Why would a supplier give any new business credit?

17. **CRITICAL THINKING** If an employee is consistently cash over when proving cash, should the employer be pleased or concerned?

Mixed Review

18. $57.5 \times 1{,}000$

19. $7.82 \div 1{,}000$

20. $78 \div 2\frac{1}{2}$

21. $\frac{1}{2} \times 9 \times \frac{2}{3}$

22. Find $\frac{1}{8}$ of $54.80.

23. Find $\frac{1}{3}$ of $81.36.

24. 36% less than $700 is what number?

25. A city has taxable property assessed at $540,000,000. To meet expenses, $28,000,000 must be raised by property tax. What is the decimal tax rate to four places?

26. A $300,000 term life insurance policy is sold at a rate of $1.18 per $1,000 of insurance. What is the annual premium?

10-2

Cash and Trade Discounts

GOALS

- Calculate cash discount and cash price
- Calculate trade discount and invoice price
- Calculate the rate of discount

KEY TERMS

- cash discount
- trade discount

Start Up

Imagine that you recently started your own business and a supplier offers you a 3% discount if you pay cash when you make a purchase. Would you pay cash for each purchase from this supplier? Why, or why not?

Stephen Coburn/ Shutterstock.com

Math Skill Builder

Review these math skills and solve the exercises that follow.

1. **Multiply** money amounts by percents.
   Find the product. 3% × $21,780 = 0.03 × $21,780 = $653.40

   1a. 1.5% × $4,692 **1b.** 35% × $129

2. Find what **percent** one number is of another number.
   Find the percent. $1.16 ÷ $5.80 = 0.2, or 20%

   2a. $52.20 ÷ $116 **2b.** $2,000 ÷ $80,000

Cash Discounts

Most business-to-business transactions begin with a *purchase order* from the buyer that identifies the goods ordered, the agreed upon price, and delivery terms. The purchase order may be written or verbal. After the order is filled, the seller issues a *sales invoice.* Both documents include the details of the purchase.

Remit payment to: Original invoice *GB4302*
Everyday Textures
2382 Bingham Road
Fremont, OH 43420

Shipping: *FOB Fremont*
Sold to: *Farwell Paints* Ship to: *Same address*
231 Grove Avenue
Clyde, OH 43410

Refer to:

Order Number	Date Entered	Date Shipped	Invoice Date	Terms of Sale
56827	06/07/20--	06/07/20--	06/10/20--	2/10, 1/30, n/60

Quantity	Unit	Description	Unit Price	Amount
36	*each*	*Brush, Nylon, 2 inch*	2.90	104.40
48	*each*	*Brush, Bristle, 3 inch*	4.52	216.96
12	*each*	*Ladder, wood, 6-foot*	42.67	512.04
50	*pkg*	*Painting gloves, latex*	3.84	192.00
		Total Due		$1,025.40

On the previous page is an invoice for Farwell Paints from Everyday Textures. While the placement of items on an invoice may vary, the information shown is typical. The main portion of the invoice contains detailed information on the order. This includes the unit price of each item, the price extension, and the total of the extensions, or *total due.*

Notice that the top portion of the invoice includes information about when the order was made. The top portion of the bill also identifies the terms of the sale. *Terms of sale* specify how and when an invoice will be paid. Usually businesses sell to other businesses *on account.* This means the customer will be billed later for purchases. The time a purchaser has to pay a bill, usually 30 to 90 days, is called the *credit period.*

Business *Tip*

Buyers who do not pay their bills by the due date will most likely be charged interest on the amount owed and a late payment fee.

Businesses frequently offer **cash discounts** as a way to encourage their business customers to pay their invoices early. Cash discounts are figured on the *invoice price* that is shown on the invoice as the *total due.* The terms of sale tell how much of a discount will be given.

For example, the terms of Farwell Paints' invoice are 2/10, 1/30, n/60. That means that 2% may be deducted from the invoice price if it is paid within 10 days of the invoice date. If the invoice is paid within 30 days, 1% may be deducted. The full amount is due within 60 days.

Sometimes a term such as 3/10 EOM is used. This means that the buyer can claim a 3% discount if the bill is paid within 10 days after the end of month shown on the invoice.

To find the due date of the invoice and the last day for receiving a discount, count ahead, from the date on the invoice, the number of days shown in the terms.

EXAMPLE 1

The credit terms are 2/10, 1/30, n/60, and the invoice date is June 10. Find the last date for taking each cash discount and the date on which the invoice is due.

SOLUTION

Add 10 days to the invoice date.

June 10 + 10 days = June 20 Last day for 2% discount

Count 30 days from the invoice date.

20 Days = June 10 to June 30
10 Days = July 1 to July 10

30 Days July 10 Last day for 1% discount

Count 60 days from the invoice date.

20 Days = June 10 to June 30
31 Days = July 1 to July 31
9 Days = August 1 to August 9

60 Days August 9 Due date of invoice

✔ CHECK YOUR UNDERSTANDING

A. Credit terms of 1/15, n/30 appear on an invoice dated November 28. What are the discount date and the due date of the invoice?

B. An invoice dated March 20 has credit terms of 3/15, 1/30, n/45. Find the discount dates and due date of the invoice.

If the customer pays the invoice within the discount period and deducts a cash discount from the invoice price, the amount paid is called the *cash price.*

Cash Discount = Invoice Price × Rate of Cash Discount

Cash Price = Invoice Price − Cash Discount

Sometimes shipping charges appear on the invoice. If so, they must be added to the cash price to find the total amount due to the seller. Cash discounts are not allowed on shipping charges. If the terms of sale show f.o.b. destination, or f.o.b. customer, this means the seller will pay the shipping charges and carry the liability for the shipment. Terms of sale such as f.o.b. factory, or f.o.b. shipping point, mean the buyer will pay shipping costs and carry the liability for the shipment.

Business *Tip*

The shipping charge may be identified on the invoice as shipping and handling or freight.

EXAMPLE 2

Find the amount of cash discount Farwell Paints would get if the invoice price of $1,025.40 were paid within the 2% discount period. Also find the cash price of the invoice Farwell would pay after taking the discount.

SOLUTION

Multiply the invoice price by the discount percent to find the cash discount.

$2\% \times \$1{,}025.40 = 0.02 \times \$1{,}025.40 = \$20.508$, or $20.51

Subtract the cash discount from the invoice price.

$\$1{,}025.40 - \$20.51 = \$1{,}004.89$ cash price

Photodisc/Getty Images

✔ CHECK YOUR UNDERSTANDING

C. Find the cash discount and the cash price of a $23,747 invoice if the invoice is paid within the first discount period. The credit terms are 4/5, 1/30, n/60.

D. An invoice dated July 28 totals $12,836. Credit terms are 2/15, 1/30, n/45. If the invoice totals paid on August 13, what are the cash discount and the cash price?

Trade Discounts

Many businesses offer **trade discounts**. These are reductions from the list price to their business customers. Trade discounts are usually figured on each item within the invoice since it is possible for a different trade discount to apply to different items on an invoice. Also, the trade discounts offered to customers may differ.

Trade discounts are always based on the list price of the items being purchased. To find the trade discount, multiply the list price times the rate of discount. To find the *invoice price,* or net price, subtract the trade discount amount from the list price.

Trade Discount = Rate of Discount × List Price

Invoice Price = List Price − Trade Discount

EXAMPLE 3

The list price of an air compressor is $480. The trade discount given to a retailer is 40%. What invoice price will a retailer pay for the compressor?

SOLUTION

Multiply the list price by the rate of discount.

40% × $480 = 0.4 × $480 = $192   trade discount

Subtract the trade discount from the list price.

$480 − $192 = $288   invoice price

✔ CHECK YOUR UNDERSTANDING

E. The list price of a gas grill is $249 with a 35% trade discount to retailers. What trade discount would a retailer get on the grill? What invoice price would be paid?

F. A water filter system has a list price of $172. What trade discount would a retailer get if a $37\frac{1}{2}\%$ trade discount is given? What is the invoice price of the filter system?

Rate of Discount

If you know the discount amount and the total amount on which a discount is based, you can find the *rate of discount.* For example, the cash discount divided by the invoice price (invoice amount) will give the rate of cash discount. The trade discount divided by the list price will give the rate of trade discount.

Rate of Cash Discount = Cash Discount ÷ Invoice Price

Rate of Trade Discount = Trade Discount ÷ List Price

EXAMPLE 4

By taking advantage of a cash discount, a retailer paid $2,522 to settle a $2,600 invoice. Find the amount of the cash discount and the rate of cash discount.

SOLUTION

Subtract the cash price from the invoice price.

$2,600 − $2,522 = $78   cash discount

Divide the cash discount by the invoice price.

$78 ÷ $2,600 = 0.03, or 3%   rate of cash discount

EXAMPLE 5

A garage door opener is listed in a wholesaler's catalog at $220. The opener is sold to retailers at an invoice price of $154. What is the rate of trade discount?

SOLUTION

Subtract the invoice price from the list price.

$220 − $154 = $66 trade discount

Divide the trade discount by the list price.

$66 ÷ $220 = 0.3, or 30% rate of trade discount

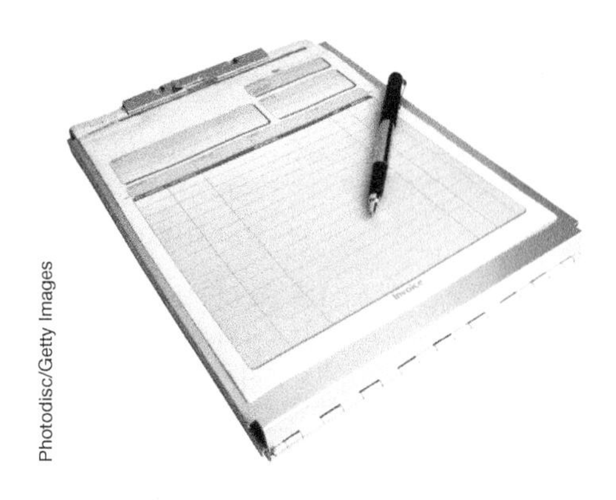

Photodisc/Getty Images

✔ CHECK YOUR UNDERSTANDING

G. An invoice totaling $5,200 was paid within the discount period. The amount paid was $5,070. What was the rate of cash discount?

H. A swimsuit with a list price of $75 was sold to a retailer for $41.25. What was the amount and rate of trade discount?

Wrap Up ►►►

Many new businesses may not have the cash available to take advantage of an immediate cash discount and will buy merchandise on credit. As they sell the stock they buy they use that money to pay the invoice when it is due. For this reason it is unlikely that a new business will choose to take an immediate cash discount.

TEAM Meeting

If you could buy an item directly from a factory, you would save part of the markup that distributors, such as wholesalers and retailers, charge. Distributors get a markup on the products they sell because they perform an important economic function of creating time utility, place utility, and form utility.

Form a team and do library or Internet research to find the meanings of these terms. Then give an example of how the terms apply to distributors. Present your findings to the class.

Exercises

Find the difference.

1. $7,117 − $213.51
2. $99.98 − $29.99

Find the product.

3. $1\frac{1}{2}\% \times \$6,027.45$
4. $33\frac{1}{3}\% \times \$480$

Find the percent.

5. $18.20 ÷ $910
6. $0.91 ÷ $2.60

Find the date on which each invoice must be paid.

	Invoice Date	Terms
7.	August 16	10 days
8.	March 5	30 days
9.	November 14	90 days

	Invoice Date	Terms
10.	April 24	60 days
11.	October 28	75 days
12.	January 7	45 days

Solve.

13. An invoice with a total due of $3,400 is paid within the discount period. Credit terms are 2/30. Find the cash discount and the cash price.

14. The credit terms of an invoice dated June 23 are 2/10, 1/20, n/30. The total amount due on the invoice is $3,106. Find the cash discount and cash price if the invoice is paid July 5.

15. The usual cost to ship men's shirts is $16 a dozen. A retailer in Peoria bought 6 dozen men's shirts on March 16 from a wholesaler in Chicago at $16.32 per shirt. The terms of the sale were 2/15, n/30; f.o.b. Chicago. The invoice was paid by check on March 29. What was the amount of the check?

16. How much would a retailer pay for 30 dozen work gloves if the wholesaler's list price is $62 a dozen, less 28%?

17. Josh Hill, the owner of Hill's Auto Parts, paid $6,873.75 to settle Invoice #B-2826 after taking a cash discount. The original invoice amount was $7,050. What rate of cash discount was given?

Carrie Osterman, a storeowner whose shop is on a boardwalk by the Atlantic Ocean, buys 250 beach towels with a list price of $18.20 each. She receives a 40% trade discount.

18. What amount of discount will she get on the entire order?

19. What will be the invoice price of the order?

Photodisc/Getty Images

An invoice of $1,230 was paid within the discount period with a check written for $1,211.55.

20. What cash discount was received?

21. What was the cash discount rate?

Carlos Durbin paid an invoice on February 16. The invoice for $783.20 was dated January 7 with credit terms of 3/20 EOM.

22. What was the last day on which a discount could be taken?

23. If a discount could have been taken, what amount was sent when the invoice was paid?

24. A rocking chair with a list price of $245 is offered to a retailer for $98. What rate of trade discount is offered?

A retailer ordered 8 dozen casual slacks for $14.50 each. The list price of the slacks is $25.

25. What trade discount was given per slack?

26. What was the rate of trade discount?

27. What will be the amount of the invoice when it is received?

28. **BEST BUY** Pearl Keating, a storeowner, gets prices on a curio cabinet from two wholesale firms. The Trill Company offers a cabinet for $800, less 40%. Pender Products offers the same cabinet for $650, less 30%. From which firm should Pearl buy the cabinet in order to get the lowest price? How much less will the lower price be?

29. **FINANCIAL DECISION MAKING** Name three things that you consider when you buy a product and rank them in order of importance. Next list three things that you think a business considers when it buys a product and rank them in order of importance. What are the similarities and differences between the lists and the rankings?

STRETCHING YOUR SKILLS On November 6, Owen Tormo bought $28,000 worth of goods with terms of 3/10, n/30. To pay the invoice on November 16 and get the 3% discount, Owen borrowed the cash price of the invoice at his bank for 20 days at 12%, ordinary interest.

30. What was the cash price of the invoice?

31. What was the cost of interest on the loan?

32. What cash discount did Owen receive?

33. How much did Owen save by borrowing money to take the cash discount?

Mixed Review

34. Write 1.93 as a percent.

35. Find $\frac{2}{5}$ of 550.

36. Estimate: 3,016 ÷ 18

37. $\frac{2}{3} \div 48$

38. Find the quotient of 9.5 ÷ 2.4 to the nearest hundredth.

39. The Bremmers want to buy a home. They estimate these home operating expenses: property tax, $3,400; insurance, $485; utilities, $1,480; maintenance, $1,200; mortgage interest, $6,400; and lost interest on down payment, $420. Their estimated income tax savings are $1,800. What will be the net cost of owning the home in the first year?

40. Seletha Payne bought 150 shares of Regal Building Materials preferred stock. The stock paid a quarterly dividend of 1.25% on a par value of $100. What total amount in dividends did Seletha receive from owning this stock for one year?

41. Nanno Metal Products estimated that the fixed costs of producing filing cabinets are $111,300 and their variable costs are $65 a cabinet. The filing cabinets will sell for $118. How many cabinets must be sold to break even, to the nearest unit?

42. Ridgeway Industries buys a planer for $3,500. The estimated life is 8 years, and the annual depreciation rate used is 12%. What will the book value of the planner be at the end of 3 years?

43. What is the girth of a box that is 40 inches long, 10 inches wide, and 5 inches deep?

10-3

Series Trade Discounts

GOALS

- Calculate the invoice price for a series of discounts
- Calculate the single discount equivalent to a series of discounts

KEY TERMS

- list price
- invoice price
- discount series

Start Up

Have you ever seen or been offered a product at a price less than the suggested retail price printed on the package? How can a retailer afford to sell for less than the suggested price?

ifong/Shutterstock.com

Math Skill Builder

Review these math skills and solve the exercises that follow.

1 **Subtract** money amounts and percents.
Find the difference. $1,807 − $36.14 = $1,770.86
Find the difference. 82% − 8.2% = 73.8%

1a. $545.60 − $163.68 **1b.** $281.32 − $4.22

1c. 90% − 13.5% **1d.** 95% − 9.5%

2 **Multiply** money amounts by percents.
Find the product. 15% × $759.29 = 0.15 × $759.29 = $113.893, or $113.89

2a. 35% × $17.12 **2b.** 7% × $763.11

3 **Multiply** percents by percents.
Find the product. 80% × 90% = 0.8 × 0.9 = 0.72, or 72%

3a. 70% × 90% **3b.** 85% × 90%

Series Discounts

In the previous lesson, you saw that businesses frequently offer business customers trade discounts. Companies, such as wholesalers, often publish a catalog that lists items offered for sale and their prices. The price in the catalog is called the **list price**, which is the price that a consumer might pay. The price the retailer pays after the trade discount is given is called the **invoice price**.

To save the cost of reprinting their catalog when list prices change, wholesalers simply change the trade discount percent to raise or lower the prices they charge retailers.

Some wholesalers may give a trade discount that has two or more discounts, called a **discount series** or a series of discounts. For example, the trade discount may be 20%, 15%, 10%. This means that three discounts are given on an item.

To find the invoice price, the first discount is based on the list price. The second discount is based on the remainder after deducting the first discount. The third discount is based on the remainder after deducting the second discount, and so on.

EXAMPLE 1

The list price of a display case is $2,100, less 25%, 10%, and 5%. Find the invoice price and the amount of trade discount.

SOLUTION

Multiply the list price by the rate of the first discount. Subtract the amount of discount from the list price. Use each remainder you get to repeat the process until all discounts have been taken.

$2,100.00	List price
− 525.00	First discount (25% of $2,100)
$1,575.00	First remainder
− 157.50	Second discount (10% of $1,575)
$1,417.50	Second remainder
− 70.875	Third discount (5% of $1,417.50)
$1,346.625	Third remainder

$1,346.625 = $1,346.63 invoice price

Subtract the invoice price from the list price.

$2,100 − $1,346.63 = $753.37 trade discount amount

Math Tip

Note that amounts are rounded to the nearest cent ONLY after all computations are done.

Math Tip

Check your work by taking the series discounts in a different order.

✔ CHECK YOUR UNDERSTANDING

A. What are the invoice price and the amount of discount for a kitchen cabinet that has a list price of $260, with discounts of 30%, 20%, and 10%?

B. A faucet with a list price of $140 is offered to retailers with discounts of 20%, 20%, and 5%. Find the invoice price and the amount of discount.

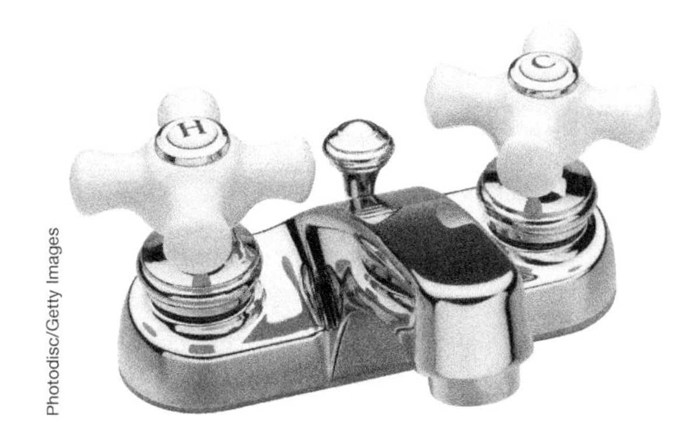

Photodisc/Getty Images

Single Discount Equivalent

If you buy regularly from one vendor and always receive the same discount series, you can calculate the invoice price faster by using one discount that is equal to the series of discounts. That one discount is called the *single discount equivalent*.

EXAMPLE 2

Use the three methods below to find the single discount equivalent and the invoice price for a pair of hiking boots that have a list price of $80. They are offered to retailers with a series discount of 20%, 10%, and 10%.

a. percent method b. table method c. complement method

SOLUTION

a. The Percent Method: To find the single discount equivalent for a series discount of 20%, 10%, and 10% follow these steps.

Step 1:	100%	List price
	− 20%	First discount (20% of 100%)
	80%	First remainder
	− 8%	Second discount (10% of 80%)
	72.0%	Second remainder
	− 7.2%	Third percent (10% of 72%)
	64.8%	Third remainder, or invoice price

Step 2: 64.8% × $80 = 0.648 × $80 = $51.84 invoice price

Step 3:	100.0%	List price
	− 64.8%	Invoice price
	35.2%	Single discount equivalent

b. The Table Method: To find the single discount equivalent for a series discount of 20%, 10%, and 10% follow these steps.

The table shows the invoice price equivalents for a variety of series discounts.

Rate	5%	10%	15%	20%	25%	30%
5%	0.9025	0.855	0.8075	0.76	0.7125	0.665
5%, 5%	0.85738	0.81225	0.76713	0.722	0.67688	0.63175
10%	0.855	0.81	0.765	0.72	0.675	0.63
10%, 5%	0.81225	0.7695	0.72675	0.684	0.64125	0.5985
10%, 10%	0.7695	0.729	0.6885	0.648	0.6075	0.567

Step 1: Locate the 20% column. Then find the 10%, 10% row.
The invoice price equivalent is 0.648, or 64.8%

Step 2: 64.8% × $80 = 0.648 × $80 = $51.84 invoice price

Step 3: 100.0% − 64.8% = 35.2% single discount equivalent

c. The Complement Method: To find the single discount equivalent for a series discount of 20%, 10%, and 10% follow these steps.

The complement of any discount rate is the difference between that rate and 100%. For example, if the discount rate is 30%, the complement of the discount rate is 70%.

Step 1: Multiply the complements of the discounts.
0.80 × 0.90 × 0.90 = 0.648 = 64.8% invoice price equivalent

Step 2: 64.8% × $80 = 0.648 × $80 = $51.84 invoice price

Step 3: 100% − 64.8% = 35.2% single discount equivalent

Look back to the previous examples, parts a, b, and c. In Step 1 you found the percent that the invoice price is of the list price. In Step 2 you multiplied the list price by the percent (invoice price equivalent) to calculate the invoice price.

In Step 3 you subtracted the percent that is the invoice price equivalent from 100%. The result is a single discount, 35.2%, which is equivalent to the series discount of 20%, 10%, and 10%. If you know the single discount equivalent, you may also calculate the invoice price this way:

$35.2\% \times \$80 = \$28.16$ discount amount

$\$80 - \$28.16 = \$51.84$ invoice price

✔ CHECK YOUR UNDERSTANDING

Find the single discount equivalent to the series discount, the amount of trade discount, and the invoice price for Problems C, D, and E, using the method indicated.

C. Percent Method: series discount, 20%, 10%, 5%; list price $900.

D. Table Method: series discount, 30%, 5%, 5%; list price, $34.16.

E. Complement Method: series discount, 10%, 25%, 5%; list price, $12.80.

Wrap Up ▶▶▶

The retailer pays less than the suggested retail price for goods. As long as the actual selling price is greater than the retailer's invoice price, a gross profit is made on the sale. Also, the "suggested price" is just that, a suggested price, not a required price.

TEAM Meeting

Some manufacturers sell only through authorized dealers. A new company that wishes to become an authorized dealer is told that there are enough dealers in the area already. The new company states that this policy is unfair, and that anyone should be able to buy and sell any product they wish.

Two teams should be formed to debate the issue that manufacturers have the right to use authorized dealers.

- One team should take the *pro position* and agree with the issue.
- The second team should take the *con position* and argue against the issue.

The debate should be held in class. The class members that are not on a debate team will listen to the debate and decide which team wins the debate.

Exercises

Find the difference.

1. $2,689 − $672.25
2. 88% − 13.2%

Find the product.

3. $12\frac{1}{2}\% \times \$250.37$
4. $45\% \times \$87.90$
5. $80\% \times 85\% \times 75\%$
6. $95\% \times 95\% \times 80\%$

Find the invoice price and the amount of trade discount. Check your work by taking the discounts in a different order.

	List Price	Trade Discount	Invoice Price	Amount of Trade Discount
7.	$120	20%, 10%		
8.	$87	20%, 10%, 10%		
9.	$315	30%, 15%, 10%		

For Exercises 10–15, use any method you choose to find the single discount equivalent for each. Use each method at least once. Show your work.

10. 10%, 10%
11. 20%, 25%
12. 20%, 12.5%
13. 10%, $33\frac{1}{3}\%$
14. 25%, 20%, 5%
15. 30%, 20%, $12\frac{1}{2}\%$

Sovle.

16. A jade bracelet has a list price of $460 and series discounts of 30%, 10%, and 5%. Find the invoice price.
17. Xavier, Inc. sells a computer chair for $238, with discounts of 20%, $12\frac{1}{2}\%$ and 7%. Find the invoice price.
18. What invoice price will a retailer pay for a book with a list price of $35 and series discounts of 20%, 20%, and 15%?
19. A garden hose with a list price of $32 is offered for series discounts of 15%, 15%, and 10%. What are the amounts of the invoice price and trade discount?

Photodisc/Getty Images

20. **FINANCIAL DECISION MAKING** Newland Distributing offers to deliver an order of 30 sets of patio furniture for $14,000 list price, less 30%, 20%, and 5%. Ingle Wholesale offers the same furniture at 25%, 20% and 10%. Which is the best offer, and how much would be saved by taking the better offer?
21. **CRITICAL THINKING** The order in which a series of discounts are taken may be changed and the answer will be the same. Why is this so?

INTEGRATING YOUR KNOWLEDGE A hotel wants to buy a new ice machine. Three vendors were asked to submit bids. The terms of their bids are shown in the chart below. Assume the hotel will take the highest cash discount offered. Copy and complete the chart by finding the invoice price and the total due. Then answer these questions:

22. Which vendor offers the best package price?

23. If the purchase is made on March 18, what is the due date of the invoice?

	Vendor A	Vendor B	Vendor C
List Price	$1,600	$1,650	$1,930
Cash Discount	1/30, n/60	2/30, n/60	2/10, 1/30, n/60
Trade Discount	5%	3%, 7%	10%
Shipping Terms	FOB Factory	FOB Factory	FOB Seller
Delivery Charges	$175	$203	$175
Invoice Price			
Total Due			

Mixed Review

24. $\frac{1}{6}$ of $468.30 is what number?

25. Write $\frac{1}{10}$% as a decimal.

26. $87\frac{1}{2}$% of $2,480

27. 30% less than $570

28. Round 0.1896 to the nearest hundredth.

29. Fiona Adams is paid a salary of $187 a week and a commission of 2% on sales. Last week her sales were $58,120. What were her total wages for the week?

30. On October 1, Dwight's check register had a balance of $943.06. His bank statement balance on the same date was $773.55. Checks outstanding were #259, $56.71; #261, $33.78. A line on the bank statement showed that the bank charged his account $260 for the October car loan payment due the bank. Reconcile the statement.

31. The preparation of a job description and job specifications for a college recruiter position took 12 hours of human resources staff time. The direct costs were: wages at $23.60 an hour, fringe benefits at 35% of wages, and 2 hours of computer time at $12.60 an hour. What was the total direct cost of this project?

32. The records of XD Products show these costs for producing 12,000 garden hoses: raw material, $14,460; direct labor, $23,760; overhead, $8,112. What was the prime cost of making the hoses? What was the manufacturing cost of a hose, to the nearest cent?

10-4

Markup and Markdown

GOALS
- Calculate cost and selling price when markup is based on selling price
- Calculate the rate of markup based on cost
- Calculate markdown and selling price

KEY TERMS
- markup
- markdown

Start Up

Often new products appear on the market and there are no competitors in sight. Because of this, a very high price may be charged for the item. Would the manufacturer or seller of the product be better off by charging a lower price so as to put it into the price range of more buyers?

Photodisc/Getty Images

Math Skill Builder

Review these math skills and solve the exercises that follow.

1. **Subtract** money amounts and percents.
Find the difference. $120 − $56.80 = $63.20
Find the difference. 100% − 37.5% = 62.5%

1a. $87 − $65.10

1b. $199.95 − $59.99

1c. $100\% - 33\frac{1}{3}\%$

1d. 100% − 38.5%

2. **Multiply** money amounts by percents.
Find the product. 60% × $118 = 0.6 × $118 = $70.80

2a. 85% × $6.24 **2b.** 55% × $18.95

2c. 34% × $15.99 **2d.** 72% × $46.55

3. **Divide** money amounts by money amounts.
Find the quotient. $63 ÷ $180 = 0.35, or 35%

3a. $16.25 ÷ $65 **3b.** $45 ÷ $50

3c. $60 ÷ $80 **3d.** $12 ÷ $30

Markup Based on Selling Price

Businesses have to decide what price to charge for an item. The price must cover the cost of the item, all expenses, including overhead, and generate a profit.

One way businesses price items is to use *markup pricing.* With markup pricing, an amount is added to the cost of the goods to cover all other expenses plus a profit. This is known as the **markup** or *margin.* The *selling price* is the price at which the item is actually sold.

Retailers often sell goods in *price lines.* For example, an auto supply store may stock three price lines of car batteries. One price line sells for $79.99; another line sells for $69.99; a third line sells for $59.99. The different price lines are expected to appeal to the different needs of buyers.

When batteries are bought by a retailer for a price line, such as $59.99, the selling price is already known. The problem is to find the highest price the retailer can pay for the batteries and still get the markup it wants. If the cost of the batteries is too high, the retailer will make too little profit in selling them.

EXAMPLE 1

Ida Schrader owns the Luggage Place and has to buy carry-on luggage for a line she sells for $90. She knows that her markup must be 40% of the selling price to cover expenses and get the net income she wants. What is the highest price she can afford to pay for each piece of luggage?

SOLUTION

Method 1:

Multiply the selling price by the markup rate to find the markup.

$40\% \times \$90 = 0.4 \times \$90 = \$36$

Subtract the markup from the selling price to find the cost:
$\$90 - \$36 = \$54$

Method 2:

Subtract the markup rate from 100% to find the cost as a percent:
$100\% - 40\% = 60\%$

Multiply the selling price by the cost as a percent to find the cost.

$60\% \times \$90 = 0.6 \times \$90 = \$54$

Algebra *Tip*

You can use the following formula when the rate of markup is based on the *selling price.*

Cost = Selling Price × (100% − Markup)

When you are trying to find the Cost, the formula can be transformed using algebra to:

Selling Price = Cost ÷ (100% − Markup)

or

$$\text{selling price} = \frac{\text{Cost}}{(100\% - \text{Markup})}$$

✔ CHECK YOUR UNDERSTANDING

A. A wholesaler wants to sell a drill to retailers for $45 and earn a markup of 20% on the selling price. What is the most the wholesaler can pay a drill manufacturer and get the desired markup?

B. Marion Cassidy, a jeweler, wants to set a price of $600 for a bracelet. The markup he wants is 60% of the selling price. What is the most he can pay for the bracelet?

Retailers also buy goods without a specific selling price in mind. However, since they know the cost of an item and the rate of markup *on the selling price* they want, they can calculate the selling price. The selling price is found by dividing the cost of an item by the percent that cost is of the selling price.

Cost as a Percent of Selling Price = 100% − Rate of Markup

Selling Price = Cost ÷ Cost as a Percent of Selling Price

EXAMPLE 2

A retailer bases its markup of shoes on their selling price. What is the selling price of a shoe with a cost of $56 and a 30% markup on the selling price?

SOLUTION

100% − 30% = 70%   cost as a percent of selling price

$56 ÷ 70% = $56 ÷ 0.7 = $80   selling price

✔ CHECK YOUR UNDERSTANDING

C. The cost of $\frac{1}{2}$ oz. of a perfume is $18. The markup is 60% of the selling price. What is the selling price?

D. A birdbath costs $20.30. A retailer wants a markup of 42% of the selling price. What is the selling price of the birdbath?

Markup Based on Cost

While many businesses use the selling price to figure markup, others use the cost. When the cost price is used, you find the rate of markup by dividing the markup by the cost.

Rate of Markup = Markup ÷ Cost

EXAMPLE 3

The cost of a rain suit that Emil Shulman bought for his store is $120. Emil wants to sell the jacket for $162. What is the rate of markup on the cost of the rain jacket?

SOLUTION

Subtract the cost from the selling price.

$162 − $120 = $42   markup

$42 ÷ $120 = 0.35, or 35%   rate of markup

✔ CHECK YOUR UNDERSTANDING

E. The cost of a comb is $0.80. A retailer wants to sell the comb for $1.28. What rate of markup based on cost will the seller have to use?

F. A grocer pays $2.50 for a half gallon of milk. The grocer wants to sell the milk for $2.70. What is the milk's rate of markup on cost?

Markdown

In some cases, retailers reduce prices by applying a **markdown** to their *marked price*, or original selling price. The marked price is the price that is marked on the item. Recall that the selling price is always the price the item actually sold for.

A markdown may be taken on the marked price at the end of a season or on items that are not selling well. Markdowns might also be done to attract customers to the store for a sale or to be more competitive.

Markdown, which is also known as *discount*, is stated as a percentage of a marked price. The markdown formulas we will use are:

Markdown = Rate of Markdown × Marked Price

Selling Price = Marked Price − Markdown

EXAMPLE 4

To move an overstocked item, a discount of 15% is given on a backpack with a marked price of $70. Find the amount of the discount and the selling price.

SOLUTION

Multiply the marked price by the rate of discount.

15% × $70 = 0.15 × $70 = $10.50   discount, or markdown

Subtract the amount of discount from the marked price.

$70 − $10.50 = $59.50   selling price

✔ CHECK YOUR UNDERSTANDING

G. A man's suit with a marked price of $270 is being offered at a 20% discount. What is the selling price of the suit?

H. At the end of the selling season a flat of flowers marked at $12.50 is discounted 50%. What is the selling price of the flat?

Photodisc/Getty Images

Consumer Alert

Dating back as far as 1954, wholesale clubs have been an option for consumers. A wholesale club is a retail store to which consumers must pay an annual membership fee so they can shop at the club's locations. The idea behind these retail clubs is that their stores appear to be more like warehouses that offer low prices on products. In many cases, products are sold in large quantities.

Consumers can save money shopping at wholesale clubs and can usually recoup their membership fees after several purchases. Beware that not all items in a wholesale club are priced cheaper than at local discount stores. Many times buyers incorrectly assume that because they are in a wholesale club the price is the lowest that can be found.

Wrap Up ▶▶▶

It is possible that the new item has a very high manufacturing cost and must be sold at a high price. It is also possible that the company has a marketing strategy of first charging the highest price possible knowing they will attract some buyers. The next step in their strategy would be to cut the price so as to bring in the most anxious buyers at that price level, and so on until the item is widely distributed.

Exercises

Find the difference.

1. $4.56 − $4.04
2. $1,276 − $982
3. 100% − $92\frac{1}{2}$%
4. 100% − 26.5%

Find the product.

5. 72.5% × $10.76
6. 28% × $1,764

Find the quotient, as a percent.

7. $0.24 ÷ $2
8. $21.09 ÷ $56.24

For Exercises 9–13, markup is based on selling price. Find the cost as a percent of selling price. Also find the selling price.

	Item	Cost	Rate of Markup	Cost as a Percent of Selling Price	Selling Price
9.	Mattress	$562.50	55%		
10.	Framed Print	$112.70	54%		
11.	Sandals	$4.51	45%		
12.	Trailer Mirror	$26.22	31%		
13.	Sleeping Bag	$46.78	$37\frac{1}{2}$%		

Solve.

14. Margaret Kogut buys golf bags for a line with a selling price of $80. What is the most she can pay for a golf bag if the markup must be at least 45% of the selling price?
15. A wholesaler has a line of rugs that sell for $129.50 each. What is the highest price the wholesaler can pay for the rugs and make a markup of 44% of the selling price?
16. What is the most that a store owner should pay per dozen for slippers with a selling price of $28.75 each if the owner needs a markup of 34% on the selling price?
17. Lori McCabe owns a hardware store and figures markup based on selling price. For a special sale, she buys 500 rolls of masking tape at 48¢ each. Lori wants a markup of 20%. What is the selling price of each roll?

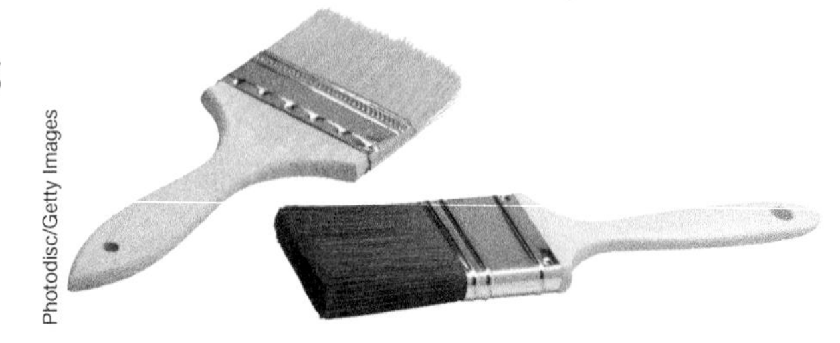

For Exercises 18–22, find the markup and the rate of markup based on cost.

	Stock Number	Cost	Selling Price	Markup	Rate of Markup On Cost
18.	501-A	\$77.25	\$101.97		
19.	501-B	\$150.00	\$225.00		
20.	503-AA	\$720.00	\$864.00		
21.	556-M	\$110.25	\$147.00		
22.	576-V	\$42.00	\$60.90		

Solve.

23. A retailer paid \$220 for an end table plus \$28 for delivery charges. What is the lowest price at which the table may be sold if a 48% markup, based on cost, is wanted?

24. Amy Hearns' bicycle shop regularly sells a racing bike for \$1,285. At the end of the season, Amy reduces the price of the bike to \$1,050. What is the rate of markdown, to the nearest percent?

25. A catalog store discounts the price of a pool liner to \$215 from \$280. The cost of shipping the liner is \$34.50. What is the rate of markdown, to the nearest tenth percent?

26. **INTEGRATING YOUR KNOWLEDGE** Barry Waldo, a retailer, purchased 25 cooler chests for \$28.10, less 30% and 10%. Barry priced the coolers at a markup of 35% on the selling price. What are the cost price and the selling price of each cooler?

27. **INTEGRATING YOUR KNOWLEDGE** A wholesaler buys a shop vacuum from a manufacturer for a 30% trade discount from its list price of \$100. What rate of markup on selling price will the wholesaler have to charge so the vacuum sells for the \$100 list price?

Mixed Review

28. $0.87 \div 10$

29. $4\frac{11}{12} + 6\frac{5}{6}$

30. $1.274 \div 4.9$

31. $856 - 0.427$

32. Write 0.034% as a decimal.

33. What is 12% more than \$842?

34. The Sorgen Delivery Company hires per diem workers during busy periods at \$96 a day. Last year, Sorgen hired 4 per diem workers for six weeks in the winter and 8 per diem workers for one week in the spring. The workweek at Sorgen is Monday through Friday. What total amount did Sorgen spend in the year on per diem workers?

35. Zeppa Electronics uses National's two-day service (see Lesson 9-4) to ship customer purchases. On Tuesday, Zeppa shipped 20 identical packages, each weighing 1.75 lb, to customers in zone 3. Insurance will cost \$2.10 per package. What is the total cost of Tuesday's shipment?

36. You can buy a TV for \$275 cash or pay \$50 down and the balance in 18 monthly payments of \$18.70. What is the installment price? By what percent, to the nearest tenth of a percent, would your installment price be greater than the cash price?

10-5

Marketing Surveys

GOALS

- Calculate the response rate of surveys
- Calculate the results of surveys

KEY TERM

- respondent

Start Up ▸▸▸

You just completed a 10-minute telephone survey covering your opinions about your bank. At the end of the survey you are asked for the amount of your total income. When you hesitate, the interviewer tells you that your survey can't be counted unless you answer all questions. Why would your income be important to the survey's results?

Monkey Business Images/Shutterstock.com

Math Skill Builder

Review these math skills and solve the exercises that follow.

1 **Add** whole numbers.
Find the sum. 1,282 + 281 = 1,563

1a. 238 + 94 | **1b.** 3,879 + 1,082

2 **Divide** whole numbers to find a percent.
Find the percent. 84 ÷ 560 = 0.15, or 15%

2a. 180 ÷ 720 | **2b.** 2,883 ÷ 4,650

2c. 189 ÷ 420 | **2d.** 452 ÷ 565

Response Rate

Businesses survey customers to find what they think about products and services. Mail, Internet, telephone, or personal contact may be used. The total potential number of people or organizations that may be surveyed is called the *population.* Organizations such as the Bureau of Census try to survey every person in the U.S. To save time and money, most organizations survey only part of the total population, called the *sample,* or *sample population.*

To insure that the people in the sample represent the people in the whole population, demographic data about the people in the sample are collected. *Demographic data* may include age, sex, education, marital status, occupation, and income.

People who complete surveys are called **respondents**. Surveys are complete and *valid* when all questions are answered and demographic data are collected. Some responses may not be counted if a quota for respondents with certain characteristics has been filled. The response rate to a survey is found by dividing the number of responses by the number of surveys attempted.

Response Rate = Responses ÷ Surveys Attempted

EXAMPLE 1

The Danton Company mailed 25,000 surveys to users of its Silk Smooth Soap. Of the 1,257 returned surveys, 57 were discarded as incomplete. What was the survey's response rate?

SOLUTION

Subtract the discarded surveys from the total surveys returned.

1,257 − 57 = 1,200 valid surveys

Divide the valid surveys by the total surveys.

1,200 ÷ 25,000 = 0.048, or 4.8% response rate

✔ CHECK YOUR UNDERSTANDING

A. An auto research firm mailed questionnaires to the 240,000 buyers of a certain car model. Replies were received from 149,050 buyers, but 250 incomplete surveys were not counted. What was the response rate to the mailing?

B. A manufacturer surveyed 80,000 buyers who registered their new printers online. There were 34,387 respondents to the initial e-mail survey request and 28,113 respondents to an e-mail reminder. Of that total, 4,100 surveys did not have enough information and were not counted. How many valid surveys were there? What was the response rate?

Survey Results

The results of surveys are tabulated to provide information about the respondents. Companies use this information to improve their products or to plan more effective ways to sell products or services. The results of a product satisfaction survey are shown.

Danton Company Silk Smooth Satisfaction Survey					
	Product Rating				
Ages	Excellent	Good	Average	Poor	Total
18–30	78	185	118	15	396
31–45	145	190	31	37	403
46–60	203	135	52	11	401
Totals	**426**	**510**	**201**	**63**	**1,200**

Business *Tip*

Business people often tabulate data to make them easier to understand. *Tabulate* means to place data in columns and rows and take totals of the columns and rows.

EXAMPLE 2

What percent of all respondents, from the survey results on the previous page, gave Silk Smooth soap a rating of better than average?

SOLUTION

Add the totals of the excellent and good rating categories.

$426 + 510 = 936$ total respondents, better than average rating

Divide the above total by the total number of respondents.

$936 \div 1{,}200 = 0.78$, or 78% rating, better than average

Photodisc/Getty Images

✔ CHECK YOUR UNDERSTANDING

C. What percent of all respondents rated Silk Smooth as average, to the nearest tenth of a percent?

D. What percent of respondents in the 31–45 age group gave Silk Smooth a poor rating?

Wrap Up ▸▸▸

Income, along with other demographic data, gives the bank a profile of your personal characteristics to see if your views about the bank are typical of other people with the same characteristics. However, you are under no obligation to provide such personal information.

TEAM Meeting

You and your team members are to conduct a survey of 200 people to determine what they think about a movie. First, select a relatively new movie that would appeal to a wide audience. Then construct a simple questionnaire to determine whether the respondents like or do not like the movie, or have not seen the movie. Demographic data should be the age of the respondent in five-year segments, such as to 17 and 18 to 22, and so on. Also ask for the number of movies seen in a year within these categories: once a week, once a month, less than once a month.

Tabulate the data from the survey of your 200-person population. Then randomly select 20 surveys and tabulate their results. Compare the results of the population and sample surveys. Discuss the reasons for any differences you find in the results.

Exercises

Find the sum.

1. $738 + 183$

2. $1{,}583 + 192 + 3{,}192$

Find the percent, to the nearest tenth percent.

3. $237 \div 1{,}815$

4. $32{,}717 \div 187{,}512$

Solve.

5. A researcher at a shopping mall stopped 214 young people who wore certain brands of clothing. Fifty-four people refused to be interviewed and 27 people did not answer enough questions to complete the survey. The rest answered all the questions. What was the total number of respondents? What was the response rate, to the nearest tenth of a percent?

6. A mail survey of 8,400 households received a 8.5% response rate. How many households responded to the survey?

7. A car manufacturer invited a group of customers to evaluate a new shade of red paint. The results showed that 500 people liked the new color, 315 liked the old shade better, 975 thought both colors looked alike, and 613 did not like the color red at all. To the nearest tenth of a percent, what percent of the respondents could not tell the difference between the old and new shades of red?

The Yard Pal Mower Company asked 550 customers to rate its newest lawnmower in three categories on a scale of 1–5, with 5 being the highest. The chart shown below summarizes the results of the survey.

Yard Pal Mower Company: Ratings of Selected Features						
Feature	Rating (5 = highest; 3 = average; 0 = lowest)					
	5	4	3	2	1	Totals
Easy to start?	350	117	40	14	29	550
Easy to change oil?	240	130	64	46	70	550
Easy to operate?	364	92	54	31	9	550

8. Which feature received the greatest number of average ratings?

9. What percent of respondents gave the "easy to change oil" category a rating of average or better, to the nearest percent?

10. What percent of respondents gave the "easy to operate" category a below-average rating, to the nearest tenth of a percent?

11. **CRITICAL THINKING** How do companies doing surveys select the people they survey to make sure they get accurate data?

12. **FINANCIAL DECISION MAKING** Drew Enterprises mailed a 90-question survey to 6,000 businesses in 12 states. Drew selected its sample population by taking every fifth business listed in a directory. They received 135 responses, mainly from small businesses. What was the response rate? Can the new company make any business decisions based on the survey?

Mixed Review

13. $\frac{4}{9} \div \frac{5}{20}$

14. 920×400

15. $3\frac{1}{2} \times 4\frac{2}{3}$

16. $5.41 \times 1{,}000$

17. Lance Bader earned $6,200 in January. He paid an overall FICA tax rate of 7.65% that consists of a 6.2% Social Security tax and a 1.45% Medicare tax. Federal income taxes withheld from his wages were $1,512. What total amount of taxes was deducted from his pay in January?

10-6

Sales Forecasts

GOALS
- Calculate future sales using trend data
- Calculate future sales using forecast methods

KEY TERMS
- forecast
- trend

Start Up ▶▶▶

During winter break week, Darlene Silvin, a high school student, earned $232 in gross wages at her part-time job. At that gross wage rate, Darlene figures that she will earn $12,064 in wages for a full year. She plans to buy a car with the money she will make. Do you agree with Darlene's forecast and plans?

Photodisc/Getty Images

Math Skill Builder

Review these math skills and solve the exercises.

1. **Multiply** money amounts by percents.
Find the product. $2,340,500 × 13% = $2,340,500 × 0.13 = $304,265

1a. $60,100,000 × 3.5%　　**1b.** $381,000 × 7.7%

2. **Multiply** whole numbers by whole numbers and money amounts.
Find the product. 12 × 1,800 = 21,600
Find the product. $9.12 × 58,000 = $528,960

2a. 48 × 6,780　　**2b.** 35 × 2,736

2c. $24.05 × 12,200　　**2d.** $7.34 × 31,900

Sales Trends

Many companies try to estimate their future sales. They may call these estimates sales **forecasts**, sales predictions, or sales projections. The sales forecast allows them to plan for future expansion and prepare annual budgets based on expected sales.

One way that companies may forecast future sales is by analyzing sales trends. A **trend** is a historical relationship between sales and time. For example, a trend may show that over 12 years sales have increased 10% each year, on average. A forecast based on this trend is that sales will be 10% higher next year than they are this year.

Using trends to forecast sales assumes that all the factors that influence sales remain the same. Some factors that may influence sales include consumer spending habits, new competition, and the general condition of the United States and world economies. Also, the longer the time period analyzed, the more accurate is the forecast.

EXAMPLE 1

Sales of Morning Treat cereal increased an average of 20% a year over the past six years. This year's sales of Morning Treat were $6,500,000. Based on this trend, what are the forecasted sales for next year?

SOLUTION

Multiply current sales by the average percent of sales increase.

20% × $6,500,000 = 0.2 × $6,500,000 = $1,300,000 forecasted sales increase

Add the forecasted sales increase to current sales.

$1,300,000 + $6,500,000 = $7,800,000 forecasted next year's sales

✔ CHECK YOUR UNDERSTANDING

A. The sales of Dentmile, a new toothpaste produced by Universal Brands, were $2,820,000 last year. Based on Universal's historical pattern of new product sales, Dentmile's sales are projected to be $2\frac{1}{2}$ times greater this year. What is the projection for this year's sales?

B. Last year's sales of an antivirus software package were 215,000 units to the business market and 98,000 units to the school market. Software sales to both markets are expected to increase by an average of 8% this year. What are the predicted total unit sales of the software to businesses and schools combined this year?

Forecasting Methods

In addition to analyzing sales trends, companies use other methods to forecast sales. These include market tests, surveys, sales projections, and management opinion.

Market tests involve the actual sale of a new product in a small geographic area, such as a large city. Based on the success of the product in the test market, a projected sales figure for all markets is calculated.

Surveys may ask consumers to examine a product and state whether or not they intend to buy the product.

Sales force projections are made from reports filed by sales staff about their customer's future spending plans.

Management opinion may be used to forecast sales for new products that have no sales history. Whatever method is used, the data you get do not guarantee actual sales.

EXAMPLE 2

The Useful Products Company developed a clip-on carrying case that holds two pairs of eyeglasses. In one test market, Useful Products sold 7,200 cases at \$3.80 a case. Useful Products now wants to sell the cases nationally in all 60 of its markets, all of which are similar in size. Based on sales in the test market, what total unit and dollar sales are forecasted for the first year?

SOLUTION

Multiply the test market unit sales by the number of markets.

$60 \times 7{,}200 = 432{,}000$ unit sales forecast for first year

Multiply the unit sales forecast by the unit price

$\$3.80 \times 432{,}000 = \$1{,}641{,}600$ dollar sales forecast for first year

✔ CHECK YOUR UNDERSTANDING

C. Sales of House-Glow dish soap increased by 2.5% in a test market due to a new bottle color. Total sales in all markets are now \$16,115,000. House-Glow plans to change the color permanently and predicts total sales to increase by the same percent as in the test market. By what amount are sales predicted to increase? What are the total predicted sales once the color change is made?

D. The sales staff of an office products dealer expects to sell 31 more copiers next year than this year. This year 176 copiers were sold at an average price of \$16,400. Assume the sales projections are correct and the average price will remain the same. What is the total projected copier sales next year, in dollars?

Wrap Up ▶▶▶

Darlene may have worked more hours than usual during winter break week because she didn't have to go to school. If so, her estimate of earning \$12,064 in annual wages is wrong. She should calculate her annual wages on a typical week or on her wages over a longer period, such as two months. She may have to reconsider her car buying plans.

Exercises

Find the product.

1. $13.5\% \times \$346{,}700$
2. $4.25\% \times \$41{,}200$
3. $15 \times \$12{,}800$
4. $42 \times \$117{,}985$

Total sales of the Worland Markets were \$8,200,000 two years ago and \$9,020,000 last year.

5. By what percent did sales increase last year?
6. If the same trend continues, what are sales likely to be this year?

Solve.

7. Sales of a wood picnic table have dropped an average of 5% each year for the past 6 years. Last year, 220 tables were sold. Based on the trend, how many tables will be sold this year?

8. Based on past sales records, the sales of Uncle Stan's Original Sauce are expected to increase by 6% a year for the next two years from $270,000 this year. What are sales forecasted to be in each of the next two years?

9. A child's toy was recalled as being unsafe. Before the recall, sales of the toy were 16,760 units. The toy has been redesigned, and the company expects unit sales to be only 40% of what they were before the recall. What unit sales should the toy maker expect?

10. Canfield Medical Supply sold 2,312 units of a new blood pressure reader last year. Canfield's management expects sales of the reader to increase 250% this year. What are the projected sales for this year?

A survey showed that 22% of the people in a test market who sampled Sweet Delite ice cream said they preferred Sweet Delite to other brands and would buy it. In the test market area, 68,000 gallons of ice cream were sold last year.

11. How many gallons of ice cream might the makers of Sweet Delite expect to sell in the test market area based on the survey results?

12. Assume that only 25% of the respondents who said they would purchase Sweet Delite actually did so. How many gallons of Sweet Delite will be sold?

13. **CRITICAL THINKING** Because of a business slowdown that lasted three years, the sales of replacement batteries for older cars increased by 17% a year during that time. The battery production manager wants to build his budget based on battery sales for the past three years. Higher-level management wants budgets to be developed based on average sales for the past 12 years. Select the plan to be used and give reasons for your choice.

14. **INTEGRATING YOUR KNOWLEDGE** A survey of 260 store customers yielded 78 respondents who said they would buy cut flowers more often if the price was 25% lower. The store estimates they have 10,000 different customers who buy 2,200 bouquets of flowers a year. What would bouquet sales be if prices were cut by 25% and the survey results were true?

Mixed Review

15. Find the average of 119, 203, 89, and 417.

16. What is $8.19 ÷ 10 rounded to the nearest cent?

17. The used car that Billie purchased was $8,100. Other costs were registration fees of $97 and a sales tax of 6%. She paid $2,100 down. What was the delivered price of the car and the balance due?

18. A golf course owner bought a mower for $12,619. The mower has a class life of 5 years. Using the MACRS schedule find the first year's depreciation and the book value of the mower at the end of the first year.

Photodisc/Getty Images

10-7

Market Share

GOAL
- Calculate market share

KEY TERMS
- market
- market share

Start Up

Two stores each with local owners shared equally the market for clothing sales in Green City. A large chain store that sells everything at discount prices will open soon. What is likely to happen to the total market for clothing and to market share of each store when the chain store opens?

Photodisc/Getty Images

Math Skill Builder

Review these math skills and solve the exercises that follow.

1. **Calculate** a percent.
   Find the percent. 8,820 ÷ 49,000 = 0.18, or 18%

   1a. 17,360 ÷ 56,000 **1b.** $96,000 ÷ $2,400,000

   1c. $26,026 ÷ $100,100 **1d.** 44,700 ÷ 74,500

2. **Multiply** whole numbers or money amounts by percents.
   Find the product. 12% × 136,000 = 0.12 × 136,000 = 16,320

   2a. 37% × 43,200 **2b.** 12% × $13,600,000

   2c. 28% × 56,000 **2d.** 19% × 2,400,500

Market Share

A **market** is the total of all the persons or organizations that are potential customers. The market for corn seed, for example, consists of all farmers and home gardeners who grow corn. **Market share** tells what percentage of the total market's sales one seller has.

An individual company's market share may be very high. For example, a local bakery in a very large city where few people drive cars may have 80–90% of the market for baked goods in the few blocks surrounding the bakery.

Other companies may have a relatively small market share, yet be financially successful. For example, a pharmaceutical company that sells medical equipment may have only a 5% share of the worldwide export market. However, if the worldwide market is $5 billion a year, the 5% market share equals $250,000,000 in annual export sales.

Market share is usually stated as a percent and may be based on the number of units sold or their dollar value.

EXAMPLE 1

In one year, an airport served 450,000 passengers. One airline served 315,000 of the passengers. What market share did the airline have?

SOLUTION

Divide the airline's passengers by the total passengers.

$315,000 \div 450,000 = 0.7$, or 70% market share

EXAMPLE 2

The retail stores in Hartwell had estimated total annual sales of $3,180,000 last year. Stengel's Department Store is estimated to have a 44% share of Hartwell's retail market. What amount of sales is Stengel's estimated to have?

SOLUTION

Multiply the total annual sales by the market share.

$44\% \times \$3,180,000 = 0.44 \times \$3,180,000 = \$1,399,200$ Stengel's estimated sales

✔ CHECK YOUR UNDERSTANDING

A. Two publishers sell daily newspapers in the City of Wilson. Each day, the *Wilson Post* sells 16,240 papers and the *Wilson Daily News* sells 11,760 papers. What market share of daily newspaper sales does the *Wilson Post* have?

B. A chain of tire stores is estimated to have 21% of the new tire market in a county. If new tire sales in the county are 126,000 tires, how many new tires are likely to have been sold by the chain?

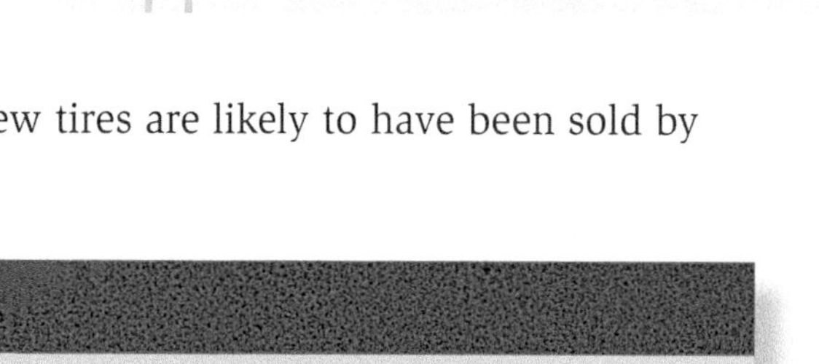

Use the Internet to research the market share held by the top three companies in the industries of personal desktop computers, refrigerators, and breakfast cereals.

Wrap Up ▶▶▶

Because of its lower prices it is likely that the chain store will gain the greatest share of the clothing market in Green City. Since the chain store may draw customers from a wider area, it is possible that total clothing sales in Green City may increase. However, even though total sales may increase, the two stores that are locally owned are likely to have lower sales because of their reduced market share.

Exercises

Find the percent.

1. 1,330 ÷ 26,600
2. \$125,000 ÷ \$625,000

Find the product.

3. 21.5% × 86,000
4. 27% × \$1,327,000

Solve.

5. The industry forecast for next year is that 2,800,000 pairs of leather slippers will be sold. The Comfort-Wear Company expects to sell 1,008,000 pairs of leather slippers. What market share does Comfort-Wear expect to have?
6. The annual sales of dishwasher soap in an eleven-state region are estimated to be \$90,000,000 a year. The Blair Company expects its brand of dishwasher soap to have annual sales of \$8,670,000. What market share, to the nearest tenth of a percent, does Blair expect to have?
7. The manufacturers of replacement car floor mats expect to sell 120,000 mats in a certain state. Of that total, the Binko Company expects to sell 46,500 mats and Solwell Products expects to sell 31,000 mats. What market share does the Binko Company expect to have, to the nearest tenth of a percent?
8. There are six companies that sell and clean work uniforms. Their combined revenues for the year are \$4,120,500. Two of the companies have combined annual revenues of \$2,940,000. What market share do the four remaining companies have, to the nearest percent?
9. A new company hopes to get a 15% share of a \$2,400,000 school supply contract. What is the dollar value of the contract the new company hopes to get?
10. The Perton Company increased its market share from 15% two years ago to 16% last year. Perton expects to gain another 1% of market share this year. What total sales can it expect if total industry sales are \$48,500,000?
11. Sales of used ski equipment through Internet ads are expected to reach a 7% market share this year. The value of used ski equipment sold through all types of ads is estimated to be \$7,000,000. What is the value of the ski equipment sold through Internet ads?
12. The total market for specialty woodworking tools sold through catalogs is \$14,000,000 annually. Wilkin Wood Works is the industry leader with a 45% market share in catalog sales. What annual catalog sales does Wilkin have?
13. A publisher of dictionaries plans to get 18% of the year's sales in the new markets they enter. The projected combined sales of dictionaries in these markets total \$19,425,000. What is the planned sales amount of the publisher?

Photodisc/Getty Images

14. **CRITICAL THINKING** The executives of Landry Enterprises believe they can start a new division that within five years will have a 10% share of the one billion dollars a year modular office furniture market. Is this a realistic goal?

15. **FINANCIAL DECISION MAKING** A company's share of a $40,000,000 market has remained at 6.1% for the past 11 years. The company's planners estimate they can gain an additional 0.1% market share by spending $150,000 each year to promote aggressively their products. Should the company go ahead with the promotion plan?

Mixed Review

16.
$$\begin{array}{rcrcrcl} 345 & + & 254 & + & 378 & = & \\ 692 & + & 235 & + & 758 & = & \\ \underline{843} & + & \underline{718} & + & \underline{698} & = & \underline{\quad\quad} \\ & & & & & = & \end{array}$$

17.
$$\begin{array}{r} 3{,}007 \\ -\ 1{,}849 \\ \hline \end{array}$$

18. $\frac{3}{4} \times \frac{7}{12} \times \frac{8}{21}$

19. $14 - 4\frac{3}{8}$

20. Find the number of days from January 13 to February 27.

21. Round 42.95¢ to the nearest cent.

22. Darla Runnels began the day with a bank balance of $356.32. Her employer direct deposited her weekly pay of $512.10. After work she made an ATM withdrawal of $75 and made two debit purchases for $61.15 and $119.04. What was Darla's bank balance at the end of the day?

23. Merle Croswell's employer provides a health insurance plan that covers him and three family members. Merle pays 15% of the $238 monthly premium. How much does Merle pay in one year for health insurance?

24. Connie Alpern drove her truck 21,470 miles last year. She uses her truck 40% of the time on her job, installing dishwashers and gas ranges. Her employer reimburses business mileage at the rate of 32.8¢ a mile. What total reimbursement for mileage did Connie receive last year?

25. Stock paying quarterly dividends of $1.80 is bought for a total cost of $45 per share. What rate of income is earned on the investment?

The results of a survey asking people whether they liked the fragrance of a new shampoo are shown. Show any percent answers correct to the nearest tenth percent.

	Yes	No	No opinion
Female	570	211	109
Male	270	386	212

26. What was the total number of respondents?

27. What percent of the respondents had no opinion?

28. What percent of the females liked the new fragrance?

10-8

Advertising

GOALS

- Calculate the cost of advertising in print media
- Calculate the cost of advertising in other media
- Calculate the advertising cost per person reached

KEY TERM

- reach

Start Up

Your friend, Sheila, just opened a sunglass stand in a shopping mall. Sheila thinks that she ought to advertise her business, but doesn't know where she should advertise. What advice might you give Sheila?

Photodisc/Getty Images

Math Skill Builder

Review these math skills and solve the exercises that follow.

1. **Add** money amounts.
   Find the sum. $3,120 + $790 = $3,910

   1a. $458 + $1,235 **1b.** $774 + $2,560

2. **Multiply** money amounts by whole numbers.
   Find the product. 12 × $62 × 3 = $2,232

   2a. 7 × $34.50 × 5 **2b.** 14 × $468

3. **Divide** money amounts by whole numbers.
   Find the quotient to the nearest tenth of a cent. $846 ÷ 1,980 = $0.427

   3a. $56,300 ÷ 980,000 **3b.** $615 ÷ 46,000

Cost of Print Ads

Advertising is a way of communicating a message about an organization, or its products and services, to a target audience through advertising media. A *target audience* consists of all possible customers for a product or service. *Media* refers to the different forms of advertising.

Print ads are ads that appear in printed publications such as newspapers. Most advertisers purchase *general ads,* also known as *display ads.* General ads usually contain large type and an illustration.

General ads are often sold by column inch. A *column inch* is a space one column wide by one inch deep.

The illustration shows one column inch on a newspaper page that is 6 columns wide and 21″ deep.

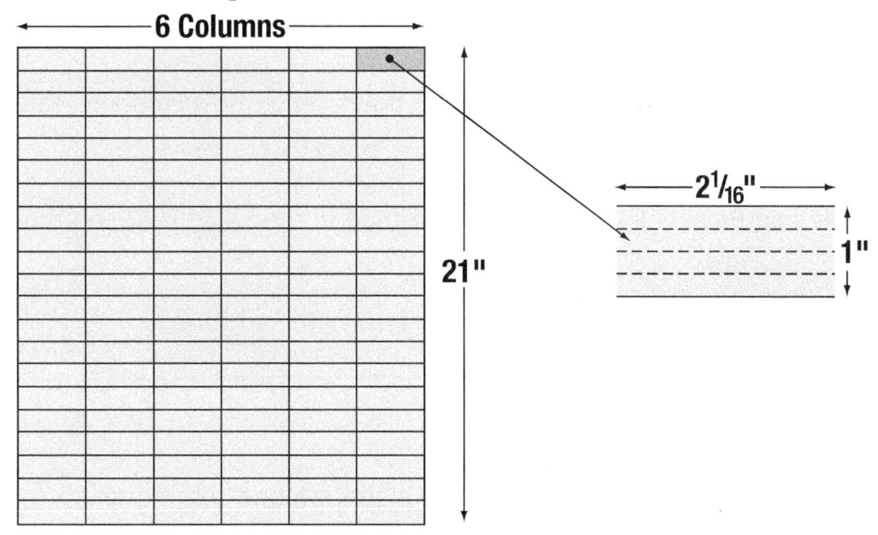

The formula for determining an ad's cost based on column inch is:

Ad Cost = Column Inches × Cost per Inch × Number of Columns Width

EXAMPLE 1

A newspaper charges $237 per column inch for a one-column width general ad. What is the cost of an ad that is 4 column inches long and one column width wide?

SOLUTION

Multiply the cost per column inch by the number of column inches and by the number of column widths.

4″ × $237 × 1 = $948   cost of ad

Print ads may also be sold by *page size,* such as a full page or part of a page. An example of an advertising rate card shown below gives the per-issue costs for a magazine that sells ads by page size. A discount is given when the advertiser agrees to run ads in more than one issue. The 1X, 3X, etc. refer to the *frequency rate,* or the number of times the ad will run.

Advertising Rate Card				
Size	**1X**	**3X**	**5X**	**12X**
Full page	$3,095	$2,940	$2,725	$2,570
$\frac{2}{3}$ page	$2,045	$1,945	$1,800	$1,700
$\frac{1}{2}$ page	$1,555	$1,475	$1,370	$1,290
$\frac{1}{3}$ page	$1,085	$1,030	$955	$900
Special Charges: Add to page rate				
4-color	$1,460	$1,390	$1,285	$1,210
2-color	€525	$500	$465	$435
Inside Cover	€3,875	$3,680	$3,410	$3,215

To calculate the total cost of an ad, find the base price by matching the ad size with the frequency rate. Then add any special charges for color or a preferred location, such as the inside cover. Multiply the total by the number of times the ad will run.

EXAMPLE 2

Find the cost of a full-page, 4-color ad that runs 5 times using the advertising rate chart.

SOLUTION

Add the base rate and the 4-color rate of a full-page ad in the 5X (times) column.

$2,725 + $1,285 = $4,010 cost of single issue ad

Multiply the single-issue ad cost by the number of times the ad runs.

5 × $4,010 = $20,050 cost of ad

✔ CHECK YOUR UNDERSTANDING

A. Your local newspaper charges $135 a column inch for a one-column width general ad. What would you pay for an ad that is 3 column inches long and 2 column widths wide?

B. Find the total cost of a $\frac{1}{3}$ page, two-color ad that runs 1 time.

Cost of Other Ads

In addition to print media, advertisers frequently use TV, radio, direct mail, billboards, and the Internet to reach their target audience. The total cost of an ad is found by multiplying the cost of the ad by the number of times the ad will run.

EXAMPLE 3

An Internet service provider charges $5,000 a week for an ad on its home page. What would be the total cost of an Internet ad that runs for 6 weeks?

SOLUTION

Multiply the weekly cost of the ad by the number of weeks it runs.

6 × $5,000 = $30,000 total cost of ad

✔ CHECK YOUR UNDERSTANDING

C. The cost of a 20-second, afternoon ad on a local radio station is $120 per ad, per day. What will be the ad's total cost if it runs once a day for 10 weekdays?

Photodisc/Getty Images

D. A network charges $875,000 for a 30-second ad at a major sports event televised nationally. The advertisers must agree to purchase a minimum of four ads during the telecast. What is the advertiser's total cost of buying the minimum number of ads?

Cost per Person Reached

Advertising rates are based on an ad's reach through a particular media. **Reach** refers to how many people see or hear an ad. The number of people reached by an ad is usually a measurable number. It may be the number of subscribers, readers, listeners, households in a geographic area, or cars passing a certain location. Internet sites may define reach as the number of hits, registrations, or searches beyond the home page.

The cost per person of an ad is found by dividing the ad's cost by the number of people reached. The per person cost of the ad may be impressively small. However, the ad is likely to reach people who are not interested in the product. So, the actual cost of the ad, per person, to reach the target audience is likely to be much greater.

EXAMPLE 4

The cost of an advertising insert in a daily newspaper delivered to one zip code is $1,600 per edition. There are 7,920 households that receive the newspaper. What will be the cost of the ad per household, to the nearest tenth of a cent?

SOLUTION

Divide the cost of the ad by the number of households.

$1,600 ÷ 7,920 = $0.202 cost per household

✔ CHECK YOUR UNDERSTANDING

E. A direct mailer inserts 30 ads in an envelope and bulk mails them to 24,300 households in a city. The cost of each ad is $825. To the nearest tenth of a cent, what is the cost per household reached to an advertiser who places one ad?

F. A furniture store spent $3,640 on a newspaper ad featuring a sale on leather sofas. In the next two days, 130 people who looked at the sofas said they saw the ad. What was the cost of the ad for each shopper who was attracted by the ad?

Wrap Up ▶▶▶

Since the sunglass stand is likely to be fairly small, it probably wouldn't pay to place ads in the usual places. Sheila might consider placing an ad in the local Yellow Pages or getting her stand's name listed on an electronic message board at the mall, if there is one. She may also print one-page flyers and hand them to shoppers entering the mall if she is allowed to do so by the mall's management.

TEAM Meeting

Contact your local newspaper and find the cost of a full-page daily ad and the daily circulation of the paper. Also contact a local or nearby radio station to find the cost of 30-second ad in the most expensive part of the day. Determine how many listeners the radio station has at that time of day.

For each type of media, calculate the cost per person reached. Write a short paragraph describing any differences you find.

Exercises

Find the sum or product.

1. $238 + $784
2. $990 + $1,360
3. 14 × $183
4. 8 × $23.25 × 14

Find the quotient, to the nearest tenth of a cent.

5. $230 ÷ 78,000
6. $554,000 ÷ 12,000,000

Use the advertising rate card to solve Exercises 7–8.

7. Find the cost of a $\frac{1}{3}$ page ad with four colors that runs 12 times.
8. How much more would it cost to buy four sets of $\frac{2}{3}$ page ads that run 3 times each than to buy the same sized ad to run 12 times?

Solve.

9. A newspaper charges $148.75 for a one-column inch, one-column wide display ad that runs one day. Ads that run for more than four consecutive days receive a 5% discount. What is the cost of a 4-column inch, one-column ad that runs from Monday through Friday?
10. The *Sunshine Journal's* weekday ad rates are $98.50 per one-column inch, per one-column width. The rates increase 40% on Saturdays and Sundays. An 8-column inch, 3-column width ad runs for 7 consecutive days. What is the cost of this ad space?
11. Nextbid.com, an Internet auction site, pays $3,270 a week to an ISP for a banner ad that will automatically link to Nextbid's web site. Nextbid signed a contract with the ISP that runs for 13 weeks. What is the cost of the ad?
12. The Internet edition of the *Marval County Daily Tribune* charges $75 a day for display ads on its home page. What income will *Marval* receive from Internet ads in June if it sells 3 ads daily?
13. A billboard rents for $128 a day with a 30-day minimum rental. What is the cost of renting the billboard for one rental period?

Photodisc/Getty Images

14. An online magazine charges $45,000 for a 3-month contract for a banner ad displayed above its weekly feature article. The magazine has 120,000 subscribers and gets 195,000 hits per day. What is the banner's cost per subscriber if a banner ad was taken for a year?

The *Daily Press* sells 230,000 newspapers daily. Through surveys it found that 2.3 persons read each newspaper sold.

15. What total number of people read the newspaper daily?
16. What is the cost per reader, to the nearest tenth of a cent, of an ad that sells for $3,200?

17. **CRITICAL THINKING** An Internet web site claims it gets 240,000 hits per day and wants to charge an advertiser a daily advertising rate based on 3¢ for each hit the web site records. What questions might the advertiser ask before agreeing to the rate?

INTEGRATING YOUR KNOWLEDGE Stephanie Hagan, an artist, has a successful business making ink sketches of infants for $80 a sketch. To expand her business she plans to mail an advertisement to the families of the 200 children born each month in areas near her home. She estimates that 10% of the families will respond to the ad and that 50% of the respondents will order a sketch. Stephanie estimates the total annual cost of preparing, printing, and mailing the announcements to be $4,680.

18. How much will Stephanie spend on her ads per family in a year?
19. How many orders does Stephanie expect to receive each year?
20. Will Stephanie make a profit or loss on her advertising campaign, and how much will her profit or loss be?

Mixed Review

21. $\frac{3}{4} - \frac{1}{2}$
22. 72 + 83 + 46 + 54
23. Rewrite 0.007 as a percent.
24. 78 ft @ $0.01
25. $5\frac{1}{4}$% of $729
26. $\frac{1}{8} \times$ $15,000
27. Estimate the quotient of 9,217 ÷ 8.9.
28. Estimate the product of 675 × 23.
29. Estimate the sum of $23,578 + $8,209 + $12,207.
30. Estimate the difference of 35,813 and 19,775.
31. Lonnie Erhardt borrowed $4,000 at 7% exact interest for 170 days. What amount of interest did he pay?
32. Rodger Cleary bought 120 shares of Fleider Oil at $78.12. His broker charged $176 commission. Fleider Oil pays an annual dividend of $1.35 a share. What is the annual yield to the nearest tenth percent?
33. A wireless carrier offers a wireless phone plan that includes 450 minutes per phone for $65.45 a month. Each extra minute costs $0.14. The plan allows unused minutes on one phone to cover extra minutes on other phones under contract. A business owner leased a phone for himself and three employees. The minutes used in March on all four phones are 468, 564, 390, and 420. What should be the amount of the owner's monthly bill for March?
34. An office printer uses 26 inkjet cartridges a year @ $26.50 each. The annual maintenance contract on the printer is $195. Annual depreciation on the printer is $125. What is the total cost of operating the printer for a year?
35. What property tax is due on a home assessed for $213,000 if the tax rate is $2.85 per $100?

Chapter *Review*

Vocabulary Review

Find the term, from the list at the right, that completes each sentence. Use each term only once.

cash discount
credit memo
discount series
forecast
invoice price
list price
markdown
market
market share
markup
on account
prove cash
reach
respondent
sales invoice
trade discount
trend

1. A reduction of the invoice price given for early payment is known as __?__.
2. The number of people who see or hear an ad is called __?__.
3. The relationship between time and data, such as sales, is referred to as a(n) __?__.
4. The selling price of an item less its cost is called the __?__.
5. A price for goods that is published in a catalog is called the __?__.
6. Price reductions, such as 10%, 10%, and 5%, are called a(n) __?__.
7. All the people interested in buying a certain product are referred to as the product's __?__.
8. A form sent to retailers showing the value of returned merchandise is called a(n) __?__.
9. A prediction, projection, or estimate of future sales is also called a(n) __?__.

10-1 Cash Sales and Sales on Account

10. A cashier began the day with a $140 change fund. His cash register totals were $4,529 cash received and $389.15 in cash paid out. There was $4,278.15 cash in the register. How much was cash over or short?
11. Mendrin Markets bought these items from a wholesaler: 24 cases of canned vegetables @ $80.16 and 120 bags of sugar @ $2.31. What was the total of the sales invoice?
12. Mendrin Markets found that four of the bags of sugar purchased in Problem 11 were ripped and unusable. What should be the amount of the credit memo the wholesaler issues?
13. A wholesaler recorded these transactions in April for Rasher Hardware's account: April 1, balance forward, $3,482; April 7, Invoice #358 for $4,233; April 12, payment of $3,482; April 13, Credit Memo #98 for $78.40; and April 28, Invoice #427 for $1,873. What was the account's balance on April 30?

10-2 Cash and Trade Discounts

14. An invoice dated March 11 has terms of sale of 1/10, n/45. What are the discount date and the due date of the invoice?
15. An invoice for $3,410 dated November 6 has terms of 2/10, 1/20, n/60. What is the amount of the cash discount and cash price if the invoice is paid on November 23?

16. A straw hat has a list price of $26 and is offered at a 45% trade discount. What is the amount of the trade discount and the invoice price?

10-3 Series Trade Discounts

17. A box of party streamers has a list price of $15.70. The series discount offered is 20%, 15%, and 5%. What is the invoice price of the box of streamers?

18. What are the invoice price and single discount equivalents written as percents for a series discount of 10%, 15%, and 20%?

10-4 Markup and Markdown

19. A store carries women's suits in a $400 price line. The markup is 47% of the selling price. What is the highest amount the store may pay for suits to get the desired markup?

20. Multi-Sports sold a hockey jersey for $112. The cost of the jersey was $85. What rate of markup on cost did Multi-Sports earn on the jersey, to the nearest percent?

21. At the end of a season a ski sweater with a marked price of $140 was offered at a 60% discount. What is the selling price of the sweater?

10-5 Marketing Surveys

22. A research firm mailed 12,500 surveys to a company's shareholders. The number of valid surveys returned was 4,872. What was the response rate, to the nearest percent?

23. The respondents to a theater lobby survey rated a movie as follows: 4 stars, 125; 3 stars, 86; 2 stars, 31; and 1 star, 47. What average star rating did the movie get? What percent of the raters gave the movie 3 or more stars, to the nearest percent?

10-6 Sales Forecasts

24. A four-year-old company's sales last year were $3,120,000. The average sales increase over the past four years is 32.5%. Based on that rate, project this year's sales.

25. Sales of a security scanner were 240 in a test market. There are an estimated 54 similar sized markets in the United States and another 280 in foreign countries. What are the expected unit sales for the scanner when it is marketed worldwide?

10-7 Market Share

26. A hardware chain's total annual sales were estimated to be $5,200,000 out of a total of $18,000,000 in estimated sales of all hardware stores in an area. What market share does the chain have, to the nearest tenth percent?

27. Of the 600,000 gallons of gasoline sold annually in a town, 34% are sold by Blue's Garage. What are Blue's gasoline sales for a year?

10-8 Advertising

28. A weekly newspaper charges $82 a column inch and one column width. What is the cost of an ad that runs 2 column inches and 2 column widths?

29. The cost of advertising during a fifteen-minute radio news program is $150 a day. What amount would an advertiser pay for one ad daily during the month of March?

30. A magazine has 45,000 subscribers. What is the cost per subscriber of an ad that costs $3,890, to the nearest tenth of a cent?

Technology Workshop

Task 1 Forecasting Sales

This chapter showed how business firms forecast future sales by various methods. Another method used is a statistical method called "least squares." The FORECAST function in Excel calculates trends using the least squares method. It produces different projections of future sales than you would get by calculating a simple average.

Enter data into a template that uses the FORECAST function to project sales for Year 10 based on sales in Years 1–9. In statistical work, time in years is often identified by a numeral instead of the actual year.

Open the spreadsheet for Chapter 10 (tech10-1.xls) and enter the data shown in blue (cells B5-13) into the spreadsheet. The spreadsheet will calculate the amount and percent of sales increase for each year based on a year-to-year comparison. The FORECAST function will project sales for Year 10 based on the value of sales for the entire nine previous years. Your computer screen should look like the one below.

	A	B	C	D
1	**Buchanan Products, Inc.**			
2	**Sales Data, Electric Motor Division**			
3			**Sales Increase From Previous Year**	
4	**Year**	**Sales**	**Dollar Increase**	**Percent Increase**
5	1	568,000	---	---
6	2	637,864	69,864	12.30%
7	3	687,043	49,179	7.71%
8	4	695,287	8,244	1.20%
9	5	791,724	96,437	13.87%
10	6	875,647	83,923	10.60%
11	7	939,043	63,396	7.24%
12	8	1,019,332	80,289	8.55%
13	9	1,068,565	49,233	4.83%
14	10	???		
15	**Average of Percent Increases, Years 1-9**			8.29%
16	**Sales Forecast for Year 10**			$1,128,419

Task 2 Analyze the Spreadsheet Output

Answer these questions about the sales forecast calculations.

1. In which year from 1–9 did the lowest percent year-to-year sales increase occur?
2. In which year from 1–9 was the year-to-year percent increase in sales the greatest?
3. What was the average of the percent increases for Years 1–9?
4. What spreadsheet functions were used to calculate the average of percent increases?

5. What amount of sales was forecast for Year 10?
6. In which years was the year-to-year percent increase within 1% plus or minus of the average of the percent increases for Years 1–9?

Now assume that sales in Year 1 were $818,000. Move the cursor to cell B5, which holds the sales for Year 1. Enter sales of 818,000.

7. What is the new figure for the average of the percent increases?
8. What was the new figure for the Year 10 sales forecast?
9. Did the year-to-year sales percent change for any year other than Year 2? Why?
10. In which year did sales recover to equal or exceed the Year 1 sales?
11. What might be some reasons why sales decreased $180,136 in Year 2 and did not recover for some time?

Task 3 Design a Spreadsheet to Calculate Series Discounts

Design a spreadsheet that will calculate series discounts. The spreadsheet should allow a list price and up to three series discounts to be entered. The spreadsheet should calculate the single discount equivalent to the series discounts, the amount of the trade discount, and the invoice price.

If there are less than three discounts in the series, enter 0 for the missing discount. Calculate the single discount equivalent as a percent to four decimal places.

SITUATION: A dishwasher with a list price of $480.54 is offered to retailers with series discounts of 15%, 10%, and 5%. Find the single discount equivalent, trade discount, and invoice price of the dishwasher.

Task 4 Analyze the Spreadsheet Output

Answer these questions about your completed spreadsheet:

12. What is the single discount equivalent to the 15%, 10%, and 5% series discount?
13. What is the amount of the trade discount?
14. What is the invoice price of the dishwasher?
15. Enter the discounts in a different order. Do you get the same results?
16. How can you use the spreadsheet's results to calculate the invoice price equivalent of the series discounts?

You may use the series discount spreadsheet to verify the examples, and exercises in Lesson 10-3 that required the use of series discounts.

Chapter *Assessment*

Chapter Test

Answer each question.

1. $3,412.80 + $1,815.09 + $212.18
2. $16,090.18 − $436.77
3. $1\frac{1}{2} \times$ $978,000
4. Rewrite 3.25% as a decimal.
5. 250 @ $26.12
6. 85% × 90% × 72%
7. 1 − 0.64125
8. 4.65% × $2,367
9. Find the percent: $234.90 ÷ $522
10. Round $2,838,922 to the nearest thousand.

Solve.

11. What is the invoice price for the purchase of a tractor with a list price of $18,560 and terms of 10%, 15%, and 4%?
12. An advertiser bought space for 12 ads at a list price of $680 per ad, less a 7% discount. What was the cost of the ads?
13. After two markdowns a pair of boots has a marked price of $38.95. A retailer marks down the boots another 15%. What is the marked price of the boots after the third markdown?
14. An invoice for 56 shirts @ $22.46 dated August 2 was paid on September 6. Credit terms were 1/10 EOM, n/60. What amount should be sent to pay the invoice?
15. A survey of a sample population of 1,800 households out of a population of 75,000 households resulted in 1,247 valid responses. What was the response rate, to the nearest tenth percent?
16. A cashier began a work period with a $180 change fund. At the end of the work period, the register showed cash received of $8,359.29 and cash paid out of $400.16. Actual cash in the register was $8,141.85. Was cash over or short?
17. The overall trend of sales increases is 4.2% a year. Forecast next year's sales to the nearest ten thousand dollars based on last year's sales of $76,428,500.
18. A nursery is estimated to have a 35% market share of the landscaping market. Total market sales are $1,000,000. What is the nursery's market share in dollars?
19. The cost of a desk telephone is $56.12. A retailer plans a markup of 36% on cost. What is the selling price of the telephone?
20. What is the single discount equivalent to a discount series of 20%, 10%, and 6.5%?
21. A customer's account had a balance of $2,680.65 on September 1. Transactions recorded in September are: Invoice #1211, $3,583; Credit Memo #129, $23.80; Payment, $656.85; Invoice #1387, $1,452; and Payment, $3,196. What was the account's balance on September 30?
22. A company's owner set a sales goal for next year that each of 3 salespersons would sell 20% more lawn watering systems than they did this year. This year's sales averaged 75 systems per salesperson. How many watering systems are projected to be sold next year?

Planning a Career in Marketing, Sales, and Service

Career choices in marketing, sales and service include advertising, public relations, direct selling, promoting, communications, Internet marketing, design, and management. The need for people who can effectively and efficiently sell products or services exists in every area of the market place. If you work well under pressure, are an excellent communicator, and can work independently, this could be a career path for you.

Jaimie Duplass 2008/Shutterstock.com

Job Titles

- Market research
- Advertising manager
- Sales agent
- Copywriter
- Interactive marketer
- Media planner
- Public relations manager

Needed Skills

- exceptional business, organizational and leadership skills
- written and verbal communication skills
- ability to scrutinize and evaluate market data
- ability to communicate in persuasive fashion

What's it like to work in Real Estate Sales?

Real estate agents guide people who are buying or selling residential, commercial, or investment real estate. They have a thorough knowledge of the neighborhoods, the legal requirements in buying and selling properties, and the financial aspects of mortgages in their localities. For buyers, agents show properties and assist with inspections and bank loans. For sellers, the agent lists the property, locates potential buyers, and shows the property. A real estate agent must be able to effectively communicate and have an exceptional eye for detail.

What About You?

Can you see yourself helping others find a new home? What types of skills do you need to work in sales?

How Times Have Changed

For Questions 1–2, refer to the timeline on page 417 as needed.

1. If NCR sold 800,000 brass-encased cash registers in 1910, how many brass-encased registers did all of their competitors combined sell?
2. In a 12-digit UPC code, the first 6 digits are the manufacturer's ID number, the next 5 digits are the item number, and the last digit is the check number that confirms the UPC code is correct. To determine a check digit, add the digits in the odd positions and multiply that sum by 3. Then add the remaining digits (in the even positions) and add it to the product. To calculate the check digit, find the single digit number that when added to the final sum will be a multiple of 10.What is the check number for the UPC code with the manufacturer's ID 045893 and item number 09648?

Chapters 9–10 *Cumulative Review*

MULTIPLE CHOICE

Select the best choice for each question.

1. The monthly rent on a 4,000 sq. ft. office is $12,000. The sales area is 1,400 sq. ft. Rent is allocated on the basis of space used. What amount of monthly rent is charged to sales?

   A. $350 **B.** $4,850 **C.** $4,200 **D.** $8,570 **E.** $3,600

2. An invoice dated April 13 has terms of 3/12, 1/30, n/60. What is the invoice's due date?

   A. June 12 **B.** May 12 **C.** May 13 **D.** April 25 **E.** June 14

3. Stellar Sports has a price line for a bowling ball of $120 with a 35.5% markup on selling price. What is the most that Stellar can pay for the bowling ball?

   A. $74.40 **B.** $42.60 **C.** $84.50 **D.** $78 **E.** $77.40

4. A commercial-grade riding lawnmower that costs $10,800 is estimated to depreciate 13% a year. What is the book value of the mower at the end of the first year?

   A. $1,404 **B.** $9,396 **C.** $8,619 **D.** $12,204 **E.** $3,348

5. A hunting cap originally priced at $18.29 was marked down 15% for a preseason sale. What is the selling price of the cap?

   A. $18.14 **B.** $13.72 **C.** $2.74 **D.** $15.55 **E.** $21.03

6. Rita used her van at work for 25% of the 26,112 miles she drove last year. Her employer reimbursed work mileage at 28¢ a mile. What reimbursement did Rita receive?

   A. $1,958.40 **B.** $6,528 **C.** $5,483.52 **D.** $7,311.36 **E.** $1,827.84

7. The MACRS depreciation rates for 7-year class life equipment are 14.29% the first year and 24.49% the second year. What total depreciation would there be for the first two years on a copier that costs $34,500 and has a 7-year class life?

   A. $12,171.73 **B.** $21,120.90 **C.** $4,930.05 **D.** $26,050.95 **E.** $8,449.05

8. What is the single discount equivalent to a series discount of 12%, 20%, and 5%?

   A. 66.88% **B.** 33.12% **C.** 32.5% **D.** 37% **E.** 63%

9. For one package weighing 6 lb, the shipping charge is $11.75 to Zone 3 and $19.25 to Zone 6. Ten, 6 lb packages are to be shipped, 8 to Zone 3 and 2 to Zone 6. Insurance costs are $1.45 a package. What is the total cost of shipping the packages?

   A. $133.95 **B.** $132.50 **C.** $192 **D.** $147 **E.** $184.25

10. Five percent of the 620 survey returns to a mail survey were not valid. If 3,400 surveys were mailed, what was the response rate, to the nearest percent?

    A. 5% **B.** 18% **C.** 17% **D.** 9% **E.** 13%

OPEN ENDED

11. Trexx Products wants to produce camera cases that will sell at $16 a unit. The fixed costs are $19,200. The variable costs of producing each case are estimated to be $6. How many cases must be sold by Trexx to break even?

12. A picnic basket with a list price of $46 is sold to a retailer for $29.90. What rate of trade discount was the retailer given?

13. The freight charge for a 786 lb shipment is $185 per hundredweight. What is the freight charge?

14. Wedamor Products had these manufacturing costs in March: raw materials, $265,190; direct labor, $717,080; and overhead, $241,800. What were Wedamor's prime cost and total manufacturing costs for March?

15. What is the invoice price of an item with a list price of $520 and a discount series of 15%, 15%, and 10%?

16. The usual rate of a one-half page ad is $1,800. An advertiser signs a contract to run a one-half page ad weekly for 52 weeks and receives a $12\frac{1}{2}$% discount. What annual amount will the advertiser pay for the ads?

17. Ben's Hardware bought a hand vacuum for $16.70 and marked it for sale at $23.38. What rate of markup based on cost did Ben's Hardware use?

18. A building with 50,000 sq. ft. is divided into a production area, warehouse, and offices. Production occupies 60% of the total building. The rent, maintenance, insurance, power, and other costs are $1,460,000 a year for the entire building. How many square feet does production occupy? What is the cost of the production area per square foot?

19. The Kite House had sales of $192,300 this year. The Kite House's 6.1% average annual sales growth is expected to continue. What is the forecast for next year's sales, to the nearest thousand?

20. A cash proof form shows these facts: change fund, $110; cash receipts, $2,986.20; cash paid out, $112.45; and cash in the drawer, $2,981.90. Was cash short or over, and by how much?

21. Cecile Doran attended a trade show for her employer. Her expenses for a four-day, three-night trip were: airfare, $528; hotel, $139 per night; meals, $327; show registration, $250, taxis, $86; and other expenses, $190. What was the amount of her reimbursement?

22. A retailer bought 240 knit shirts @ $19.84. Because the shirts arrived late for a special sale, the wholesaler gave the retailer a 3% credit on the total order. What is the amount of the credit memo the retailer received?

23. Out of 312 people surveyed, 79 gave a product a superior rating, 207 gave it an average rating, and 26 gave it a poor rating. What percent of the ratings were average or better, to the nearest tenth percent?

CONSTRUCTED RESPONSE

24. The owner of a company tells you to cut each expense item by 5% next year. You tell him that he has given you an impossible assignment. He threatens to fire you unless you give a good explanation. Explain to the owner in writing why the plan will not work.

Chapter 11

Manage People and Inventory

11-1 Employee Recruitment Costs
11-2 Wage and Salary Increases
11-3 Total Costs of Labor
11-4 Tracking and Reordering Inventory
11-5 Inventory Valuation
11-6 Ordering and Carrying Inventory

Statistical Insights

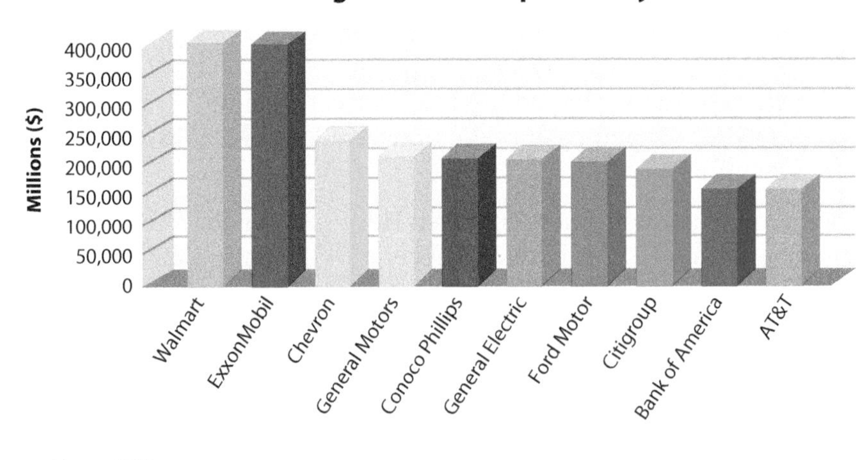

Source: Money.CNN.com

Companies in the United States are ranked by revenue, growth, number of employees, return to investors, or profit. As the economy changes, companies pursue inventive approaches to remain financially viable in the marketplace. Use the bar graph to answer Questions 1–4.

1. How many of these companies had revenues under $100,000,000,000?
2. Which one of these 10 top companies had revenues closest to $250,000 million?
3. What is the approximate range of revenue of the top 10 businesses?
4. **Describe** the revenues of Bank of America compared to the revenues of ExxonMobil.

How Times Have Changed

A Pennsylvania company introduced the first profit sharing plan in 1797; however it was not until the early twentieth century that profit sharing plans gained wide acceptance. Today, profit sharing plans play a part in the recruitment of employees. One of these plans, the 401(k), allows employees to contribute money to a retirement fund, and also allows companies to contribute a matching amount.

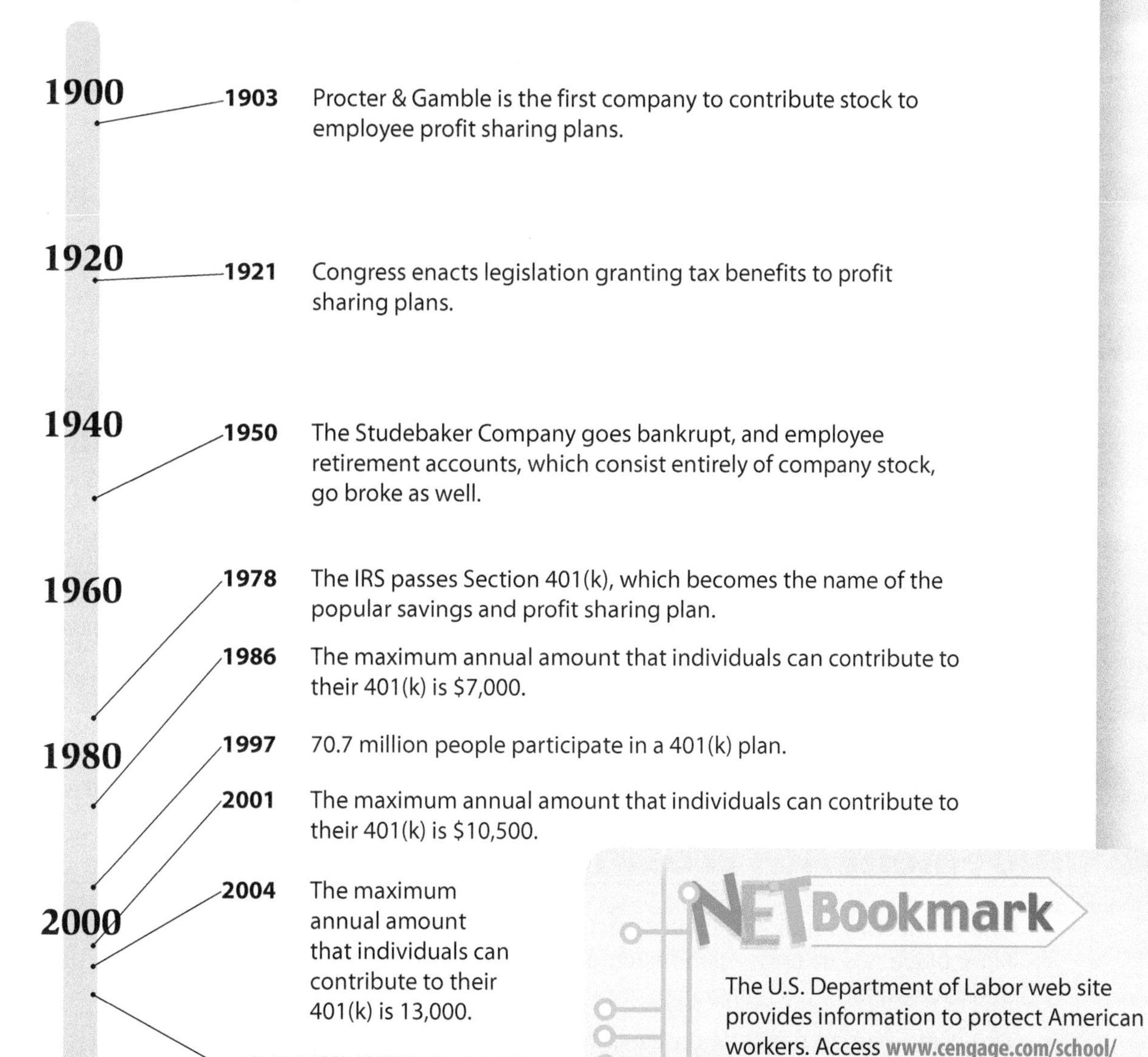

1903 Procter & Gamble is the first company to contribute stock to employee profit sharing plans.

1921 Congress enacts legislation granting tax benefits to profit sharing plans.

1950 The Studebaker Company goes bankrupt, and employee retirement accounts, which consist entirely of company stock, go broke as well.

1978 The IRS passes Section 401(k), which becomes the name of the popular savings and profit sharing plan.

1986 The maximum annual amount that individuals can contribute to their 401(k) is $7,000.

1997 70.7 million people participate in a 401(k) plan.

2001 The maximum annual amount that individuals can contribute to their 401(k) is $10,500.

2004 The maximum annual amount that individuals can contribute to their 401(k) is 13,000.

Research to find the significant event that occurred in 2008 with 401k consumer protection.

NET Bookmark

The U.S. Department of Labor web site provides information to protect American workers. Access www.cengage.com/school/business/businessmath and click on Chapter 11. On what type of topics does the Department of Labor provide information on their web site? What is the FLSA?

11-1

Employee Recruitment Costs

GOALS

- Calculate the cost of employment advertising
- Calculate the cost of hiring an employee
- Calculate the cost of using an employment agency

KEY TERMS

- exempt employee
- nonexempt employee
- executive recruiter
- contract employee

Start Up ▸▸▸

You read in the newspaper that a company paid $500,000 to an executive recruiting firm to search for a new CEO. Was the search worth the cost considering the number of people who are probably interested in the CEO position?

Stephen coburn/ shutterstock.com

Math Skill Builder

Review these math skills and solve the exercises that follow.

1. **Add** money amounts.
   Find the sum. $1,238 + $156 + $442 = $1,836

   1a. $218 + $176 + $39 **1b.** $83,100 + $2,176

2. **Multiply** money amounts by whole numbers and percents.
   Find the product. 12 × $125 = $1,500
   Find the product. 38% × $1,200 = 0.38 × $1,200 = $456

   2a. 6 × $134 **2b.** 26 × $157

   2c. 24.6% × $36,100 **2d.** 41.3% × $184,000

Business Tip

CEO stands for chief executive officer. In some firms the CEO's title is president.

Employment Advertising

One of the ways that employers recruit future employees is through advertising. They place ads in *print media* such as local newspapers, national publications, or trade magazines. Ads may also be posted on Internet job sites.

The cost of an ad depends on its size and the number of days it will run. Other costs include the costs of preparing the ad.

EXAMPLE 1

The XTR Company placed an ad for quality control employees for its new plant. The cost of designing the ad was a one-time charge of $230. The ad appeared in a local newspaper for two days during the week and again on Sunday. The newspaper's charge for the ad was $680 a day during the week and $810 on the weekend. What amount did XTR spend for advertising?

SOLUTION

Multiply the daily charge by the number of days.

$2 \times \$680 = \$1,360$ cost of weekday ad

Add charges.

$\$1,360 + \$810 + \$230 = \$2,400$ total spent on ad

Photodisc/Getty Images

✔CHECK YOUR UNDERSTANDING

A. Fenton Financial Advisors plans to use print media to recruit new employees. They decide to buy one ad each month in a financial publication at a monthly cost of $5,400. Because of the number of ads Fenton will run, they receive a 25% discount per ad. What is the cost of running one employment ad per month for a year?

B. For employment ads, a newspaper's daily charges are $56 for one inch of a column for weekday and $82 for one inch of a column for weekend. What is the cost of a 4-inch weekday ad for two days?

Hiring Costs

There are two basic classifications of employees: exempt and nonexempt. **Exempt employees** are usually paid a salary and do not qualify for overtime pay. **Nonexempt employees** are usually paid by the hour and do get paid overtime pay.

The hiring costs for exempt employees are greater than those for nonexempt employees. Also, the hiring costs for high-level executives are greater than those for lower-level exempt employees.

Labor costs are a large part of the hiring costs for many employers. Human resource staff members process applications, verify prior employment, check references, conduct interviews, and complete the paperwork necessary to add an employee to the payroll.

Employers also spend money to recruit employees when they hold job fairs or interview on college campuses. Employers may also pay relocating costs for a new employees.

Business *Tip*

In many companies a Human Resource (HR) department hires employees. The HR department may also be known as the Personnel Department.

Business *Tip*

Overhead costs are not directly related to a specific activity, such as interviewing an applicant. They include such costs as a manager's time, heat, power, light, office space, and depreciation on furniture and equipment.

EXAMPLE 2

The Human Resource department hired a new factory worker. The hiring costs included: 0.5 hours, application processing; 1.5 hours, reference checking; 0.6 hours, interviewing; and 1.4 hours, budget approval and benefit processing. These costs were charged at $80 an hour. General or *overhead* costs of $250 apply to each new worker hired. What was the total cost of hiring this worker?

SOLUTION

Add the hours spent hiring: $0.5 + 1.5 + 0.6 + 1.4 = 4$ hours

Multiply the cost per hour by the hours. Add overhead costs.

$(4 \times \$80) + \$250 = \$320 + \$250 = \$570$ total cost of hiring

✔ CHECK YOUR UNDERSTANDING

C. Five applicants went through the first round of screening for one opening as an accounts payable supervisor at Wilkins Aircraft Products. The average amount spent by Wilkins of bringing each applicant in for interviews was $1,610 per applicant. Managers spent two hours interviewing each applicant at a cost of $65 an hour. Background and reference checks took 12 hours per applicant at a cost of $41 an hour. Overhead costs were $500 per applicant. What total amount did Wilkins spend on first-round screening for the supervisor job?

Photodisc/Getty Images

D. A large company received 150 applications for two receptionist job openings. Bobbie Elder took two, 8-hour days to review and sort the applications into the 15 best applicants. The cost of Bobbie's time is $28 an hour. Bobbie spent 45 minutes per applicant to contact and briefly interview each of the 15 applicants. She then spent 1.5 hours per applicant checking the references of each of the 5 applicants who appeared to be best suited for the job. What was the cost of Bobbie's time on this project, to date?

Employment Agency Recruitment

Some employers use an employment agency to do some employee recruitment. For example, some employers may use **executive recruiters**, often called *head-hunters,* to find full-time employees for management or specialized technical positions.

Many executive recruiters specialize in one particular industry so that they can become familiar with the best businesses and talent available in that industry. Some recruitment agencies also specialize in one industry. When a recruiter finds talent for only one field, they will often have contact information for the top individuals nationwide that work in that field whether those individuals are seeking a new position or not.

Executive recruiters make all the contacts with potential employees, screen their suitability for the position, determine their level of interest in a new position, and identify their salary and benefit requirements. These recruiters are paid either a contingency fee or a retainer fee.

Recruiters who are paid a *contingency fee* get paid only if they find a suitable employee. The recruiters who are paid a *retainer fee* get paid even for an unsuccessful search.

Instead of recruiting permanent employees, some firms use temporary help agencies. These are companies that have a pool of people who will accept work as contract employees. **Contract employees** are temporary employees who receive their paychecks and benefits from the temporary help agency. The employer who hires contract employees pays a fee to the agency for each contract employee.

EXAMPLE 3

The Tonnel Company hired a contract accountant from the Triple-Star Employment Agency at a total cost of \$224 a day. A regularly employed accountant at Tonnel is paid \$182 a day. What is the difference between the costs of the contract and regular accountants in wages only for a five-day workweek?

SOLUTION

Multiply the daily rate for each employee by 5 days.

$5 \times \$224 = \$1,120$ contract employee wages

$5 \times \$182 = \$910$ regular employee wages

Subtract employee wage from contract wage.

$\$1,120 - \$910 = \$210$ wage difference

Business Tip

One of the benefits of using outside agencies is that employers pay for the recruiting services of executive recruiters or for the cost of contract employees only when they are needed.

EXAMPLE 4

An executive recruiting firm recruited a new accounting manager for the Tonnel Company. The recruiter's contingency fee was 35% of the manager's annual salary of \$91,500. The recruiter also charged for \$1,900 in travel expenses. What total amount did Tonnel pay the recruiting firm?

SOLUTION

Multiply the annual salary by the contingency fee percent.

$35\% \times \$91,500 = \$32,025$ contingency fee

Add the fee and the travel expenses.

$\$32,025 + \$1,900 = \$33,925$ total amount paid

✔ CHECK YOUR UNDERSTANDING

E. Use the facts from Example 3. Assume the regular employee receives benefits of 27.5% of wages. For which employee, contract or regular, will it cost more per day considering wages and benefits? How much more?

F. A headhunter, Boyd Carron, is recruiting a plant manager for Exwell Manufacturing on a contingency fee basis. Exwell will pay Boyd a 22% fee based on the expected \$150,000 salary the manager will earn plus another 3.2% of the manager's salary to cover recruiting expenses. What total fee will Exwell pay Boyd if the search is successful?

Wrap Up ▶▶▶

Although many people may be interested in the CEO position, the recruiting firm would likely have a list of highly qualified people to consider because of the high-level contacts it can make. The recruiting firm is paid for its expertise and its discretion. From the company's viewpoint, the \$500,000 expense is small compared to the financial growth a company can make by getting the ideal person for the CEO position.

TEAM Meeting

Each class member must join a team of 4–6 students. Each team member needs to interview five employed people in the general community to determine how they got their current job.

Then the team should look at the responses received and summarize them into 6–10 categories, such as want ads or referred by a friend or relative.

Teams are to present their findings to the class. The class should identify patterns and trends in how people got their jobs and the job types. Summarize all the team findings into 6–10 categories.

Exercises

Find the sum.

1. \$8,500 + \$1,280 + \$575
2. \$1,110 + \$355 + \$720

Find the product.

3. 5 × \$127.50
4. 52 × \$876
5. 65% × \$240,000
6. 20% × \$35,550

Solve.

7. A trade journal prints four editions monthly, one each for four different parts of the United States. The cost of a small ad in the job postings section of one of the editions is \$1,400 per month. A 17% discount is given if an ad is placed in all editions in a month. What is the total cost of the ad if it is placed in all editions in the same month?
8. A newspaper charges \$67.20 a line per day for employment advertising. The Color-True Paint Company placed a 13-line ad for a polymer chemist in the newspaper for two days. What was the total cost of the ad?
9. The Grix spent \$108,000 on employment advertising for technical positions. Grix hired 25 technical employees as a direct result of the ad. How much did Grix spend on employment advertising for each technical employee hired?
10. Trex Graphics estimates it spends these amounts to hire each new employee: recruiting, \$1,750; 15 hours personnel staff time at \$31 an hour; 8 hours interview time at \$84 an hour; and 5 days employee orientation and training at \$360 a day. What is the total cost of hiring an employee?
11. A law firm spent \$18,000 to screen and interview recent college graduates for one job opening. The company hired Gabriella Coates, the top student in her class. Gabriella received an annual salary of \$85,000, 41% in benefits, payment of her college student loan of \$29,000, and a \$10,000 signing bonus. What is the total cost of hiring and employing Gabriella for the first year?

12. A plastic molding company with 300 employees has to replace 5% of its employees each year. The average cost of hiring one new employee is $3,900. What total amount does the company pay a year to hire new employees?

13. Mayfair Industries paid Rosman Recruiting a retainer fee of $114,000 to recruit a chief financial officer who will be paid a salary of $235,000 a year. Rosman was also used to recruit two purchasing agents, each of whom will be paid an annual salary of $49,000. Rosman's contingency fee for recruiting each purchasing agent was 23% of the position's annual salary. What total amount in recruiting fees did Mayfair pay Rosman?

14. **CRITICAL THINKING** A job fair for a new store had 500 applicants for 10 jobs. The one recruiter assigned had hoped to interview applicants at the job fair, but was overwhelmed by the response. The recruiter now has to review all the applications and call back certain applicants for an interview. How could the recruiting have been handled better?

To meet a deadline, Ronan Products hired four contract employees for 80 total hours at $27.50 an hour. Ronan estimates that regular employees could have done the same work in 60 hours by working overtime at time-and-a-half pay. Regular employees receive $19.20 an hour and benefits of 25% of their wages. By hiring the contract employees, Ronan did not have to pay 60 hours of overtime.

15. What was the cost of using contract employees?

16. What would have been the cost if regular employees had been used instead of contract employees?

17. **INTEGRATING YOUR KNOWLEDGE** Conrad Harris took a new job 700 miles from where he now lives. Conrad's new employer agreed to pay his relocation expenses which were: $3,800 moving bill, $4,200 in closing costs on a new home purchase, and the 7% agent's commission on the sale of his old home at $156,000. What total relocation costs did Conrad's employer pay?

Mixed Review

18. Rewrite $\frac{1}{2}$ as a decimal.

19. $3\frac{3}{5} \times 1\frac{7}{8}$

20. $2,142 \div 9$

21. Find 8% of $23.50.

22. Inez Folz agrees to a purchase price of $23,050 for a new car. The sales tax is 3.5% of the purchase price and registration fees are $118. Inez made a 20% down payment of the purchase price. What is the delivered price of the car?

23. The CPI for the Recreation category was 102.0 in 1999 and 103.7 in 2000. By what percent did Recreation prices increase from 1999 to 2000, to the nearest tenth percent?

24. We Have Puppets, a retail store, plans to sell its puppets online. E-Market will design an e-business web site for $14,800. Server use and storage space will cost $410 a month while high-speed Internet access will cost $275 a month. E-Market will maintain the web site for $780 a month. What is the first-year cost of the site to We Have Puppets?

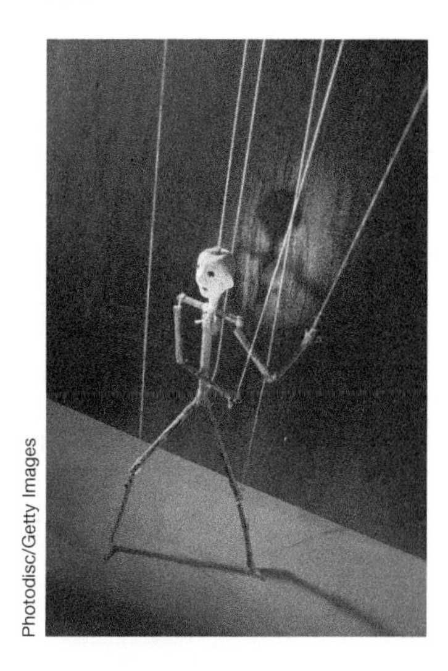
Photodisc/Getty Images

11-2

Wage and Salary Increases

GOALS

- Calculate Cost-of-Living Adjustments (COLA)
- Calculate bonuses
- Calculate profit sharing

KEY TERMS

- Cost-of-Living Adjustment (COLA)
- bonus
- profit sharing

Start Up

An employer offers to pay a wage increase of 4% in one payment at the end of a year. The employees want a 1% wage increase on the first day of each quarter starting at the beginning of the year. The employer argues that 1% for 4 quarters is 4%. Is there any difference between the wage plans?

Andresr/Shutterstock.com

Math Skill Builder

Review these math skills and solve the exercises that follow.

1. **Add** money amounts.
   Find the sum. $0.86 + $17.41 = $18.27

   1a. $0.39 + $10.64 **1b.** $1,806 + $31,098

2. **Multiply** dollar amounts by percents. Round to the nearest cent.
   Find the product. 1.9% × $18.10 = 0.019 × $18.10 = $0.343, or $0.34

   2a. 3.8% × $14.39 **2b.** 0.6% × $38,400 **2c.** 15% × $3,100,000

3. **Divide** money amounts by whole numbers.
   Find the quotient. $390,000 ÷ 600 = $650

   3a. $86,000 ÷ 215 **3b.** $28,000 ÷ 40

Cost-of-Living Adjustments

Some employers pay their employees a **Cost-of-Living Adjustment (COLA)**, which is a wage increase based on changes in the Consumer Price Index (CPI). COLA pay increases may be written into wage contracts made between a company's management and a labor union.

COLA is designed to help employee wages keep up with inflation. COLA wage adjustments are usually made annually, semiannually, or quarterly. Wage adjustments often become a permanent part of the regular hourly pay or salary earned by employees.

Since the Consumer Price Index reports price increases that have already happened, COLA adjustments often lag behind inflation. Because of this, some employers give *retroactive pay*. They compute the amount employees should have been paid since the last COLA wage adjustment.

EXAMPLE 1

The Skrynn Company gives its employees a COLA equal to the CPI's 2.6% increase in the previous year. Its employees now earn an average rate of $14.50 an hour. What will be the average hourly pay rate after the COLA is given?

SOLUTION

Multiply the hourly rate by the percent increase. Round to the nearest cent.

2.6% × $14.50 = 0.026 × $14.50 = $0.377, or $0.38 COLA

Add the COLA to the old hourly pay.

$0.38 + $14.50 = $14.88 new pay rate

✔ CHECK YOUR UNDERSTANDING

A. Omer Brombach earns $11.80 an hour. His employer gives him a COLA based on the 0.8% the CPI rose last quarter. What amount of raise will Omar get, to the nearest cent? What will be his new hourly pay rate?

B. Marsha Ahlquist earned a salary of $56,200 last year. Her employer gave her a COLA pay raise equal to the 4.8% increase in the CPI. What is the amount of the COLA? What is her salary after the COLA?

Photodisc/Getty Images

Bonuses

Some companies may give their employees a bonus. A **bonus** is pay given to reward employees who make a significant contribution to the success of the company or whose work record is exceptional. A rating system may be used to determine which employees deserve bonuses.

Bonuses may be a stated dollar amount or a percentage of an employee's regular pay. They may also be given in addition to other forms of pay raises, such as COLA. Bonuses may become a permanent part of an employee's pay or they may be only a one-time payment.

EXAMPLE 2

The TSU Company decides to give a bonus to Lowell Hurst. The bonus is 3.5% of his annual earnings of $38,500 a year. What amount of bonus will Lowell receive?

SOLUTION

Multiply the annual earnings by the bonus percent.

3.5% × $38,500 = 0.035 × $38,500 = $1,347.50 bonus

✔CHECK YOUR UNDERSTANDING

C. Althea Gannon received a bonus of 2% of total sales for being the top salesperson in the company. Her sales for the year were $850,000. What bonus did Althea receive?

D. Four team members received a bonus of $2,000 each for finishing a project two months ahead of schedule. Twelve support team members each received a $1,200 bonus. What total bonus did the employer pay to all team members?

Profit Sharing

Some companies offer profit sharing to their employees. **Profit sharing** means that employees get part of the profit earned by a company. An employee's share is usually based on the amount of profit and the number of employees eligible for it.

Many profit sharing plans make payments in cash to employees. Employees owe income tax on those payments. Others allow employees to place their shares of the profits in pension programs. When the pension plan is used, the payment is not taxable until it is taken out at retirement.

The idea behind profit sharing plans is to motivate employees to be productive, cut costs, and make suggestions for improvements because they benefit directly from higher profits. Employees share only in the profits of a company, not its losses.

NET Bookmark

Use the Internet or do library research to find a company that pays profit sharing. You could use General Motors, Ford, Daimler-Chrysler, or any other company you find. Find the average amount of profit sharing each hourly-rated employee received for each of the last five years. Find the percent that profit sharing was of the average pay of hourly-rated employees.

EXAMPLE 3

A company plans to share 20% of its annual profit of $5,000,000 with 250 employees. What profit sharing amount will each employee get?

SOLUTION

Multiply the annual profit by the profit sharing percent.

$20\% \times \$5,000,000 = 0.2 \times \$5,000,000 = \$1,000,000$ profit to be shared

Divide the profit to be shared by the number of employees.

$\$1,000,000 \div 250 = \$4,000$ profit sharing amount per employee

✔CHECK YOUR UNDERSTANDING

E. A total of $13,500 profit is to be shared equally by a company's 18 purchasing and warehouse staff. What amount will each employee receive?

F. The management of a company shared 20% of its $511,875 profit with its 65 employees. How much did each employee receive?

Wrap Up ▶▶▶

By getting the 1% wage increase each quarter, employees would have use of part of their raise throughout the year. The wage increase could be spent or invested. Also, if the wage increase becomes part of regular pay, the 1% raise per quarter would be compounded. So, the actual percent increase would be 4.06% instead of 4%.

Exercises

Find the sum.

1. $0.57 + $14.83
2. $1,847 + $38,141

Find the product, to the nearest cent.

3. 0.9% × $18.72
4. 2.4% × $950,000

Find the quotient.

5. $17,920,000 ÷ 28,000
6. $18,512 ÷ 26

Solve.

7. Viola Forberg earns $19.50 an hour. On April 1, she received an 0.8% COLA. What is her new hourly pay rate, to the nearest cent?
8. The total payroll of a company whose workers receive COLA is $11,200,000. The CPI is expected to rise 3.46% this year. What total amount of payroll should the company expect to pay this year after COLA, to the nearest ten thousand dollars?
9. The four top-rated employees in six departments of the Roth Company each received a $1,250 bonus. What total amount of bonus did the Roth Company pay?
10. Cromwell Electronics set aside 12% of its annual profit of $2,250,000 to share with its 118 employees. What profit sharing check did each employee receive, to the nearest cent?
11. All the workers in a factory received a bonus of 2% of their annual pay for finishing a project on time and below expected cost. The average annual pay of the workers is $49,200. What average bonus was paid per employee?
12. A company issued profit sharing checks averaging $3,180 per employee to its 218 employees. The largest check was for $5,008. The smallest check was for $2,117. What total amount of profit did the company share with its employees?

Photodisc/Getty Images

A company's annual payroll for 400 hourly workers is $14,000,000 and for its 40 salaried workers, $1,800,000. Each worker gets a 3% COLA raise that will be paid at one time as retroactive pay.

13. What retroactive pay will each hourly worker get on average?

14. What retroactive pay will each salaried worker get on average?

15. What total retroactive payment will the company make to all workers?

Photodisc/Getty Images

16. **CRITICAL THINKING** A company has paid COLA for the past 10 years. The company expects a slight loss this year and an even greater loss the next year. The company asks its employees to give up the COLA increase until business improves. What should the employees do?

17. **FINANCIAL DECISION MAKING** You have a choice between two similar jobs, both paying $20,000 a year. One job offers you automatic pay raises of $500 every six months until you reach the maximum pay of $25,000 for the job's classification. The other job does not offer automatic pay increases, but will give increases up to $2,000 a year to employees who do good work. Which job would you take? Why?

18. **INTEGRATING YOUR KNOWLEDGE** An employee was paid for an average of 44.5 hours a week for 52 weeks last year. His regular-time pay rate was $15.10 an hour. Time-and-a-half is paid for hours worked over 40 hours a week. His profit sharing check was 4% of his regular time wages for the year. What was his total gross pay for the year from all sources?

Mixed Review

19. $6.6 + 1.76 + 0.9 + 1.875$

20. $12 \times 5{,}400$

21. $\frac{3}{4} - \frac{5}{8}$

22. $3\frac{1}{2} \div 4\frac{2}{3}$

23. Find the estimated product of 989×62.

24. Find the average of 414, 917, 582, 336, and 712.

25. Flavian Gvendron's check register showed a balance of $782.29 at the beginning of the day. During the day he wrote checks for $89.39, $585.87, and $193.92. He also transferred $200 from savings to checking. What was the balance of Flavian's check register after these transactions?

26. What annual interest is earned by 6, $1,000, 8.25% interest bonds?

27. The Fleider Company is replacing its 76 desktop computers. The company plans to lease new desktop computers for 36 months at $41.80 a month for each computer. What will be the total cost of leasing per month and over the life of the lease?

11-3

Total Costs of Labor

GOALS
- Calculate cost of full-time employees
- Calculate cost of part-time employees

KEY TERM
- part-time employee

Start Up

A company restricts the hours that several of its employees work to a maximum of 34 hours a week. If the employees worked 40 hours a week, they would be full-time employees and receive the benefits of paid medical insurance and tuition reimbursement. Is this business policy of limiting the hours worked fair to the employees who work less than 40 hours?

Photodisc/Getty Images

Math Skill Builder

Review these math skills and solve the exercises that follow.

1. **Add** money amounts.
   Find the sum. $12,345 + $1,888 + $210 = $14,443

   1a. $27,458 + $5,292 **1b.** $312.47 + $34.87

2. **Multiply** money amounts by percents and whole numbers.
   Find the product. 7.65% × $45,600 = $3,488.40
   Find the product. 32 × $14.10 = $451.20

   2a. 0.8% × $1,245 **2b.** 4.3% × $6,500

   2c. 16 × $9.45 **2d.** 52 × $269.34

Cost of Full-Time Employees

The cost of full-time employees consists of the wages they are paid and the benefits they receive. Employees may consider only their wages to be income and benefits as only a non-cash incentive for working. Employers consider benefits to be costs, or expenses, because they actually pay the benefits.

Some of the benefits paid by employers are required to be paid by law. For example, employers must match employee payments of FICA taxes that include a 6.2% Social Security tax on annual wages up to $102,000 and a 1.45% Medicare tax on all wages.

In addition, the Federal Unemployment Tax Act (FUTA) requires employers to pay a FUTA tax on the first $7,000 of earnings of each employee.

Each state has an unemployment tax law, called the State Unemployment Tax Act, or SUTA. SUTA also levies taxes to be paid by the employer.

SUTA pays cash to workers who are temporarily unemployed. These cash payments are called unemployment compensation. FUTA taxes are primarily used to pay the cost of running the unemployment compensation program.

Other employee benefits commonly provided by employers include retirement plans, medical and related insurance, life insurance, vacation pay, holiday pay, and sick leave. Still other benefits may include a company car, tuition reimbursement, child care, paid parking, employee discounts, and legal services.

EXAMPLE 1

The Todd Production Company pays Charlene Mitchell a salary of $47,000 a year. The benefits that Todd pays include a 7.65% FICA tax, an 0.8% FUTA tax and 4.5% SUTA tax on the first $7,000 of wages, medical insurance of $4,120 a year, and a retirement contribution of 5% of Charlene's salary. What was the total cost of the benefits the Todd Company provided to Charlene? What total amount did it cost the Todd Company to employ Charlene for the year?

Math *Tip*

Round to the nearest cent the results of all tax and benefit calculations involving percents.

SOLUTION

Multiply the annual salary by the FICA and retirement percents.

$7.65\% \times \$47,000 = 0.0765 \times \$47,000 = \$3,595.50$ FICA taxes paid

$5\% \times \$47,000 = 0.05 \times \$47,000 = \$2,350$ retirement contribution

Multiply the first $7,000 of wages by the FUTA and SUTA percents.

$0.8\% \times \$7,000 = \$56$ FUTA tax paid

$4.5\% \times \$7,000 = \$315$ SUTA tax paid

Add benefits.

$\$3,595.50 + \$2,350 + \$56 + \$315 + \$4,120 = \$10,436.50$ total benefits

Add the annual salary and the total benefits.

$\$47,000 + \$10,436.50 = \$57,436.50$ total cost of employment

✔ CHECK YOUR UNDERSTANDING

A. Tek-Power pays its employees an average of $540 a week. It also provides these benefits at no cost to employees: medical insurance worth $73.07 a week, life insurance worth $2.80 a week, and a pension plan contribution equal to 3% of wages. FICA taxes of 7.65% and unemployment taxes of 4.7% of wages are paid by Tek-Power. What is the average cost of the benefits of one employee for one week?

B. The total wages paid by Rizzo Restaurant Supply Company for the second week in January are $10,350. Total benefits paid to employees average 32% of wages. What was the total cost of labor for Rizzo for the week?

Photodisc/Getty Images

Cost of Part-Time Employees

Part-time employees are those employees who work less than 40 hours a week. Part-time employees are often paid the same rate of pay as full-time employees, but do not receive the same benefits. At a minimum, part-time employees receive the benefits required by law.

EXAMPLE 2

Quintin Napier, a college student, works 25 hours a week during evening hours at the help desk of the Renner Software Company. Renner pays Quintin $11.25 an hour and provides him with tuition reimbursement benefits of $500 a year. The Renner Company's benefits include the 7.65% FICA tax and total unemployment taxes of 4.2% on the first $7,000 of earnings. What total benefits does Renner provide? What is the total cost of Quintin's employment for 52 weeks?

SOLUTION

Multiply the weekly hours worked by the hourly wage and by 52 weeks.

$25 \times \$11.25 \times 52 = \$14,625$ annual earnings

Multiply the annual earnings by the FICA tax rate and the $7,000 of earnings by the unemployment tax rate

$7.65\% \times \$14,625 = \$1,118.81$ FICA tax

$4.2\% \times \$7,000 = 0.042 \times \$7,000 = \$294$ unemployment tax

Add the annual earnings, FICA and unemployment taxes, and the tuition reimbursement.

$\$14,625 + \$1,118.81 + \$294 + \$500 = \$16,537.81$ total cost of employment

✔ CHECK YOUR UNDERSTANDING

C. A restaurant pays Carrie $7.90 an hour and gives her a daily $3.10 meal allowance. Taxes paid by the restaurant include a 7.65% FICA tax and 5.1% in unemployment taxes. What is the maximum amount the restaurant will spend for Carrie's benefits if she works seven hours a day, four days a week?

D. The Boot Fitter Shop filled a vacancy by hiring two part-time employees who each work 20 hours a week at the regular pay of $10.85 an hour. As a result, the shop will not have to pay medical benefits that average 14% of employee wages for full-time employees. For each 40-hour week, how much less will the Boot Fitter Shop spend by hiring the two part-time employees than it would have spent by hiring one full-time employee at the same rate?

Wrap Up

One method employers use to cut costs is reducing the amount they spend on employee benefits. Employers avoid paying employee benefits by hiring two part-time employees to do the work of one full-time employee. Each full-time employee that is not hired can potentially save the employer $4,000 a year. While such a policy appears to be unfair to employees, the policy is legal.

Financial Responsibility

Personnel Marketing Firms

When you are searching for employment, numerous avenues are available. You may read employment advertisements in newspapers, professional journals, and online job posting web sites. Another option is to seek assistance through an employment agency. Basically, there are two types of employment agencies. One type of agency is hired by the company with the open position and that company pays the agency fees. Another type charges the fee to the person looking for a job and in return they help you prepare a resume, cover letters, and identify possible positions for which you qualify.

Before you sign a contract with a personnel marketing firm, read the contract thoroughly to be completely aware of the services for which you are paying and any guarantees the firm offers. The individuals trying to sell you their services are salespeople.

Depending on the length of time you have been unemployed or looking for a new job, you may be vulnerable to the sales pitches promising you the job of your dreams. Personnel marketing firms often have fees ranging from $5,000 to $7,000 that are paid by the individual seeking employment. Research the background of any company and be sure that they have a proven track record for finding employment for their clients before you hand over any money.

Exercises

Find the sum.

1. $12,451 + $781 + $26.02
2. $68,260 + $12,573

Find the product to the nearest cent.

3. 2.06% × $762
4. 3.98% × $1,071
5. 29 × $14.43
6. 52 × 23 × $8.92

Solve.

7. Eldon Schaff earns wages of $780 a week. His non-tax benefits include medical, dental, and vision insurance worth $4,110 a year; employer pension plan contribution of $2,250 a year; three weeks of vacation and holiday pay; and a paid annual physical exam worth $800. What total benefits does Eldon receive?
8. Erika Fogarty is allowed to contribute 6% of her annual pay of $73,200 to a retirement savings program. Her employer adds to the retirement program $1 for every $2 that Erika contributes. What is the employer's cost for this benefit?
9. Edgar Henneman earned $104,000 in wages last year. What was the cost to Edgar's employer of providing these legally required benefits: net FUTA tax of 0.8% on the first $7,000 of wages, SUTA tax of 7.6% of the first $9,500 of wages, Social Security tax of 6.2% on the first $102,000 of wages, and a Medicare tax of 1.45% on all wages?
10. The Sure-Freeze Company pays its full- and part-time employees the same hourly pay, an average rate of $15.12 an hour. The cost of benefits per full-time employee is 28% of hourly pay and 9% of hourly pay for part-time employees. To the nearest cent, what is the total labor cost per hour for full-time employees? For part-time employees?

11. A company employs 60 part-time workers and pays them $12.40 an hour for working an average of 30 hours in a 5-day week. To keep its employees, the company plans to raise the pay rate to $12.95 an hour. It will also give each employee a $4 a day parking allowance and free coffee or juice at an estimated cost of $0.90 a day per employee. What total amount more will it cost the company per week when these pay and benefit changes are made?

Photodisc/Getty Images

The Lentag Company spent $3,750 per employee this year to provide a medical insurance benefit for its 1,100 employees, at no cost to the employees. The insurance company notified Lentag that medical insurance costs would increase 12% next year. Lentag's management decides to keep the same medical coverage but will require employees to pay 45% of the increase.

12. What will be the new cost of medical insurance per employee?
13. What amount will each employee pay a year for medical insurance?
14. What total amount more will Lentag pay for the medical insurance benefit next year than it did this year for all employees?
15. **FINANCIAL DECISION MAKING** You are the owner of a small company with six part-time employees who each work 36 hours a week. To improve staff morale you would like to have full-time employees even though the benefits would cost $2,000 a year per employee. To keep costs in line you would have to let one person go. The five employees would have to work 43 hours a week to get the same work done. What should you do?
16. **CRITICAL THINKING** To reduce labor costs, a company offers a retirement incentive for employees who are at least 59 years of age and have 30 years of service with the company. A 45-year old employee complains that this program favors older workers and is an example of age discrimination. Do you agree or disagree? Why?

Mixed Review

17. $862.7 - 149.34$
18. Write 0.12 as a fraction.
19. $\frac{3}{14} \div \frac{5}{7}$
20. $1{,}000 \times 2.07$
21. 10¢ ÷ 1,000
22. 35 is what percent of 50?
23. Find the number of days from July 9 to October 15.
24. Find the average: $129.43, $241.89, $851.13, and $197.43
25. Katrina Stovall needs a $60,000 mortgage loan to buy a home. Her bank gives her a 7%, 25-year mortgage. Use the Amortization Table in Chapter 6 to find Katrina's monthly payments. Then find the total interest she will pay on the loan over 25 years.
26. Chuck Boyce signed a two-year cell phone lease with Vega NationWide. He estimates he will use 500 peak minutes of airtime a month. Using the Wireless Phone Service Plans Table in Chapter 8, find the estimated cost of the lease over two years, including the activation fee.

11-4

Tracking and Reordering Inventory

GOALS

- Calculate inventory balances
- Calculate the stock reorder point

KEY TERMS

- stock
- stock record
- perpetual inventory
- reorder point

Start Up

Assume you are responsible for buying food and household items for your home. Some items, such as fresh fruit, will be used quickly. Other items, such as a bottle of furniture wax, are likely to be used over a long period of time. What method would you use to make sure you did not run short of any items needed to feed a family or use around the home?

Andre Blais/ Shutterstock.com

Math Skill Builder

Review these math skills and solve the exercises that follow.

1 **Subtract** whole numbers.
Find the remainder. 592 − 187 = 405

1a. 83 − 29

1b. 118 2 47

1c. 1,276 − 345

1d. 63,782 − 4,674

2 **Divide** whole numbers by whole numbers.
Find the quotient. 6,270 ÷ 22 = 285

2a. 527 ÷ 31

2b. 4,290 ÷ 30

2c. 5,382 ÷ 6

2d. 1,440 ÷ 45

2e. 1,380 ÷ 15

2f. 2,835 ÷ 81

3 **Multiply** whole numbers by whole numbers.
Find the product. 120 × 7 = 840

3a. 230 × 13

3b. 520 × 3

3c. 55 × 21

3d. 117 × 59

Inventory Records

All businesses keep a record of the **stock** they have on hand. The stock on hand may be merchandise that is resold to others or a part that will be used in the making of a product. Stock is often referred to as *inventory*.

Stock Record			
Item: Digital Camera Stock No. DC-106		Reorder Point: 20 Unit: Each	
Date	Quantity Received	Quantity Issued	Balance
March 1			53
March 2		31	22
March 4		18	4
March 6	51		55
March 10		36	19

Businesses are able to keep track of stock on hand by using a stock record form. A **stock record** shows how much of an item has been received, issued, and how much remains. A sample stock record is shown.

The Togan Company is a wholesaler and sells only to retail stores. It does not sell to the general public. The sample stock record shows the number of digital cameras the Togan Company received from a manufacturer and the number of digital cameras it issued, or sold, to retailers. The unit for digital cameras is "each." This means that each camera is sold individually.

Photodisc/Getty Images

In the sample stock record, a balance was found after each transaction was entered in the stock record. The system of keeping a running balance of stock on hand is called **perpetual inventory**. The perpetual inventory system allows you to maintain an up-to-date record of stock on hand. The method used to find stock balances is similar to the one you used to find check register balances.

As stock arrives from suppliers, it is inspected, counted, and stored by receiving department employees. Records of stock issued may include sales slips or other records that show the quantity and description of items issued from stock.

EXAMPLE 1

Refer to the sample stock record given. If 36 more digital cameras were received on March 15 and 17 more cameras were issued on March 17, what was the balance in the stock record on March 15 and March 17?

SOLUTION

Add quantity received on March 15 to the stock balance.

$19 + 36 = 55$ stock balance on March 15

Subtract the quantity issued on March 17 from the balance on March 15.

$55 - 17 = 38$ stock balance on March 17

✔ **CHECK YOUR UNDERSTANDING**

A. A wholesaler's stock record for garage doors showed a balance of 38 doors on June 5. The numbers of garage doors issued in June were: June 8, 6; June 14, 28; June 30, 8. Eighteen garage doors were received on June 16. What was the stock record balance on June 30?

B. A manufacturer showed a bicycle tire's stock record balance to be 1,389 on April 1. These tire sales were made: April 12, 417 tires; April 26, 874 tires. 900 more bicycle tires were produced on April 29 and added to stock. What were the balances in the stock record after each transaction was recorded?

Reorder Point

When the stock of an item runs low, it must be reordered. On the Sample Stock Record the reorder point is 20 units. The **reorder point** is the minimum stock level at which an order must be placed. When the balance on the stock record form is equal to or is less than 20 digital cameras, stock must be reordered.

Companies designate a person, usually called a buyer, to order stock. A document that lists the items to be ordered and their prices is sent to the supplier. The document is called a purchase order.

For items in regular demand, most companies know the average number of units they sell or use per day. This average number is called *daily usage.* Companies also know the *lead time,* or the average number of days it will take for ordered stock to arrive. To reduce the possibility of having no stock left at all because of delays, most companies will order additional stock, called *safety stock.*

The reorder point formula shows how the daily usage, lead time, and safety stock are related:

Reorder Point = (Daily Usage × Lead Time) + Safety Stock

Business *Tip*

By calculating a realistic reorder point, companies hope to avoid a *stockout* condition. This happens when there is no inventory left to fill orders or meet production needs.

EXAMPLE 2

The Appland Company, a wholesaler, sells 5,280 dishwashers each month in 22 working days. A supplier's lead time for delivery of dishwashers is 5 working days from the time an order is received. Appland wants to keep a safety stock of 50 dishwashers. What is Appland's reorder point for dishwashers?

SOLUTION

Divide the number of dishwashers sold by the number of working days.

$5,280 \div 22 = 240$ daily usage

Multiply daily usage by the lead-time. Then add the safety stock quantity.

$(240 \times 5) + 50 = 1,200 + 50 = 1,250$ reorder point

Photodisc/Getty Images

✔ **CHECK YOUR UNDERSTANDING**

C. The Floor Care Place sells 90 floor polishers every 30 days. It takes 10 days' lead time to replace floor polisher stock. Safety stock is 2 polishers. What is the reorder point?

D. A lamp manufacturer uses 8,280 of an electrical component every 45 days. The lead time for the component is 13 days. Safety stock is three days' usage. What is the reorder point?

Wrap Up ▶▶▶

A number of methods may be used. The simplest is to write down an item on a shopping list when you first notice that it is completely used or soon will be. You could also use your computer and print a list of all the essential food and household items and check off needed items.

Stores also have shopping lists that you could bring home. You could also buy two of everything you need and put an item on your list when you reach one item left. In any case, all the methods suggested require that you make a written record of your needs.

Consumer Alert

Rain Checks

When a sale price is advertised, a store's buyers should have projected how many consumers they expect to buy the sale item and have enough inventory to meet the demand. If the number of people that come to buy the product exceeds the number of products the store has in stock, there will be unhappy customers. They expected to be able to buy the item at the reduced price.

To keep customers happy, many stores offer *rain checks* when a sale item is out of stock. A rain check is a coupon that states when the item is back in stock, the buyer can present the coupon and still pay the reduced price for the item. Buyers have to go to the customer service counter to have a rain check issued.

Exercises

Find the difference.

1. $1{,}805 - 687$

2. $58 - 25$

Find the product or quotient.

3. 39×7

4. 31×18

5. $6{,}030 \div 5$

6. $810 \div 30$

A manufacturer buys and stores engines to be used in assembling lawn mowers. On Monday, March 6, 800 engines were in stock at the start of the day. From March 6–10, these quantities of engines were taken from storage to a production line: March 6, 315; March 7, 308; March 8, 318; March 9, 298; and March 10, 285. These numbers of engines arrived from a supplier: March 7, 750; and March 10, 750. How many engines were on hand at the end of the day:

7. on Wednesday?

8. on Friday?

Solve.

9. A truck supply store had 18 bed liners in stock on February 1. These numbers of bed liners were sold in four weeks: Week 1, 11; Week 2, 6; Week 3, 9; and Week 4, 16. A shipment of 34 bed liners was received on February 8. What was the stock record balance of bed liners at the close of business on February 28?

10. An auto parts store sells 10,350 bottles of windshield washer fluid during the first 90 days of winter. The store does not let stock of washer fluid get below 50 bottles during this period. New stock can be obtained within two days of ordering. What is the reorder point?

11. A manufacturer uses 16,800 fasteners in six days. Safety stock is 8 days' usage. Fasteners usually arrive within three days of ordering. What is the reorder point for fasteners?

A supermarket carries four brands of general-purpose flour. It sells 120 bags of Winnie's unbleached flour every 30 days. More bags of Winnie's flour can be ordered and received in two days but only in cases of 12 bags. The supermarket carries no safety stock of Winnie's flour.

12. What is the reorder point for Winnie's flour?

13. How many bags must be ordered once the reorder point is reached?

14. **CRITICAL THINKING** Lorna Vogt owns a small store and believes that she doesn't need any written perpetual inventory system. She claims she has memorized all the inventory records and can walk down each aisle and know what to order. Should Lorna keep the inventory and ordering system she claims to be working or change to another system?

15. **CRITICAL THINKING** Rank these products sold by retailers according to the lead time required to obtain them from a supplier: digital camera, motor oil, sofa, refrigerator. Rank the product with the shortest lead time as "1." Give reasons for your rankings.

Mixed Review

16. Rewrite 1.8 as a percent.

17. $35 \times 10¢$

18. Rewrite $\frac{5}{8}$ as a decimal.

19. $8 \times 2\frac{1}{4}$

20. Find the quotient to the nearest tenth: $2{,}045 \div 9$

21. What number is $1\frac{1}{2}$ of $218.96?

22. Ila Behrle bought a car for $11,200. She drove the car 12,000 miles last year. Her car expenses for the year were insurance, $786; gas and oil, $960; repairs, $125; license plates, $72; depreciation, 14%; and loss of interest on original investment, $442. What was Ila's operating cost per mile of owning this car, to the nearest cent?

23. Nettie Tindle bought 400 shares of Exoway stock at $54.37. Her broker charged a commission of $186. What was the total cost of the stock purchase?

11-5

Inventory Valuation

GOALS
- Calculate inventory value on FIFO basis
- Calculate inventory value on LIFO basis
- Calculate inventory value on weighted average basis

KEY TERMS
- First In, First Out (FIFO)
- Last In, First Out (LIFO)
- weighted average

Start Up

Suppose you found a package of saw blades with a price of $5.95 on them. You know the current price of similar blades is about $17. If you were asked to state the value of the blades you found, what would you say and why?

Gwimages/ Shutterstock.com

Math Skill Builder

Review these math skills and solve the exercises that follow.

1. **Add** whole numbers and money amounts.
Find the sum. $2,450 + $886 = $3,336

1a. $763 + $827 **1b.** 490 + 53 **1c.** 215 + 85

2. **Multiply** money amounts by whole numbers.
Find the product. 180 × $18 = $3,240

2a. 218 × $7.50 **2b.** 187 × $12.36

3. **Divide** money amounts by whole numbers.
Find the quotient, to the nearest cent. $18,300 ÷ 650 = $28.15

3a. $1,280 ÷ 58 **3b.** $46,370 ÷ 1,570

First In, First Out

Companies usually calculate the value of their inventory at a certain time, such as the end of a quarter or end of a year. The inventory's value is often needed for financial records and reports. An inventory's value may also be the basis for calculating the amount of personal property tax due to a city or state. The exact value of inventory is found by multiplying the quantity of items in stock by their cost.

There are several methods of finding the value of the ending inventory. Three widely used methods are: First In, First Out; Last In, First Out; and weighted average.

The **First In, First Out (FIFO)** method assumes that goods purchased first are sold or used first and the value of inventory is based on the cost of the most recently purchased items. The FIFO method is widely used because the value of the inventory using FIFO is close to the cost of replacing the inventory.

Photodisc/Getty Images

ZPM Business Products			
Stock Item: Paper Shredders			Stock No. PS108
Inventory Period: January 1, 20— to March 31, 20—			
	Units	Unit Cost	Total Value
Beginning Inventory			
January 1	40	$20.00	$800
Purchases			
January 13	100	$21.00	$2,100
February 7	80	$22.00	$1,760
March 1	160	$22.50	$3,600
March 20	120	$24.00	$2,880
Total	500		$11,140
Ending Inventory			
March 31	160		???

The illustration above shows an inventory record for paper shredders. In the FIFO method, the 40 shredders in the beginning inventory would be issued first to fill orders. The 100 units from the January 13 order would be issued next, and so on. Since the units bought first are issued first, the 160 units in the ending inventory must be part of the units purchased last.

Business *Tip*

The unit cost of an item often changes. The item may be bought from different suppliers who have different prices. Also, a discount may be given. The way in which goods are shipped also affects unit costs.

EXAMPLE 1

Use the FIFO method to find the value of the ending inventory of paper shredders shown in the illustration above.

SOLUTION

The number of units in the ending inventory: 160 units

Starting with the last purchase, find the quantities and the cost of the 160 units. Multiply the units by their unit cost and add their products.

March 20 = 120 units at $24 = $2,880
March 1 = 40 units at $22.50 = $900
160 units — $3,780 value of ending inventory

Using the FIFO method, the value of 160 shredders is $3,780.

✔ CHECK YOUR UNDERSTANDING

These units were purchased by the OPC Company in February.

2-3 140 units @ $36.50 2-14 170 units @ $37.00 2-26 110 units @ $36.00

2-7 110 units @ $38.00 2-18 60 units @ $37.25

A. Assume the beginning inventory was 112 units and the ending inventory is 142 units. Find the ending inventory's value on the FIFO basis.

B. Now assume the beginning inventory was 234 units and the ending inventory is 196 units. Find the ending inventory's value on the FIFO basis.

Last In, First Out

The **Last In, First Out (LIFO)** method assumes that goods purchased last are issued first. The value of the ending inventory is based on the cost of goods purchased first. Some companies use LIFO especially when prices are rising. In that case, LIFO results in the lowest value for inventory and that leads to a lower net income for the company. The company generally pays less in income taxes by using LIFO.

Look back at the illustration at the top of the previous page. Using the LIFO method, the last shredders purchased, which are the 120 units bought on March 20, would be the first to be issued. The 160 units bought on March 1 would be issued next, and so on. Since the units bought last are issued first, the 160 units in the ending inventory must be made up of the units in the beginning inventory and the units purchased in January and February.

Business *Tip*

The LIFO method is used only to find the value of the ending inventory. This does not mean that the newest stock is sold first. In fact, where possible companies rotate stock, and use the oldest stock first in the same way that the newest canned goods are moved to the back of the shelf in a supermarket.

EXAMPLE 2

Use the LIFO method to find the value of the ending inventory of paper shredders shown in the illustration at the top of the previous page.

SOLUTION

The number of units in the ending inventory: 160 Units

Starting with the beginning inventory and the first purchases, find the quantities and the costs of 160 units. Multiply the units by their costs and add the products.

January 1 =	40 units at $20 =	$800	
January 13 =	100 units at $21 =	$2,100	
February 7 =	20 units at $22 =	$440	
	160 units	$3,340	value of ending inventory

Under the LIFO method, the value of 160 shredders is $3,340.

✔ CHECK YOUR UNDERSTANDING

The Davis Alam Company purchased smoke detectors in November as follows.

11-5 32 units @ $14.60

11-11 40 units @ $14.85

11-18 20 units @ $14.90

11-25 25 units @ $15.10

C. The beginning inventory was 12 units valued at $14.20 per unit. The ending inventory is 32 units. Find the ending inventory's value on the LIFO basis.

D. The beginning inventory was 6 units valued at $14.55 per unit. Use the LIFO basis to find the value of the ending inventory of 49 units.

Photodisc/Getty Images

Weighted Average

In the **weighted average** method, the ending inventory is valued at the average cost of the beginning inventory plus the cost of all purchases during the time period.

The weighted average method is useful when the cost of stock varies a lot and frequently. In that case, the weighted average method may represent the cost of inventory better than some other methods.

EXAMPLE 3

Use the weighted average method to find the value of ending inventory of ZPM Business Products paper shredders.

SOLUTION

Multiply the units in the beginning inventory and those purchased during the quarter by their unit cost. Add the products. Also add the number of units.

$$\begin{array}{rl} 40 \times \$20.00 = & \$\ \ \ 800 \\ 100 \times \$21.00 = & \$\ 2{,}100 \\ 80 \times \$22.00 = & \$\ 1{,}760 \\ 160 \times \$22.50 = & \$\ 3{,}600 \\ \underline{120} \times \$24.00 = & \underline{\$\ 2{,}880} \\ 500 \qquad\qquad & \$11{,}140 \end{array}$$

Total units = 500; Total value of 500 units = $11,140

Divide the total value by the total units.

$11,140 ÷ 500 = $22.28 average unit cost

Multiply the ending inventory by the average unit cost.

$22.28 × 160 = $3,564.80 value of ending inventory

Using the weighted average method, the value of the ending inventory of 160 shredders is $3,564.80.

✔CHECK YOUR UNDERSTANDING

The July purchases of tractor tires by the Holwith Manufacturing Company follow. Use these data for Problems E and F. For both problems, use the weighted average method to find the unit cost of the beginning inventory and July's purchases, to the nearest dollar.

7-1	312 units @ $208	7-18	940 units @ $199
7-10	786 units @ $202	7-29	515 units @ $205

E. The beginning inventory of 607 tires was valued at $201 per tire. What is the weighted average unit cost of the beginning inventory and July purchases, to the nearest dollar? What is the value of the ending inventory of 308 units?

F. Using the weighted average method, find the unit cost of the beginning inventory of 416 units valued at $205 per tire and the July purchases, rounded to the nearest dollar. Also find the value of the 702 units in the ending inventory.

Wrap Up ▶▶▶

Assuming the old saw blades were not used, their value to you is probably slightly less than their replacement cost of $17. They would have a lower value if they were sold to someone else because they would be thought of as being old blades. They may also be worth less than new blades because of design changes that make the new blades better than the old blades.

Communication

Another way of assigning a value to the ending inventory is to use the *specific identification method*. Refer to a college accounting textbook or do an Internet search to find the information about this method. Keywords for your search include inventory and valuation and method.

Specifically, find how unit costs are assigned in the specific identification method, the types of goods the method is best used for, and how it compares with the three methods you studied in this lesson. Prepare a slide show on your findings.

Exercises

Find the sum.

1. 564 + 183
2. $8,016 + $2,917
3. 1,354 + 688
4. $5,008 + $1,871

Find the product or quotient to the nearest cent.

5. 97 × $18.20
6. $168,000 ÷ 424
7. $513 ÷ 12
8. $23.65 × 72

The Zonex Plumbing and Supply Company purchased these quantities of water heaters from April 1 to June 30: April 28, 41 @ $146; May 19, 53 @ $139; June 23, 21 @ $149. The April 1 beginning inventory of 27 units was valued at $141 each. Find the value of the ending inventory of 35 heaters:

9. using the FIFO method
10. using the LIFO method
11. using the weighted average method, to the nearest cent

Jessup's Trailer Sales purchased utility trailers from a manufacturer at these quantities and unit costs over a two-month period: April 4, 20 at $376; May 2, 34 at $340; May 29, 18 at $381. The beginning inventory of 9 utility trailers had a total value of $3,159. Find the value of the 31 utility trailers in the May 31 ending inventory:

12. using the FIFO method
13. using the LIFO method
14. using the weighted average method (round the unit cost of the beginning inventory and purchases to the nearest dollar.)

15. **CRITICAL THINKING** The weighted average method of valuing inventory is often used to find the value of a product such as gasoline that is mixed together and whose costs cannot be individually identified. What other products would be suited for being valued by the weighted average method?

INTEGRATING YOUR KNOWLEDGE A wholesaler had an inventory of 850 fishing rods valued at $49,946 on May 1. Shipments of fishing rods received from a manufacturer were: May 15, 2,300 rods at $52.50; June 12, 3,400 rods at $47.75. During May and June 6,100 rods were issued from stock and shipped to retailers. There were no other transactions in May and June.

16. What was the wholesaler's ending inventory of fishing rods on June 30?

17. What was the value of the fishing rod inventory on June 30 using the LIFO method?

Photodisc/Getty Images

Mixed Review

18. Find the quotient, to the nearest tenth: 15.015 ÷ 0.7

19. What number is 30% more than $81.19?

20. 9,184 − 4,079

21. $\frac{2}{3} \times \frac{5}{6}$

22. $4\frac{1}{3} - 1\frac{2}{3}$

23. $\frac{4}{5} \div \frac{15}{2}$

24. Find 15% of $17.80.

25. The owner of five vending machines collected these amounts of money from the machines while restocking them: $118, $138, $180, $110, and $129. What average amount was collected from the five machines?

26. The Carlisle family borrowed $2,000 on a one-year simple interest installment loan at 18%. The monthly payments were $183.48. For the first monthly payment, find the amount of interest, amount applied to the principal, and the new balance.

27. Brightman Inc. purchased a 50-workstation license for DollarSafe financial software for $4,200. In a survey of software on all its workstations, a software publisher's association audit found that DollarSafe was installed on 86 computers. The penalty is $125 for each unlicensed installation of DollarSafe. What total amount did Brightman Inc. pay to use DollarSafe software?

11-6

Ordering and Carrying Inventory

GOALS
- Find the ordering costs of inventory
- Find the carrying costs of inventory

KEY TERMS
- ordering costs
- carrying costs

Start Up

On five days a week you drive to a store three miles away to do 15 minutes of grocery shopping on each trip. What factors should you consider to determine what the cost is of each shopping trip?

Photodisc/Getty Images

Math Skill Builder

Review these math skills and solve the exercises that follow.

1. **Multiply** dollar amounts by whole numbers.
   Find the product. $8 \times \$35.50 = \$284$

   1a. $52 \times \$5,600$ **1b.** $12 \times \$3,765$

2. **Find a percent** of a number.
   Find the product. $12\% \times \$56,000 = \$6,720$

   2a. $6\% \times \$156,000$ **2b.** $8.5\% \times \$236,800$

3. **Divide** money amounts by whole numbers.
   Find the quotient, to the nearest cent. $\$268,000 \div 33,000 = \$8.12$

   3a. $\$1,458,290 \div 980,000$ **3b.** $\$180,304 \div 16,000$

Ordering Costs

The **ordering costs** of inventory include the expenses connected with creating and sending a purchase order to a supplier and handling stock.

Costs of creating purchase orders include the cost of personnel in the purchasing department, office costs, and overhead costs. Office costs include the costs of forms, envelopes, telephones, faxes, copiers, and computers. Overhead costs include such items as heat, power, light, and maintenance.

The costs of handling stock include unloading and checking stock shipments and placing stock in storage areas. Handling costs are usually stated as a cost per hour. Office costs and overhead costs may be stated as a percentage of total office costs or total overhead costs.

Many organizations calculate ordering costs as a cost per purchase order. The average ordering cost does not include the value of the items bought. So, it may cost the same to order a box of printer paper worth $36 as it does to order a $16,000 computer workstation. The cost per order increases if specifications have to be written, bids have to be compared, products have to be tested before purchase, or if the product purchased has special handling or storage requirements.

EXAMPLE 1

The Linkman Company issues 500 purchase orders a month. The company estimated these monthly costs of issuing purchase orders: 60% of the company's office personnel costs of $4,600; $150 an hour for 12 hours for handling incoming stock; 10% of overhead costs of $1,400. What is the average monthly cost of issuing a purchase order?

SOLUTION

Multiply the monthly cost for each item by the percent rate or the number of hours.

60% × $4,600 = 0.6 × $4,600 = $2,760 cost of office personnel

12 × $150 = $1,800 cost of handling incoming stock

10% × $1,400 = 0.1 × $1,400 = $140 overhead costs

$2,760 + $1,800 + $140 = $4,700 total monthly cost of purchasing

Divide the monthly purchase cost by the number of purchase orders issued.

$4,700 ÷ 500 = $9.40 average cost of issuing purchase orders

✔ CHECK YOUR UNDERSTANDING

A. A retail chain of drugstores has two purchasing agents and one assistant who work full time on stock orders. Their annual wages total $128,100. These employees process 15,000 orders a year. Of the total warehouse cost of $250,000 a year, 60% is charged to ordering. The total office costs are $54,000 a year, 35% of which is allocated to ordering. What is the average cost of each stock order?

B. The ExaSol Company estimates the basic cost of processing any purchase order to be $12 per order. Each year 125 special orders must be handled at these extra costs: 5.5 hours of staff time per order to write specifications at $43 an hour; 2.5 hours per order at $32 an hour for staff to review the bids received. What are the total and average costs of handling these special purchase orders?

Carrying Costs

Carrying costs include all the costs of holding inventory until it is issued. You usually show carrying costs by how much it costs to hold one unit of inventory for a year. A large part of carrying costs is interest on money borrowed to buy the inventory. Other costs include personal property taxes on the inventory's value, insurance, storage costs, and loss from theft, damage, or obsolescence.

EXAMPLE 2

The Turned Page carries an inventory of 40,000 books valued at $22.50 per book. It pays 11% annual interest on its inventory and an annual property tax of 1% of inventory value. Each year, about 0.5% of the inventory is damaged and must be discarded. The total of other carrying costs is $8,000 a year. What is the average carrying cost, per book?

Photodisc/Getty Images

SOLUTION

Multiply the number of books by the value per book.

$40,000 \times \$22.50 = \$900,000$ annual value of inventory

Multiply the inventory's value by the interest rate, the property tax rate, and the damage rate.

$11\% \times \$900,000 = 0.11 \times \$900,000 = \$99,000$ annual interest

$1\% \times \$900,000 = 0.01 \times \$900,000 = \$9,000$ annual property tax

$0.5\% \times \$900,000 = 0.005 \times \$900,000 = \$4,500$ annual damage

Add the annual amounts for interest, property tax, and damage.

$\$99,000 + \$9,000 + \$4,500 = \$112,500$ annual carrying costs

Divide annual carrying costs by the number of books.

$\$112,500 \div 40,000 = \$2.812$, or $2.81 average carrying cost per book

✔ CHECK YOUR UNDERSTANDING

C. A food wholesaler's cost for Happy House frozen vegetables is $1.05 per bag for the 600,000 bags handled each year. The wholesaler pays 8% interest on one-half of the frozen vegetables inventory value for a year. Each year, 0.25% of the bags tear and must be thrown out. The cost of freezers and electricity to run the freezers is $90,000 a year. What is the carrying cost for one bag of frozen vegetables, to the nearest tenth of a cent?

D. A fruit retailer's average weekly inventory is valued at $7,200. Annual interest of 9% is paid on one-fourth of the inventory's annual value. About 4% of the annual inventory spoils and must be thrown out. Other costs related to inventory total $2,000 a year. What is the carrying cost for each one dollar of annual inventory, to the nearest tenth of a cent?

Wrap Up ▶▶▶

The most visible cost per trip would be the operating cost per mile of your car. Another cost would be the value of the time you spend in driving and shopping and the alternative cost of what else you could be doing with your time.

Communication

The cost of borrowing money is one of the costs of carrying inventory. Organizations with good credit ratings or who are good customers of a lender may borrow money at or near the *prime rate of interest.*

Through research, find a definition for the prime rate of interest, how the prime rate is determined, and what the current prime rate is. Write a one-page summary of your research.

Exercises

Find the product.

1. 12 × $34,580
2. 26 × $78,192
3. 7.3% × $124,600
4. 0.75% × $1,300,000

Find the quotient.

5. $186,200 ÷ 9,500
6. $54,316 ÷ 734,000

Solve.

7. These percents are charged to ordering: 100% of the purchasing department's annual wages of $86,400 and 40% of warehouse costs of $285,000 a year. Other ordering costs amount to $27,000 a year. If 13,000 purchase orders are placed each year, what is the cost per order to the nearest cent?
8. The Sensun Fabric Company finds its average stock order costs $16. Sensun plans to allow some departments to place supply orders through the online ordering system of various suppliers instead of through the Sensun purchase order system. The company expects to save 23% of order costs on an estimated 4,100 supply orders. What will be the total estimated annual savings of changing the ordering procedure?
9. A company issues 65,000 stock orders a year. The total costs assigned to handling are: $450,000, warehouse employee wages; depreciation on equipment, $29,000; and 13% of overhead warehouse costs of $260,000. To the nearest cent, what is the average cost of handling per stock order?
10. Brent Custom Products pays 7.6% annual interest on one-half of its $3,000,000 annual inventory value. It also pays 0.65% personal property tax on the total value of inventory. Other carrying costs include $26,000 insurance, $51,000 in labor costs, and $19,000 in overhead costs. What is the carrying cost per $1 of annual inventory, to the nearest tenth of a cent?
11. Gerald's Blooms carries a weekly inventory valued at $26,000. Due to spoilage, 15% of the stock of flowers must be thrown out each week. What is the carrying cost of spoilage per dollar of weekly inventory?

The Kroban Company's cost breakdown for the average purchase order is: 15 minutes at $28 an hour, 6 minutes stocking at $96 an hour; office costs of $1.18 per order; overhead costs of $0.86 per order.

12. What àmount does Kroban spend per order?
13. What total amount is spent if 7,600 purchase orders are issued each year?

The Stone Heating Company has an annual inventory value of $1,500,000. Because of financial problems, Stone has to finance its entire inventory at 17% annual interest. If Stone had been able to maintain an excellent credit rating, it would have had to borrow money for only 40% of the inventory's annual value. The interest rate charged by a lender would have been 8.2%.

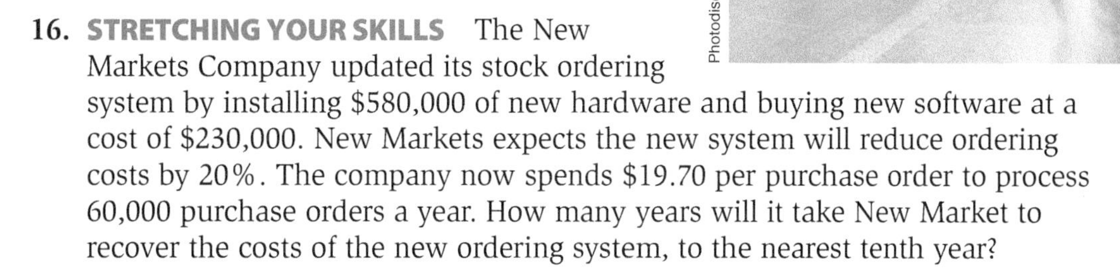

Photodisc/Getty Images

14. What amount must Stone now pay for the interest portion of carrying costs per each dollar of inventory?

15. To the nearest cent, what amount could Stone have paid for the carrying charges of interest for each dollar of inventory if it had an excellent credit rating?

16. **STRETCHING YOUR SKILLS** The New Markets Company updated its stock ordering system by installing $580,000 of new hardware and buying new software at a cost of $230,000. New Markets expects the new system will reduce ordering costs by 20%. The company now spends $19.70 per purchase order to process 60,000 purchase orders a year. How many years will it take New Market to recover the costs of the new ordering system, to the nearest tenth year?

17. **CRITICAL THINKING** A painting company uses 3,000 gallons of paint a year. If the company bought all the paint at one time, it would have high storage, or carrying costs. If it bought paint, say every working day, the time and effort put into frequent ordering would result in higher ordering costs. How do companies such as this strike a balance between the two costs?

Mixed Review

18. Rewrite 0.73 as a percent.

19. 62.5 ft @ 10¢ a foot

20. 252 yd @ $1.35 a yard

21. $0.724 + 8.516 + 15.73$

22. $\frac{4}{5} - \frac{3}{8}$

23. Round these to the nearest cent: 7.68¢ and 39.15¢

24. The Estes are a married, retired couple. Harold Estes earned $4,387.50 last year working as a crossing guard. His wife, Eva, earned $8,100 as a greeter at a chain store. They file a joint city income tax return and are allowed deductions that reduce their taxable income by $9,750. On the remainder of their income they pay a city income tax rate of $1\frac{1}{2}$%. What was their city income tax?

25. Adrian Parnell borrowed $6,000 for 18 months to buy materials to remodel his house. Adrian signed a promissory note at his bank to repay the loan at 13.5% interest. Find the amount of interest Adrian must pay. Then find the amount Adrian must pay his bank when the note is due.

Chapter Review

Vocabulary Review

Find the term, from the list at the right, that completes each sentence. Use each term only once.

bonus
carrying costs
contract employee
Cost of Living Adjustment (COLA)
executive recruiter
exempt employee
First In, First Out
Last In, First Out
nonexempt employee
ordering costs
part-time employee
perpetual inventory
profit sharing
reorder point
stock
stock record
weighted average

1. The expenses of creating purchase orders and handling stock are called __?__.
2. Payment that an employee gets for exceptional job performance is known as a(n) __?__.
3. A listing of the units received, issued, and on hand for a specific item of merchandise is called a(n) __?__.
4. Someone who works for one organization but is paid by another is called a(n) __?__.
5. A type of employee who is not usually paid overtime is called a(n) __?__.
6. A pay system that gives wage increases based on inflation is known as __?__.
7. An inventory valuation system that uses the unit cost of goods bought first to calculate the value of ending inventory is called __?__.
8. The expenses involved in storing inventory until it is issued are called __?__.
9. A person who tries to find full-time employees for an employer and gets paid only for successful searches is known as a(n) __?__.

11-1 Employee Recruitment Costs

10. The Keller Company runs a newspaper employment ad each Sunday. The ad should cost $1,400 a day. Because of the number of ads it runs, Keller receives a 22% discount off the usual price. What is the cost for Keller to run ads for one year?
11. A company hired Joanna Neech through an executive recruiter. The recruiter's fee was 26% of all money paid to Joanna in her first year of employment. Joanna agreed to a salary of $65,000 a year. She also was paid an $11,000 signing bonus. What was the cost of using the recruiter?

11-2 Wage and Salary Increases

12. Helmut Nemzek received a COLA of 2.3% on his weekly gross wages of $760. How much will Helmut earn per week after the pay increase?
13. Kruse Products, Inc. shared 15% of its $2,000,000 profit with its 120 hourly employees. What was the total amount of profit shared? What was each employee's share?

11-3 Total Costs of Labor

14. Becky Dressler earns $42,000 a year. The benefits her employer pays in taxes include 7.65% FICA tax and a combined FUTA and SUTA tax of 7.1% of the first $7,000 of income. Becky also receives medical benefits worth $3,890 a year. Her employer pays 3.4% of her annual pay into a retirement program. What is the employer's total cost of providing these benefits?

15. Lyle LeBlanc completed his first month of work at a restaurant. He was paid $7.75 an hour for the 118 hours he worked. Lyle received a $25 loyalty bonus for staying with the restaurant for at least one month. The restaurant paid FICA taxes of 7.65%, FUTA taxes of 6.2%, and SUTA taxes of 2.1% on Lyle's wages. What was the total cost of employing Lyle for one month?

11-4 Tracking and Reordering Inventory

16. A wholesaler showed a stock balance of 812 units of WT-45 electric shavers on Nov. 1. WT-45 shavers were shipped to retailers on these dates: Nov. 2, 415 units; Nov. 12, 570 units; and Nov. 28, 605 units. The wholesaler received 900 shavers from a manufacturer on Nov. 10. What was the stock record balance of shavers on Nov. 30?

17. On average, Parts Source, a wholesaler, sells 28 electric motors a day to repair shops. It takes six days for Parts Source to get new motors from a manufacturer. Safety stock is $1\frac{1}{2}$ times the daily motor sales. What is the reorder point?

11-5 Inventory Valuation

Davis & Simpson, a distributor, purchased kitchen shears from different manufacturers.

September 3	700 @ $7.30	September 18	300 @ $7.62
September 12	1,200 @ $6.90	September 18	500 @ $7.41

18. The unit value of the beginning inventory of 600 shears was $7.28. Find the value of the ending inventory of 700 shears using FIFO.

19. The beginning inventory was 160 units with a unit value of $6.85. Use LIFO to find the value of the ending inventory of 300 units.

20. The total value of a beginning inventory of 450 units was $3,253.50. Use the weighted average method to find the value of the ending inventory of 610 units.

11-6 Ordering and Carrying Inventory

21. The Piper Company issues 120 stock orders a week. A purchasing agent earning $700 a week spends 100% of her time on ordering. Her assistant earns $112 a day and spends 10% of his time entering orders into a computer system. Warehouse and office costs are calculated at $5.17 per order. What is the cost of issuing each stock order?

22. A retailer's annual inventory value is $720,000. He borrows money at 17% for three-tenths of the inventory's annual value. Customer theft is 3% of inventory. Other costs related to holding stock are $28,000. What is the carrying cost per $1 of inventory?

Technology Workshop

Task 1 Calculating COLA

Enter data into a template that calculates Cost of Living Adjustment (COLA) for employees of the Canton Manufacturing Company. You may use the template to study the effect of wage increases on a company's total wage costs.

Open the spreadsheet for Chapter 11 (tech11-1.xls) and enter the data shown in blue (cells C3-4) into the spreadsheet. The percent change in the CPI, which is calculated for you, is also known as the rate of inflation. A new hourly wage rate for employees, total annual wages based on the new wage rate, amount of wage increase for each employee, and the total of the wage increases are also calculated. Your computer screen should look like the one shown when you are done.

	A	B	C	D	E
1	Canton Manufacturing Company				
2	Projected COLA Wage Increase, January, 20--				
3	CPI Year-End Data:	Old CPI	179.9		
4		New CPI	184.0		
5		% Change	2.3%		
6		Old Hourly	New Hourly	New Annual	Amount of
7	Employee	Pay Rate	Pay Rate	Cost of Wages	Wage Increase
8	Bealer, Timothy	16.75	17.13	35630.40	790.40
9	Creslin, Ronald	17.20	17.59	36587.20	811.20
10	Eggert, Diane	16.90	17.29	35963.20	811.20
11	Fujiwara, Ritsuko	18.34	18.76	39020.80	873.60
12	Liston, Skip	17.05	17.44	36275.20	811.20
13	O'Brien, Sean	17.56	17.96	37356.80	832.00
14	Quincey, Roxanne	17.86	18.27	38001.60	852.80
15	Ramirez, Maria	19.45	19.89	41371.20	915.20
16	Zollig, Conrad	18.90	19.33	40206.40	894.40
17	TOTAL				7592.00

Task 2 Analyze the Spreadsheet Output

Answer these questions about the COLA wage calculations.

1. What was the CPI at the end of the year?
2. By what percent did the CPI change during the year to the nearest tenth percent?
3. How was the percent change in the CPI calculated?
4. By what amount will wages increase in the coming year due to the change in the CPI?
5. The annual cost of wages calculation assumes that employees work 52 weeks, 40 hours a week. How many total hours are used in the calculation?

6. Which employee will have the lowest hourly pay rate after the raise? What is the rate?
7. Which employee will have the greatest annual earnings at the new pay rate? What are the earnings?

Now move the cursor to cell C4, which holds the new CPI index number. Enter a new index number of 187.5.

8. What is the total percent change in the CPI with the new data?
9. Compare the total wage increase with the CPI at 187.5 to your total from Question 4. How much more will Canton have to pay its employees at the higher CPI?
10. Now enter a new CPI of 179.9 in cell C4. Will the employees get COLA? What is the reason?
11. Now enter a new CPI of 177.2 in cell C4. What does the lower CPI figure mean? Explain what happened to the hourly pay rate of employees.

Task 3 Design a Spreadsheet to Find an Ending Inventory's Value

You are to design a spreadsheet to find the value of the ending inventory using the weighted average method. Your spreadsheet should include columns for the date, the number of units, their unit cost, and total value. Where possible, use spreadsheet functions to calculate the ending inventory's unit cost and value.

SITUATION: A wholesaler, Storage Products, Inc., keeps an inventory record for steel utility shelving. Use the weighted average method to calculate the value of the ending inventory. Inventory and purchase data are to the right.

Storage Product, Inc.
Inventory Period: April 1, 20— to June 30, 20—
Stock Item: Steel Utility Shelving

April 1 beginning inventory, 58 units with a unit cost of $48.12.

June 30 ending inventory, 31 units.

Purchases: April 10, 35 units @ $53.90
April 23, 56 units @ $50.10
May 11, 21 units @ $58.12
June 1, 65 units @ $46.05
June 20, 44 units @ $52.80

Task 4 Analyze the Spreadsheet Output

Answer these questions about your spreadsheet output.

12. What was the unit cost of the ending inventory?
13. What was the value of the ending inventory?
14. How did you calculate the value of the ending inventory?
15. What functions did you use to calculate the unit cost of the ending inventory?

Chapter Assessment

Chapter Test

Answer each question.

1. Add: $45,384 + $3,482 + $2,005.34
2. Subtract: 1,748 − 985
3. Multiply: 0.6% × $1,480
4. Divide $1,479,400 by 260
5. $\frac{3}{4} \div 1\frac{1}{2}$
6. 4.8 ÷ 0.006%
7. What percent of $50,400 is $1,260?
8. $225 + (11% × $225)

Solve.

9. The board game Skribin has daily sales of 12 units. It takes 28 days to receive a new shipment. No safety stock is purchased. What is the reorder point for the game?
10. Five percent of a company's profits of $472,000 are to be shared with 4 supervisors. Thirty percent of the profits are to be shared with 48 hourly employees. What profit sharing amount did each of the supervisors and hourly employees get?
11. A contract employee is paid $24 an hour for the same work done by a full-time employee who earns $19 an hour plus 37% in benefits. Which employee costs more to hire per hour, and how much more?
12. Bear County will pay a COLA equal to the 0.7% rise in the CPI last quarter. If Tabatha is now paid $18.70 an hour by the county, what will her new hourly rate be when the raise is given?
13. The unit cost of a beginning inventory of 405 units is $15.16. The unit cost of the last 600 units purchased is $16.81. Use FIFO to find the value of the ending inventory of 170 units.
14. MXV Products uses a computer program to screen job applications at a cost of $1.40 per application. Screening used to be done by an employee who took ten minutes at an average cost of $36 an hour. How much will computer screening save MXV for the 426 applications it expects to receive this month?
15. A weekend employment ad costs $36 a column inch per day. Find the cost of a 3-inch ad that runs on Saturday and Sunday.
16. A vegetable stand at a farmer's market carries a weekly inventory valued at $3,000. Due to spoilage, 5% of the stock must be thrown out each week. What is the carrying cost of spoilage per dollar of weekly inventory?
17. Marta Lind received a notice showing that her employer paid these benefits based on her $53,000 annual salary: 8.7%, required tax benefits; 6%, medical insurance; 3%, pension plan contribution; and 5%, vacation and holiday pay. What total benefits did Marta's employer provide?
18. The total value of 680 units of beginning inventory and purchases in a quarter is $28,900. What is the value of an ending inventory of 76 units using the weighted average method?

Planning a Career in Transportation, Distribution, and Logistics

Edwin Verin 2008/Shutterstock.com

The arrangement and movement of materials, goods, and people by car, truck, train, airplane, and ship is the job of a worker in the transportation, distribution, and logistics field. Other career pathways in this field include related services like infrastructure planning and development, construction, and facility maintenance. If you enjoy the area of transportation, are fascinated by roadways or trains, like the challenge of creating detailed plans, a career in transportation, distribution, and logistics might be for you.

Job Titles

- Locomotive engineer
- Bus driver
- Facility manager
- Construction manager
- Distribution clerk
- Construction equipment operators

Needed Skills

- creative and detailed oriented
- problem solving and decision making skills

- strong mathematical and science background
- computer and technologically savvy
- ability to work independently, and with others

What's it like to work in Construction Equipment?

Construction equipment operators run machines to move dirt and building materials at construction sites. A construction equipment operator may dig trenches for water and sewer or lay asphalt to the ground for the construction of roadways. Some construction equipment operators use cranes to lift heavy materials, while others use machines that drive wood into the ground for securing piles for the support of bridges, oil rigs, and skyscrapers.

What About You?

Can you see yourself operating construction equipment? Which area of construction is most interesting to you?

How Times Have Changed

For Questions 1–2, refer to the timeline on page 471 as needed.

1. Douglas Kerr's employer will match his contribution to a 401(k). The matching amount will be a dollar-for-dollar match not to exceed 6% of his salary. In 2004, Douglas's salary was $36,470. If he contributed $2,000, what would be the total amount contributed to the plan? What would be the total amount contributed to the plan if he contributed $4,000?
2. The maximum annual amount that individuals could contribute to their 401(k) in 2008 was $15,500. Yolanda Washington's employer matches 25% of every dollar that she contributes, up to 9% of her salary. Yolanda earned $63,800 in 2008 and she contributed only the amount her employer would match. What was the total amount contributed to her plan? How much more could Yolanda have contributed?

Chapter 12

Business Profit and Loss

12-1 Preparing Income Statements
12-2 Analyzing Income Statements
12-3 Partnership Income
12-4 Preparing Balance Sheets
12-5 Analyzing Balance Sheets
12-6 Bankruptcy

Statistical Insights

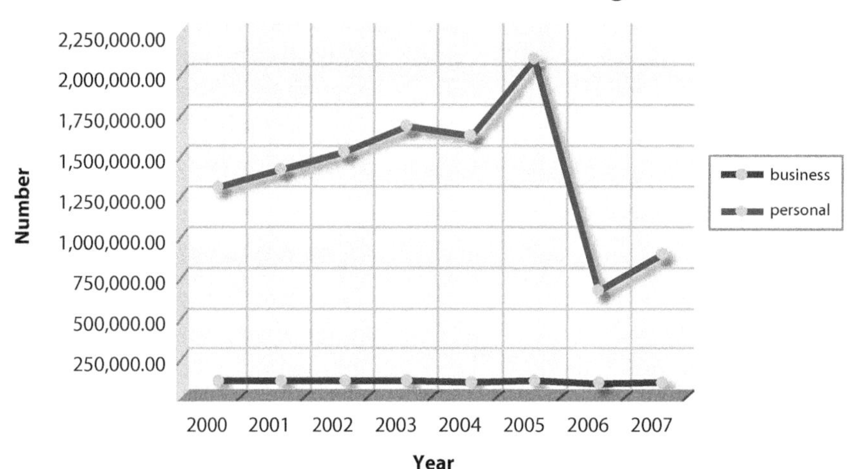

Source: UScourts.gov

There are different chapters of bankruptcies that discharge all or part of the qualified debt. The number of bankruptcies filed depends partially on the divorce rate, bankruptcy laws, consumer debt, and recession.

1. What is the overall trend of bankruptcy filings?
2. What year had the greatest increase in personal bankruptcy filings?
3. What year had the greatest decrease in filings?
4. The passing of the Bankruptcy Abuse Prevention and Consumer Protection Act (BAPCPA) made discharging personal debt through bankruptcy more difficult. **Explain** why you think this would affect the number of bankruptcy filings. Use the graph and tell when you think the BAPCPA went into effect.

How Times Have Changed

Work for wages—for thousands of years, people have done work for payment; either a barter for goods or for an exchange of currency. Employees and employers agree on the amount of goods or currency to be traded for the labor. In 2600 BC clay tablets were used in Babylon to record the time laborers worked and the wages the laborers received. In 200 BC, Roman soldiers were paid in salt, a precious preservative and seasoning. The word *salary* comes from the word *salarium* which means *salt-money*.

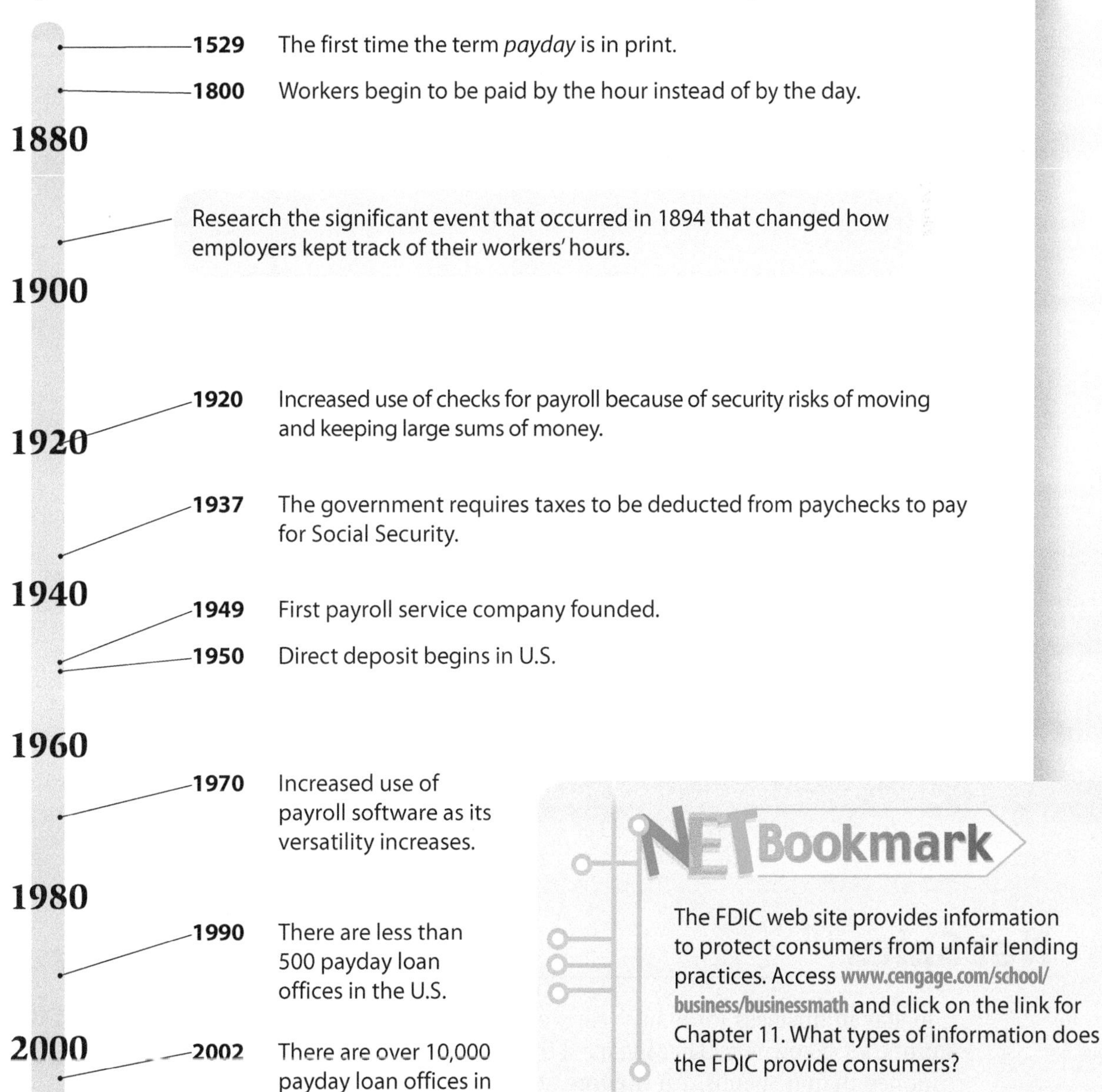

1529 The first time the term *payday* is in print.
1800 Workers begin to be paid by the hour instead of by the day.
1880
Research the significant event that occurred in 1894 that changed how employers kept track of their workers' hours.
1900
1920 Increased use of checks for payroll because of security risks of moving and keeping large sums of money.
1920
1937 The government requires taxes to be deducted from paychecks to pay for Social Security.
1940
1949 First payroll service company founded.
1950 Direct deposit begins in U.S.
1960
1970 Increased use of payroll software as its versatility increases.
1980
1990 There are less than 500 payday loan offices in the U.S.
2000
2002 There are over 10,000 payday loan offices in the U.S.
NET Bookmark
The FDIC web site provides information to protect consumers from unfair lending practices. Access www.cengage.com/school/business/businessmath and click on the link for Chapter 11. What types of information does the FDIC provide consumers?

12-1

Preparing Income Statements

GOALS

- Calculate net sales
- Calculate cost of goods sold
- Calculate gross profit
- Calculate net income

KEY TERMS

- income statement
- cost of goods sold
- gross profit
- net income
- net loss

Start Up

If you are the owner of a small business, how do you know if you are making a profit? Would looking at the amount of cash in the cash register at the close of a business day tell you if you are making a profit?

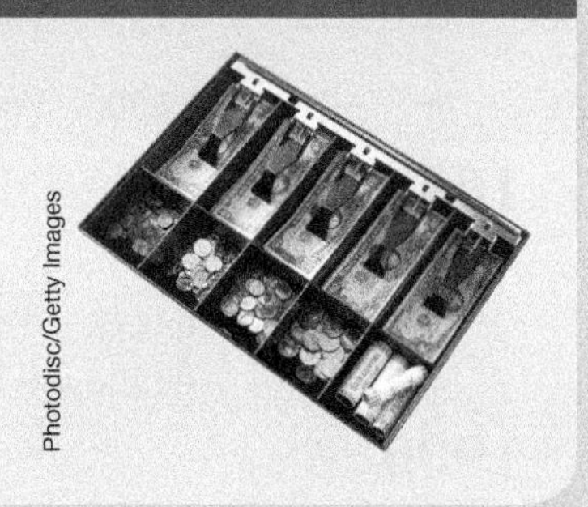
Photodisc/Getty Images

Math Skill Builder

Review these math skills and solve the exercises that follow.

1. **Add** money amounts.
   Find the sum. $25,149 + $4,509 + $108,558 = $138,216

   1a. $145,345 + $87,981 + $16,129 **1b.** $340,288 + $72,465 + $1,598

   1c. $38,265 + $45,107 + $4,967 **1d.** $16,983 + $8,297

2. **Subtract** larger money amounts from smaller money amounts.
   Show negative differences in parentheses.
   Find the difference. $45,608 − $64,733 = ($19,125)

   2a. $84,891 − $101,397 **2b.** $12,793 − $24,197

   2c. $735,978 − $1,000,783 **2d.** $72,197 − $98,288

Net Sales

To stay in business for many years, every business must make a profit for its owners. That means a business must bring in more money than it spends. An **income statement** shows how much money is earned and spent during a period of time, such as a month, quarter, or year. If more money is earned than spent, there is a profit. If more money is spent than earned, there is a loss.

Business Tip

An income statement is also called a *profit and loss statement* and an *earnings report.*

There are five major sections to an income statement: revenue, cost of goods sold, gross profit, operating expenses, and net income or loss.

The revenue section starts with *sales,* or the total value of goods sold in the period covered by the statement.

Some goods may be returned for refunds by customers, and allowances or price reductions may be given to customers for damaged goods.

These ***sales returns and allowances*** decrease sales, so they are subtracted from revenue. The amount left is called *net sales* (see A).

Net Sales = Sales − Sales Returns and Allowances

Two income statements are shown below, Petrie's Gift Shop and Cepeda Manufacturing.

Petrie's Gift Shop
Income Statement
For the month ended July 31, 20—

Revenue		
Sales	75,952	
Less sales returns and allowances	1,900	
Net sales		(A) 74,052
Cost of goods sold		
Beginning inventory, July 1	211,638	
Purchases	31,548	
Goods available for sale	(B) 243,186	
Ending inventory, July 31	196,920	
Cost of goods sold		(C) 46,266
Gross profit on sales		(D) 27,786
Operating expenses		
Salaries and wages	10,567	
Rent	3,194	
Taxes	1,864	
Utilities	1,678	
Advertising	780	
Depreciation of equipment	726	
Insurance	253	
Other expenses	854	
Total operating expenses		19,916
Net income		(E) 7,870

Cepeda Manufacturing
Income Statement
For the month ended July 31, 20—

Revenue		
Sales	205,000	
Less sales returns and allowances	8,552	
Net sales		(A) 196,448
Cost of goods sold		
Beginning inventory, July 1	21,148	
Add costs of goods manufactured	145,280	
Goods available for sale	(B) 166,428	
Ending inventory, July 31	24,511	
Cost of goods sold		(C) 141,917
Gross profit on sales		(D) 54,531
Operating expenses		
Salaries	18,250	
Office expenses	2,800	
Payroll taxes	2,452	
Depreciation	2,688	
Total operating expenses		26,190
Net income		(E) 28,341

Business Tip

Revenue is income from all sources. For most businesses, the main revenue is income from the sale of goods produced, services rendered, or goods resold.

EXAMPLE 1

Regal Book Shop's sales for the year were $156,790. Sales returns and allowances for the same year were $2,398. What were Regal's net sales for the year?

SOLUTION

Subtract the sales returns and allowances from sales.

$156,790 − $2,398 = $154,392   net sales

✔ **CHECK YOUR UNDERSTANDING**

A. The Cycle Shop's sales for the first quarter of the year were $307,892. Sales returns and allowances for the quarter were $14,640. What were the net sales for the quarter?

B. Crestwood Products had sales of $45,240 in May. Sales returns and allowances in the same month were $1,125. What were Crestwood's net sales?

Cost of Goods Sold

The **cost of goods sold** is expenses directly related to buying or producing the goods sold. For a retailer, these are the costs of purchasing merchandise. For a manufacturer, the cost of goods sold is the prime cost, or the sum of the raw materials costs and direct labor costs.

Each income statement has a *beginning inventory,* or the dollar value of all goods on hand at the beginning of the period. Companies add the cost of the goods purchased or made during the time period to the beginning inventory to find the goods available for sale (see B) for the period. The value of the inventory on hand at the end of the period is called the *ending inventory.*

The ending inventory is subtracted from goods available for sale. The difference is the cost of goods sold (see C) for that period. For a retailer, the formula is

Cost of Goods Sold = Beginning Inventory + Purchases − Ending Inventory

For a manufacturer, the formula is

Cost of Goods Sold = Beginning Inventory + Cost of Goods Manufactured − Ending Inventory

EXAMPLE 2

On May 1, Agle Farm Implements had an inventory of merchandise costing $1,895,200. During May, goods costing $497,800 were bought. Merchandise inventory at the end of May was $1,038,700. What was the cost of goods sold for May?

SOLUTION

Add the beginning inventory and the purchases for May.

$1,895,200 + $497,800 = $2,393,000 goods available for sale in May

Subtract the ending inventory from the goods available for sale.

$2,393,000 − $1,038,700 = $1,354,300 cost of goods sold for May

✔ **CHECK YOUR UNDERSTANDING**

C. On January 1, The Shoe Palace had an inventory of merchandise costing $317,800. During the quarter, merchandise costing $219,900 was bought. Merchandise inventory on March 31 was $275,300. Find the cost of goods sold for the quarter.

D. Rogers Products Company records show these amounts for the year: beginning inventory, January 1, $630,750; cost of goods produced during the year, $1,682,900; ending inventory, December 31, $673,580. Find the cost of goods sold for the year.

Gross Profit

Gross Profit (see D) is the difference between the net sales and the cost of goods sold.

Gross Profit = Net Sales − Cost of Goods Sold

EXAMPLE 3

Main Street Appliances had net sales of $34,590 in July. The cost of goods sold in July was $19,670. What was Main Street's gross profit on its sales?

SOLUTION

Subtract the cost of goods sold from the net sales.

$34,590 − $19,670 = $14,920 gross profit for July

✔ CHECK YOUR UNDERSTANDING

Dex Image/Jupiter Images

E. The Tee's net sales of golf equipment for the year were $589,280. The cost of the goods sold for the same period was $298,100. What is the Tee's gross profit for the year?

F. On July 1, Robles Metal Products had a beginning inventory of $505,700. During the quarter, goods costing $290,900 were manufactured. The inventory at the end of September was $485,300. During the quarter, Robles' net sales were $759,200. What was the cost of goods sold for the quarter? What was Robles' gross profit for the quarter?

Net Income

Operating expenses are the costs of running a business and may include salaries and wages, taxes, advertising, rent, depreciation of equipment, utilities, and insurance. Operating expenses are usually listed and totaled on the income statement.

Total operating expenses are deducted from gross profit to determine a company's **net income** or net loss (see E). Net income may also be called *net profit*. This is the actual profit earned after all costs are accounted for. Both Petrie's Gift Shop and Cepeda Manufacturing reported a net income at the bottom of their statements.

Net Income or Loss = Gross Profit − Operating Expenses

If operating expenses are greater than gross profit on sales, the result, or difference, is a **net loss**.

For example, suppose your business had a gross profit of $15,000 for a month, and the operating expenses were $20,000 that month. You had a net loss of $5,000. When a net loss occurs, write Net Loss instead of Net Income at the bottom of the income statement. Put parentheses around the amount ($5,000) to show it is a loss, or negative amount.

Business Tip

Utilities may include many different commodities or services, such as gas, electricity, telephone, cable, and online services.

EXAMPLE 4

A store had a gross profit of $37,200 and total expenses of $28,500 in October. What was the store's net income or loss?

SOLUTION

Because the gross profit is greater than the total expenses, there will be a net income.

Net Income = $37,200 − $28,500 = $8,700

✔CHECK YOUR UNDERSTANDING

G. The Pet Corner had a gross profit of $29,400 in June. It also had these expenses for the month: salaries and wages, $7,400; rent, $1,300; taxes, $550; utilities, $390; advertising, $125; depreciation, $270; insurance, $105; and other expenses, $420. What were the store's total expenses? What was the store's net income or loss?

H. Leo's Card Shop had a gross profit of $3,300 in April. It also had these expenses for the month: salaries and wages, $1,340; rent, $1,570; taxes, $450; utilities, $290; advertising, $175; depreciation, $170; insurance, $155; and other expenses, $552. What were the shop's total expenses? What was the shop's net income or loss?

Wrap Up ▶▶▶

The amount of cash in the cash register tells you only the amount of cash receipts you had in one day. That is important information. However, to figure whether you made a profit or loss, you must find all the money your business earned and all the money your business spent during a time period, such as a month. If what the business earned was more than what it spent, you made a profit.

Communication

Use the Internet to find the income statements for a retail firm and a manufacturing firm. Most large companies maintain a web site that contains what are called the annual reports to stockholders. The annual reports contain the current income statement for the companies. Identify and list the major sections of the income statement for each company and compare them to the major income statement sections used in the chapter. What differences did you find? What sections were the same? Prepare a brief oral presentation of your findings.

Exercises

Find the sum.

1. $19,803 + $28,809 + $12,490
2. $389,122 + $15,967 + $1,488

Find the difference.

3. $89,220 − $67,208
4. $108,893 − $82,415
5. $45,280 − $65,820
6. $348,199 − $399,286

Solve.

7. Fleigle Supplies, Inc. had sales of $258,440 in February. Sales returns and allowances in the same month were $2,256. What were Fleigle's net sales?

8. T-R Roofing Manufacturing Company's beginning inventory on January 1 was $1,530,470. The cost of the goods it produced during the year was $3,828,200. Its ending inventory on December 31 was $1,463,850. What was the cost of goods sold for the year?

Photodisc/Getty Images

9. A company's net sales for the month were $1,498,780. The cost of goods sold for the same period was $849,400. What is the company's gross profit for the month?

Kemper Foundry had a beginning inventory of $355,700 on July 1. During the quarter, goods costing $1,124,300 were manufactured. The inventory at the end of September was $385,600. During the quarter, Kemper's net sales were $1,675,600.

10. What was the cost of goods sold for the quarter?

11. What was Kemper's gross profit for the quarter?

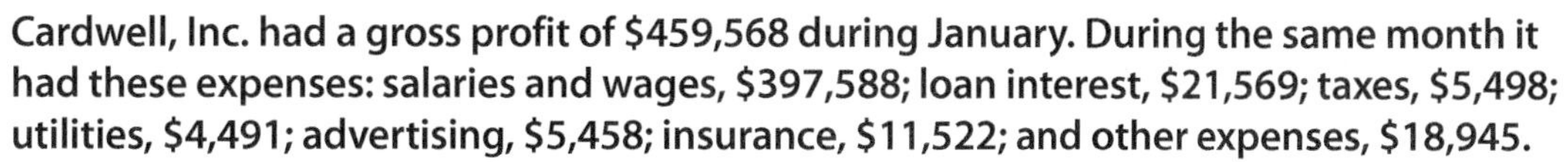

Cardwell, Inc. had a gross profit of $459,568 during January. During the same month it had these expenses: salaries and wages, $397,588; loan interest, $21,569; taxes, $5,498; utilities, $4,491; advertising, $5,458; insurance, $11,522; and other expenses, $18,945.

12. What was the total of the expenses?

13. What was the net income or loss?

14. **CRITICAL THINKING** Rob Dougherty listened to a local TV news program that claimed that a new shoe store on Main Street had made a net income of 12% last year. Rob thought that was high because most of the local banks only paid 1.5% on interest-bearing accounts and corporate bonds were only paying 7.8% interest. Is Rob right? Why might a net income of 12% from the store be reasonable?

Mixed Review

15. What amount is 150% of $84?

16. What percent of 96 is 144?

17. What amount is 1% of $1,000,000?

18. $298.80 is what percent less than $360?

19. Write 57.3% as a decimal, rounded to the nearest hundredth.

20. Last year, Chandna Venkatraman was paid a salary of $700 a week for 52 weeks. Her federal taxable income for the same period was $21,200. She paid to the city of Carthage in which she worked an income tax of 3.5% on her taxable income. How much did Chandna pay in city income taxes?

21. A company spent $785,600 on computer support for their 460 desktop, notebook, and handheld computers last year. What was the per computer cost of the computer support?

12-2

Analyzing Income Statements

GOALS
- Use percentages to analyze income statement items
- Use ratios to analyze income statement items

KEY TERMS
- gross profit margin
- net profit margin
- merchandise turnover rate

Start Up

Jason told Juanita about a company listed on the stock market in which he had decided to invest. Jason was excited because the company had earned a 15% net income for last year. Juanita didn't seem too thrilled, at the net income percentage. Why might Juanita be skeptical of the net income percentage?

mdd/Shutterstock.com

Math Skill Builder

Review these math skills and solve the exercises that follow.

1. **Divide** dollar amounts by dollar amounts.
Find the quotient. $420,000 ÷ $60,000 = 7

1a. $55,352 ÷ $6,512 **1b.** $461,900 ÷ $74,500

2. **Divide** dollar amounts to find percents.
Find the quotient. $70,000 ÷ $350,000 = 0.2, or 20%

2a. $35,235 ÷ $652,500 **2b.** $3,588 ÷ $89,700

3. **Average** dollar amounts.
Find the average of $38,500 and $36,300.
$38,500 + $36,300 = $74,800; $74,800 ÷ 2 = $37,400

3a. Average $12,080 and $14,280. **3b.** Average $58,280 and $63,120.

Percentage Analysis of Income Statements

Businesses use income statements to report their income. They also use them to analyze a business. One way to analyze a business using income statements is to compare each item on the statement to net sales using percentages. That means dividing each item by the net sales amount. Cepeda Manufacturing (see Lesson 12-1) net income, $28,341, as a percent of net sales, $196,448, is found this way:

Photodisc/Getty Images

$28,341 ÷ $196,448 = 0.1443, or 14.4%

Two important percentage comparisons are the comparisons of gross profit to net sales and net income to net sales. These comparisons are called the **gross profit margin** and the **net profit margin**.

Converting income statement items to percents lets you compare a company's income statements from one year to the next. It also lets you compare a company's income statement to the statements of other companies.

For example, a percentage comparison of key items on the Petrie's Gift Shop (see Lesson 12-1) income statements for two years is shown in the table.

Item	Amount This Year	Percent of Net Sales	
		This Year	Last Year
Net Sales	$74,052	100.0%	100.0%
Cost of Goods Sold	$46,266	62.5%	65.2%
Gross Profit	$27,786	37.5%	34.8%
Operating Expenses	$19,916	26.9%	28.9%
Net Income	$7,870	10.6%	5.9%

The comparison shows that Petrie's net income as a percent of net sales has risen from 5.9% to 10.6%. There appear to be two reasons for this increase. Petrie has lowered the price his merchandise costs him from 65.2% to 62.5% of net sales and lowered the expenses of operating his business from 28.9% to 26.9% of net sales.

EXAMPLE 1

Landau Mechanics, Inc. had gross sales of $543,370, sales returns and allowances of $5,370, and net income of $28,640 last year. What was the company's net profit margin to the nearest whole percent?

SOLUTION

Subtract sales returns and allowances from gross sales.

$543,370 − $5,370 = $538,000 net sales

Divide net income by net sales.

$28,640 ÷ $538,000 = 0.053, or 5% net profit margin

Math Tip

The relationship of two numbers such as 1 and 2 can be shown as 1:2, $\frac{1}{2}$, or 50%.

✔CHECK YOUR UNDERSTANDING

A. A vacuum cleaner dealership's net sales were $238,180 last year. Its operating expenses for the same period were $17,700. What were the operating expenses as a percent of net sales, to the nearest tenth percent?

B. Cliver Telecommunications, Inc. had net sales of $1,589,300 and cost of goods sold of $1,032,850 this year. Last year the amounts were $1,453,200 and $960,300, respectively. What is the company's gross profit margin, to the nearest tenth percent, for each year? Is the percent rising or falling, and by how much?

Ratio Analysis of Income Statements

Ratio analysis compares items on an income statement. One ratio, the merchandise turnover rate, can be helpful in analyzing a business.

The **merchandise turnover rate** is the number of times per period that a store replaces, or turns over, its average stock of merchandise. The period is often a month, quarter, or year. Different kinds of businesses have different acceptable turnover rates.

Businesses that sell durable and expensive goods such as furniture or large appliances have low turnover rates. Businesses that sell perishables such as food have high turnover rates.

$$\textbf{Merchandise Turnover Rate} = \frac{\textbf{Cost of Goods Sold for Period}}{\textbf{Average Merchandise Inventory for Period}}$$

The turnover rate may be used not only on an entire inventory but also on each item of inventory. Doing so lets you identify goods that are moving more slowly than others or moving more slowly than they have in the past. To prevent these items from becoming over-stocked, you may decide to discount them.

EXAMPLE 2

Calculate the merchandise turnover rate for Cepeda Manufacturing based on the July income statement.

SOLUTION

Find the average of the July 1 inventory ($21,148) and the July 31 inventory ($24,511).

($21,148 + $24,511) ÷ 2 = $22,829.50   average merchandise inventory for July

Divide the cost of goods sold for July by the average inventory for July.

$141,917 ÷ $22,829.50 = 6.22, or 6.2   merchandise turnover rate for July

✔ CHECK YOUR UNDERSTANDING

C. Rock Creek Stores had a beginning inventory on January 1 of $168,350. The ending inventory on December 31 was $155,490. The stores' cost of goods sold for the year was $945,280. What was its merchandise inventory turnover rate, to the nearest tenth, for the year?

D. A fabric manufacturer had an inventory of a stock item costing $34,200 on July 1 and $39,400 on September 30. The cost of goods sold for that inventory item for the quarter was $295,200. What was the turnover rate for the item for the quarter, to the nearest tenth?

Wrap Up ▶▶▶

Juanita might be skeptical of the net income percentage for many reasons. One reason is that there is no information against which the percentage can be compared. Several questions should be asked. What has the company's net income percentage been for the last 5 or 10 years? What is the net income of similar companies? What are the industry's forecasts for growth? Company income statements from a single period need to be compared to other years and other firms.

Financial Responsibility

Examine the income statement for Cepeda Manufacturing in Lesson 12-1. Identify steps the firm might consider taking to increase its net income. Detail the steps you have identified and how each step will affect net income.

Exercises

1. What percent is $9,216 of $76,800?

2. Find the average of $62,780 and $58,270.

Prepare a percentage analysis of the items below to net sales using Cepeda Manufacturing's income statement. Find each percent to the nearest tenth.

3. Cost of Goods Sold
4. Gross Profit on Sales
5. Operating Expenses
6. Net Income

A business had net sales of $140,000. The cost of goods sold was $84,000.

7. The cost of merchandise sold was what percent of net sales?

8. What was the gross profit margin?

The records of the Carlson Company last year showed these data: net sales, $220,000; cost of merchandise sold, $143,000; and operating expenses, $52,800.

9. Find the gross profit and the net income for last year.

10. Find the gross profit margin for last year.

11. Find the net profit margin for last year.

From January through June, Dollar Appliance Center took 3 inventories of merchandise: $88,000, $188,200, and $124,600. The cost of merchandise sold during the 6-month period was $601,200.

12. What was Dollar's average merchandise inventory?

13. What was the merchandise turnover rate for the 6 months?

14. What is the equivalent merchandise turnover rate for a year?

INTEGRATING YOUR KNOWLEDGE ReAlarm Sales purchased alarm systems from a wholesaler at these quantities and unit costs during June: June 5, 20 at $3,760; June 10, 24 at $4,752; June 29, 28 at $5,572. The beginning inventory of 9 systems had a total value of $1,683.

15. Find the value of the 26 alarm systems in the June 30 ending inventory using the FIFO method.

16. If the cost of the alarm systems sold was $10,593, what was the turnover rate, to the nearest tenth, for the alarms for the month?

Mixed Review

17. $0.45 \times 1{,}000$

18. $3.108 \div 100$

19. Rewrite $\frac{7}{8}$ as a percent.

20. How many days are there between May 5 and July 16?

21. The senior class of Ekersville High School ran a dart game at the town's 3-day homecoming event. The class collected $250.50 on Friday, $242.75 on Saturday, and $224.25 on Sunday. What was the average amount collected by the class each day?

12-3

Partnership Income

GOALS

- Distribute partnership income in proportion to investments
- Distribute partnership income in a fixed ratio or fixed percent
- Distribute partnership income after paying interest on investments

KEY TERM

- partnership

Start Up ▶▶▶

Sally's friend, Ralph, told her he was forming a partnership with another friend to start a photography business. Sally asked Ralph which lawyer was drawing up the partnership agreement. Ralph said he had known his friend for many years and trusted him. Sally was concerned. Is Sally right to be concerned?

Photodisc/Getty Images

Math Skill Builder

Review these math skills and solve the exercises that follow.

1. **Subtract** money amounts.
   Find the difference. $135,800 − $75,000 = $60,800

   1a. $84,500 − $15,400   **1b.** $138,540 − $63,470

2. **Multiply** money amounts by percents and fractions.
   Find the product. $120,000 × 8% = $120,000 × 0.08 = $9,600
   Find the product. $350,000 × $\frac{3}{5}$ = $210,000

   2a. $560,800 × 7%   **2b.** $280,500 × 9.5%

   2c. $\frac{4}{5}$ × $750,000   **2d.** $\frac{\$20{,}000}{\$50{,}000}$ × $280,000

In Proportion to Investments

Firms owned by one person are called *sole proprietorship.* A firm owned by two or more people that is not a corporation or other form of business is called a **partnership**. When you form a partnership, you may sign a partnership agreement.

The agreement tells how much money the partners invested and how they will share net income or net loss. Ways to distribute income or loss are:

- Equally between the partners
- In proportion to the partners' investments
- Paying interest to the partners on their investments
- In a fixed ratio or percent
- Combining two or more of the above methods

If no way is specified, income or loss may be distributed equally.

To find how much each partner receives when net income is shared in proportion to investments, show each partner's investment as a fractional part of the total investment. Then, multiply the total partnership net income by each partner's fraction.

Business *Tip*

The IRS expects profits and losses to be distributed according to each partner's percentage interest in the business. Even if there is a written agreement, the partners may have to justify how they divide profits and losses.

EXAMPLE 1

Matsumi and O'Conner invest $300,000 and $200,000, respectively, in a partnership. They agree to share net income in proportion to their investments. At the end of the first year, the partnership earns a net income of $52,000. Find each partner's share of the net income.

SOLUTION

Add the investments of each partner.

$\$300{,}000 + \$200{,}000 = \$500{,}000$ total investment

Multiply the net income by the fraction that shows each partner's investment of the total investment.

$\frac{\$300{,}000}{\$500{,}000} \times \$52{,}000 = \$31{,}200$ Matsumi's share

$\frac{\$200{,}000}{\$500{,}000} \times \$52{,}000 = \$20{,}800$ O'Conner's share

Math *Tip*

Check your work by adding each partner's share of net income together. They should equal the total net income from the partnership.

✔ CHECK YOUR UNDERSTANDING

A. Aguilar and Trent share their partnership net income in proportion to their investments. Aguilar has $180,000 and Trent $120,000 invested. What is each partner's share of this year's $64,000 net income?

B. Silverstein, Clark, and Salinas have invested $250,000, $200,000, and $150,000 in a partnership that had a net income of $90,000 last year. If the partners share the net income in proportion to their investments, what is each partner's share, to the nearest dollar?

Fixed Ratio or Fixed Percent

In some partnerships, net income is shared in a fixed ratio, such as 5 to 4. In this case, the net income is divided into nine equal parts with five parts going to one partner and four parts to the other partner. In other partnerships, each partner's share may be a certain percent of the net income, such as 55% to one partner and 45% to the other.

EXAMPLE 2

Yeater and Wilson, business partners, agree to divide a net income of $120,000 in the ratio of 5 to 3. What is each partner's share?

SOLUTION

Add the shares.

$5 + 3 = 8$

Photodisc/Getty Images

Form fractions for each partner that show their part of the total shares. Multiply the fractions by the net income.

$\frac{5}{8} \times \$120{,}000 = \$75{,}000$ Yeater's share

$\frac{3}{8} \times \$120{,}000 = \$45{,}000$ Wilson's share

✔CHECK YOUR UNDERSTANDING

C. Tandjung and Sikorski share the net income of $24,000 from their partnership in the ratio of 6 to 5. What is each partner's share?

D. Cerkez and White share their partnership income 58% and 42%, respectively. If their net income this year is $130,500, what is each partner's share?

Interest on Investments

When partners invest different amounts, they may be paid interest on their investments. The rest of the net income may then be divided equally, or by other means.

EXAMPLE 3

Thome and Dulay form a partnership, investing $170,000 and $160,000, respectively. They agree to pay each partner 8% interest on their investment and to divide the remainder of the net income equally. The net income for the first year is $40,000. Find each partner's share of the net income.

SOLUTION

Multiply the interest rate by each partner's investment.

$8\% \times \$170{,}000 = \$13{,}600$ Thome's interest $8\% \times \$160{,}000 = \$12{,}800$ Dulay's interest

Add each partner's interest payment.

$\$13{,}600 + \$12{,}800 = \$26{,}400$ total interest paid

Subtract the total interest from the net income.

$\$40{,}000 - \$26{,}400 = \$13{,}600$ remainder of net income

Divide the remainder of net income by the number of partners.

$\$13{,}600 \div 2 = \$6{,}800$ each partner's share of remainder

Add each partner's share of remainder to their interest payment.

$\$13{,}600 + \$6{,}800 = \$20{,}400$ Thome's total share of net income

$\$12{,}800 + \$6{,}800 = \$19{,}600$ Dulay's total share of net income

✔ CHECK YOUR UNDERSTANDING

E. Zeigler and Soga are partners with investments of $250,000 and $375,000, respectively. The partners receive 10% interest annually on their investments. The rest of the net income is divided equally. What is each partner's total share of a net income of $96,000?

F. Volmer and Ramos invested $400,000 and $500,000 in a partnership, respectively. The partners agreed to pay 8% interest on their investments and to divide the remainder of any net income equally. The first year's net income was $240,000. What was each partner's share?

How are the net incomes from partnerships taxed? Are partnerships taxed differently from sole proprietorships? If one partner dies, how are the partnership properties distributed according to the partnership laws in your state? Is one partner responsible for the business debts made by the other partner? Research the answers to these questions using the library, Internet sources, or talking to an attorney. Prepare a brief report containing the answers you found to each question. List the sources you used, including Internet sources.

Wrap Up ▶▶▶

Sally is right to be concerned. Many problems can ruin friendships in a partnership. Most states have laws governing partnerships that take effect if no partnership agreement exists. Ralph and his partner might not like the way that those laws affect their partnership arrangements. The best advice is to agree on who will do what and who will get paid what at the beginning and put those conditions in a written partnership agreement.

Exercises

1. $45,689 − $23,185
2. $134,288 − $53,718
3. $375,500 × 7%
4. $745,850 × 9.5%
5. $\frac{3}{5} \times \$37,500$
6. $\frac{5}{8} \times \$91,000$

For Exercises 7–8, the partners distribute income in proportion to their investments. Find the net income each partner will get to the nearest dollar.

	Investments			
	Partner X	Partner Y	Partner Z	Net Income
7.	$260,000	$180,000		$68,400
8.	$280,000	$270,000	$86,000	$162,600

Solve.

9. In the partnership of Mallory and Reese, Mallory's investment is $225,000 and Reese's is $130,000. Net income is divided in proportion to their investment. Their net loss for the first year is $18,900. What is each partner's share of the loss?

10. The partnership agreement of Premmel and Tratia states that Premmel should receive 40% of any net income and Tratia should get 60%. The net income last year was $98,500. Find each partner's share of the net income.

11. Berrios and Gruber formed a partnership. They agreed to divide net income in the ratio of 7 to 4, with Berrios receiving the larger share. Last year's net income was $97,400. Find each partner's share, to the nearest dollar.

12. Riccio, Delgado, and Saburo invested $150,000, $160,000, and $190,000, respectively, in a partnership. For the first year, their gross profit was $208,400 and their expenses were $86,300. The partners got 7% on their investment, and the rest of the net income was shared equally. What was each partner's total share of the net income for the first year, to the nearest dollar?

Find each partner's share of the net income in each problem to the nearest dollar. Interest is paid on investments. The rest of the net income is divided equally.

	Investments			Interest on Investment	Net Income for Year
	Partner A	Partner B	Partner C		
13.	$50,000	$30,000		7%	$83,000
14.	$22,000	$37,000	$46,000	9%	$137,500
15.	$82,000	$58,000	$69,000	12%	$195,000

16. **INTEGRATING YOUR KNOWLEDGE** The investments of 3 partners are: Medina, $120,000; Douglass, $136,000; and Sanchez, $144,000. Each year net income is distributed in proportion to the partners' investments. Last year, the firm's net sales were $660,400. The cost of merchandise sold was $420,800, and the operating expenses were $129,600. The partners estimate that the business could be sold for $850,000. How much of last year's net income did each partner get?

17. **CRITICAL THINKING** When the partners' shares of net income are written as ratios or as percents, what should the sum of the shares be?

Mixed Review

18. Round: $567,398.45 to the nearest thousand dollars.

19. Rewrite $1\frac{3}{5}$ as a percent.

20. Rewrite $2\frac{5}{8}$ as a decimal.

21. Rewrite 305% as a decimal.

22. Tom Raymond bought 1,000 shares of a load fund, International Health Fund, at its offering price of $35.87. What was his total investment in the fund?

12-4

Preparing Balance Sheets

GOAL

- Calculate total assets, liabilities, and equity

KEY TERMS

- balance sheet
- assets
- liabilities
- equity

Start Up ▶▶▶

"I don't understand why the bank won't lend me the money," complained Tom Westman. Tom had asked a bank for a business loan. He had two other business loans from two other banks. "I always pay my bills as soon as I can!" Can you think of any reasons why Tom could not get a bank loan for his business?

khz/Shutterstock.com

Math Skill Builder

Review these math skills and solve the exercises that follow.

1. **Add** money amounts.
   Find the sum. $24,809 + $17,362 + $54,672 = $96,843

   1a. $10,689 + $21,765 + $28,510 + $3,592 + $5,221

   1b. $74,562 + $52,214 + $79,603 + $22,132 + $582

2. **Subtract** money amounts.
   Find the difference. $107,329 − $45,228 = $62,101

   2a. $274,656 − $132,109 **2b.** $92,176 − $53,737

   2c. $198,341 − $127,335 **2d.** $452,867 − $227,186

Business Tip

The balance sheet is often called a "snapshot" of a business because it reports data on the business for one specific day.

Calculate Total Assets, Liabilities, and Equity

A **balance sheet** is prepared at least once a year and shows a company's financial status on a specific day, usually the end of the month, quarter, or year.

Notice that the balance sheet shown on the next page is divided into three categories: assets, liabilities, and equity.

Assets are the things that are owned by a business that have value. Assets may be classified as current or long term. *Current assets* can be turned into cash or used within a year. Current assets include cash, accounts receivable, inventory, and supplies.

Business Tip

Long-term assets are also called *fixed assets.*

Photodisc/Getty Images

Petrie's Gift Shop
BALANCE SHEET
July 31, 20--

Assets		
Cash	$ 16,300	
Merchandise Inventory	196,920	
Store Supplies	5,380	
Store Equipment	24,500	
Total Assets		$243,100
Liabilities		
Accounts Payable	$ 51,560	
Bank Loan	45,200	
Total Liabilities		$ 96,760
Equity		
Tom Petrie, Equity		146,340
Total Liabilities and Equity		$243,100

Accounts receivable are the accounts of customers who owe the business money for the merchandise that was sold to them on credit. Petrie's Gift Shop's current assets included cash, merchandise inventory, and store supplies.

Long-term assets are property, such as machinery, land, and buildings that have a useful life of more than 1 year. Petrie's Gift Shop's long-term asset is store equipment.

The total assets of a business are the sum of its current and long-term assets.

Total Assets = Current Assets + Long-Term Assets

Liabilities are the debts of a business. Businesses often get some of their assets by buying them on credit and promising to pay later. The persons to whom the money is owed are called the *creditors* of the business.

Like assets, liabilities may be *current* or *long term.* An example of a current liability is *accounts payable.* Accounts payable are the companies that sold Petrie merchandise on credit. An example of a long-term liability would be a 25-year mortgage on a building or a three-year bank loan.

Petrie's current liabilities were accounts payable. The store's long-term liability is the bank loan.

The total liabilities of a business are the sum of its current and long-term liabilities.

Total Liabilities = Current Liabilities + Long-Term Liabilities

Business *Tip*

Long-term liabilities are also called fixed liabilities.

Business *Tip*

Equity may also be called *net worth.*

The owner's share of the business is called **equity**. If all the assets of a business are owned free of debt, the owner's share of the business is equal to the total value of the assets. If there are liabilities, the value of the owner's share is found by subtracting the liabilities from the assets. Since Tom Petrie is the owner of the gift shop, the equity account is written in his name. The total equity of $146,340 is Petrie's claim against the assets of the business and is found this way:

Equity = Assets − Liabilities

EXAMPLE 1

Petrie's Gift Shop has assets worth $243,100. It also owes creditors $96,760. Find the owner's equity of the business.

Photodisc/Getty Images

SOLUTION

Subtract the total liabilities from the total assets.

$243,100 − $96,760 = $146,340 equity

✔ CHECK YOUR UNDERSTANDING

A. TriCity Jewelry Suppliers had these assets on April 30, 20—: Cash, $14,746.10; Accounts Receivable, $63,754.65; Merchandise Inventory, $537,562.50; Supplies, $8,196.00; Equipment, $39,163.00; and Building, $167,470.00. The company owed $310,799 in accounts payable and $75,000 to the Luzerne County Bank. Find each of the following: 1) company assets, 2) liabilities, and 3) owner's equity.

B. On October 31, 20—, Tremont Brothers, Inc. had these assets: Cash, $5,850; Accounts Receivable, $14,750; Store Equipment, $7,430; Delivery Equipment, $32,540; and Supplies, $1,150. The company owed $16,790 in accounts payable. The brothers also owed $7,500 to a local bank for a 4-year truck loan. What were Tremont's 1) assets, 2) liabilities, and 3) equity?

Algebra *Tip*

A formula that represents the arithmetic of the balance sheet is
$A = L + E$;
A = Assets,
L = Liabilities,
E = Equity

Wrap Up ▸▸▸

Tom needs to make sure that he makes his loan and bill payments on time to receive the confidence of banks. Tom may also have total liabilities that are getting close to or exceeding the total of his assets. Tom may need to pay off some of his current debt before he asks for another loan.

Exercises

1. $38,297 + $67,110 + $3,207 + $763
2. $257,383 − $131,105

John Graber's business had these assets on December 31, 20—: Cash, $2,482; Accounts Receivable, $5,275; Merchandise Inventory, $87,650; Store Supplies, $3,732; Store Equipment, $22,647; and Delivery Equipment, $21,582.

3. What were John's current assets?
4. What were John's long-term assets?
5. What were John's total assets?

A business owed the following creditors for merchandise and store supplies: Richard's Landing, Inc., $3,489; Tree-Top Supplies, $4,121; Denson Wholesale Distributors, $8,522; and Abner Products, $3,971. The business also owed Macon Bank $78,300 on a 30-year mortgage, and Crevor National Bank $12,670 on a 3-year loan for store equipment.

6. What were the business's current liabilities?
7. What were the business's long-term liabilities?

Tal Benrique owns a flower shop with these assets: Cash, $4,200; Merchandise, $103,400; Store Equipment, $15,600; and Store Supplies, $1,860. He has accounts payable of $10,050 and a bank loan of $3,500.

8. What are Tal's total assets?
9. What are his total liabilities?
10. What is his equity?

Ama Rawini owns a tire store. She has these assets: Cash, $5,775; Merchandise, $63,000; Store Supplies, $780; Store Equipment, $13,800; Delivery Van, $12,600; and Land and Building, $90,000. She owes the State Bank $12,900 and the Logan Manufacturing Company $35,700.

11. What are Ama's total current assets?
12. What are Ama's total assets?
13. What are Ama's total liabilities?
14. What is her equity?

Vicente Lopez has an art store with these assets: Cash, $3,500; Accounts Receivable, $4,600; Merchandise, $73,500; Store Equipment, $8,700; Land and Buildings, $125,500. His liabilities are accounts payable, $13,200 and Lassiter Bank, $89,400.

Photodisc/Getty Images

15. What are Vicente's total assets
16. What are Vicente's total liabilities
17. What is Vicente's equity?

Kassie Burns owns a delivery company. She has 6 employees whose total annual wages are $114,000. On December 31 last year, her assets were: Cash, $3,620; Office Supplies, $2,185; and Delivery Equipment, $32,950. On that date, she owed $6,358 in accounts payable and $14,280 to Valley Bank.

18. What were Kassie's total assets?

19. What were Kassie's total liabilities?

20. What was Kassie's equity?

21. On December 31, Anthony Fouts, a small engine repair shop owner, had the following assets and liabilities. Complete his balance sheet by calculating his total assets, total liabilities, equity, and total liabilities and equity.

Photodisc/Getty Images

Downtown Engine Repair, Inc.
Balance Sheet, December 31, 20-

Cash	$ 2,834	Liabilities:	
Inventory	43,713	Accounts Payable	$ 7,530
Shop Supplies	2,570	Reston National Bank	12,780
Shop Equipment	28,800	Total Liabilities	
		Chieko Kimura, Equity	
Total Assets		Total Liabilities and Equity	

22. **CRITICAL THINKING** Look at the balance sheet for Petrie's Gift Shop. Suppose that every amount was the same except for the bank loan, which was changed to $200,000. How would Petrie's credit be affected? How would you show the new equity on the balance sheet?

23. **FINANCIAL DECISION MAKING** Why might a bank or other lender want to know a firm's current assets as well as the total assets?

Mixed Review

24. 3.6×15

25. $260 \times 1\frac{3}{8}$

26. $1,486.72 − $978.37

27. 40×275

28. Estimate the product: $875 × 48

29. What part of 360 is 90?

30. Find the average, to the nearest tenth: 8, 9, 7, 8, 10, 7.

31. Luis Bartolemo wants to earn 12% annual net income on his $60,000 cash investment in a business property. His annual expenses of owning the property are $12,000. What monthly rent must Luis charge?

32. The inflation index for 1996 was 156.9 and 160.5 for 1997. What was the rate of inflation for 1997, to the nearest tenth percent?

33. Towson Enterprises, Inc. decided to lease 20 notebook computer systems for 3 years for their technical staff. The monthly lease for each system was $62.88. What was the annual cost of leasing one notebook? All the notebooks?

34. The Clearview Window and Door Company increased its market share from 6% to 8% in one year's time. Clearview expects to gain another 3% next year. What is the company's sales goal next year if the window and door industry sales per year are $44 billion?

12-5

Analyzing Balance Sheets

GOALS
- Calculate the current ratio
- Calculate the debt to equity ratio
- Calculate return on equity

KEY TERMS
- current ratio
- debt-to-equity ratio
- return on equity

Start Up

Patricia was thinking about starting her own business and was asked by a friend how much she expected to make on her investment in the business. She was not sure how to answer the question. What might Patricia use as a guide?

Blend Images/Jupiter Images

Math Skill Builder

Review these math skills and solve the exercises that follow.

1. **Divide** money amounts and show as a decimal, to the nearest tenth.
   Find the quotient. $155,600 ÷ $85,400 = 1.82, or 1.8

   1a. $42,500 ÷ $102,840 **1b.** $5,308 ÷ $2,408

2. **Divide** money amounts and show as a ratio, to the nearest tenth.
   Find the ratio. $450,000 ÷ $200,000 = 2.3:1

   2a. $65,750 ÷ $31,200 **2b.** $245,900 ÷ $120,400

3. **Divide** money amounts and show as a percent, to the nearest tenth.
   Find the quotient. $24,500 ÷ $130,000 = 18.8%

   3a. $84,560 ÷ $956,300 **3b.** $45,870 ÷ $325,560

Current Ratio

Like the income statement, items on the balance sheet can be compared to one another as ratios or percents. There are a number of ratios and percents that can be used: the current ratio, the debt-to-equity ratio, and return on equity.

The **current ratio** compares current assets to current liabilities. It is a means of testing whether the business has enough cash, or items that can be turned into cash quickly, to pay its short-term debts.

The formula for finding the current ratio is:

$$\text{Current Ratio} = \frac{\text{Current Assets}}{\text{Current Liabilities}}$$

A current ratio of 2:1 means you have twice as many current assets as current liabilities. A current ratio of 2:1 or better is usually considered good. The ratio may be expressed as 2:1, or simply 2.

EXAMPLE 1

Petrie's Gift Shop has current assets of $218,600 and current liabilities of $51,560. What is the shop's current ratio?

SOLUTION

Divide the current assets by the current liabilities, to the nearest tenth.

$218,600 ÷ $51,560 = 4.2, or 4.2:1 current ratio

Photodisc/Getty Images

✔ CHECK YOUR UNDERSTANDING

A. A wholesale firm has $289,130 in current liabilities and $428,410 in current assets. What is its current ratio, to the nearest tenth?

B. A camping equipment store has the following current assets: cash, $2,580; accounts receivable, $2,380; merchandise inventory, $87,450; store supplies, $4,730. It owes the following current liabilities: accounts payable, $35,260; 30-day promissory note to the bank, $4,500. What is the store's current ratio, to the nearest tenth?

Debt-to-Equity Ratio

The **debt-to-equity ratio** is found by dividing the total debt of a firm by the firm's equity, or net worth. The ratio helps measure the amount of financial risk a business faces.

A debt-to-equity ratio shows the level of debt the firm is carrying. That allows others to determine if the level is appropriate for the business considering such factors as the level of debt similar firms are carrying and the overall condition of the economy at the time. The formula for the ratio is:

$$\text{Debt-To-Equity Ratio} = \frac{\text{Total Debt}}{\text{Equity}}$$

Business *Tip*

Some people use long-term liabilities rather than total debt to find debt-to-equity ratio.

Debt-to-equity ratio is often expressed as a percent. Thus, a debt-to-equity ratio of 1:1 would be shown as 100%. Whether a given debt-to-equity ratio is high or low depends on the industry within which the firm operates. In some industries, such as leasing and office technology, debt-to-equity ratios of 1:1 or higher are common. In others, ratios of 1:2, or 50% are more frequently found.

EXAMPLE 2

Petrie's Gift Shop has $45,200 in total debt in the form of a bank loan. Petrie's equity is $146,340. What is its debt-to-equity ratio, shown as a percentage, to the nearest tenth?

SOLUTION

Divide the total debt by the equity.

$45,200 ÷ $146,340 = 0.3089, or 30.9% debt-to-equity ratio

✔ CHECK YOUR UNDERSTANDING

C. Johnson's Creamery has debts of $356,890 and equity of $745,200. What is its debt-to-equity ratio, shown as a percent, to the nearest tenth?

D. Kriege's Steel Products has a mortgage on its factory for $450,300 and a 3-year bank loan for $250,000. Its total net worth is $1,205,500. What is its debt-to-equity ratio, shown as a percent, to the nearest tenth?

Return on Equity

Some ratios use information from both the balance sheet and income statement. One such ratio, **return on equity**, compares net income, taken from the income statement, to equity, or owner's equity taken from the balance sheet. The ratio lets you know what you've earned on your investment in a business during a particular period. Return on equity is usually shown as a percent.

Business *Tip*

Return on equity is often called return on investment, ROI.

$$\textbf{Return on Equity} = \frac{\textbf{Net Income}}{\textbf{Equity}}$$

You can compare the return on equity from a business to what the same money might have made in other investments during the same time period. For example, you can compare your return on equity with what a savings account, bond, stock, or money market fund would have paid you. If you are making less by owning and operating a business than you could make in less risky investments, such as savings accounts or bonds, you should question if continuing to operate the business is the best investment of your time and money.

However, new businesses may take some time before they begin to "pay off." Also, many businesses are seasonal. That is, they make much greater net incomes in some months than others. A gift shop, for example, may make much more net income during the winter holiday season than it does do in the summer months.

EXAMPLE 3

Petrie's Gift Shop had net income of $7,870 and equity of $146,340 on July 31, 20—. What was its return on equity, to the nearest tenth percent?

SOLUTION

Divide net income by equity.

$7,870 ÷ $146,340 = 0.054, or 5.4% return on equity

✔CHECK YOUR UNDERSTANDING

E. Gulliver Travel Agency had a net income of $67,280 on its December 31 income statement. The balance sheet on the same day showed equity of $121,500. What is its return on equity, to the nearest tenth percent?

F. Driger Sales, Inc.'s income statement showed these amounts on December 31: net sales, $248,400; cost of goods sold, $145,650; and operating expenses of $65,800. Its balance sheet showed a net worth of $246,300. What was its return on equity, to the nearest tenth percent?

Wrap Up ▶▶▶

There are many ways to evaluate what a person makes on an investment, including an investment in one's own business. One guideline is to compare the return from the business to the return on other investments that could be made. Starting a business is risky. Many new businesses fail. Patricia could lose all the money she put into it. So, she should expect a return that is greater than investments with less risky investments, such as savings accounts, CDs, or money market funds.

Exercises

1. Add $67,308 + $10,845 + $93,070 + $23,189
2. Divide $45,688 by $23,418 and show as a decimal, to the nearest tenth.
3. Divide $58,320 by $28,590 and show as a ratio, to the nearest tenth.
4. Divide $194,335 by $352,187 and show as a percent, to the nearest tenth.

Solve.

5. The balance sheet of Corida Manufacturing, Inc. shows $1,168,560 in current liabilities and $2,486,150 in current assets. What is its current ratio, to the nearest tenth?
6. A golf equipment store has the following current assets: cash, $1,958; accounts receivable, $1,080; merchandise inventory, $27,560; and store supplies, $1,350. It owes these current liabilities: accounts payable, $15,890 and a 90-day promissory note for $2,400. What is the store's current ratio, to the nearest tenth?

Photodisc/Getty Images

7. Tane's Nursery has debts of $136,910 and equity of $241,020. What is its debt-to-equity ratio, shown as a percent, to the nearest tenth?
8. Nieves Foundries has a mortgage on its factory for $845,400 and a 3-year bank loan for $1,500,000. Its total equity is $2,278,150. What is its debt-to-equity ratio, shown as a percent, to the nearest tenth?

Vitaleze Markets' income statement showed these amounts at the end of a year: net sales, $324,700; cost of goods sold, $214,530; and operating expenses of $98,400. Its balance sheet showed a net worth of $362,800.

9. Fiddler Toy Store had a net income of $176,780 on its December 31 monthly income statement. The balance sheet on the same day showed a net worth of $712,800. What is its return on equity, to the nearest tenth percent?

Photodisc/Getty Images

10. What was its return on equity, to the nearest tenth percent?

11. Vitaleze's owner can earn 2.5% in a one-year CD. Is this more or less than the business's return on equity for the year? How much more or less?

12. **FINANCIAL DECISION MAKING** A business earns a net income of $345,288 on owner's equity of $3,785,240. What is its return on equity, to the nearest tenth percent? Is this return a good one for the owner? Justify your answer.

INTEGRATING YOUR KNOWLEDGE During one month, a firm had net sales of $1,358,000, cost of goods sold of $831,070, and operating expenses that were 30% of net sales. The firm's balance sheet showed total assets of $4,568,000 and total liabilities of $3,000,300.

13. What was the firm's gross profit margin, to the nearest whole percent?

14. What was the firm's return on equity, to the nearest whole percent?

Mixed Review

15. $\frac{3}{4} + \frac{4}{5}$

16. $49.08 - 23.4$

17. $400.7 \div 1{,}000$

18. 56.08×0.003

19. Estimate the product of $432.10 × 328.

20. $120 is 60% of what amount?

21. Ted Wallach earns $12.50 an hour. His employer gives him a COLA based on the 2.8% the CPI rose last year. What is his new hourly pay rate?

22. Trish Rothstein drove her car 2,356 miles in October, 22.5% of which were for her job as a salesperson for a firm. The firm reimbursed Trish $0.35 a mile for business use. What did Trish receive as reimbursement for the use of her car in October?

23. The sales of a product were $11,268,000 last year. Sales are projected to be $1\frac{1}{2}$ times greater this year. What is the forecast for this year's sales?

24. Reece and Nitobe share partnership net income in proportion to their investments. Reece invested $351,000 and Nitobe $429,000 in the partnership. What is each partner's share of a net income of $160,000, to the nearest cent?

12-6

Bankruptcy

GOAL
- Calculate the percent and amount of bankruptcy claims

KEY TERM
- bankrupt

Start Up

What do you think happens to businesses that become bankrupt? Do they ever have to repay their debts? If bankrupt businesses don't have to repay their debts in full, is it fair to others who make all of their payments?

Photodisc/Getty Images

Math Skill Builder

Review these math skills and solve the exercises that follow.

1. **Subtract** money amounts.
   Find the difference. $154,780 − $85,259 = $69,521

   1a. $256,733 − $192,076 **1b.** $34,199 − $18,228

2. **Multiply** money amounts by percents.
   Find the product. $3,540 × 41% = $3,540 × 0.41 = $1,451.40
   Find the product. 100¢ × 35% = 100¢ × 0.35 = 35¢

   2a. $52,368 × 23.5% **2b.** $1,809 × 15.9%

   2c. 100¢ × 12.5% **2d.** 50¢ × 75%

3. **Divide** money amounts to get a percent.
   Find the percent, to the nearest tenth. $134,500 ÷ $245,600 = 0.5476, or 54.8%

   3a. $456,200 ÷ $1,500,200 **3b.** $39,980 ÷ $86,630

Bankruptcy Claims

When a business keeps operating at a loss, the amounts it owes may become more than its assets are worth. When this happens, the business becomes *insolvent* and a court may declare it **bankrupt**. The court then appoints a trustee or receiver to sell all the assets and pay the debts.

Business Tip

People, as well as businesses, can declare bankruptcy. States usually have personal bankruptcy laws to cover this situation.

After selling the assets, the trustee must pay the legal costs of the bankruptcy, any other claims that the law says must be paid first, and *secured creditors,* such as banks that hold a mortgage. Then the money that is left is paid to all other creditors in proportion to their claims. The percent to be paid to each creditor is found by dividing the total cash available for the creditors by the total of all creditors' claims.

$$\textbf{Claim Percent Paid} = \frac{\textbf{Cash Available for Creditors}}{\textbf{Total Creditors' Claims}}$$

EXAMPLE 1

The court declared a landscaping firm bankrupt. The trustee sold the firm's assets for $23,000. Legal costs of bankruptcy and other claims the court required to be paid first totaled $8,000. Creditors' claims totaled $37,500. What percent of the creditors' claims can the trustee pay? How many cents on the dollar will creditors get on their claims? How much will a creditor get who has a claim of $3,200?

SOLUTION

Subtract the legal costs and claims that must be paid first from the proceeds of the sale of the firm's assets.

$23,000 − $8,000 = $15,000 amount available to creditors

Divide the amount available to creditors by the total of the creditors' claims.

$15,000 ÷ $37,500 = 0.40, or 40% percent paid on each claim

Multiply 100 cents by the percent paid on each claim.

100¢ × 0.4 = 40¢ cents on the dollar paid to each creditor

Multiply the creditor's claim by the percent paid on each claim.

$3,200 × 0.4 = $1,280 amount creditor will receive

✔ CHECK YOUR UNDERSTANDING

A. A store is declared bankrupt. Creditors' claims total $420,000. After the assets are sold and bankruptcy costs are paid, $150,800 is left for creditors' claims. What percent of their claims will the creditors get?

B. Fairmount, Inc. was declared bankrupt, and its assets sold for $261,900. Legal costs of bankruptcy and other claims that must be paid first totaled $71,600. The total of creditors' claims was $420,000. Tricol, Inc. had a claim for $2,700. How many cents on the dollar will creditors get? How much will Tricol get?

Research the personal bankruptcy laws in your state to answer these questions. Write your answers in complete sentences.

1. Under what conditions is a person allowed to declare bankruptcy?
2. What kinds of creditors might have secured claims on the assets of the bankrupt person?
3. Who would be notified about a personal bankruptcy?
4. What property, if any, is a bankrupt person allowed to keep?
5. What are the disadvantages to a person who declares bankruptcy?

Wrap Up ▸▸▸

When a firm goes bankrupt, creditors take the firm's assets, the bankrupt person's credit rating is impaired, and the owners may be held criminally liable if fraud was intended.

Exercises

1. \$158,929 − \$45,216 **2.** \$187,225 × 19.4% **3.** 100¢ × 28%

Find what percent of the creditors' claims the trustee can pay.

	Total Creditors' Claims	Cash Available for Creditors
4.	\$ 68,000	\$ 27,200
5.	37,200	11,160
6.	146,700	51,345
7.	288,000	187,200

Find how many cents on the dollar can be paid to the creditors.

	Total Creditors' Claims	Cash Available for Creditors
8.	\$ 83,600	\$ 53,504
9.	71,760	25,116
10.	196,000	82,320
11.	54,000	8,688

Solve.

12. A shoe discount store is declared bankrupt. Creditors' claims total \$230,000, but only \$85,100 is available to pay claims. What percent of their claims will the creditors get?

13. A bankrupt wholesaler has debts totaling \$124,250. The cash available for the creditors is \$57,155. How many cents on the dollar will creditors get?

14. The creditors of a bankrupt manufacturer are paid at the rate of 16.3¢ on the dollar. There are 40 creditors with total claims of \$94,557. How much will a creditor receive on a claim of \$2,560?

Trail Bikes, Inc. was declared bankrupt, and its assets sold for \$174,600. Legal costs and claims paid first totaled \$47,625. The total of creditors' claims was \$370,000. Klesko, Inc. had a claim for \$750. Torrant Corp. had a claim for \$23,000.

15. How much money was available for all creditors?

16. What percent of each creditor's claim was paid?

17. How much did the Klesko receive? Torrent receive?

18. **FINANCIAL DECISION MAKING** Can you think of reasons why personal bankruptcy has increased in the last few years?

Mixed Review

19. What part of \$1,500 is \$75? **20.** Find 3.7% of \$43,930.

21. The monthly charge for leasing a small luxury car is \$379. For all miles driven over 10,000 miles in 1 year, there is a 20¢ per mile charge. What is the yearly cost of leasing the car if it is driven 21,000 miles in a year?

Chapter Review

Vocabulary Review

Find the term, from the list at the right, that completes each sentence. Use each term only once.

- assets
- balance sheet
- bankrupt
- cost of goods sold
- current ratio
- debt-to-equity ratio
- equity
- gross profit
- gross profit margin
- income statement
- liabilities
- merchandise turnover rate
- net loss
- net profit margin
- net income
- partnership
- return on equity

1. A report that shows how much money has been earned and spent during a period of time is called a(n) __?__.
2. Expenses directly related to producing or buying the goods sold are called __?__.
3. Total operating costs are deducted from gross profit to determine a company's __?__.
4. The percentage that gross profit is of net sales is called the __?__.
5. The number of times per period that a store replaces its average stock of merchandise is called the __?__.
6. Things that are owned by a business that have value are called __?__
7. The owner's share of the business is called __?__.
8. The __?__ compares current assets to current liabilities.
9. A comparison of net income to equity is called __?__.
10. When a business becomes *insolvent* a court may declare it __?__.

12-1 Preparing Income Statements

11. A small business's sales for the year were $56,980. Sales returns and allowances for the same year were $1,308. What were the net sales for the year?
12. Tom's TV Lot had an inventory of merchandise costing $96,300 in April. During April, goods costing $147,200 were bought. Merchandise inventory at the end of April was $88,500. What was the cost of goods sold for April?
13. Lee's Interiors had net sales for the year of $459,840. The cost of the goods sold for the same period was $268,700. What was Lee's gross profit for the year?
14. The Furniture Factory had a gross profit of $368,220 and operating expenses of $293,100 in October. What was its net income?

12-2 Analyzing Income Statements

15. Eddars Bottled Gas Company had net sales of $592,300 and cost of goods sold of $332,450 this year. What is Eddars' gross profit margin, to the nearest tenth percent?
16. The Bookshelf had net sales of $287,900 and net income of $38,220 last year. What was its net profit margin, to the nearest tenth percent?

17. Central Marts had a beginning inventory on Jan. 1 of $1,668,450. The ending inventory on Dec. 31 was $1,655,490. The Mart's cost of goods sold for the year was $9,945,160. What was its merchandise inventory turnover rate, to the nearest tenth?

12-3 Partnership Income

18. Byrd and Tyne invest $200,000 and $400,000, respectively, in a partnership. They agree to share net income in proportion to their investments. The partnership's net income for the year is $82,000. Find each partner's share of the net income.

19. The partners Merk and Guterrez agree to divide net income of $270,000 in the ratio of 4 to 5. What is each partner's share?

20. The partners Rich and Matsui agree to share their net income of $350,000 by 45% and 55%, respectively. What is each partner's share?

21. Bril and Sosa are partners with investments of $150,000 and $300,000, respectively. The partners receive 12% annually on their investments. The rest of the net income is divided equally. What is each partner's total share of a net income of $89,000?

12-4 Preparing Balance Sheets

22. Bevo Discount Mart had these assets on June 30, 20—: Cash, $28,574.10; Accounts Receivable, $123,574.85; Merchandise Inventory, $1,257,612.80; Supplies, $17,936.00; Equipment, $79,637.00; and Building, $362,670.00. The company owed $631,939 in accounts payable and $150,000 to the Maritime Bank. What were the company's assets, liabilities, and equity?

12-5 Analyzing Balance Sheets

23. A store has current assets of $428,800 and current liabilities of $115,760. What is the store's current ratio?

24. Watkin Manufacturing has long-term debts of $813,900 and equity of $1,425,800. What is its debt-to-equity ratio, shown as a percent, to the nearest tenth?

25. A business had net income of $127,820 on its December 31 income statement. The balance sheet on the same day showed equity of $241,600. What is its return on equity, to the nearest tenth percent?

12-6 Bankruptcy

26. A wholesaler was declared bankrupt, and its assets sold for $252,600. Legal costs of bankruptcy and other preferred claims totaled $71,600. The total of creditors' claims was $420,000. What percent, to the nearest tenth, of their claims will the creditors get?

27. A business is declared bankrupt. Creditors' claims total $240,000. After the assets are sold and bankruptcy costs are paid, $100,320 is left for creditors' claims. Dunne, Inc. had a claim for $4,800. How much will Dunne get?

28. The cash available for creditors from a bankruptcy is $340,000. The total claims of creditors are $720,000. How many cents on the dollar will creditors receive?

Technology Workshop

Task 1 Enter Data in a Balance Sheet Analyzer Template

Complete a template that calculates the current ratio, debt-to-equity ratio, and return on equity.

Open the spreadsheet for Chapter 12 (tech12-1.xls) and enter these amounts in the blue cells from the Wehling Wholesale Company: current assets, $250,000; current liabilities, $95,000; long-term liabilities, $100,000; equity, $455,000; net income, $75,000; and capital, $455,000. Your computer screen should look like the one shown below when you are done.

	A	B	C
1	**Balance Sheet Analyzer**		
2			
3	**Current Ratio**		
4	Current Assets	250,000.00	
5	Current Liabilities	95,000.00	
6	Ratio	2.6	:1
7			
8	**Debt-to-Equity Ratio**		
9	Long-term Liabilities	100,000.00	
10	Equity	455,000.00	
11	Debt-to-Equity Ratio	1:	4.6
12			
13	**Return on Equity**		
14	Net Income	75,000.00	
15	Equity	455,000.00	
16	Return on Equity	16.5%	

Notice that the spreadsheet does not automatically calculate equity from the asset and liability data entered. This feature lets you use the spreadsheet to find the debt-to-equity ratio for one company and the return on equity for another at the same time.

Task 2 Analyze the Spreadsheet Output

Calculate or find the amounts needed to fill the blue cells using the data from the income statement and balance sheet for Petrie's Gift Shop shown in Lessons 12-1 and 12-4. Make sure that you show Petrie's bank loan of $45,200 as a long-term liability.

Answer the following questions.

1. What is Petrie's current ratio?
2. What is Petrie's debt-to-equity ratio?
3. What is Petrie's return on equity?
4. What spreadsheet function is being used in cell B6?
5. What math is being used in cell B16?

Task 3 Design an Income Statement Analyzer Spreadsheet

Design a spreadsheet that will calculate the major items on an income statement and also calculate the percent each item on the income statement is of net sales.

Use the income statement for Petrie's Gift Shop to provide you with the headings, item labels, and amounts. Use one column for the item labels, one column for the amounts, and one column for the percentages.

Petrie's Gift Shop
Income Statement
For month ended July 31, 20—

	Amounts	**Percent**
Revenue		
Sales	75,952	102.6%
Less sales returns and allowances	1,900	2.6%
Net sales	74,052	100.0%
Cost of goods sold		
Beginning inventory, July 1	211,638	285.8%
Purchases	31,548	42.6%
Goods available for sale	243,186	328.4%
Ending inventory, July 31	196,920	265.9%
Cost of goods sold	46,266	62.5%
Gross profit on sales	27,786	37.5%
Operating expenses		
Salaries and wages	10,567	14.3%
Rent	3,194	4.3%
Taxes	1,864	2.5%
Utilities	1,678	2.3%
Advertising	780	1.1%
Depreciation of equipment	726	1.0%
Insurance	253	0.3%
Other expenses	854	1.2%
Total operating expenses	19,916	26.9%
Net income	7,870	10.6%

Task 4 Analyze the Spreadsheet Output

Answer these questions about your completed spreadsheet:

6. How did you calculate cost of goods sold?
7. How did you display percentages that had multiple decimal places?
8. How did you display percentages in general?
9. What sections of the spreadsheet would have to change if you used it to display Cepeda Manufacturing's income statement from Lesson 12-1?

Chapter *Assessment*

Chapter Test

Answer each question.

1. 3,856 ÷ 1,000

2. $\frac{2}{3}$ × $20,000 to the nearest cent

3. $299 is what percent of $4,600?

4. 52 × $77.78

5. 250% × $98

6. 75% × $3,200

7. $21,524 − $16,800

8. 2,080 × $0.29

9. $172 + $329 + $136 + $473

10. 2.62 × $562

Solve.

11. A discount store took three inventories of merchandise during a quarter: $176,000, $178,820, and $246,600. The cost of merchandise sold during the 3-month period was $1,360,200. What was the merchandise turnover rate, to the nearest tenth?

12. A firm had net sales of $280,000. The cost of goods sold was $168,000. What was the gross profit margin?

13. Miskai invests $425,000 and Donald invests $260,000 in a partnership. Net income is divided in proportion to their investment. Their net loss for the first year is $38,600. What is each partner's share of the loss?

14. The partners Herrera and Rothenberg share net income at 40% and 60%, respectively. Their net income last year was $248,500. Find each partner's total share of the net income.

15. A small store has assets worth $124,800. It also owes creditors $65,860. Find the equity of the business.

16. Jamask, Inc. has $589,300 in current liabilities and $856,410 in current assets. What is its current ratio, to the nearest tenth?

17. Billows Stores had net income of $158,060 and equity of $1,863,450 on October 31, 20—. What was its return on equity for October, to the nearest tenth percent?

18. A firm is declared bankrupt. Creditors' claims total $443,000. After the assets are sold and bankruptcy costs are paid, $178,529 is left for creditors' claims. How much will a creditor with a claim of $14,500 get?

Kline Products had a beginning inventory of $635,500 on January 1. During the quarter, goods costing $2,245,600 were manufactured. The inventory at the end of March was $785,400. During the quarter, Kline's net sales were $3,265,800.

19. What was the cost of goods sold for the quarter?

20. What was the gross profit?

Planning a Career in Business, Management, and Administration

Dmitriy Shironosov 2008/Shutterstock.com

Employees in business, management, and administration work in corporate offices, at universities, hospitals, and in food service. People in these careers might manage people, projects, or both, as well as plan business operations. Others might be the support staff for those in management positions. If you enjoy planning projects, a career in business, management, and administration might be for you.

Job Titles

- Financial analyst
- Project manager
- Assistant
- Recruiter
- Entrepreneur
- Operating officer

Needed Skills

- outstanding leadership skills
- excellent decision making skills

- computer and technology skills
- creativity and talent for problem solving
- strong communication and negotiation skills

What's it like to work in Administrative Support?

In larger organizations, an administrative support manager may supervise an area or department, such as human resources, administration, or payroll. These managers report to mid-level managers who then work with the top executives. In smaller firms, an administrative support manager might oversee all support services and work directly with top management.

What About You?

Can you see yourself managing people and helping a business succeed and grow?

How Times Have Changed

For Questions 1-2, refer to the timeline on page 511 as needed.

1. Ursula's time card is shown below. She is paid to the nearest quarter hour. She earns $11.85 an hour. What are her gross wages for this week of work? If this time card was dated 1935, what would be one difference in Ursula's net pay compared to a time card dated 1940?

	M	T	W	TH	F
IN	7:58	8:00	7:55	8:10	8:02
OUT	12:01	11:55	12:00	11:59	12:05
IN	1:05	1:00	12:59	12:58	
OUT	4:30	4:28	4:32	4:29	

2. Describe benefits to both an employer and employees to utilizing electronic transfer of paychecks instead of paper checks.

Chapter 13

International Business

Statistical Insights

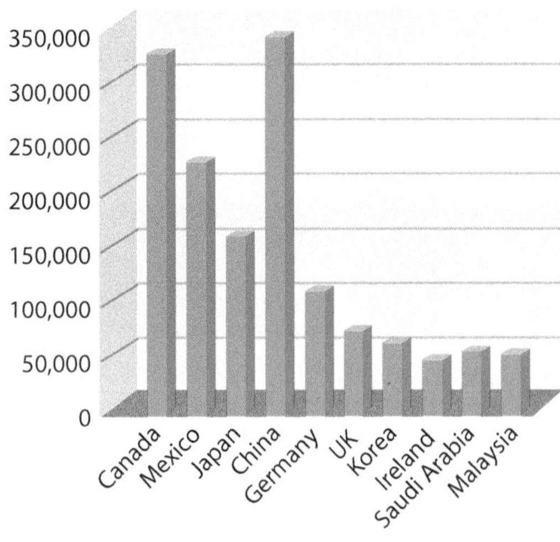

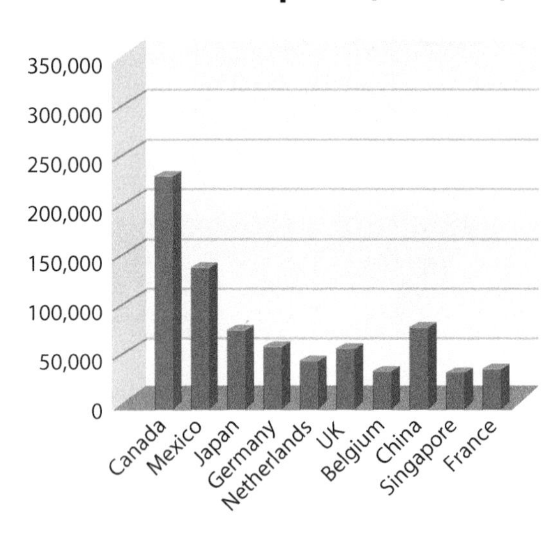

Source: census bureau

The U.S. sells its goods to other countries and other countries sell their goods to the U.S. Use the bar graphs to answer Questions 1–4.

1. Which country has the least difference between U.S. exports and imports?
2. About how much greater are U.S. imports from China than U.S. exports to China?
3. What is the approximate total of exports to the countries shown?
4. A trade deficit exists when the U.S. imports more than it exports. A trade surplus is when the U.S. exports more than imports. **Explain** whether the U.S. has trade deficits or surpluses with Canada, Mexico, Japan, and China.

How Times Have Changed

For most of history time was not an international concern. Each local area set its own times based on the position of the sun and kept by a well-known clock in the community. When railroad travel became popular in the nineteenth century, a standardized time system became necessary. This system was the beginning of the international time zones used today.

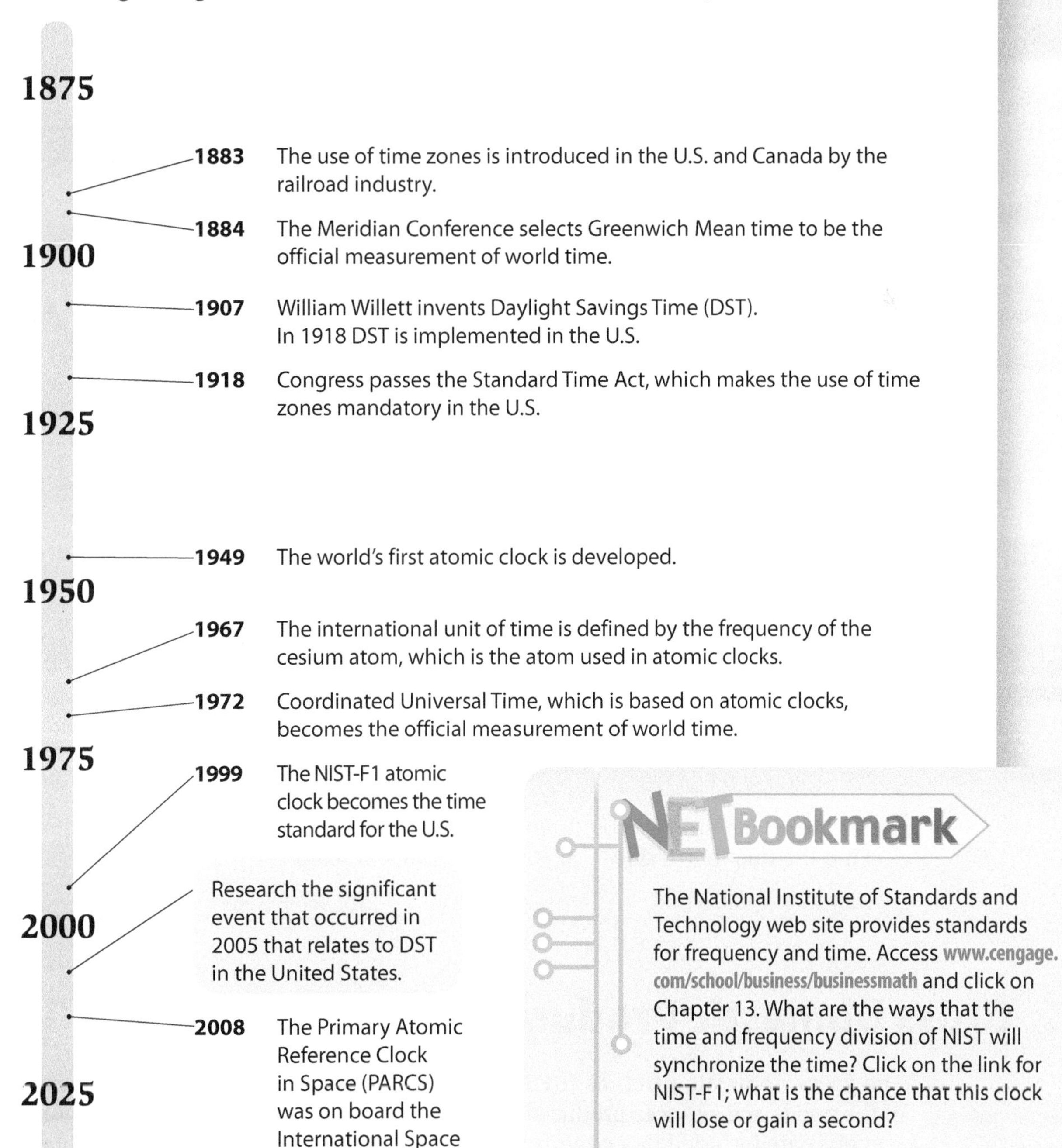

1875
1900
1925
1950
1975
2000
2025
1883 The use of time zones is introduced in the U.S. and Canada by the railroad industry.
1884 The Meridian Conference selects Greenwich Mean time to be the official measurement of world time.
1907 William Willett invents Daylight Savings Time (DST). In 1918 DST is implemented in the U.S.
1918 Congress passes the Standard Time Act, which makes the use of time zones mandatory in the U.S.
1949 The world's first atomic clock is developed.
1967 The international unit of time is defined by the frequency of the cesium atom, which is the atom used in atomic clocks.
1972 Coordinated Universal Time, which is based on atomic clocks, becomes the official measurement of world time.
1999 The NIST-F1 atomic clock becomes the time standard for the U.S.
Research the significant event that occurred in 2005 that relates to DST in the United States.
2008 The Primary Atomic Reference Clock in Space (PARCS) was on board the International Space Station.
NET Bookmark
The National Institute of Standards and Technology web site provides standards for frequency and time. Access www.cengage.com/school/business/businessmath and click on Chapter 13. What are the ways that the time and frequency division of NIST will synchronize the time? Click on the link for NIST-F I; what is the chance that this clock will lose or gain a second?

13-1

Production, Trade, and Finance

GOALS

- Calculate a nation's per capita production
- Calculate trade surpluses or trade deficits
- Convert currency using exchange rates

KEY TERMS

- gross domestic product (GDP)
- per capita GDP
- domestic business
- international business
- exporting
- importing
- trade surplus
- trade deficit
- foreign exchange rate

Start Up

Eric, a professional hockey player, signs a contract that pays $1.8 million a year in Canadian dollars. Marcel's contract is for $1.25 million a year payable in U.S. dollars. Which player is paid more per year?

Photodisc/Getty Images

Math Skill Builder

Review these math skills and solve the exercises.

1 **Divide** money amounts by whole numbers.
Find the quotient. $1,200,000,000,000 ÷ 80,000,000 = $15,000

1a. $940,000,000 ÷ 160,000 **1b.** $1,910,260,000 ÷ 190,000

2 **Multiply** whole numbers by decimals; **round** to the nearest hundredth.
Find the product. 368 × 0.1208 = 44.454, or 44.45

2a. 1,200 × 0.004373 **2b.** 1,714 × 2.657

3 **Divide** whole numbers by decimals; **round** to the nearest hundredth.
Find the quotient. 1,800 ÷ 3.672 = 490.196 = 490.20

3a. 35 ÷ 0.8975 **3b.** 96 ÷ 2,302

Gross Domestic Product

The **gross domestic product (GDP)** of a country measures the total market value of the goods and services produced within its borders. Goods sold to other countries (exports) are included in a country's GDP. The goods bought from other countries (imports) are not.

GDP is usually reported annually in dollars. GDP is an indication of the size of a country's economy. In this book, GDP will be shown in United States dollars.

One way to compare GDP among countries is to calculate per capita GDP. *Per capita* means per person. So **per capita GDP** shows the amount of goods and services produced per person by a country. Per capita GDP is often used to measure a country's standard of living.

Math *Tip*

Large money amounts may be written in this way:
$1.4 billion for $1,400,000,000
$1.12 trillion for $1,120,000,000,000
$0.45 million for $450,000

EXAMPLE 1

In a recent year, Norway had a GDP of $171,700,000,000 ($171.7 billion) and a population of 4,500,000 (4.5 million) people. Find the per capita GDP for Norway in that year to the nearest hundred dollars.

SOLUTION

Divide GDP by the number of people: $171,700,000,000 ÷ 4,500,000 = $38,155.56

To the nearest hundred dollars, Norway's per capita GDP is $38,200.

✔ CHECK YOUR UNDERSTANDING

A. Suppose Japan had a GDP of $3.6 trillion and a population of 127 million. What was Japan's per capita GDP, to the nearest dollar?

B. The GDP of Zimbabwe is $24 billion. Its population is 12.7 million. Find Zimbabwe's per capita GDP, to the nearest dollar.

Balance of Trade

Domestic business is the manufacturing, buying, and selling of goods and services within a country. Many companies conduct **international business** that includes all the business activities necessary to manufacture, buy, and sell goods and services across national borders. When people talk about *foreign trade*, they are talking about international business.

The two most important international business activities are exporting and importing. **Exporting** is the selling to other countries of the goods and services produced within your country. **Importing** is the buying of products produced outside your country. A measure of a country's international business activity is its balance of trade. *Balance of trade* is the difference between a country's exports and imports.

Photodisc/Getty Images

When foreign exports exceed imports, a country has a **trade surplus**. When foreign imports exceed exports, a country has a **trade deficit** and may have to borrow money or ask for credit from other countries, thus establishing *foreign debt*. As with individuals, governments must pay interest on money they owe another country.

If the foreign debt is too large, it can affect the country's economy by limiting the amount of money available for necessary work in that country. For example, a large debt could mean a country could not improve public buildings or roads, or provide some services for its citizens.

EXAMPLE 2

Recently, Argentina had exports of $29.57 billion and imports of $13.27 billion. Did Argentina have a trade surplus or trade deficit? How much?

SOLUTION

Subtract the smaller number from the larger number.

$29,570,000,000 − $13,270,000,000 = $16,300,000,000

Argentina's exports exceeded imports, so there was a trade surplus.

✔ CHECK YOUR UNDERSTANDING

C. In one year, Germany's imports were $585 billion and its exports were $696.6 billion. Did Germany have a trade surplus or trade deficit, and what was the amount?

D. Find the amount of trade surplus or trade deficit for the United Arab Emirates in a year when its exports were $56.73 billion and its imports were $37.16 billion.

Exchange Rates

Most nations have their own kind of money or *currency*. Before a business can buy something in another country, it has to buy that country's money. How much the business must pay for another country's currency is based on the **foreign exchange rate**, or simply the *exchange rate*.

Foreign exchange is the process of changing or converting the currency of one country into the currency of another country. The exchange rate is the amount of your money you must trade for the currency of another country.

The value of currency like most things in the market is affected by supply and demand. Currency exchange rates are also affected by the country's economic condition and political stability. Exchange rates change daily. The table below shows the exchange rates for sample currencies for one day in a recent year.

Country	Currency	Symbol	Value in U.S. Dollars	Units per U.S. Dollar
Brazil	real	R$	$0.430104	2.32502 R$
Canada	dollar	$	$0.949938	$1.0527 Canadian
Great Britain	pound	£	$1.82640	£ 0.547525
Hong Kong	dollar	HK$	$0.128093	$7.8068 HK
India	rupee	Re	$0.022909	43.65 Re
Japan	yen	¥	$0.009127	109.56 ¥
Mexico	peso	MXN	$0.098005	10.2035 MXN
South African	zar	R	$0.129701	7.71 R
Various European	euro	€	$1.46849	0.680967 €

The unit of currency in Brazil is the *real*. The last column shows there are 2.32502 *reais* (plural of real) in one U.S. dollar. The adjacent column states that each *real* is worth $0.430104 in U.S. dollars.

Since January 1, 1999, twelve European countries have replaced their national currencies with a single currency called the *euro*. The euro is now used to settle inter-bank and international commerce payments. Euro currency replaced national currencies on January 1, 2002. This means, for example, that the French franc is no longer used.

There are two ways to find the equivalent amount in U.S. dollars for a given foreign currency amount.

EXAMPLE 3

A businessperson hires a translator in India for 5,000 rupees. How much was the translator paid in U.S. dollars?

SOLUTION

Method 1: Use the rupee's Value in U.S. Dollars from the table.

$5{,}000 \times \$0.022909 = \$114.55$ cost in U.S. dollars

Method 2: Use the rupee's Units per U.S. Dollar from the table.

$5{,}000 \div 43.65 = \$114.547$, or $\$114.55$ cost in U.S. dollars

Since you are converting foreign currency into U.S. dollars, you round to the nearest cent and add a dollar sign to the final answer.

Another way to use the exchange rate table is to calculate how many units of foreign currency you could get in exchange for U.S. dollars.

EXAMPLE 4

On a business trip to England, Maria Cedeno exchanged $120 of U.S. currency to pay for incidental expenses while in England. She got the exchange rate shown in the table. How much money did she receive in return, and in what currency?

SOLUTION

Multiply the pound's units per U.S. dollar by the amount exchanged.

$\$120 \times 0.547525 = 65.703£$ amount received in pounds

✔ CHECK YOUR UNDERSTANDING

E. A manufacturer from Illinois took business clients in Mexico to dinner. The cost of the dinner was 9,895 pesos. Find the dinner's cost in U.S. dollars.

F. Find how many rupees would be received for an exchange of $50 in U.S. currency.

Identify the countries that use the euro as their sole currency. Find the answers to these questions and then write a summary about your findings.

1. May other countries join at a later date?
2. What are the advantages of using a common currency?
3. What are the disadvantages?
4 Is every euro bill and coin identical from country to country?

Wrap Up ▸▸▸

The only way to compare money amounts in different currencies is to adjust them for the exchange rate. Using the exchange rate table you can calculate that Eric's contract is worth about $1,709,888 in U.S. funds. So, Eric is paid more per year.

Exercises

Find the product or quotient, rounded to the nearest hundredth.

1. 23×15.736
2. 127×88.084
3. $4{,}500 \div 0.286$
4. $18.25 \div 0.467$

Solve.

5. Egypt had a GDP of $295.2 billion and a population of 76.1 million. What was Egypt's per capita GDP, to the nearest ten dollars?
6. New Zealand has a population of 3,993,817 and a GDP of $85.34 billion. What is its per capita GDP, to the nearest dollar?
7. In one year, Vietnam's GDP was 203.7 billion. Its population was 82.7 million. To the nearest dollar, what was its GDP per capita?
8. A country's total exports for a year were $3.049 billion. Its imports for the first 6 months of the year were $940 million and $1.08 billion for the last 6 months. What was the amount of the trade surplus or trade deficit for the year?

Photodisc/Getty Images

Use the Exchange Rate Table for Exercises 9–11. Round to the nearest U.S. dollar.

9. A flying service bought a used airplane in Canada for $140,000 in Canadian dollars. What was the cost of the airplane in U.S. dollars?
10. A company bought electronics equipment in Japan for 2,400,000 yen. What was the equipment's cost in U.S. dollars?
11. The purchase of a factory in Finland was settled in euros. The factory's cost was 586,000 euros. What was its cost in U.S. dollars?
12. **CRITICAL THINKING** Instead of having their own currency, some countries use U.S. dollars as their currency. Why would they want to do this?
13. **BEST BUY** You can buy a sports car in Sweden for 303,320 krona. The krona's exchange rate in U.S. dollars is $0.150707. Shipping the car to the United States would cost 39,420 krona. Once the car arrives in the United States, there would be an additional cost of $800 to get the car transported to your home. Exactly the same car could be bought in a nearby city for $51,840. Which is the better deal?

Mixed Review

14. $2\frac{1}{3} \times \frac{1}{5}$
15. $1\frac{9}{16} \div 1\frac{1}{4}$
16. Estimate, then find the actual product of 317×9.
17. Round to the nearest 10 million: 543,458,284,009
18. Eleanor Scripps has earned exactly the same weekly salary of $1,200 for two years. What is her average monthly salary?

13-2

International Time and Temperature

GOALS

- Calculate time in different time zones
- Convert Fahrenheit and Celsius temperatures

KEY TERM

- time zone

Start Up ▶▶▶

While vacationing in Toronto, Canada, you call the front desk and ask for the outdoor temperature. The clerk answers, "It's 5 degrees." You put on a heavy coat over an extra sweater. When you step outdoors, you are too warm. Why is this so?

Photodisc/Getty Images

Math Skill Builder

Review these math skills and solve the exercises.

1 **Add** or **subtract** hours to or from time.
Find the times. 12:15 P.M. + 7 hours = 7:15 P.M.
3:00 A.M. − 5 hours = 10:00 P.M.

1a. 9:15 A.M. + 4 hours　　**1b.** 7:00 P.M. + 11 hours

1c. 12:00 Noon − 6 hours　　**1d.** 4:10 A.M. − 8 hours

2 **Multiply** numbers by fractions or decimals; **round** to the nearest whole.
Find the products. $\frac{5}{9} \times 68 = 37.7$, or 38
$1.8 \times 54 = 97.2$, or 97

2a. $\frac{5}{9} \times 72$　　**2b.** $\frac{5}{9} \times 29$　　**2c.** 1.8×27　　**2d.** 1.8×-10

International Time

Dealing with international travel and customers requires knowledge of international time zones and the International Date Line.

The globe is divided into 24 standard **time zones** based on a 360° circle as shown on the next page. The center of each time zone is designated by the standard meridians of longitude. Each time zone is spaced 15° apart. Greenwich, England is designated as the starting point, 0°, or the *prime meridian*.

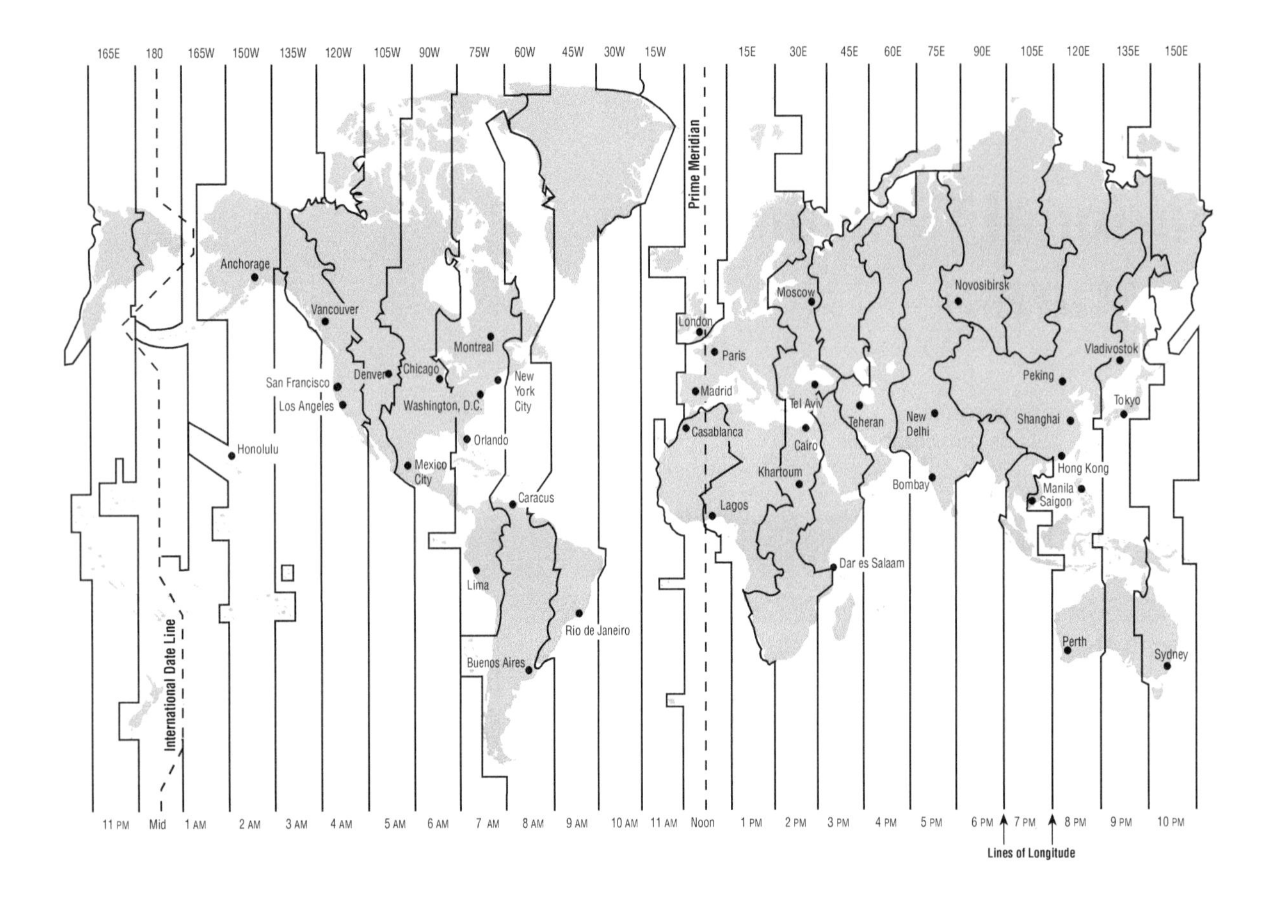

If you started in Greenwich and moved east one time zone, you would add one hour of clock time for each 15° of longitude. If you moved west one time zone from Greenwich, you would subtract one hour for every 15° of longitude.

In the United States, the contiguous land is divided into four time zones: Eastern Standard Time (EST), Central Standard Time (CST), Mountain Standard Time (MST), and Pacific Standard Time (PST).

If the time in California is 8:00 A.M. (PST), the time in New York is 11:00 A.M. (EST). If the time in Chicago (CST) is 7:00 P.M., the time in California is 5:00 P.M. (PST).

Remember that when you move across time zones to the *east* you *add* the number of time zones moved to the current time. When you move to the *west,* you *subtract* the number of time zones moved from the current time.

The International Date Line is an imaginary line located in the middle of the Pacific Ocean (west of the west coast of the United States). It is halfway around the world from Greenwich, England. When you travel east and cross the date line, you lose a day. If you are traveling east around the world and it is May 15, when you cross the International Date Line it will be May 14. If you were traveling west on May 15, when you cross the International Date Line, it would be May 16.

To determine the time and date in a particular country, first locate the place where you are starting and determine if you are traveling east or west. Count the time zones traveled to calculate the time and note if you crossed over the International Date Line. When you cross the International Date Line, you must adjust the day forward for travel to the west and backward for travel to the East.

EXAMPLE 1

It is 3 P.M., Thurs., in San Francisco. What time and day is it in Sydney, Australia?

SOLUTION

Look at the time zone map and count the number of time zones.

Sydney is 6 time zones to the west of San Francisco.

Subtract the number of time zones from the time you know.

3:00 P.M. − 6 hours = 9 A.M.

The direction of travel was to the west and the International Date Line was crossed, so add 1 day to the current day.

Thursday + 1 day = Friday

It is 9 A.M. on Friday in Sydney.

✔ CHECK YOUR UNDERSTANDING

A. It is 7:00 A.M. Thursday in Chicago, Illinois. How many time zones away is Peking, China? What is the day and time in Peking?

B. It is 1:30 P.M. Sunday in Vladivostok, Russia. What day and time is it in Anchorage, Alaska? How many time zones were counted and in which direction?

Photodisc/Getty Images

Temperature Conversion

In the United States, temperature is measured and recorded in the *Fahrenheit scale* in everyday use. Many other countries measure temperature by using the *Celsius scale.* Knowing the differences between the two scales and how to convert from one scale to the other is important to travelers and for business planning. The Fahrenheit and Celsius scales are shown.

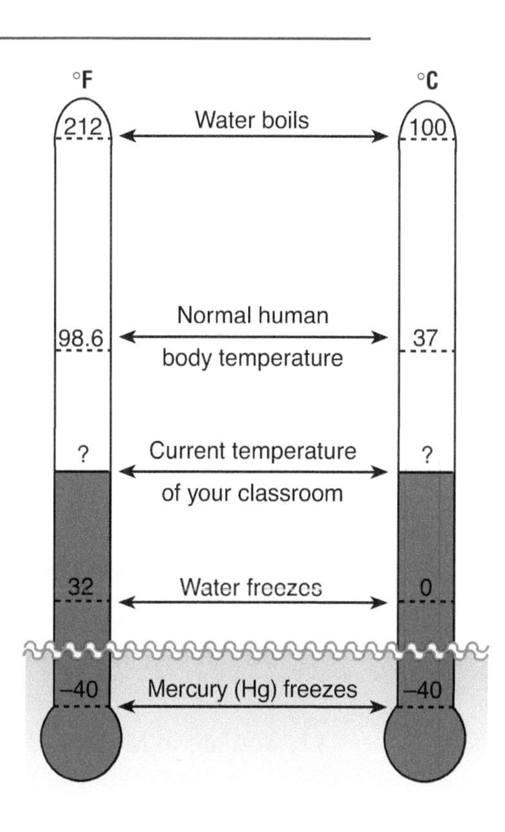

As you can see, the Fahrenheit scale is based on a scale of 32° (freezing point) and 212° (boiling point). The Celsius system is based on a scale of 0°C (freezing point) to 100° (boiling point).

To convert between Celsius (C) and Fahrenheit (F) temperatures, you can use one of these two formulas.

$$F = 1.8C + 32$$

$$C = \frac{5}{9}(F - 32)$$

EXAMPLE 2

The average body temperature in Fahrenheit degrees is 98.6. What is its Celsius equivalent?

SOLUTION

Apply the Celsius conversion formula: $C = \frac{5}{9}(F - 32)$

$\frac{5}{9}(98.6 - 32) = \frac{5}{9} \times 66.6 = 37°C$ body temperature in Celsius

Check your answer by using the Fahrenheit conversion formula: $F = 1.8C + 32$

$1.8(37) + 32 = 66.6 + 32 = 98.6°F$ body temperature in Fahrenheit

✔ CHECK YOUR UNDERSTANDING

C. The weather report for New York City forecasts a high today of 94°F. What is the equivalent temperature in Celsius, to the nearest degree?

D. A weather reporter in Naples, Italy states that an overnight low temperature of 15°C is expected. What is the equivalent temperature in Fahrenheit?

Algebra *Tip*

The temperature conversions you have done show a different number for temperatures in the two scales. However, the Fahrenheit and Celsius thermometers on the previous page shows the element mercury freezes at –40°F and –40°C.

Using the formulas for conversion, you can see that both conversions equal –40° on the other scale.

Wrap Up ▸▸▸

The clerk gave the temperature in Celsius, the measure used in Canada. The temperature of 5° Celsius is equal to 41° Fahrenheit. You are warm because your clothing is more suited to a much colder temperature.

Exercises

Find the sum or difference.

1. 7:00 P.M. + 10 hours

2. 1:47 A.M. + 11 hours

3. 3:13 A.M. – 9 hours

4. 12:01 A.M. – 4 hours

Find the product, round to the nearest whole.

5. $\frac{5}{9} \times 14$

6. $\frac{5}{9} \times 52$

7. 1.8×6

8. 1.8×30

Solve.

9. It is 5:00 A.M. Saturday in Orlando, Florida. What day and time is it in Rio de Janeiro?

10. It is 12:00 Noon Greenwich meridian time. What time is it in Khartoum?

11. You are in Vancouver, Canada. Your boss is in Bombay, India. She wants you to call Tuesday at 5:00 P.M. Bombay time. What day and time do you call?

12. You are in Moscow, Russia. You want to call someone in Mexico City at 3:30 P.M., their time. What time is it in Moscow when you place the call?

13. You imported an exotic bird for display in your pet shop. The exporter tells you the temperature in the area the bird will be housed must always be 30°C or higher. At what Fahrenheit temperature should you set your thermostat?

Photodisc/Getty Images

14. You receive a shipment of items that includes the warning "Do Not Accept Shipment if Exposed to Temperatures Below 10°C." You know that the package was stalled for several days in a snowstorm where temperatures never got above 20°F. Should you accept the shipment?

15. You are sending cargo overseas. The cargo must be kept refrigerated at 45°F. Someone tells you that if you mark the package 32°C it will be kept cool enough. Is the advice correct? Explain.

16. To conserve energy you are advised to set your thermostat to 68°F in the winter and 72°F in the summer. What are the settings in Celsius to the nearest whole degree?

17. **CRITICAL THINKING** You are flying from Los Angeles, California to Montreal, Canada. Your departure time is 9:30 A.M. PST, and the flight time is 5 hours. What time will it be in Montreal when your flight arrives?

18. **CRITICAL THINKING** The time at the International Date Line is Midnight and 11:00 P.M. one time zone to the west. The days are different. When it is 2:00 A.M. at the International Date Line and 1:00 A.M. one time zone to the west, are the days now the same? Give your reasons.

Mixed Review

19. $14 billion ÷ $500 million

20. 866 × 10¢

21. Milton Kohl's job pays $39,400 in annual salary and 28% of his salary in fringe benefits. His total job expenses are $940 plus $140 a month for commuting costs. What are Milton's net job benefits?

22. You signed a promissory note at your credit union to borrow $4,200 for 6 months at 16% annual interest. What was the amount of interest on the loan? What total amount was due at the end of 6 months?

23. The monthly payments on a $68,500, 30-year mortgage loan are $561.23. Closing costs were $5,113. What total amount of interest was paid on the loan over 30 years?

24. The CPI for 1991 was 136.2. What was the purchasing power of the dollar in 1991, to the nearest cent?

25. Investments in a partnership were: Goldberg, $84,000; and Makoski, $118,000. The partners agree to pay each partner 6.5% interest on their investment and to divide the remainder of the net income equally. The net income for the year was $117,000. Find each partner's share of the net income.

13-3

International Measures of Length

GOALS

- Convert one metric unit of length to another
- Calculate using metric units of length
- Convert metric and customary measurements of length

KEY TERM

- meter

Start Up ▶▶▶

A highway sign in Canada gives the speed limit as 100 km/h. A friend tells you that km is a Canadian expression for miles, so the sign means you can drive at 100 miles an hour. You think you have to drive about 60 miles per hour to be below the speed limit? Who is right? Why?

Elena Elisseeva/ Shutterstock.com

Math Skill Builder

Review these math skills and solve the exercises that follow.

1 **Calculate** arithmetic operations on metric lengths.
Find the sum. 123 mm + 8.7 mm = 131.7 mm

1a. 0.96 km + 2.3 km = ? km

1b. 6 cm × 54 = ? cm

1c. 79 cm − 46 cm = ? cm

1d. 340 hm ÷ 17 = ? hm

2 **Multiply** whole numbers and decimals.
Find the product. 20 × 3.28 = 65.6

2a. 6.5 × 2.54

2b. 35 × 1.61

Metric Measures of Length

The *metric system* of measurement is used in most countries. In the United States, the *customary system* is used most often. Because a growing variety of products are manufactured in the U.S. for worldwide use, you should be familiar with basic units of metric measurements. These basic units are the meter (length), the liter (capacity), and the gram (weight or mass).

The metric system is based on the decimal system. Therefore, all relationships among metric units are based on 10. Prefixes before the basic unit indicate smaller and larger units and are used with all of the basic metric measurements. The prefixes and their order are shown in the metric length place value table.

1 000 m	100 m	10 m	1 m	0.1 m	0.01 m	0.001 m
kilometer	hectometer	decameter	meter	decimeter	centimeter	millimeter
km	hm	dam	m	dm	cm	mm

The basic unit of length in the metric system is the **meter**. One meter is equivalent to 3.28 feet, or slightly more than one yard in the customary system. The metric units of length, their symbols, and their values in meters are shown.

	Unit	Symbol	Value in Meters
Parts of a meter	**millimeter**	**mm**	0.001 m (one-thousandth meter)
	centimeter	**cm**	0.01 m (one-hundredth meter)
	decimeter	dm	0.1 m (one-tenth meter)
Basic unit	**meter**	**m**	1 m (one meter)
Multiples of a meter	decameter	dam	10 m (ten meters)
	hectometer	hm	100 m (one hundred meters)
	kilometer	**km**	1 000 m (one thousand meters)

(Commonly used units are in **bold** type.)

When writing metric amounts, follow these rules:

1. Write the symbol, not the unit name, in small letters. (18 millimeters = 18 mm)
2. Use the same symbol for singular and plural values. (3 centimeters = 3 cm; 0.03 centimeters = 0.03 cm; 1 centimeter = 1 cm)
3. To break up large numbers, use a space instead of a comma. (1200 m = 1 200; 4243 km = 4 243 km)

To change or convert metric measurements is easy because the metric system is a decimal system. Each position in the metric system is either 10 times more than, or one-tenth of, the next unit. To change units, multiply or divide by 10 as many times as needed to change to the unit you want.

Another way to change units is by moving the decimal point. When changing from a larger to a smaller unit, move the decimal point to the right. When changing from a smaller to a larger unit, move the decimal point to the left.

EXAMPLE 1

Change 94 m to centimeters.

SOLUTION

Use the metric table to count the number of places to the desired unit.

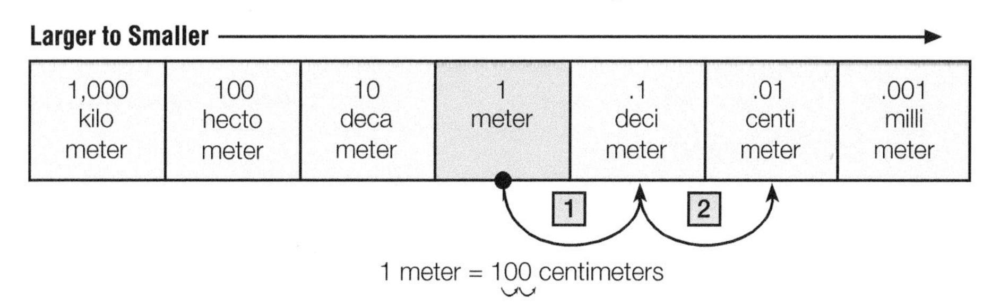

To move from meter (the larger unit) to centimeter (the smaller unit), you must move the decimal point two places to the right of meter (m).

94 m = 9 400 cm

EXAMPLE 2

Change 9 400 cm to kilometers.

SOLUTION
Use the table to count the number of places needed to move from cm to km.

Move the decimal point 5 places to the left.

9 400 cm = 0.094 km

EXAMPLE 3

Which is longer, 4 m or 485 cm?

SOLUTION
4 m = 400 cm   change 4 meters to centimeters

400 cm is less than 485 cm   compare numbers

485 cm is longer than 4 m

✔CHECK YOUR UNDERSTANDING

A. Change 16 m to mm and to hm.

B. Which is longer, 0.45 km or 450 dm?

Calculating With Metric Measures

Metric measurements are added, subtracted, multiplied, and divided in the same way as customary values. If the numbers have different units of measure, change measurements so that they are all in the same unit.

EXAMPLE 4

a. Add 6.85 m and 70 cm.

b. Multiply 3.5 cm by 800.

SOLUTION

a. 70 cm = 0.7 m   change cm to m

6.85 m + 0.7 m = 7.55 m   answer in meters

NOTE: *When computing with more than one metric measure, such as meters and centimeters, state your answer in the largest measure (the meter in this case), unless you are directed otherwise.*

b. 3.5 cm × 800 = 2 800 cm

✔CHECK YOUR UNDERSTANDING

C. Find the difference: 33 cm − 124 mm

D. Find the quotient: 2 100 km ÷ 7

Converting Metric and Customary Measures

Once in a while, you may find it necessary to convert a measurement from the metric system to the customary system or vice versa. Listed are some equivalencies between the metric and customary systems.

Metric to Customary	Customary to Metric
1 cm ≈ 0.39 inches	1 inch ≈ 2.54 cm
1 m ≈ 3.28 feet	1 foot ≈ 0.305 m
1 km ≈ 0.62 miles	1 mile ≈ 1.61 km

Math *Tip*

The symbol ≈ means approximate and is used when the equivalency of items is not exact, but nearly equal.

EXAMPLE 5

A sailboat is 8 m long. What is its length in feet?

SOLUTION

Multiply the meters by the correct equivalency.

1 m ≈ 3.28 feet

$8 \times 3.28 \approx 26.24$ feet   sailboat length

EXAMPLE 6

Elaine bicycled 7 miles. How many kilometers did she bicycle?

SOLUTION

Multiply the miles by the correct equivalency.

1 mile ≈ 1.61 km

$7 \times 1.61 \approx 11.27$ km   kilometers bicycled

Photodisc/Getty Images

✔ CHECK YOUR UNDERSTANDING

E. The length of a ribbon is 243 cm. What length is the ribbon in inches?

F. The distance from home plate to first base is 90 feet. What is the equivalent distance in meters?

Wrap Up ▶▶▶

The 100 km means 100 kilometers an hour, not miles per hour. Since a kilometer is equivalent to about 0.62 miles, the 100 km is equal to about 62 miles per hour.

TEAM Meeting

Form a group and visit retail stores, study catalogs, or do Internet research to find 10 items that are sold by metric measurements of length or that list metric measures on the items' labels or wrappings. Produce a table that gives the name of the items, the metric measures, and the customary measures. Share your table with the class.

Exercises

Find the unknown length.

1. 1 m = ? cm **2.** 1 m = ? mm **3.** 1 km = ? m **4.** 1 cm = ? m

5. 1 dm = ? m **6.** 1 mm = ? m **7.** 6 mm = ? cm **8.** 900 mm = ? m

9. 368 m = ? km **10.** 7 m = ? cm **11.** 5 m = ? mm **12.** 9 cm = ? mm

13. 3 km = ? m **14.** 700 m = ? km **15.** 290 cm = ? m **16.** 1 085 cm = ? m

Find the result.

17. 56 mm − 14.5 mm **18.** 12 m + 187 cm **19.** 4.1 cm × 4

20. 70 km × 8 **21.** 4.8 mm ÷ 12 **22.** 7 320 m ÷ 6

23. 23 × 120 cm **24.** 1 280 mm ÷ 64 **25.** (55 m + 45 m) ÷ 2

26. 3 cm + 4 cm + 1 cm **27.** 6.2 km − 4 500 m

28. 14.1 km + 930 m **29.** 345 cm − 753 mm

30. 23 m + 74 cm + 6 829 mm **31.** 115 m − 960 cm

Find the equivalent measures of length.

32. 2.5 cm ≈ ? inches **33.** 2,200 miles ≈ ? km

Which measure is longer?

34. 0.4 km or 4 100 m **35.** 0.6 m or 6 000 cm

36. 53 mm or 53 dm **37.** 1 040 mm or 1.4 cm

38. 7 832 m or 78.32 cm **39.** 4 700 m or 4.64 km

Solve.

40. Lena Hailey needed these lengths of rope: 12 m, 3.5 m, 2.6 m. What total amount of rope did Lena need?

41. A company used 312 m of fiber optic cable from a spool that originally held 1 200 m of cable. How much cable was left on the spool?

42. A trip of 600 km was estimated to take 7.5 hours. In the first 6.5 hours, 524 km were driven. How many kilometers must be driven in one hour to meet the trip's estimate?

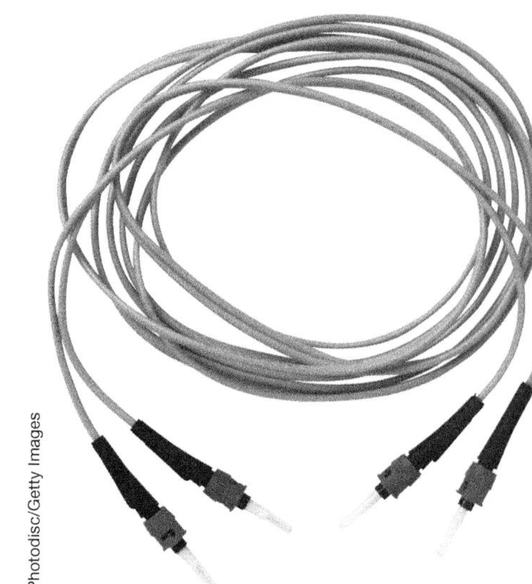
Photodisc/Getty Images

43. Four students reported the length of material they used for a project as: 0.5 m, 25 cm, 400 mm, 18.4 cm. What total length of material in meters did they use?

44. Laura Hiller is cutting pieces of tubing to fill an order. The order is for these pieces and lengths of tubing: 12 pieces, 180 cm long; 30 pieces, 54 cm long; 126 pieces, 1.2 m long. What total length of tubing, in meters, does Laura need to fill the order?

45. The length of a car is 5.5 m. What is its length in feet?

46. A circle has a diameter of 18 cm. What is the circle's diameter in inches?

47. A four-year-old car had been driven for a recorded distance of 105,000 km. What was the distance driven in miles?

On a business trip, Rudy Komiko drove these distances in three days: 216 km, 237 km, 165 km.

48. How many kilometers did Rudy drive in those three days?

49. What estimated average number of kilometers did he drive per day?

50. What actual average number of kilometers did he drive each day?

51. A box measures 14 inches long by 8 inches wide by 7 inches tall. Will a book that measures 34 cm by 20.8 cm by 3 cm fit inside the box completely?

52. **CRITICAL THINKING** Unlike other countries of the world, the United States uses primarily the customary system in everyday life. What would have to happen for this situation to change?

53. **FINANCIAL DECISION MAKING** B & K Manufacturing uses customary measures but wants to do business with a company in England that will only purchase items made to metric measure. What problems might this present to B & K?

Mixed Review

54. What is the value of X when $\frac{10}{5} = \frac{12}{X}$?

55. Rewrite $\frac{3}{4}$ as a decimal.

56. $436 - 27.4$

57. $\frac{7}{8} \div \frac{3}{4}$

58. 1,752 × 10¢

59. Round $11.947 to the nearest cent and to the nearest dollar.

60. A van that costs $24,500 and has a residual value of $16,000 can be leased for $520 monthly with a $1,500 down payment. The van may be purchased with a $6,800 down payment and $891 monthly payments. The lease and loan terms are 24 months. Which costs more, leasing or buying? How much more?

61. A company uses the sum-of-the-years-digits method of calculating depreciation. The company buys a machine for $45,000 and expects to use it for 5 years and then sell it for $21,000. What are the total depreciation and book value of the machine at the end of the first year?

62. The Easy Tread Company had sales of $83,100 in September. Sales returns and allowances for the same month were $315.40. What were Easy Tread's net sales for September?

63. Yetzin and Golub are business partners and share net income in a ratio of 7 to 5, respectively. What is each partner's share of a net income of $12,852?

64. Cheryl Moores works full time for an annual salary of $39,560. She also works part-time 5 hours each week at a job that pays $17.45 an hour. Both jobs pay bi-weekly. What are gross wages each pay period for the two jobs combined?

13-4

International Measures of Area

GOALS

- Convert one metric unit of area to another
- Calculate using metric units of area
- Convert metric and customary measurements of area

KEY TERMS

- area
- square meter

Start Up

You notice that a package of plastic food wrap you bought contains 100 square feet or 9.2 square meters of wrap. You bought the product in the United States. Why would the package give both customary and metric measures?

Ragne Kabanova/ Shutterstock.com

Math Skill Builder

Review these math skills and answer the questions that follow.

1. **Add** or **Subtract** as indicated. Write answers using the larger unit.
   Find the sum. 2 m + 100 cm = 2 m + 1 m = 3 m
   Find the difference. 12 hm − 38 dam = 12 hm − 3.8 hm = 8.2 hm

   1a. 14 cm + 1 420 mm **1b.** $8\ hm^2 + 1.6\ hm^2$

   1c. $8\ km^2 - 5\ km^2$ **1d.** 6.75 cm − 50 mm

2. **Multiply** or **Divide** as indicated.
   Find the product. $16\ m^2 \times 4 = 64\ m^2$
   Find the quotient. 96 km ÷ 4 km = 24

   2a. $12\ mm^2 \times 3$ **2b.** 35 × 4 m

   2c. 45 mm ÷ 5 **2d.** 26 cm ÷ 13 cm

3. **Multiply** whole numbers and decimals.
   Find the product. 2 × 10.8 ft = 21.6 ft

   3a. 50 × 0.16 in. **3b.** 3 × 0.4 m

Metric Measures of Area

The amount of surface an item has is called its **area**. To find the area of any rectangle, such as a desk or floor, multiply its length by its width.

Area = Length × Width **or** $A = L \times W$

Calculating area is a skill used in many businesses. For example, the owner of a paving company must find the area of a rectangular parking lot to estimate how much to charge for paving the lot. If a parking lot is 120 feet long by 60 feet wide, its area is 120 ft × 60 ft or 7,200 ft^2.

Area measures the space covered by two dimensions, so area is expressed in terms of square units. In the customary system, area may be in square inches, square feet, square yards, or square miles, depending on what is being measured. For example, a tablecloth may cover a certain area measured in square inches. The area covered by a box of floor tiles can be measured in square feet. Carpeting is sold by the square yard, while forests may be measured in square miles.

In the metric system, the basic unit of area is the **square meter**. The square meter is equivalent to approximately 1.2 square yards in the customary system. Square meter is written as m^2. The word "square" is not written. It is named by the exponent "2" that appears to the right and above the m notation for meter.

	Unit	Symbol	Value in Square Meters
Parts of a square meter	**square millimeter**	**mm^2**	0.000 001 m^2 (one-millionth square meter)
	square centimeter	**cm^2**	0.000 1 m^2 (one-ten-thousandth square meter)
	square decimeter	dm^2	0.01 m^2 (one-hundredth square meter)
Basic unit	**square meter**	**m^2**	1 m^2 (square meter)
Multiples of a square meter	square decameter	dam^2	100 m^2 (one hundred square meters)
	square hectometer or **hectare**	hm^2 or **ha**	10 000 m^2 (ten thousand square meters)
	square kilometer	**km^2**	1 000 000 m^2 (one million square meters)

(Commonly used units are in **bold** type.)

Square centimeter and square meter are the measures of area used most often. Large areas are measured in either hectares (ha) or square kilometers (km^2). Hectare is another name for square hectometer. Hectare is used because it is easier to say and write.

In metric measures of area, each square unit is *100 times* the next smaller unit, or *one hundredth* of the next larger unit. This is shown in the illustration at the right. Remember that to determine area, multiply length by width.

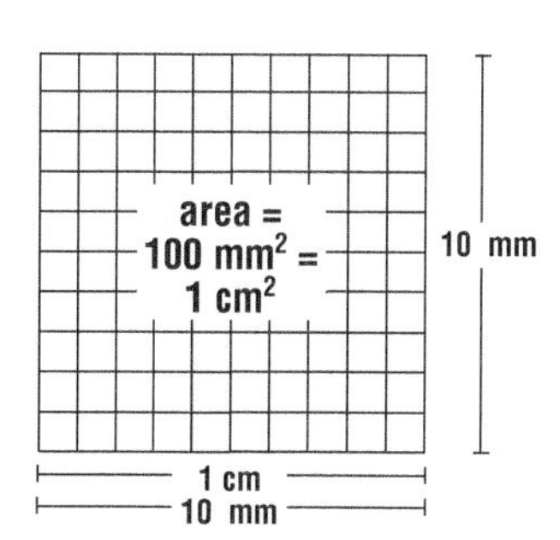

1 cm × 1 cm = 1 cm^2

Because 1 cm = 10 mm, the area can also be expressed as:

10 mm × 10 mm = 100 mm^2.

So 1 cm^2 = 100 mm^2.

Metric measures of area may be changed by moving the decimal point in a way similar to changing metric measures of length. A metric area place value table is shown on the next page.

$1\ 000\ 000\ m^2$	$10\ 000\ m^2$	$100\ m^2$	$1\ m^2$	$0.01\ m^2$	$0.000\ 1\ m^2$	$0.000\ 001\ m^2$
kilo-meter²	hectare or hectometer²	deca-meter²	meter²	deci-meter²	centi-meter²	milli-meter²
km^2	ha or hm^2	dam^2	m^2	dm^2	cm^2	mm^2

To change from a metric unit of area to the next smaller unit, you move *one place* on the area place value table and move the decimal point *two places* to the right. To change to the next larger unit of area, move the decimal two places to the left.

EXAMPLE 1

Change the metric measures of area shown below to the unit indicated.

a. $1\ cm^2 = ?\ mm^2$ **b.** $1\ m^2 = ?\ cm^2$ **c.** $100\ mm^2 = ?\ cm^2$ **d.** $10\ 000\ cm^2 = ?\ m^2$

SOLUTION

To change to a smaller unit, move the decimal point two places to the right for each place you move on the metric area place value table.

a. $1\ cm^2 = 100\ mm^2$ The decimal point is moved 2 places to the right.

b. $1\ m^2 = 10\ 000\ cm^2$ The decimal point is moved 4 places to the right.

To change to a larger unit, move the decimal point two places to the left for each place you move on the metric area place value table.

c. $100\ mm^2 = 1.00\ cm^2$ The decimal point is moved 2 places to the left.

d. $10\ 000\ cm^2 = 1\ m^2$ The decimal point is moved 4 places to the left.

✔ CHECK YOUR UNDERSTANDING

A. Change to the smaller measure indicated: $1\ km^2 = ?\ m^2$

B. Change to the larger measure indicated: $1\ 000\ m^2 = ?$ ha

Calculating With Metric Measures

When you add or subtract square units, the unit in the measure remains the same. Measurements must have the same units in order to be added or subtracted.

When two measures in millimeters are multiplied, the result is mm^2. When a square unit, such as cm^2, is multiplied by a number, the answer is given in cm^2. When square units, such as km^2, are divided by a measurement, such as km^2, the quotient is in units, not square units.

Math *Tip*

When calculating with metric measures, such as km^2 and m^2, state your answer in the largest measure (km^2 in this case), unless you are otherwise directed.

EXAMPLE 2

Solve each problem.

a. $12\ km^2 + 600$ ha **b.** $18\ cm^2 - 700\ mm^2$

c. $8\ cm \times 4\ cm$ **d.** $6\ km^2 \times 3$

e. $48\ mm^2 \div 12$ **f.** $48\ km^2 \div 24\ km^2$

SOLUTION

a. $12 \text{ km}^2 + 600 \text{ ha} = 12 \text{ km}^2 + 6 \text{ km}^2 = 18 \text{ km}^2$

b. $18 \text{ cm}^2 - 700 \text{ mm}^2 = 18 \text{ cm}^2 - 7 \text{ cm}^2 = 11 \text{ cm}^2$

c. $8 \text{ cm} \times 4 \text{ cm} = 32 \text{ cm}^2$

d. $6 \text{ km}^2 \times 3 = 18 \text{ km}^2$

e. $48 \text{ mm}^2 \div 12 = 4 \text{ mm}^2$

f. $48 \text{ km}^2 \div 24 \text{ km}^2 = 2$

✔ CHECK YOUR UNDERSTANDING

C. What is the sum of 7 cm² and 400 mm²? What is the difference of $13 \text{ km}^2 - 1\ 200 \text{ ha}$?

D. What is the product of $9 \text{ mm} \times 2 \text{ mm}$? What is the product of $5 \text{ ha} \times 11$? What is the quotient of $25 \text{ km}^2 \div 5$?

Algebra *Tip*

Two related conversion formulas are shown below.

$1 \text{ cm}^2 = 0.16 \text{ in}^2$

$1 \text{ in}^2 = 6.5 \text{ cm}^2$

If the accuracy of the formulas is not as precise as you might want, search the Internet for equivalencies that show more decimal places.

Converting Metric and Customary Measures

You may occasionally find it necessary to convert from the metric system to the customary system or vice versa.

Metric to Customary	Customary to Metric
$1 \text{ cm}^2 \approx 0.16 \text{ in.}^2$	$1 \text{ in.}^2 = 6.5 \text{ cm}^2$
$1 \text{ m}^2 \approx 10.8 \text{ ft}^2$	$1 \text{ ft}^2 \approx 0.09 \text{ m}^2$
$1 \text{ ha} \approx 2.5 \text{ acres}$	$1 \text{ acre} \approx 0.4 \text{ ha}$

Photodisc/Getty Images

EXAMPLE 3

Alvin bought a roll containing 7 m² of gift-wrap. How many square feet of gift-wrap are in the roll?

SOLUTION

Multiply the number of square feet in one square meter by the number of square meters.

$1\text{m}^2 \approx 10.8 \text{ ft}^2$ equivalency for one square meter

$7 \times 10.8 \text{ ft}^2 \approx 75.6 \text{ ft}^2$ paper in one roll

EXAMPLE 4

A stock clerk found a carpet remnant that covers 360 square feet of floor space. How many square meters will the remnant cover?

SOLUTION

Multiply the number of square meters in one square foot by the number of square feet.

$1 \text{ ft}^2 \approx 0.09 \text{ m}^2$ equivalency for one square foot

$360 \times 0.09 \text{ m}^2 \approx 32.4 \text{ m}^2$ size of remnant

✔CHECK YOUR UNDERSTANDING

E. Maria Blasko bought a can of spray paint that will cover a 1.8 m^2 surface. How many square feet will the paint cover?

F. A computer monitor occupies a space of 272 square inches. This space is equal to how many square centimeters?

Wrap Up ▶▶▶

Although you purchased the product in the United States, it is possible the product is sold in other countries that use the metric system. Many products sold in the U.S. and other countries include labeling in more than one language for those people that are bilingual. Instead of printing two different packages, the manufacturer places all the information on one package and saves the expense of keeping the packages separate.

Exercises

Find the missing area.

1. 1 cm^2 = ? mm^2
2. 1 m^2 = ? cm^2
3. 86 km^2 = ? ha
4. 960 cm^2 = ? m^2
5. 26 ha = ? m^2
6. 2 000 mm^2 = ? m^2

Find the sum or difference.

7. 3 cm^2 + 900 mm^2 + 1.5 m^2
8. 6 km^2 + 120 ha
9. 9 m^2 + 7 000 cm^2
10. 10 000 m^2 + 6 ha
11. 4.5 km^2 − 18 ha
12. 5 cm^2 − 100 mm^2
13. 4 km^2 − 1 000 000 m^2
14. 1 ha − 6 000 m^2

Find the product or quotient.

15. 827 mm^2 × 5
16. 100 m × 100 m
17. 3 km^2 × 0.5
18. 0.4 × 58 cm^2
19. 240 km^2 ÷ 60 km^2
20. 0.75 m^2 ÷ 3
21. 156 mm^2 ÷ 39 mm^2
22. 920 cm^2 ÷ 115 cm^2

Find the equivalent measure of area.

23. 25 ha ≈ ? acres
24. 16 ft^2 ≈ ? m^2
25. 140 cm^2 ≈ ? $in.^2$
26. 2,000 acres ≈ ? ha

Solve.

27. The Liverpool Home Building Company bought these tracts of land: 9 ha, 140 ha, 73 ha, 1 168 ha. Find the total amount of land they purchased.
28. Roland wants to buy carpeting for a room that measures 3 m × 4.1 m. What amount of carpeting will he need?
29. Keisha Beard's yard is 44 m long by 25 m wide. She turned part of her yard into a garden 12 m long by 5 m wide. She also has a 3 m by 3 m shed on her lawn. She has to mow the rest of the lawn. How much lawn does she mow?

30. Umberto inherited a 450 ha farm in Argentina. What is the equivalent size of the farm in acres?

31. A paper mill produced 190 000 m^2 of card stock last month. What was last month's card stock production in square feet?

32. Rhonda Issel plans to make and sell decorative pillows that are about 1.5 ft^2. The pattern for the pillow gives the area of a pillow in centimeters. To the next highest 100, how many cm^2 are equal to 1.5 ft^2?

Photodisc/Getty Images

INTEGRATING YOUR KNOWLEDGE A box holds 45 floor tiles. Each tile measures 30.5 cm by 30.5 cm.

33. What does a tile measure, in inches, to the nearest inch?

34. Estimate the area, in cm^2, covered by a tile?

35. What area, to the nearest square meter, will a box of tiles cover?

36. What area, to the nearest ft^2, will a box of tiles cover?

37. **FINANCIAL DECISION MAKING** A carpet company quotes an installed price of $5 a square foot for a type of carpeting. Another company quotes an installed price of $50 a square meter for the same type of carpeting. The room to be carpeted measures 24 feet by 15 feet. Which is the better offer? How much will you save by taking the better offer?

Mixed Review

38. $\frac{7}{8} \times \frac{16}{35} \times \frac{5}{6}$

39. $6\frac{1}{2} - 3\frac{1}{2}$

40. How many days are there between June 27 to August 14?

41. $27.9 - 15$

42. $0.035 \times \$516.12$

43. What percent of 90 is 27?

44. Write 1.2 as a fraction.

45. Round $872,598 to the nearest ten thousand.

46. Valerie Clay-Gelen's credit card statement for April showed these items: 4/1, previous balance, $218.06; 4/12, purchase, $119; and 4/18, payment, $200. Valerie's credit card company uses a 1.5% monthly periodic rate and the average daily balance method including new purchases. What is the finance charge for April? What is Valerie's new balance?

47. Zetex provides its employees with a disability policy. The policy pays a benefit of 5% of wages plus $1\frac{1}{4}$% for each of the 14 years that Max Randolph has been employed with Zetex. What is his benefit percent?

48. An invoice for $2,050 dated December 12 has credit terms of 1.5/10, n/30. The invoice is paid on December 21. What amount should have been paid?

13-5

International Measures of Capacity and Weight

GOALS

- Convert one metric unit of capacity to another
- Convert one metric unit of weight to another
- Calculate using metric units of capacity and weight
- Convert metric and customary measures of capacity and weight

KEY TERMS

- liter
- gram

Start Up

You pick up a suitcase in a store and notice its label gives the weight of the case in kilograms. The label on a second suitcase of the same size states its weight in pounds. The second suitcase feels lighter than the first when you pick it up. Might this be your imagination at work? How might you compare the weights?

Photodisc/Getty Images

Math Skill Builder

Review these math skills and solve the exercises.

1. **Convert** one measure of metric length to another measure of metric length. Find the capacity. 21 mm = ? m = 0.021 m

   1a. 134 mm = ? dm **1b.** 1.3 km = ? m

2. **Calculate** arithmetic operations on metric capacity and weight. Find the sum. 17 cg + 17 cg = 34 cg

   2a. 9 L + 200 L **2b.** 9 kg − 2.8 kg

   2c. 960 L ÷ 12 **2d.** 15 000 g ÷ 600

Metric Measure of Capacity

The basic metric measure of capacity is the **liter**. A liter is slightly larger than a quart. The symbol for the liter is a capital L. Most small capacity measures are shown in milliliters or as decimal parts of a liter, such as 400 mL or 0.4 L. To change from one measure of capacity to another you move the decimal point as you have done with metric lengths.

	Unit	Symbol	Value in Liters
Parts of a liter	**milliliter**	**mL**	0.001 L (one-thousandth liter)
	centiliter	cL	0.01 L (one-hundredth liter)
	deciliter	dL	0.1 L (one-tenth liter)
Basic unit	**liter**	**L**	1 L (one liter)
Multiples of a liter	decaliter	daL	10 L (ten liters)
	hectoliter	hL	100 L (one hundred liters)
	kiloliter	**kL**	1 000 L (one thousand liters)

(Commonly used units are in **bold** type.)

1 000 L	100 L	10 L	1 L	0.1 L	0.01 L	0.001 L
kilo-liter	hecto-liter	deca-liter	liter L	deci-liter	centi-liter	milli-liter
kL	hL	daL		dL	cL	mL

EXAMPLE 1

Change the metric measures of capacity to the unit indicated.

a. 38 daL to liters

b. 36 000 mL to liters

SOLUTION

To change to a smaller unit, move the decimal point to the right.

a. 38 daL = 380 L   The decimal point is moved one place to the right.

To change to a larger unit, move the decimal point to the left.

b. 36 000 mL = 36 L   The decimal point is moved three places to the left.

✔ CHECK YOUR UNDERSTANDING

A. Change 4 L to milliliters.

B. Change 180 cL to liters.

Metric Measure of Weight

In this text and in everyday conversation, *weight* and *mass* are used interchangeably. Technically, however, mass is a measurement of the quantity of matter. Weight is a force exerted by gravitational pull against a mass. That is why astronauts on the moon weigh approximately $\frac{1}{6}$ of their weights on Earth. An astronaut's mass, however, is the same on Earth and in space.

	Unit	Symbol	Value in Grams
Parts of a gram	**milligram**	**mg**	0.001 g (one-thousandth gram)
	centigram	cg	0.01 g (one-hundredth gram)
	decigram	dg	0.1 g (one-tenth gram)
Basic unit	**gram**	**g**	1 g (one gram)
Multiples of a gram	decagram	dag	10 g (ten grams)
	hectogram	hg	100 g (one hundred grams)
	kilogram	**kg**	1 000 g (one thousand grams)
	metric ton	**t**	1 000 kg (one thousand kilograms)

(Commonly used units are in **bold** type.)

In the metric system, the **gram** is the basic unit of weight. One kilogram, or 1 000 grams, is equivalent to approximately 2.2 customary pounds. A *gram* is equivalent to about 0.04 of a customary ounce.

Usually grams or milligrams are used to measure items with a small weight or mass. Kilograms or metric tons are used to measure large items. A metric ton, *t*, is equal to 1 000 kg.

Use the metric weight place value table shown to count the number of places you need to move the decimal point to change from one weight measure to another.

1 000 g	100 g	10 g	1 g	0.1 g	0.01 g	0.001 g
kilo-gram	hecto-gram	deca-gram	gram	deci-gram	centi-gram	milli-gram
kg	hg	dag	g	dg	cg	mg

EXAMPLE 2

Change the metric measures of weight to the unit indicated.

a. 3 g to centigrams **b.** 762 dag to kilograms

SOLUTION

To change to a smaller unit, move the decimal point to the right.

a. 3 g = 300 cg The decimal point is moved two places to the right.

To change to a larger unit, move the decimal point to the left.

b. 762 dag = 7.62 kg The decimal point is moved two places to the left.

✔CHECK YOUR UNDERSTANDING

C. Change 5 cg to milligrams.

D. Change 19 000 kg to metric tons

Photodisc/Getty Images

Calculating With Metric Measures

You can add, subtract, multiply, and divide capacity and weight measures in the same way as other numbers. When measures are in different units, convert them to the same unit. Remember to state your answer in the larger unit.

EXAMPLE 3

Perform the indicated operation.

a. 18 L + 250 cL **b.** 56 kL − 2 000 L **c.** 24 L × 15

d. 2 300 g × 1.4 **e.** 620 mg ÷ 0.5 **f.** 625 mg ÷ 4 mg

SOLUTION

a. 18 L + 250 cL = 18 L + 2.5 L = 20.5 L

b. 56 kL − 2 000 L = 56 kL − 2 kL = 54 kL

c. 24 L × 15 = 360 L

d. 2 300 g × 1.4 = 3 220 g

e. 620 mg ÷ 0.5 = 1 240 mg

f. 625 mg ÷ 4 mg = 156.25

✔CHECK YOUR UNDERSTANDING

E. Find the sum: 2 L + 400 mL

F. Find the difference: 19 g − 20 mg

Converting Metric and Customary Measures

As you did with length and area, use the appropriate conversion figure to convert metric capacity and metric weight measures to customary measures, and vice versa. Commonly used capacity and weight conversions are shown below.

Metric to Customary	Customary to Metric
1 L ≈ 1.06 quarts	1 quart ≈ 0.95 L
1 L ≈ 0.26 gallons	1 gallon ≈ 3.79 L
1 g ≈ 0.035 ounce	1 ounce ≈ 28.3 g
1 kg ≈ 2.2 pounds	1 pound ≈ 0.45 kg

EXAMPLE 4

What is the equivalency in liters of a bottle of bleach whose capacity is 0.75 gallons?

SOLUTION

Multiply the gallons by the correct equivalency.

0.75 gallons × 3.79 ≈ 2.84 L capacity of bottle

EXAMPLE 5

A jar of olives has a net weight of 70 g. What is the equivalent weight in ounces?

SOLUTION

Multiply the grams by the correct equivalency.

70 g × 0.035 ≈ 2.45 oz   net weight of olive jar

✔CHECK YOUR UNDERSTANDING

G. While traveling, you buy 54 L of gasoline. This purchase is equivalent to how many gallons of gasoline?

H. You calculate from a European recipe that you need 240 kg of flour to bake a certain type of bread for a town fundraiser. How many pounds of flour must you buy?

Wrap Up ▶▶▶

If there is a sizable difference in the weight of the cases, you may be correct in saying one case is lighter than the other. The best way to make comparisons is to change the weights to the same measures, either metric or customary.

Exercises

Change the metric measures to the unit indicated.

1. 1 kL = ? L
2. 1 mL = ? kL
3. 600 mL = ? L
4. 470 L = ? kL
5. 1 900 mg = ? g
6. 740 g = ? kg
7. 0.65 kg = ? g
8. 500 kg = ? t
9. 1 mL = ? L

Perform the indicated operation.

10. 400 mL + 34.6 mL
11. 3 t + 1.5 t + 54 kg
12. 35.23 kL − 35 000 L
13. 7 450 mg − 2.09 g
14. 22 × 23.5 kL
15. 4% × 1 800 kg
16. 100 mL ÷ 8
17. 9 t ÷ 0.2
18. 2 004 t ÷ 12 t
19. 913.5 mL ÷ 7 mL

Find the equivalent measures of capacity and weight.

20. 6 quarts ≈ ? L
21. 12 pounds ≈ ? kg
22. 2 L ≈ ? quarts
23. 24 ounces ≈ ? g

Solve.

24. A tank holds 150 kL of fuel oil when full. On Monday, the tank was 90% full. These amounts of fuel were pumped from the tank in the next three days: 33 500 L, 27 000 L, and 22 450 L. How many kL of fuel oil were left in the tank at the end of the third day?

25. Jacob Stearns is providing refreshments for the 12 cast members of a dance group. If Jacob buys 15 liters of juice, how many liters of juice will there be for each cast member?

26. A box contains 240 writing pads. The pads alone weigh 31.2 kg and the box weighs 3 kg. How many grams does each steno pad weigh?

27. A full box of crackers weighed 0.410 kg. The box is now $\frac{1}{2}$ full. What is the weight in grams of the crackers left?

28. Regal hand soap is sold in 8-bar packs. Each pack weighs 1.136 kg. Hand soap made by Rose Mist is sold in 3-bar packs with a total weight of 405 g. What is the average weight in grams of one bar of each brand of soap?

29. To keep hydrated during a bike race, racers were advised to drink 2.5 L of water during the race. There are 32 ounces in a quart. How many ounces of water should each racer drink?

A 354 mL can of frozen concentrate makes 0.94 liters of lemonade.

30. How many liters of lemonade can you make from 6 cans?

31. How many cans of concentrate would you need to make 4.7 liters of lemonade?

Northern Products produced 6 kiloliters of syrup. The syrup is poured into one-fourth liter size bottles and is packed 48 bottles to a case.

32. What total number of bottles may be filled with the syrup produced?

33. If all the bottles are packed in cases, how many cases will there be?

34. **CRITICAL THINKING** Since many products now show customary and equivalent metric measures, do you think that the people who use such products gradually learn metric measures? What are your reasons?

Mixed Review

35. What number is $\frac{3}{8}$ less than $205.68?

36. Cedric Marshall spends 8.5% of his $46,000 annual income on commuting costs. He wants to cut that cost to $40 a week by using the subway instead of driving. How much less will Cedric spend in a year on commuting by making this change?

Photodisc/Getty Images

Chapter *Review*

Vocabulary Review

Find the term, from the list at the right, that completes each sentence. Use each term only once.

- area
- domestic business
- exporting
- foreign exchange rate
- gram
- Gross Domestic Product (GDP)
- importing
- international business
- liter
- meter
- per capita GDP
- square meter
- time zone
- trade deficit
- trade surplus

1. A(n) __?__ results when a country buys more goods from other countries than it sells to other countries.
2. The basic metric measure for mass is known as the __?__.
3. The purchase of goods and services among countries is referred to as __?__.
4. The basic unit of length or distance in the metric system is called the __?__.
5. The amount of currency of one country that can be exchanged for another country's currency is determined by the __?__.
6. Buying goods from other countries is called __?__.
7. All the trade that takes place within a country's borders is called __?__.
8. In the metric system, the basic unit of capacity is known as the __?__.
9. The result you get by multiplying a rectangle's length by its width is called __?__
10. The result of dividing a country's Gross Domestic Product by its population is known as __?__.

13-1 Production, Trade, and Finance

11. The GDP of Mongolia is $5.8 billion. Its population is 2.62 million. What is its per capita GDP, to the nearest dollar?
12. In one year, Saudi Arabia had imports of $30.4 billion and exports of $86.5 billion. What was the amount of their trade surplus or trade deficit?
13. The exchange rate for the Chilean peso is $0.0019117. What is the equivalent in U.S. dollars of 4,000 Chilean pesos, to the nearest cent?

13-2 International Time and Temperature

14. A business call is made on Wednesday from Los Angeles, California to Tokyo, Japan at 5:00 P.M. Los Angeles time. Find what time it is in Tokyo.
15. The average low temperature in January in Green Bay, Wisconsin is 7°F. What is the temperature in Celsius, to the nearest whole degree?
16. The high temperature for the day in Dublin, Ireland was reported as 17°C. To what Fahrenheit temperature is this equivalent, to the nearest degree?

13-3 International Measures of Length

17. How many kilometers are in 5,000 m?
18. 530 mm is equal to how many centimeters?
19. Find the value of a in this equation: $a = 5(2 \text{ m} + 125 \text{ cm})$
20. Find the value of b in this equation: $b = (23 \text{ km} - 19\ 000 \text{ m}) \div 2.5$
21. Find the number of meters in 25 ft.
22. Find the number of inches in 46.1 cm.
23. The Just Right Tailor shop buys twelve 500-m spools of black thread every three months. How many kilometers of thread does the shop buy in one year?

13-4 International Measures of Area

24. A farm measures 40 000 m^2. What is its measure in hectares?
25. The inland waterways of a European country measure 19 km^2. What is this measure in square meters?
26. Find the value of c in this equation: $c = 3(49 \text{ m}^2 - 15\ 000 \text{ cm}^2)$
27. Find the value of d in this equation: $d = 160 \text{ ha} - (6 \text{ km}^2 \div 4)$
28. Find the number of hectares in 840 acres.
29. Find the number of square inches in 540 cm^2.
30. Pam's living room is 14 feet by 16 feet. She has a grand piano that covers an area of about 4 square meters. Her entertainment center covers about 2.16 square meters. About how many square feet of floor space remains for Pam to place other furniture in her living room?

13-5 International Measures of Capacity and Weight

31. How many kiloliters are in 24 000 L?
32. What number of centiliters is 212 mL equal?
33. A shipment weighing 6 700 kg is equal to what number of metric tons?
34. What weight in grams is a packet of spice that weighs 40 000 mg?
35. Find the value of e in this equation: $e = 36(22 \text{ L} - 1\ 200 \text{ cL})$
36. Find the value of f in this equation: $f = (62 \text{ g} + (10 \times 400 \text{ mg})) \div 4$
37. Find the number of liters in 4 quarts and the number of gallons in 12 L.
38. Find the number of grams in 16 oz.
39. Find the number of pounds in 250 kg.
40. The Zealan Company is shipping 1,500 roasting timers. The weight of one timer and its shipping box is 730 g. What is the total weight of this shipment in kilograms?

A certain model of a car used to weigh 1 200 kg. The car has been redesigned so that its total mass is now 1 170 kg.

41. By how many kg was the weight of the car reduced?
42. What percent of the car's original weight is this reduction?

Technology Workshop

Task 1 Currency Exchange

Enter data into a template that will calculate the U.S. currency equivalent to a foreign currency, given an exchange rate in dollars.

Open the spreadsheet for Chapter 13 (tech13-1.xls) and enter the data shown in blue (cells B2-3) into the spreadsheet. The spreadsheet assumes that a hotel room was rented in Sydney, Australia for four days and the total bill came to 685.16 AUD. Your computer screen should look like the one shown below when you are done.

	A	B
1	**Currency Exchange Calculator**	
2	Units of Foreign Currency	685.16
3	Exchange Rate in U.S. Dollars	$0.676221
4	U.S. Currency Equivalent	$463.39

Task 2 Analyze the Spreadsheet Output

Answer these questions about the spreadsheet.

1. Why was the hotel bill entered as 685.16 instead of 685.16 AUD?
2. What does the exchange rate mean?
3. To what amount in U.S. currency was the hotel bill equal?
4. If the hotel room rate does not change during the next year, would the U.S. currency equivalent remain the same for a four-day stay? Give your reasons.
5. Why was the U.S. currency output rounded to the nearest cent and not the nearest dollar, as $463?

Now assume that a currency trader exchanged AU $300,000,000 AUD for U.S. dollars on July 1 at the exchange rate of $0.676221. Move the cursor to B2 and enter 300,000,000 without the commas.

6. What amount of U.S. dollars did the trader receive from this exchange?
7. Now assume that by one week later, on July 8, the value of a Australian Dollar dropped by 1.5¢ in relation to the U.S. dollar. What would be the new exchange rate?
8. Enter the new exchange rate to find the amount in U.S. dollars that could be purchased with $300,000,000 AUD. What is the amount?
9. What do the answers to Problems 6 and 8 suggest?

Task 3 Design a Spreadsheet to Convert Temperatures

Design a spreadsheet that will convert Fahrenheit temperature to Celsius, and Celsius to Fahrenheit. The spreadsheet should show the equivalent temperature at the right of each temperature entered. Use the conversion formulas that appear in Lesson 13-2.

SITUATION: The low temperature for a day in January in Duluth, Minnesota is expected to be 21°F. You receive an email from a friend who lives in Geneva, Switzerland. She reports that the low temperature in Geneva for the same day is expected to be –2°C. Compare the two temperatures.

Task 4 Analyze the Spreadsheet Output

Answer these questions about your completed spreadsheet.

10. To what Celsius temperature is 21°F equivalent?

11. What is the equivalent Fahrenheit temperature for –2°C?

12. In which city will it be warmer, at the low temperature for the day?

13. The temperature during the day in Deluth is supposed to increase 20°F, while in Geneva, the temperature is supposed to increase 14°C. Which city is supposed to have a greater temperature increase during the day?

You may want to use the spreadsheet to verify the temperature conversions you have already solved in Lesson 13-2.

Photodisc/Getty Images

Chapter *Assessment*

Chapter Test

Answer each question.

1. Write out $1.072 million as a numeral.
2. Multiply 55×0.3642 and round to the nearest tenth.
3. Solve $2{,}900 \div 0.3642$ and round to the nearest hundredth.
4. Add 7 hours to 10:14 A.M.
5. 15 cm + 900 mm + 0.014 m
6. 60 km ÷ 0.4 km
7. 24% of 6.5 kL
8. What number is $\frac{1}{8}$ more than 160 L?
9. Find what percent 340 mg is of 200 mg.
10. $1\frac{3}{4} \times 76.2$ kg

Solve.

11. In a recent year, a country had a Gross Domestic Product of $4,200,000,000 and a population of 1,200,000. What was its per capita GDP?
12. Last year a country had exports of $1.04 billion and imports of $924 million. What was the amount of the trade surplus or deficit?
13. Thailand's currency, baht (B), has a value in U.S. dollars of $0.02893. There are 34.571 B per U.S. Dollar. What amount in U.S. dollars is equivalent to 60,000 B?
14. A phone call is made at 6:00 A.M. across 6 time zones to the east and across the International Date Line. What time is it in the place where the call is received?
15. What temperature in Celsius equals a temperature of 50°F?
16. A store sold 218 cm of gold chain from a roll that originally held 12.5 m of chain. How much chain is left on the roll in centimeters?
17. One inch ≈ 2.54 cm, and 1 foot ≈ 0.305 m. How many meters are there in a board 7.5 ft long, to the nearest tenth meter?
18. Michelle cut 2 boards for shelving out of a 2.88 m^2 sheet of wood. One board was cut 20 cm by 120 cm; the other was cut 30 cm by 120 cm. How much wood in m^2 was left of the larger sheet?
19. A cabinet top measures 20 inches × 32 inches. If 1 $in.^2 \approx 6.5\ cm^2$, what is the area of the cabinet top in cm^2?
20. The Hilton family saves 25¢ per liter by buying orange juice in bulk. The family uses 2.5 L of orange juice a week. How much will the Hiltons save in a year by buying juice in bulk?
21. A deli restaurant chain sells 572 lb of potato salad a week. How many kg of potato salad will the chain sell in 1 year at this rate, to the nearest kg?

Planning a Career in Hospitality and Tourism

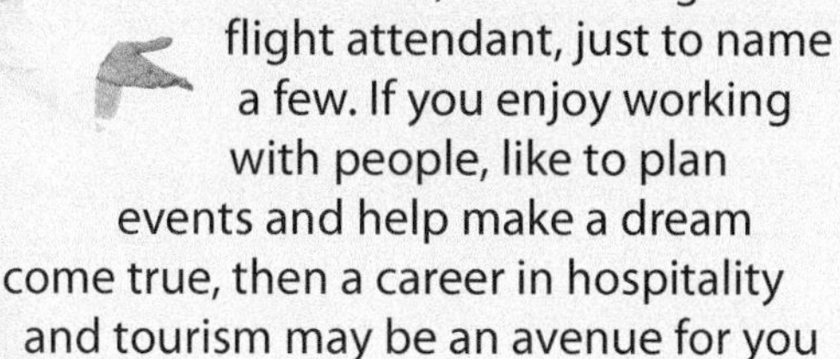

Photodisc/Getty Images

Opportunities in hospitality and tourism include management of hotels and restaurants, planning weddings, working at a resort, assisting people with their vacation plans, being a chef in a fine dining establishment or cook at a local diner, and serving as a flight attendant, just to name a few. If you enjoy working with people, like to plan events and help make a dream come true, then a career in hospitality and tourism may be an avenue for you to pursue.

Job Titles

- Chef
- Restaurant manager
- Travel agent
- Airline attendant
- Tour guide
- Event planner
- Amusement park worker
- Cruise ship activities director
- Front desk clerk
- Concierge

Needed Skills

- enthusiasm and an upbeat personality
- an eye for detail
- excellent communication skills

- strong management skills
- math and computer skills
- ability to work independently and with other, as well as under pressure
- good physical stamina and strength

What's it like to work in Food Service Management?

Food service managers supervise businesses like restaurants, clubs, banquet halls, and dining rooms that make and serve food to customers. Food service managers are responsible for coordinating activities in the kitchen and dining room to create a total dining experience for their customers. Food service managers may keep track of inventory and order food, supplies and equipment to ensure that enough product is on hand. Managers typically handle all of the administrative and staffing issues including hiring, training, and firing. They usually work closely with a restaurant owner or corporate supervisor to ensure sales goals are met and operations run smoothly.

What About You?

Can you see yourself in a career that serves people and brings happiness to others? Which aspect of hospitality is most interesting to you?

How Times Have Changed

For Question 1–2, refer to the timeline on page 547 as needed.

1. Daylight savings time was implemented during WWI to save energy. It's estimated that using DST saves about $180 million dollars in energy costs. This is thought to be 1% of the country's energy costs. What would the total energy costs be?
2. The NIST-F1 atomic clock that was put into use in 1999 was the most accurate clock ever. In 20 million years the clock would be off at most by one second. In how many years would it be possible that the clock was off by one hour?

Chapters 11–13 *Cumulative Review*

MULTIPLE CHOICE

Select the best choice for each question.

1. A day care center has current assets of $60,100 and current liabilities of $32,100. What is the center's current ratio, to the nearest tenth?

   A. $28.000 **B.** 1.9:1 **C.** 1.3:1 **D.** $46,100 **E.** 0.5:1

2. What is the area of a building that is 15 m × 30 m?

   A. 450 m **B.** 90 m^2 **C.** 327 m^2 **D.** 90 m **E.** 450 m^2

3. On Tuesday, you flew west from Chicago to Rio de Janeiro, which is 3 time zones to the east of Chicago. If you land in Rio de Janeiro at 2:00 in the afternoon, what time is it in Chicago?

   A. 11:00 A.M. **B.** 12:00 P.M. **C.** 1:00 P.M. **D.** 3:00 P.M. **E.** 5:00 P.M.

4. Lowen's Books had net sales of $541,290, gross profit of $162,300, and net income of $42,460. What was Lowen's net profit margin, to the nearest tenth percent?

   A. 7.9% **B.** 30% **C.** 7.8% **D.** 37.8% **E.** 26.2%

5. A product's beginning inventory on January 1 was $8,400, and its ending inventory on March 31 was $3,600. The cost of goods sold for the product in the quarter was $14,500. What was the product's turnover rate, to the nearest tenth?

   A. 3.0 **B.** 2.4 **C.** 1.3 **D.** 1.7 **E.** 6.4

6. A business had a net income for a year of $86,580 and equity of $619,900. What was its return on equity, to the nearest tenth percent?

   A. 6.2% **B.** 17.3% **C.** 7.2% **D.** 14% **E.** 11.4%

7. The exchange rates in U.S. Dollars for Israel's shekel and Poland's zloty, are $0.2246 and $0.2980, respectively. What would a 149 zloty purchase cost in U.S. dollars?

   A. $71.04 **B.** $500 **C.** $33.47 **D.** $44.40 **E.** $663.40

8. Ninety employees share a company's profit of $325,350. The share is to be paid in two equal payments. What amount will each employee receive in the first payment?

   A. $3,615 **B.** $361.50 **C.** $18,075 **D.** $1,807.50 **E.** $162,675

9. The partners, Boyd and Blake, agree to share equally a $145,000 annual net income after Boyd is paid a salary of $12,000 plus 7.2% interest on his $115,000 investment in the business. What is Boyd's total share of the net income?

   A. $20,280 **B.** $68,360 **C.** $68,720 **D.** $78,500 **E.** $82,640

10. If there are 32 oz. and 0.95 L in a quart, how many liters would there be in a can that holds 48 oz. of paint thinner?

    A. 15.2 L **B.** 45.6 L **C.** 1.425 L **D.** 30.4 L **E.** 50.5 L

11. On Elmer Byrnes' first $7,000 of earnings, his employer must pay FICA tax of 7.65%, FUTA tax of 6.2%, and SUTA tax of 0.5%. How much does the employer pay in legally required taxes on Byrnes' $7,000 of earnings?

    A. $535.50 **B.** $434 **C.** $1,004.50 **D.** $35 **E.** $6,027

OPEN ENDED

12. Two partners agree to share profits in proportion to their investments. Partner A invested $120,000, and Partner B invested $160,000. What share of a $214,970 annual profit does each partner receive, to the nearest dollar?

13. A bankrupt firm has $42,500 to pay $250,000 of creditors' claims. How many cents on the dollar will each creditor receive?

14. A country with a population of 109,000 has a Gross Domestic Product of $156,960,000. What was the per capita GDP in this country?

15. Pruitt Furnishings had sales of $218,400 and sales returns and allowances of $1,572 in March. The cost of goods sold in March was $130,789. What was the gross profit for March?

16. A wholesaler had a beginning inventory of 683 blenders on February 1. A total of 1,340 blenders were added to stock in February. Stock shipments of blenders were: Feb. 5, 280; Feb 13, 459; Feb 19, 580; and Feb 26, 387. What was the ending inventory?

17. U.S. exports to Brazil were $13,202,198,000. U.S. imports from Brazil were $11,313,062,000. Did the U.S. have a trade surplus or trade deficit with Brazil, and what was the amount?

18. On a trip to Canada your car breaks down and you are towed 60 kilometers to the Canadian border. The towing service charged you $1.85 per kilometer plus $0.18 for each kilogram of your car's 2,200 lb weight. What was the total towing charge?

19. LeMoyne's Department Store had these assets on December 31: Cash, $34,288.92; accounts receivable, $123,573.65; store and warehouse, $435,200. The store owed $85,323.05 in accounts payable and a $178,000 loan to Second Street Bank. What were the store's assets, liabilities, and equity?

20. One day in a recent year, the euro exchange rate in U.S. dollars was $1.2198. Also, the euro converts as 0.8201 euros per U.S. Dollar. A sale of computers was priced at 2,500,000 euros. What was the sale amount in U.S. dollars, to the nearest thousand dollars?

21. A store sells 45 boxes of Crunch cereal daily. More cereal will arrive within three days of ordering. Safety stock is 16 boxes. What is the reorder point for Crunch cereal?

22. An appliance repair company carries a $41,000 annual parts inventory. The company pays 8.2% interest on 50% of the inventory and a 0.4% personal property tax on the inventory's full annual value. The annual cost of storage and insurance is $2,665. What is the carrying cost of inventory per $1 of inventory?

CONSTRUCTED RESPONSE

23. You read an article distributed on the Internet discussing how important it is for the United States to have a strong dollar. Explain the effect that you think a strong dollar will have on imports and exports.

Selected Answers

Chapter 1: Gross Pay

Lesson 1-1, pages 4–9

Exercises
1. 36.5 **3.** $2,223 **5.** $456 **7.** $10,300
9. $3,239.50 **11.** 40.25 **13.** 43.5 **15.** $298.99
17. $112 **19.** 5.5 **21.** $98.59 **23.** 7.3
25. $131.40 **27.** $17.055; $22.74 **29.** $646.39
31. $548.10 **33.** $1,337.35
Mixed Review
35. $982 **37.** $312 **39.** $34.50

Lesson 1-2, pages 10–13

Exercises
1. $1,250 **3.** $45,000 **5.** $1,435.78
7. $736 **9.** $1,570 **11.** $40,820 **13.** $1,846.15
15. $3,000
Mixed Review
19. $43,420 **21.** $1,250 **23.** $175

Lesson 1-3, pages 14–19

Exercises
1. 0.0925 **3.** 0.005 **5.** $975 **7.** 18%
9. 4.5% **11.** 11.625% **13.** $600; $592.25
15. $403.34 **17.** $632.30 **19.** $6,144
21. 3.2% **23.** $2,100
Mixed Review
29. $973.98 **31.** $21.60 **33.** $141.55

Lesson 1-4, pages 20–25

Exercises
1. 203 **3.** $97.44 **5.** $225 **7.** $37.40
9. $650 **11.** $2,090 **13.** $477.75
15. 140; $392.00 **17.** 354; $417.72 **21.** $580.75
23. $96; $116.15; $480; $580.75; $24,960; $30,199
Mixed Review
25. 24 **27.** $2,100 **29.** $840
31. $1,935

Lesson 1-5, pages 26–31

Exercises
1. $15.50 **3.** $51 **5.** $10.41 **7.** $19,990
9. $132.40 **11.** $90 **13.** $3,245.83; $749.04
15. $77 **17.** $116 **19.** $9/hour
21. $395; 44 hours **23.** $1,808
25. $13.26
Mixed Review
29. $940 **31.** $70,304

Chapter Review, pages 32–33

1. gross pay **3.** tip **5.** straight commission
7. overtime **9.** per diem **11.** $26.445, $35.26
13. 9.3 **15.** $55,200 **17.** $1,560 **19.** $2,250
21. $4,228 **23.** $5,950 **25.** $1,872 **27.** $3,230
29. 49

Chapter 2: Net Pay

Lesson 2-1, pages 40–46

Exercises
1. $138.09 **3.** $416.77 **5.** $309.76
7. 0.08 **9.** 0.1206 **11.** 0.89145
13. $27.89 **15.** $34 **17.** $47
19. $3 **21.** $43 **23.** $29.45, $6.89
25. $15.46, $3.62 **27.** $46.45, $10.86
29. $17.95, $4.20 **31.** $45.00; $28.97; $6.78; $162.87; $304.42 **33.** $22.00; $35.85; $8.38; $97.74; $480.47 **35.** $25.00; $33.40; $7.81; $114.43; $424.33 **37.** $473.66 **39.** $539.40
41. $1,513.80 **43.** $651
Mixed Review
47. $32 **49.** $1\frac{3}{4}$ **51.** 50

Lesson 2-2, pages 47–53

Exercises
1. $8,496 **3.** $3,493 **5.** $57,599 **7.** $21,238
9. $15,281 **11.** $39,712.27 **13.** $7,500
15. $16,500 **17.** $2,724 **19.** $3,006
21. $172 **23.** $34,781 **25.** $32,762
27. $3,183 **29.** $850 **31.** $5,175 **33.** $240
35. $263
Mixed Review
37. $27.50 **39.** 25% **41.** 0.432 **43.** 4

Lesson 2-3, pages 54-57

Exercises
1. $1,254 **3.** 0.072 **5.** 0.2498 **7.** $2,160.43
9. $58.63 **11.** $814.50 **13.** $667.60
15. $850 **17.** $2,872.65 **19.** $23,450
21. Tax due: $178

Mixed Review

23. $10\frac{11}{12}$ **25.** 2,667.50

Lesson 2-4, pages 58–62

Exercises

1. $67,969 **3.** $4,186 **5.** $4,629
7. 0.564 **9.** 0.236 **11.** $11,333.19
13. $6,956.30 **15.** $6,301 **17.** $42,776
19. $5,461.56 **21.** B-Tree

Mixed Review

23. 0.375 **25.** 0.005 **27.** $\frac{1}{4}$ **29.** $\frac{1}{10}$

Lesson 2-5, pages 63–69

Exercises

1. $64.40 **3.** $557.59 **5.** $1,415.95
7. 84% **9.** 20% **11.** $31.45
13. $1,130 **15.** 72% **17.** $3,042.31 **19.** 71%
21. $44; $30.23; $298.19; 75% **23.** $0; $34.60; $332.29; 73% **25.** $34.68 **27.** $481.69
29. $67.66 **31.** $14.23; $739.96 **33.** 74%
35. 79% **37.** 81%

Mixed Review

39. $1,809.50 **41.** $17\frac{19}{24}$ **43.** $33.85
45. $2,167.11 **47.** last week, $7.17

Chapter Review, pages 70–71

1. deduction **3.** taxable income **5.** net pay
7. adjusted gross income **9.** $26
11. $34.47; $8.06; $153.31; $402.69
13. $29,050 **15.** $318 **17.** $1,894 **19.** $41,075
21. $37,281 **23.** $24.50; $18.65

Chapter 3: Banking

Lesson 3-1, pages 80–86

Exercises

1. $874.89 **3.** $16.08 **5.** $5.60
7. $3.75 **9.** $0.82 **11.** $1,050.95; $50.95
13. $955.44 $55.44 **15.** $676.49; $76.49
17. $865.45 **19.** $\$1,600 \times 3\% = \$48$ estimated; $50.18 actual **21.** $613 **23.** $817.14

Mixed Review

27. 8 **29.** $\frac{1}{28}$ **31.** $3,218,000 **33.** 7.5%
35. $0.70 **37.** $428.48 **39.** 5%

Lesson 3-2, pages 87–92

Exercises

1. $770.62 **3.** $1,260.93 **5.** $140
7. $680.95 **9.** $981.75 **11.** $299.94
13. $1,354.52 **17.** $593.25
19. $1,199.47

Mixed Review

21. $\frac{23}{24}$ **23.** $5\frac{1}{12}$ **25.** 14.859
27. 66% **29.** $36,920

Lesson 3-3, pages 93–99

Exercises

1. $15,371.61 **3.** $641.49 **5.** $856.80
7. $644 **9.** $24.82 **11.** $252.35

Mixed Review

15. $19\frac{1}{9}$ **17.** $702.80

Lesson 3-4, pages 100–108

Exercises

1. $1,266.49 **3.** $14,906.25 **5.** $2,257.84
7. $972.37 **9.** $210.77

Mixed Review

13. $\frac{2}{5}$ **15.** $\frac{4}{5}$ **17.** 150 **19.** 5,440; 5,400
21. $49,200 **23.** $41,930 **25.** $47.30
27. $7,150, $196.63

Lesson 3-5, pages 109–114

Exercises

1. $15.92 **3.** $7,500 **5.** $30.67
7. $13.92 **9.** $11.25 **11.** $108.40

Mixed Review

15. 1,429 **17.** 3.45% **19.** 188

Lesson 3-6, pages 115–119

Exercises

1. 0.25 **3.** $839.36 **5.** $398.54 **7.** $185.06
9. $580.41 **11.** $4,314.47; $314.47
13. $308.39; $8.39 **15.** $1,852.71; $52.71
17. $1,780.73; $219.27 **19.** $65,018.50; $4,981.50
21. $4,142.84 **23.** $24,443.19 **25.** $5,640.72
27. $359.28 **29.** $283.02
31. José, $901.31 more

Mixed Review

33. $468.41

Chapter Review, pages 120–121
1. annuity **3.** compound amount
5. certificate of deposit
7. Electronic Funds Transfer
9. \$4,032 **11.** \$3,868.09 **13.** \$1,609.01
15. \$269.47 **17.** \$615.34 **19.** \$181.33
21. \$30,886.92

Chapter 4: Credit Cards

Lesson 4-1, pages 128–136

Exercises
1. \$25.38 **3.** \$321.76 **5.** \$9.01 **7.** \$53
9. \$1,711.74 **11.** 0.000316 **13.** \$103.86
15. \$3.74, \$267.88, \$29 **17.** \$72.65 **19.** \$338.11
21. \$391,200.40
Mixed Review
25. 3.0225 **27.** \$33 **29.** \$851.08 **31.** \$54,128

Lesson 4-2, pages 137–142

Exercises
1. \$944.51 **3.** \$4.63 **5.** 1.25%; \$2.67 **7.** 1; \$8.67
9. 0.0575%; \$3.75 **11.** 3.07; \$162.22 **13.** \$6.96, \$244.76 **15.** \$3.49, \$479.98
Mixed Review
19. $\frac{4}{59}$ **21.** 1.75 **23.** \$708.27

Lesson 4-3, pages 143–147

Exercises
1. \$55.61 **3.** \$0.57 **5.** \$7.45 **7.** \$4.26
9. \$5.56, \$371.18
Mixed Review
13. 0.0584 **15.** \$4,500

Lesson 4-4, pages 148–152

Exercises
1. 1.5833 **3.** 0.018745 **5.** \$33 **7.** \$22.31
9. \$13.03 **11.** \$346.93 **13.** \$158.24
15. \$18.90 **17.** \$14.43
Mixed Review
21. \$2,500 **23.** \$54.30 **25.** \$100 **27.** \$10
29. \$3.11 **31.** \$137.50 **33.** \$2,083.33

Lesson 4-5, pages 153–159

Exercises
1. \$80.98 **3.** \$6 **5.** \$11 **7.** \$6.75 **9.** \$26
11. \$4.17; \$554.17; \$11; \$543.17
13. \$585.70; \$4.88; \$640.58; \$13; \$627.58
15. \$668.81; \$5.57; \$724.38; \$14; \$710.38
17. \$30.30 **19.** 39%
21. \$486.25; \$6.08; \$492.33; \$20; \$472.33
23. \$459.23; \$5.74; \$464.97; \$19; \$445.97
25. \$433.54; \$5.42; \$438.96; \$18; \$420.96
27. \$34.96 **29.** 69%
Mixed Review
33. \$104.40 **35.** 25% **37.** \$120

Chapter Review, pages 160–161
1. debt-to-income ratio
3. adjusted balance method
5. periodic rate **7.** annual percentage rate
9. credit score **11.** \$105 **13.** \$98.45
15. \$10 **17.** \$2.95, \$322.25
19. \$2.99, \$320.67 **21.** \$14.68 **23.** \$252.50

Chapter 5: Loans

Lesson 5-1, pages 170–177

Exercises
1. \$93.78 **3.** \$660 **5.** 12.5% **7.** \$750
9. \$20 **11.** \$6.75; \$156.75 **13.** \$29,370
15. 10.5% **17.** \$29.00, \$29.40 **19.** \$47.34, \$48
21. \$21.60, \$21.90 **23.** \$75 **25.** 11%
27. Estimates will vary; \$600 **29.** \$10,591.78
33. 93%
Mixed Review
35. 3.4113 **37.** \$1,230.77 **39.** \$27,677
41. \$337.37 **43.** 1.25%

Lesson 5-2, pages 178–183

Exercises
1. 22 **3.** 28 **5.** \$4.18 **7.** \$3,107.90
9. \$512.80 **11.** \$2.19 **13.** \$37.91 **15.** \$4.68
17. \$15.98 **19.** \$33.53 **21.** \$0.3842
23. \$0.1375 **25.** \$0.5548 **27.** \$0.4167
29. \$3.50 **31.** \$2.44 **33.** \$23.41 **35.** \$554.79
37. \$9,144
Mixed Review
39. $\frac{7}{24}$ **41.** $8\frac{3}{4}$ **43.** \$67.20 **45.** \$335.25

47. \$336.39

Lesson 5-3, pages 184-190

Exercises
1. \$2,589.15 **3.** \$575 **5.** \$441 **7.** \$1,413
9. 0.017 **11.** \$1,793.68; \$293.68
13. \$2,280.60; \$130.60 **15.** 6%
17. \$6.25, \$80.77, \$419.23
19. \$12, \$61.34, \$738.66 **21.** \$60.88
23. \$72.54 **25.** \$24.86 **27.** \$63.81
Mixed Review
31. 7.082 **33.** \$18 **35.** 0.365 **37.** \$602.80
39. \$2.82; \$250.15 **41.** \$50 **43.** 14 months

Lesson 5-4, pages 191–194

Exercises
1. 2.456 **3.** \$719 **5.** \$2,278.44 **7.** \$1,076.23
9. \$2,750.40 **11.** \$2,632.57 **13.** \$3,429.38
Mixed Review
17. 29,500 **19.** \$34.47 **21.** \$18.90; \$25.20

Lesson 5-5, pages 195–199

Exercises
1. \$4.56 **3.** \$9,820,800 **5.** \$10,784.30
7. 0.225 **9.** 0.635 **11.** \$15 **13.** $13\frac{1}{2}$% **15.** 13%
17. $15\frac{1}{4}$% **19.** \$184 **21.** $12\frac{3}{4}$% **23.** \$7.08
25. $13\frac{1}{4}$% **27.** $14\frac{1}{4}$% **29.** $14\frac{1}{4}$% **31.** \$4,320
33. \$8 **35.** \$320; it is the same amount of finance charges as the installment loan
Mixed Review
37. \$9,750 **39.** \$3.97 **41.** 1,262.5%
43. \$1,810.18 **45.** \$50.75 **47.** \$485.64
49. \$26,820

Chapter Review, pages 200–201

1. exact interest method **3.** finance charge
5. total amount due **7.** interest
9. down payment
11. \$250.50 **13.** \$192; \$3,392 **15.** 15%
17. a) \$1.16, b) \$5.15 **19.** \$83.11 **21.** \$40.60
23. \$30 **25.** \$1,514.30; \$314.30 **27.** \$1,503.29
29. \$9.75, $17\frac{1}{2}$% **31.** 13% **33.** $13\frac{1}{2}$%

Chapter 6: Own a Home or Car

Lesson 6-1, pages 208–215

Exercises
1. \$1,095 **3.** 180 **5.** \$7,846.44 **7.** \$36,920
9. \$85,874 **11.** \$20,355 **13.** \$4,071
15. \$2,933 **17.** \$7,033 **19.** \$501.86; \$60,446.40
21. \$598.77; \$125,557.20 **23.** \$733.76; \$164,153.60
25. \$1,718.64
Mixed Review
29. \$ 211,862.73 **31.** 5,600 **33.** $5\frac{1}{2}$ **35.** \$5,082
37. 74% **39.** \$6.75

Lesson 6-2, pages 216–222

Exercises
1. \$13,879 **3.** \$3,361 **5.** \$3,507.50 **7.** \$6,936
9. \$20,528 **11.** \$4,068 **13.** Buying saves \$1,022.
15. Renting costs \$124 more.
Mixed Review
19. 130,000 **21.** 13% **23.** \$949.20 **25.** 58 days

Lesson 6-3, pages 223–228

Exercises
1. \$6,174 **3.** 1,253 **5.** 0.0284 **7.** 0.0183
9. \$1,344.15 **11.** \$9,886.59 **13.** \$902,000; 0.033
15. \$314,300; 0.014 **17.** \$936 **19.** \$1,295
21. \$2,805.48 **23.** \$2,576.56
Mixed Review
27. 0.33708 **29.** 0.45 **31.** 2,400; 2,330.15
33. \$24.74 **35.** \$7.17

Lesson 6-4, pages 229–236

Exercises
1. \$899 **3.** \$26,620 **5.** \$56,950 **7.** \$58,500
9. 0.706 **11.** 1,505 **13.** 1,893.5 **15.** \$287
17. \$1,970 **19.** \$142 **21.** \$936 **23.** \$6,000
25. \$11,000 **27.** \$17,500 **29.** \$6,400; \$11,000, \$18,000
Mixed Review
33. \$14.29 **35.** $2\frac{1}{2}$ **37.** \$134.03 **39.** $\frac{11}{25}$
41. \$2.07 **43.** \$173.38 **45.** \$605; 25%
47. \$4,403.80, \$403.80

Lesson 6-5, pages 237–243

Exercises
1. \$18,697 **3.** \$21,259.50 **5.** \$1,471.36
7. \$23,929 **9.** \$24,912 **11.** Yes, \$203
13. \$21,130.50; \$17,960.50 **15.** \$5,189.10
Mixed Review
19. 10.1; 10 **21.** 238 **23.** 0.0578 **25.** \$14.60

Lesson 6-6, pages 244–249

Exercises
1. \$690 **3.** \$1,517 **5.** \$32,904 **7.** \$3,018.40
9. \$14,953 **11.** Leasing costs \$1,973 more.
13. \$29,112 **15.** \$28,514.08

Mixed Review

19. 3,000; 3,046 **21.** $9\frac{1}{10}$ **23.** $42.28 **25.** $11.67 **27.** 46%

Lesson 6-7, pages 250–254

Exercises

1. $14,173 **3.** $2,351 **5.** 13% **7.** $12,231 **9.** $12,627 **11.** $7,700 **13.** $4,075 **15.** $2,030 **17.** $4,815 **19.** 15% **21.** 13% **23.** $1,675 **25.** $12,400

Mixed Review

27. 12.8 **29.** $\frac{5}{6}$ **31.** $18 **33.** $4,775 **35.** Carthage **37.** $54

Lesson 6-8, pages 255–260

Exercises

1. 0.05 **3.** 0.1035 **5.** 1.05 **7.** $11.28 **9.** $829.27 **11.** $980.12 **13.** $1,454.44 **15.** $800.25 **17.** $2,113 **19.** $2,992.92

Mixed Review

25. 48 **27.** $1,927.96 **29.** $459; $10,659 **31.** $9.87; $1,032.75 **33.** $15,664.20 **35.** $757.39 **37.** $164

Chapter Review, pages 261–263

1. premium **3.** bodily injury **5.** depreciation **7.** renters policy **9.** security deposit **11.** $89,845, $18,280 **13.** $53 **15.** $21,442 **17.** 0.00074 **19.** $18,721.80 **21.** $728 **23.** $210 **25.** $63,000 **27.** $32,553.72; $25,053.72 **29.** $9,368.64; $1,241.64 **31.** Buying is $2,410 less expensive. **33.** 10.3% **35.** $2,678.75 **37.** $3,176

Chapter 7: Insurance and Investments

Lesson 7-1, pages 272–278

Exercises

1. $104.58 **3.** $7 **5.** $67.80 **7.** 25% **9.** $1,784 **11.** $596 **13.** $807 **15.** $6,720 **17.** $2,244.10 **19.** $9,800 **21.** 59% **23.** $526,000 **25.** $3,150 **27.** $4,000 **29.** $5,831.45

Mixed Review

33. $23 **35.** $39.26 **37.** $3\frac{1}{5}$ **39.** $0 **41.** $13\frac{1}{4}$% **43.** $2,1121 **45.** $23,750

Lesson 7-2, pages 279–284

Exercises

1. $16,742 **3.** $378.96 **5.** $4,921 **7.** $533 **9.** $6,583 **11.** $67.96 **13.** 39.6% **15.** $3,070 **17.** $106 **19.** $1,030.50 **21.** $20,391.50 **23.** $200

Mixed Review

25. $3,015.46 **27.** 13.957 **29.** $2\frac{3}{8}$ **31.** $1,645 **33.** $624.09

Lesson 7-3, pages 285–289

Exercises

1. $47,450 **3.** $4,999 **5.** $19,062 **7.** 27% **9.** $3,615.63 **11.** $486.25 **13.** $1,487.50 **15.** 15 **17.** $50,600

Mixed Review

21. 12.5% **23.** 23.1% **25.** $\frac{15}{16}$ **27.** $567.60 **29.** $1,700 **31.** $44.45 **33.** $3,085.84 **35.** $452.74; $727.09

Lesson 7-4, pages 290–294

Exercises

1. $6,290 **3.** $1,027 **5.** $1,185 **7.** $928.77; disc. **9.** $1,030.88; prem. **11.** $1,098.36; prem. **13.** $4,572.25 **15.** $21,258.40 **17.** $14,174.10 **19.** $13,760.76 **21.** $4,766.82 **23.** $7,085.04 **25.** $30 **27.** $58.74 **29.** $7.10

Mixed Review

33. 57.78 **35.** $1\frac{5}{24}$ **37.** $6\frac{2}{7}$ **39.** $24,136.80 **41.** $3,672 **43.** $93 **45.** $140,000

Lesson 7-5, pages 295–299

Exercises

1. $840 **3.** $960 **5.** $900 **7.** $462.50 **9.** $165 **11.** $4,962.30; $450 **13.** $32,175.90; $3,750 **15.** 10.9% **17.** 8.5% **19.** $20 **21.** 10.2% **23.** 9.4% **25.** $20,090 **27.** $19,417.80

Mixed Review

29. $\frac{2}{45}$ **31.** 60.5

Lesson 7-6, pages 300–307

Exercises

1. $5,534.88 **3.** $37.50 **5.** $12,706.55 **7.** $1,791.20 **9.** $22,259.90 **11.** $2,394.20 **13.** $1,589.90 **15.** $46.80 **17.** $6,400 **19.** $1,500 **21.** $1,400 **23.** 5% **25.** 5% **27.** 6.9% **29.** 10.5% **31.** 2.9% **33.** 400 **35.** +$687 **37.** −$444.87 **39.** +$544.83 **41.** $1,248 **43.** $2,343

Mixed Review

45. 1.9% **47.** $1,600

Lesson 7-7, pages 308–313

Exercises
1. $674 **3.** $1,771.54 **5.** 23.4% **7.** $4,384 **9.** $9,904 **11.** $727 **13.** $5,235 **15.** $7,464 **17.** 5% **19.** $2.64; 4.9% **21.** $3.20; 4.5% **23.** $60,774.40 **25.** $24,330
Mixed Review
29. 8.75% **31.** $1.58 **33.** $2,002.50 loss **35.** $8,217.75 **37.** $11,169.50

Lesson 7-8, pages 314–320

Exercises
1. $1,584 **3.** $2,375 **5.** $309 **7.** $3,375.47 **9.** $742 **11.** $670 **13.** $1,110 **15.** 10% **17.** –2% **19.** 11.7% **21.** 7.3% **23.** 26.3% **25.** $1,697.67 **27.** $369.17 **29.** $985 **31.** $221.67
Mixed Review
35. 126 **37.** 32% **39.** $14

Lesson 7-9, pages 321–326

Exercises
1. $274,327 **3.** 0.4% **5.** $26,880 **7.** 52.5% **9.** $1,920 **11.** 2.5% **13.** $24,054.55 **15.** $435
Mixed Review
17. 35.6 **19.** $28.86

Chapter Review, pages 327–729

1. disability insurance **3.** major medical insurance **5.** cash value **7.** current yield on bonds **9.** mutual fund **11.** capital investment **13.** premium **15.** coinsurance **17.** $583.20; $536.49 **19.** $2,100 **21.** $1,395.43 **23.** $2,893.33 **25.** $1,013.79 **27.** $985.40, discount **29.** $4,125 **31.** $21,105 **33.** $4,837.13 **35.** 5.0% **37.** $11,823 **39.** $505.90 profit **41.** 11.2% **43.** 0.6% **45.** $2,200

Chapter 8: Budgets

Lesson 8-1, pages 336–341

Exercises
1. $103.42 **3.** $183.29 **5.** $3,662.33 **7.** $1,110.72 **9.** $561.03 **11.** $410.67 **13.** $62.13 **15.** $104.13 **17.** $199.21 **19.** $220.39 **21.** $195.80 **23.** $49.67 **25.** $250 **27.** $164.99 **29.** $60.83 **31.** $100 **33.** $75
Mixed Review
37. $\frac{3}{4}$ **39.** $304.14 **41.** $487.20 **43.** $24.20 **45.** $884.25

Lesson 8-2, pages 342–346

Exercises
1. 0.19 **3.** 39% **5.** 70.5% **7.** $4,380 **9.** 7% **11.** 2% **13.** 71% **15.** $450 **17.** $90
Mixed Review
23. $3\frac{5}{8}$ **25.** $5\frac{2}{5}$ **27.** $486 **29.** 66%

Lesson 8-3, pages 347–352

Exercises
1. $37.52 **3.** $153.24 **5.** $30.60 **7.** $10.46 **9.** $4.33 **11.** $0.44 **13.** $0.46 **15.** $46.51 **17.** larger; $0.01 **19.** $55.96 **21.** 6 days **23.** Brand B **25.** $13.61
Mixed Review
29. $\frac{3}{8}$ **31.** $618.75 **33.** $303.67

Lesson 8-4, pages 353–358

Exercises
1. $106.72 **3.** $308 **5.** $169.76 **7.** $6.50 **9.** $2.99 **11.** $131.87 **13.** $189.92 **15.** $185.03 **17.** $43.49 **19.** $696.82 **21.** satellite; $10.01
Mixed Review
23. $0.96 **25.** $73.01

Lesson 8-5, pages 359–364

Exercises
1. 6% **3.** 34% **5.** $308 **7.** Entertainment, Miscellaneous **11.** 7% **13.** $516; $6,192 **15.** $774; $9,288 **17.** $430; $5,160 **19.** $258; $3,096 **21.** $4,300; $51,600
Mixed Review
25. 119.625 **27.** $\frac{3}{7}$ **29.** 0.0064 **31.** 120 **33.** $31,650

Lesson 8-6, pages 365–371

Exercises
1. 32.9 **3.** 4.1% **5.** 0.423 **7.** 145.2 **9.** Apparel **11.** 104.8% **13.** Food and Beverages, 4.8% **15.** $0.565; $0.016 **17.** $0.193 **21.** $10.37.
Mixed Review
25. $20.67 **27.** $244.56 **29.** 2% **31.** $20.00 **33.** $8,027.50 **35.** $1,456; $6,656 **37.** $2,550.80

Chapter Review, pages 372–373

1. inflation **3.** Consumer Price Index **5.** purchasing power of the dollar **7.** unit price **9.** $1,680 **11.** $80.50 **13.** $9,240; $770 **15.** $112.50; $5,850 **17.** $1.77 **19.** 6.2 weeks **21.** $790.38 **23.** 17%,4% **25.** 115.3% **27.** $0.581

Chapter 9: Business Costs

Lesson 9-1, pages 382–386

Exercises

1. $365.25 **3.** $910.48 **5.** $445,150 **7.** $1,286,342 **9.** $464,462 **11.** $150 **13.** $162,691 **15.** $722.50; $1,734; $433.50 **17.** $1,153; $637.60 **19.** 246 **21.** $87.63

Mixed Review

25. $\frac{8}{15}$ **27.** $58 **29.** 30% **31.** 39.9 cents

Lesson 9-2, pages 387–390

Exercises

1. $31.68 **3.** $292,740 **5.** 12,800 **7.** 10,000 **9.** $1,170,000 **11.** 1,200 **13.** $18.75

Mixed Review

17. 1,763 **19.** 0.125 **21.** 206 **23.** $2,187.50 **25.** $328,404 **27.** $19,200

Lesson 9-3, pages 391–395

Exercises

1. 21 **3.** $96,680 **5.** $47,174.40 **7.** $1,288; $1,090.16 **9.** $7,568; $5,860.66 **11.** $19,836; $15,013.87 **13.** $48,576.92 **15.** $250,000 **17.** $120,000

Mixed Review

21. $36.40

Lesson 9-4, pages 396–401

Exercises

1. $929.81 **3.** $48.10 **5.** 32,900 **7.** yes **9.** yes **11.** no **13.** $33.25 **15.** $12.00 **17.** $27.75 **19.** $15.25 **21.** $25.00 **23.** $63.10 **25.** $45.45 **27.** $193 **29.** $2,137.20 **31.** $144 **33.** $1,366.56 **35.** $2,062.26 **37.** 1,200 lb **39.** $427.57

Mixed Review

43. $47.60 **45.** 9.75% **47.** $2,019.64 **49.** $2,600 **51.** $378.15

Lesson 9-5, pages 402–405

Exercises

1. $11.29 **3.** $26.72 **5.** $16.80 **7.** 1.3 **9.** 1,275 **11.** 3 **13.** $33\frac{1}{3}$% **15.** $621 **19.** $150

Mixed Review

21. $193\frac{1}{2}$ **23.** 7 **25.** 57

Lesson 9-6, pages 406–409

Exercises

1. $314.42 **3.** $708.50 **5.** $192.97 **7.** $1,744.22 **9.** $2,024 **11.** $5,160 **13.** $0.455

Mixed Review

17. 8%

Chapter Review, pages 410–411

1. total manufacturing cost **3.** breakeven point **5.** variable costs **7.** MACRS **9.** f.o.b. **11.** $3,840,000; $4,524,850 **13.** 2,400 cowlings; $48,000 **15.** $116,600 **17.** $12,696 **19.** $8,984.56, Belcro **21.** $21; $630 **23.** $0.068 **25.** $466.25

Chapter 10: Sales and Marketing

Lesson 10-1, pages 420–426

Exercises

1. $57,464.91 **3.** $4,053.77 **5.** $2,864.40 **7.** 18¢ short **9.** $2.32 short **11.** $4,354.20 **13.** $12,619.50 **15.** $5,037.82; $4,966.37; $1,820.87; $4,395.23; $2,574.36; $3,491.12; $3,478.56; $904.20

Mixed Review

19. 0.00782 **21.** 3 **23.** $27.12 **25.** 0.0519

Lesson 10-2, pages 427–433

Exercises

1. $6,903.49 **3.** $90.41 **5.** 2% **7.** Aug. 26 **9.** Feb. 12 **11.** Jan. 11 **13.** $68; $3,332 **15.** $1,151.54 **17.** 2.5% **19.** $2,730 **21.** 1.5% **23.** $759.70 **25.** $10.50 **27.** $1,392 **31.** $181.07 **33.** $658.93

Mixed Review

35. 220 **37.** $\frac{1}{72}$ **39.** $11,585 **41.** 2,100 **43.** 30 inches

Lesson 10-3, pages 434–439

Exercises

1. $2,016.75 **3.** $31.30 **5.** 51% **7.** $86.40; $33.60 **9.** $168.68; $146.32 **11.** 40% **13.** 40% **15.** 51% **17.** $154.94 **19.** $20.81, $11.19 **23.** May 17; $1,520; $1,488.47; $1,737; $1,679.80; $1,661.71; $1,877.26

Mixed Review

25. 0.001 **27.** $399 **29.** $1,349.40 **31.** $407.52

Lesson 10-4, pages 440–445

Exercises

1. \$0.52 **3.** $7\frac{1}{2}$% **5.** \$7.80 **7.** 12% **9.** 45%; \$1,250 **11.** 55%; \$8.20 **13.** $62\frac{1}{2}$; \$74.85 **15.** \$75.52 **17.** \$0.60 **19.** \$75; 50% **21.** \$36.75; $33\frac{1}{3}$% **23.** \$367.04 **25.** 23.2% **27.** 30%

Mixed Review

29. $11\frac{3}{4}$ **31.** 855.573 **33.** \$943.04 **35.** \$217

Lesson 10-5, pages 446–449

Exercises

1. 921 **3.** 13.1% **5.** 133; 62.1% **7.** 40.6% **9.** 79%

Mixed Review

13. $1\frac{7}{9}$ **15.** $16\frac{1}{3}$ **17.** \$1,986.30

Lesson 10-6, pages 450–453

Exercises

1. \$46,804.50 **3.** \$192,000 **5.** 10% **7.** 209 **9.** 6,704 **11.** 14,960

Mixed Review

15. 207 **17.** \$8,683; \$6,583

Lesson 10-7, pages 454–457

Exercises

1. 5% **3.** 18,490 **5.** 36% **7.** 38.8% **9.** \$360,000 **11.** \$490,000 **13.** \$3,496,500

Mixed Review

17. 1,158 **19.** $9\frac{5}{8}$ **21.** 43¢ **23.** \$428.40 **25.** 16% **27.** 18.3%

Lesson 10-8, pages 458–463

Exercises

1. \$1,022 **3.** \$2,562 **5.** \$0.003 **7.** \$25,320 **9.** \$2,826.25 **11.** \$42,510 **13.** \$3,840 **15.** \$529,000 **19.** 120

Mixed Review

21. $\frac{1}{4}$ **23.** 0.7% **25.** \$38.27 **27.** 1,000 **29.** \$44,000 **31.** \$130.41 **33.** \$267.68, **35.** \$6,070.50

Chapter Review, pages 462–462

1. cash discount **3.** trend **5.** list price **7.** market **9.** forecast **11.** \$2,201.04 **13.** \$6,027.60 **15.** \$34.10; \$3,375.90 **17.** \$10.14 **19.** \$212 **21.** \$56 **23.** 3 stars; 73% **25.** 80,160 **27.** 204,000 gallons **29.** \$4,650

Chapter 11: Manage People and Inventory

Lesson 11-1, pages 474–479

Exercises

1. \$10,355 **3.** \$637.50 **5.** \$156,000 **7.** \$4,648 **9.** \$4,320 **11.** \$176,850 **13.** \$136,540 **15.** \$2,200 **17.** \$18,920

Mixed Review

19. $6\frac{3}{4}$ **21.** \$1.88 **23.** 1.7%

Lesson 11-2, pages 480–484

Exercises

1. \$15.40 **3.** \$0.17 **5.** \$640 **7.** \$19.66 **9.** \$30,000 **11.** \$984 **13.** \$1,050 **15.** \$474,000

Mixed Review

19. 11.135 **21.** $\frac{1}{8}$ **23.** 60,000 **25.** \$113.11 **27.** \$3,176.80; \$114,364.80

Lesson 11-3, pages 485–489

Exercises

1. \$13,258.02 **3.** \$15.70 **5.** \$418.47 **7.** \$9,500 **9.** \$8,610 **11.** \$2,460 **13.** \$202.50

Mixed Review

17. 713.36 **19.** $\frac{3}{10}$ **21.** \$0.0001 **23.** 98 **25.** \$424.07; \$67,221

Lesson 11-4, pages 490–494

Exercises

1. 1,118 **3.** 273 **5.** 1,206 **7.** 609 **9.** 10 **11.** 30,800 **13.** 12

Mixed Review

17. \$3.50 **19.** 18 **21.** \$328.44 **23.** \$21,934

Lesson 11-5, pages 495–500

Exercises

1. 747 **3.** 2,042 **5.** \$11,765.40 **7.** \$3.76 **9.** \$5,075 **11.** \$5,000.80 **13.** \$11,359 **17.** \$26,442

Mixed Review

19. \$105.55 **21.** $\frac{5}{9}$ **23.** $\frac{8}{75}$ **25.** \$135 **27.** \$8,700

Lesson 11-6, pages 501–505

Exercise

1. $414,960 **3.** $9,095.80 **5.** $19.60 **7.** $17.49 **9.** $7.89 **11.** $0.15 **13.** $141,664 **15.** $0.033

Mixed Review

19. $6.25 **21.** 24.97 **23.** 8¢, 39¢ **25.** $1,215; $7,215

Chapter Review, pages 504–505

1. ordering costs **3.** stock record **5.** exempt employee **7.** LIFO **9.** executive recruiter **11.** $19,760 **13.** $300,000; $2,500 **15.** $1,085.36 **17.** 210 **19.** $2,118 **21.** $11.47

Chapter 12: Business Profit and Loss

Lesson 12-1, pages 512–517

Exercises

1. $61,102 **3.** $22,012 **5.** ($20,540) **7.** $256,184 **9.** $649,380 **11.** $581,200 **13.** Net loss, $5,503

Mixed Review

15. $126 **17.** $10,000 **19.** 0.57 **21.** $1,707.83

Lesson 12-2, pages 518–521

Exercises

1. 12% **3.** 72.2% **5.** 13.3% **7.** 60% **9.** $77,000; $24,200 **11.** 11% **13.** 4.5 **15.** $5,174

Mixed Review

17. 450 **19.** $87\frac{1}{2}$% **21.** $239.17

Lesson 12-3, pages 522–526

Exercises

1. $22,504 **3.** $26,285 **5.** $22,500 **7.** $40,418; $27,982 **9.** $11,978.87; $6,921.13 **11.** $61,982; $35,418 **13.** $42,200; $40,800 **15.** $66,480; $63,600; $64,920

Mixed Review

19. 160% **21.** 3.05

Lesson 12-4, pages 527–531

Exercises

1. $109,377 **3.** $99,139 **5.** $143,368 **7.** $90,970 **9.** $13,550 **11.** $69,555 **13.** $48,600 **15.** $215,800 **17.** $113,200 **19.** $20,638 **21.** $77,917; $20,310; $57,607; $77,917

Mixed Review

25. $357\frac{1}{2}$ **27.** 11,000 **29.** $\frac{1}{4}$ 31. $1,600 **33.** $754.56; $15,091.20

Lesson 12-5, pages 532–536

Exercises

1. $194,412 **3.** 2.0:1 **5.** 2.1:1 **7.** 56.8% **9.** 24.8% **11.** less; 0.7% **13.** 39%

Mixed Review

15. $1\frac{11}{20}$ **17.** 0.4007 **19.** $120,000 **21.** $12.85 **23.** $16,902,000

Lesson 12-6, pages 537–539

Exercise

1. $113,713 **3.** 28¢ **5.** 30% **7.** 65% **9.** 35¢ **11.** 16¢ **13.** 46¢ **15.** $126,975 **17.** $255; $7,820

Mixed Review

19. $\frac{1}{20}$ **21.** $6,748

Chapter Review, pages 540–541

1. income statement **3.** net income **5.** merchandise turnover rate **7.** equity **9.** return on equity **11.** $55,672 **13.** $191,140 **15.** 43.9% **17.** 6.0 **19.** $120,000; $150,000 **21.** $35,500; $53,500 **23.** 3.7:1 **25.** 52.9% **27.** $2,016

Chapter 13: International Business

Lesson 13-1, pages 548–552

Exercises

1. 361.93 **3.** 15,734.27 **5.** $3,880 **7.** $2,463 **9.** $132,991 **11.** $860,535 **13.** buying the car in the United States

Mixed Review

15. $1\frac{1}{4}$ **17.** 543,460,000,000

Lesson 13-2, pages 553–557

Exercises

1. 5:00 A.M. **3.** 6:13 P.M. **5.** 8 **7.** 11 **9.** 2 time zones east; it is 7:00 A.M. Saturday. **11.** Tuesday at 4:55 A.M. **13.** 86°F **15.** No, 32°C equals 89.6°F

Mixed Review

19. 28 **21.** $47,812 **23.** $133,542.80 **25.** $57,395; $59,605

Lesson 13-3, pages 558–563

Exercises

1. 100 **3.** 1 000 **5.** 0.1 **7.** 0.6 **9.** 0.368 **11.** 5 000 **13.** 3 000 **15.** 2.9 **17.** 41.5 mm **19.** 16.4 cm **21.** 0.4 mm **23.** 2 760 cm **25.** 50 m **27.** 1.7 km **29.** 269.7 cm **31.** 105.4 m **33.** 3 542 km **35.** 6 000 cm **37.** 1 040 mm **39.** 4 700 m **41.** 888 m **43.** 1.334 m **45.** 18.04 ft **47.** 65,100 mi **49.** 200 km **51.** No

Mixed Review

55. 0.75 **57.** $1\frac{1}{6}$ **59.** $11.95; $12 **61.** $8,000; $37,000 **63.** $7,497; $5,355

Lesson 13-4, pages 564–569

Exercises

1. 100 mm^2 **3.** 8 600 ha **5.** 260 000 m^2 **7.** 1.5012 m^2 **9.** 9.7 m^2 **11.** 4.32 km^2 **13.** 3 km^2 **15.** 4 135 mm^2 **17.** 1.5 km^2 **19.** 4 **21.** 4 **23.** 62.5 acres **25.** 22.4 in.^{2} **27.** 1 390 ha **29.** 1 031 m^2 **31.** 2,052,000 ft^2 **33.** 12 inches by 12 inches **35.** 4 m^2

Mixed Review

39. 3 **41.** 12.9 **43.** 30% **45.** $870,000 **47.** 22.5%

Lesson 13-5, pages 570–575

Exercises

1. 1 000 L **3.** 0.6 L **5.** 1.9 g **7.** 650 g **9.** 0.001 L **11.** 4.554 kg **13.** 5.36 g **15.** 72 kg **17.** 45 t **19.** 130.5 **21.** 5.4 kg **23.** 679.2 g **25.** 1.25 L **27.** 205 g **29.** 84.8 oz. **31.** 5 **33.** 500

Mixed Review

35. $128.55

Chapter Review, pages 576–577

1. trade deficit **3.** international business **5.** foreign exchange rate **7.** domestic business **9.** area **11.** $2,214 **13.** $7.65 **15.** −14°C **17.** 5 km **19.** 16.25 m **21.** 7.625 m **23.** 24 km **25.** 19 000 000 m^2 **27.** 0.1 km^2 **29.** 86.4 in.^{2} **31.** 24 kL **33.** 6.7 t **35.** 360 L **37.** 3.8 L; 3.12 gal **39.** 550 lb **41.** 30 kg

Glossary

A

Access fees Money paid to Internet service provider for usage of Internet.

Adjusted balance method The difference between payments and credits during a month from the balance at the end of the previous month.

Adjusted gross income A tax term meaning gross income less adjustments.

Airtime The minutes you spend calling on a cell phone.

Annual percentage rate (APR) A percent that shows the ratio of finance charges to the amount financed; the cost of credit for one year, expressed as a percentage.

Annuity A series of equal payments made at regular intervals of time.

Area The amount of surface an item has.

Assessed value A value put on property as a base for figuring amount of tax.

Assets Things of value owned by a person or business.

Automatic teller machine (ATM) A computer system that lets you withdraw or deposit money in your bank account without a teller's help.

Average A single number used to represent a group of numbers.

Average daily balance method The periodic rate applied to the average daily balance in the account during the billing period.

Average monthly expense The sum of the monthly amounts spent on a particular expense divided by the number of months.

B

Balance The amount of money in an account; the difference between the two sides of an account.

Balance sheet A statement showing the assets, liabilities, and capital of a person or business for a certain date.

Bank statement A monthly report to a depositor showing deposits, payments, and balance in a bank account.

Bankrupt Legally insolvent and unable to pay debts.

Base period A period in time with which comparisons are made.

Benefits An addition to wages provided by the employer that are of value.

Bodily injury insurance Auto insurance covering liability for injury to other persons.

Bond Written promise to repay the money loaned on the due date.

Bonus Pay given to reward employees who make a significant contribution to the success of the company or whose work record is exceptional.

Book value Original cost less total depreciation to date.

Breakeven point The point at which income from sales equals total costs of producing and selling goods.

Budget Future spending goals.

C

Cable television Television service by radio frequency signals transmitted through cables.

Cafeteria plan A Section 125 plan that allows employees to put pre-tax wages into an account that is used to pay for qualified expenses, such as health or life insurance premiums, child care, or health care expenses.

Capital investment The amount of cash you originally invested plus anything you spent for improvements.

Carrying costs The costs of holding inventory until it is issued.

Cash advance The cash received when money is borrowed on a credit card.

Cash discount Discount given for early payment of a bill.

Cash value The value of an insurance policy if it is canceled; cash surrender value.

Certificate of deposit (CD) A time deposit or savings certificate.

Check register A record of deposits, withdrawals, and checks.

Circle graph A circle showing how parts relate to the whole and to each other.

Closing costs Fees and expenses paid to complete the transfer of ownership of a home.

Coinsurance When the insured and the insurer share losses or costs.

Collision Auto insurance that covers damage to the insured's car.

Commission A payment amount to a salesperson that may be an amount for each item sold or it may be a percent of the dollar value of sales.

Compound amount The total in a savings account at the end of an interest period after compound interest is added.

Compound interest The difference between the original principal and the compound amount; interest figured on interest after it has been added to principal.

Comprehensive damage Insurance that covers damage or loss to your vehicle from fire, theft, vandalism, and other causes.

Consolidation loan Money that is borrowed to pay off all of your debts.

Consumer price index (CPI) A single number used to measure a change in consumer prices compared to a base year.

Contract employee- Temporary employee who receives paychecks and benefits from the employment agency.

Cost of goods sold The amount paid by the seller for the goods sold.

Cost-of-Living-Adjustment (COLA) A wage increase based on changes in the Consumer Price Index (CPI).

Credit memo A form that tells the buyer that the buyer's account has been reduced.

Credit score A number, between 300 and 850, generated by credit bureaus that is assigned to a person to rate the risk that person poses to a creditor.

Credit terms and conditions A document or disclosure box that, by law, a credit card company must supply to all customers that outline the costs associated with using the credit card.

Current ratio Compares current assets to current liabilities.

Current yield (on bonds) Found by dividing the bond's annual interest income by the bond's price.

D

Daily interest factor The interest a note is accumulating per day.

Debit card A card that lets you pay for purchases using a terminal in a store.

Debt The money that you owe because you are using someone else's money with the promise to pay it back.

Debt to equity ratio Ratio found by dividing the long term liabilities of a firm by the firm's capital. Helps measure the amount of financial risk of a business.

Debt-to-income ratio A ratio, usually expressed as a percent, that indicates the percent of one's income that is spent on housing and other debts.

Declining-balance method A way of figuring depreciation at a fixed rate on a decreasing balance; fixed-rate method.

Deductions Subtractions from gross pay.

Deposit Slip A form used to list all money deposited in a bank at one time.

Depreciation The decrease in value caused by wear and aging.

Direct Deposit Transferring funds by a company directly into their employees bank accounts without writing any checks.

Disability insurance Pays a portion of the income you lose if you cannot work due to health condition or injury.

Discount (bond) A bond with a market value that is less than par value.

Discount series A trade discount that has two or more discounts.

Domestic business The manufacturing, purchasing and selling of goods and services within a country.

Double-time pay Pay that is twice the regular pay.

Down payment The part of a price that is paid at the time of buying on the installment plan.

E

E-business Electronic business; doing business online.

Earned income Money received from working such as wages, salaries, and tips.

Electronic funds transfer (EFT) When funds are withdrawn from one account and deposited into another using computers.

Employee A person who works for others.

Employee benefit Paid vacation, sick leave, retirement plan, insurance, and other benefits beyond the wage or salary.

Employer A person or company that an employee works for.

Equity The owner's share of a business.

Exact interest method Interest based on exact time and a 365-day year.

Executive recruiter Person hired to find full-time employees for management or specialized technical positions; headhunters.

Exempt employee Paid a salary and does not qualify for overtime pay.

Exemption An amount of income that is free from tax.

Exporting Selling of goods or services produced within your country and sold to other countries.

Extension The total price of each quantity on a sales slip, found by multiplying the quantity by the unit price.

F

Factory overhead The total cost of items such as rent, depreciation, heat, light, power, insurance, supplies, and indirect labor used in a factory.

Finance charge The sum of the interest and any other charges on an installment loan or purchase.

First in, first out (FIFO) A method used to find the value of ending inventory when the exact cost is not known. Merchandise purchased first is used first and the value of the ending inventory is based on the cost of the most recently purchased items.

Fixed costs Overhead items such as rent, salaries, heat, and insurance that remain the same no matter how much is produced and sold.

Flat tax A tax rate that stays the same for every person regardless of the amount of income the yearn in a year.

f.o.b. Free on board; a term used in price quotations to tell who will pay transportation costs.

Forecast Estimate of future sales.

Foreign exchange rate The amount paid for another country's currency.

Freight A service for delivery of heavy, bulky goods.

Frequency distribution A table of numbers arranged in order that records the frequency, or how often each number occurs, tallied.

Future value of an annuity The amount of money, including interest earned, in an account after a series of equal payments are made to the account.

G

Grace period A number of days given that allows you to avoid all finance charges if you pay your balance in full by the due date.

Graduated commission A pay system in which the rate of commission increases as the base increases.

Gram The basic unit of weight in the metric system.

Grand total The corner total on a columnar table. Used to check accuracy of vertical and horizontal addition.

Gross domestic product (GDP) A system of a country that measures the total market value of the goods and services produced within its borders.

Gross income Total income in a year. Includes swages, salaries, commissions, bonuses, tips, interest, dividends, prizes, pensions, sale of stock, and profit from a business.

Gross pay The total amount that an employee is paid.

Gross profit The difference between the net sales and the cost of goods sold.

Gross profit margin The comparison of gross profit to net sales.

H

Health insurance Protects policyholder from financial loss due to illness.

Home coverage area The region of the country where you have wireless phone service.

Homeowners insurance An insurance policy that covers your home and protects you against risk.

Horizontal bar graph A graph with bars running to the left and right.

Hourly rate An amount of pay for each hour worked.

I

Importing The buying of products produced outside your country.

Income statement Shows how much money has been earned and spent during a period of time.

Individual retirement account (IRA) Retirement investment.

Inflation A rise in the prices of goods and services.

Installment loan A loan from a bank or credit union in which you repay the principal and interest in installments.

Interest Money paid to an individual or institution for the privilege of using their money.

International business All business activities necessary to manufacture, purchase, and sell goods and services across national borders; foreign trade.

Internet service provider (ISP) An organization that provides access to the Internet in your area.

Invoice price Price the retailer pays after a trade discount is given.

J

Job expense Money paid out of total job benefits for things such as travel, dues, tools.

L

Labor force All persons who are working or looking for work.

Last in, last out (LIFO) A method to find the value of ending inventory when the exact cost is not known. Merchandise purchased last is used first so inventory is based on the costs of goods purchased first.

Lease A rental agreement.

Liabilities Debts of a business.

Life insurance A way of protecting your family from financial hardship when you die.

Line graph A graph on which lines connect dots to show values.

List price A price shown in a catalog.

Liter The basic metric measure of capacity.

M

Major medical insurance Supplements basic health coverage. Designed to help pay the hospital costs or other health care expenses due to a major illness or injury.

Manufacturer's Suggested Retail Price (MSRP) The price a manufacturer suggests buyers pay.

Markdown A reduction in marked or retail price; a discount.

Market The total of all persons or organizations that are potential customers.

Market price The price at which a stock or bond is sold; the price or value of an item in the open market.

Market share Tells what percentage of the total market's sales one seller has.

Markup The amount added to the cost of the goods to cover all other expenses including the desired profit.

Maturity date The date that marks the end of the term of a loan; when the money must be repaid.

Mean The sum of the numbers divided by the number of items; arithmetic average; a measure of central tendency.

Merchandise turnover rate The number of times per period that a store replaces, or turns over its average stock of merchandise.

Meter (m) The basic unit of length in the metric system.

Modified accelerated cost recovery system (MACRS) Allows you to claim depreciation over a fixed number of years depending on the class life of the property.

Mortgage loan A paper signed by a borrower that gives the lender the right to ownership of property if the borrower does not pay the principal or interest.

Mutual fund An investment company that buys stocks and bonds of other companies.

N

Negotiation When the buyer and seller come to an agreed price.

Net asset value The value of a mutual fund share found by dividing net assets of the fund by outstanding shares.

Net income The amount left after subtracting operating expenses from gross profit; net profit.

Net job benefits The total value of the benefits received from a job less job expenses.

Net loss When operating expenses are greater than gross profit; net loss equals operating expenses less gross profit.

Net pay The remaining pay after deductions have been subtracted from total or gross wages; take-home pay.

Nonexempt employees Employees who are paid by the hour and get paid overtime.

Net profit margin The comparison of net income to net sales.

O

On account A term indication that payment is to be made at a later date; describing a partial payment on an amount owed.

Online Being connected to the Internet.

Online banking An electronic banking service that allows you to do your banking by using your personal computer and the Internet.

Ordering costs Inventory costs that include the expenses connected with creating and sending a purchase order to a supplier and handling stock.

Ordinary interest method A method used in place of the exact interest method by some businesses that uses only 360 days. The 360-day year has 12 months of 30 days each and is known as the *banker's year.*

Outstanding check A check issued but not yet received and paid by the bank.

Over-budget When you are spending more in a category or in total than was budgeted.

Overdrawn An account is overdrawn if you write a check for more money than is in the account.

Overtime Time worked beyond the regular working day or week.

P

Part-time employee An employee who work less than 40 hours per week.

Partnership A business owned by two or more persons that is not a corporation.

Per capita GDP Shows the amount of goods and services produced per person by a country.

Per diem An employee who is needed and paid by the day; a per day amount that a business reimburses an employee for expenses, regardless of how much was spent.

Periodic finance charges Interest paid for specified periods to credit card companies.

Periodic rate A daily or monthly rate found by dividing the Annual Percentage Rate (APR) by 365 or 12.

Perpetual inventory The system of keeping a running balance of stock on hand.

Piece rate A wage system in which workers are paid by the number of pieces produced.

Premium The amount paid for insurance; a bond with a market value that is for more than par value.

Prepayment penalty A fee charged if you pay the loan off early. A prepayment penalty must be disclosed in the original terms of the loan.

Present value of an ordinary annuity The balance needed in an account in order to make a series of payments from the account.

Previous balance method Charges interest on the balance in the account on the last billing date of the previous month.

Prime cost The cost of raw materials and direct labor.

Principal The one for whom an agent acts; the face of a note; the amount on which interest is paid.

Profit sharing Employees get part of the profit earned by a company.

Promissory note A written promise, or IOU, that you will repay the money to the lender on a certain date.

Property damage insurance Auto insurance covering damage to property of others.

Property tax A tax on value of real estate.

Prove cash Count cash on hand and check accuracy against the record of cash received and paid out.

Purchasing power of the dollar A measure of how much a dollar now buys compared to what it could buy during some base period.

Q

Quota A fixed amount of sales above which commission is paid.

R

Range The difference between the highest and lowest numbers in a set of data.

Rate of interest Interest shown as a percent.

Reach Refers to how many people see or hear an advertisement.

Reconcile To bring both the balances on a bank statement and a check register into agreement and to make sure the bank's records are correct.

Reconciliation form A form showing how the checkbook and bank statement balances are made to agree.

Renters policy Similar to homeowners insurance, but not covering loss of building or apartment.

Reorder point The minimum stock level at which an order must be placed.

Replacement cost policy Insurance that pays the cost of replacing property at current prices.

Resale value The amount you receive when you sell an asset, such as a car.

Respondent Person who completes a business survey.

Return on equity Compares net income to capital.

Roaming charges Charges you pay to make a call when you are outside your home coverage area.

S

Salary A fixed amount of money paid for each pay period worked.

Sales invoice Lists the goods sold and delivered to the buyer.

Sales returns and allowances The dollar amount of goods sold that were later returned for refunds or for which credit was given because of damage; a decrease in sales.

Sales tax A tax charged by a city, county, or state on the sale of items or services and collected by sellers from buyers.

Satellite television Television service via communication satellites.

Security deposit Amount of money given to guarantee a lease.

Service charge A bank charge or deduction for handling a checking account; a charge in addition to interest on an installment purchase.

Square meter The basic unit of area in the metric system.

Standard deduction A fixed amount that can be deducted from taxable income; used in place of itemized deductions.

Stock Goods or supplies on hand; shares of ownership in a corporation.

Stock record Shows how much of a particular item has been received and issued, and how much remains on hand.

Straight commission A pay system in which commission is the only pay; there is no other wage or salary earned.

Subtotal On a sales slip, the sum of the extensions before taxes are added.

Sum-of-the-years-digits method A variable-rate way of depreciating that provides decreasing amounts of depreciation as an item ages.

T

Taxable income The amount used to figure income tax with a tax-rate schedule.

Term A specified time that money is left on deposit.

Term of discount The time during which a bank holds a discounted note; date of discount to date of maturity.

Time zone One of 24 standard regions in which the globe is divided, 15° apart, along longitude lines.

Time-and-a-half-pay Pay that is one and a half times the regular pay rate.

Tip Amount of money given to someone for services he or she provides.

Total finance charges (for a cash advance) The sum of the periodic finance charges and any fee charged.

Total manufacturing cost The sum of the costs of raw materials, direct labor, and factory expense.

Trade deficit When foreign imports exceed exports.

Trade discount A reduction or discount given from a catalog or list price; a discount given" within the trade."

Trade surplus When foreign exports exceed imports.

Trade-in value The amount you get for your old car or other asset when buying a new car or other asset.

Transaction A deposit or withdrawal that is recorded.

Travel expenses Expenses incurred while traveling for business; mileage, hotel fees, meal charges, airfares, taxi fares, rental car charges, expenses for entertaining customers, and related expenses.

Trend A historical relationship between sales and time.

U

Under-budget When you are spending less in a category or in total than was budgeted.

Unearned income Money received from interest and dividends.

Unemployment rate The percentage of the total labor force that is not working.

Unit price The price of one item.

V

Variable costs Costs such as raw materials, direct labor, and energy that vary or change with the amount of goods produced and sold.

Vertical bar graph A graph with bars running up and down.

W

Wages Money paid for a job completed; the total pay for a day or week of a worker paid on an hourly rate basis.

Weighted average A method used to find the value of ending inventory when the exact cost is not known. Inventory is priced at the average price per unit of the beginning inventory plus the cost of all purchases during the fiscal year.

Withholding allowance An allowance for a person used to reduce the amount of tax withheld from pay.

Withholding tax A deduction from pay for income tax.

INDEX

A

B

C

D

E

F

G

H

I

J

K

L

M

N

O

P

Q

R

S

T

U

V

W

Y